Endocannabinoids and Lipid Mediators in Brain Functions

Miriam Melis
Editor

Endocannabinoids and Lipid Mediators in Brain Functions

Springer

Editor
Miriam Melis
Dept Biomedical Sciences
Division of Neuroscience and Clinical Pharmacology
University of Cagliari, Cagliari
Italy

ISBN 978-3-319-57369-4 ISBN 978-3-319-57371-7 (eBook)
DOI 10.1007/978-3-319-57371-7

Library of Congress Control Number: 2017942949

Printed on acid-free paper

This Springer imprint is published by Springer Nature
The registered company is Springer International Publishing AG
The registered company address is: Gewerbestrasse 11, 6330 Cham, Switzerland

To my family, for their love and support and for brightening up the day when it was needed

Endocannabinoids and Lipid Mediators in Brain Functions

The science of cannabinoids is 50 years old. These past years provided a remarkable and constant number of breakthroughs, showing that the signalling mediated by endocannabinoids and lipid mediators impacts almost every function of the body. Indeed, this represents a special field of research, which allows tackling the complexity of biological functions and provides potential therapeutic frameworks for a plethora of diseases.

Endocannabinoids and lipid mediators, acting on multiple receptors in both CNS and periphery, are involved in the control of neuronal firing, synaptic plasticity, local circuit activity and behaviour. In fact, these lipid molecules, being intimately connected with diverse metabolic and signalling pathways, differently affect various functions of neurons and glial cells through activation not only of surface receptors, such as cannabinoid and vanilloid receptors, but also of nuclear receptors such as peroxisome proliferator-activated receptors. Although the endocannabinoid system plays a fundamental role in making short- and long-term modifications to neural circuits and related behaviours, many facets of this system are still unclear.

The number of exciting discoveries brought up to the scientific community almost on a daily basis highlights the importance of an updated volume on this topic, particularly given that potential therapeutic benefits of cannabis and cannabinoids are currently under heavy analysis in many countries worldwide. Hence, the main objective of this book is to explore not only some of the many functions of endocannabinoids (and lipid mediators) in physiological control of networks at a cellular and molecular level but also to extend this knowledge for the potential use of cannabinoids and/or drugs regulating endocannabinoid levels in vivo as therapeutic target(s) in neurological and neuropsychiatric disorders. In this book, new findings and ideas about the endocannabinoid system and its roles as neuronal circuit modulator related to human brain pathologies characterized by alterations in neuroplasticity will be highlighted. Endocannabinoid roles in key systems controlling appetite, pain, learning and memory, as well as stress responses, will be presented. In addition, pathological processes associated with changes in endocannabinoid signalling will be discussed in the context of anxiety, autism, depression and addiction. Such an integrated science has important therapeutic implications because it explores the interactions among the environment, the brain and the body. Finally, given that nowadays *Cannabis* use is increasingly prevalent among adolescents and that this (mis)use interferes with endogenous endocannabinoid system, which is critically involved in both pre- and post-natal neurodevelopment, the consequences on normal development, thus predisposing to motivational, affective and psychotic disorders, will be discussed.

Given the tremendous amount of information available and the pace of research progress in this field, it is impossible for the current volume to be fully comprehensive. Nonetheless, this book will provide an excellent background to researchers looking for extending their areas of interest and to newcomers in the field.

As the book was being prepared for publication, Dr. Loren (Larry) Parsons, who contributed to this book, passed away in his sleep at only 52 years old. He brought light both to the field of endocannabinoid system, in particular with regard to the mechanisms involved in drug abuse, and to everyone who interacted with him at a personal or professional level. He will be deeply missed.

Miriam Melis

Dept Biomedical Sciences
Division of Neuroscience and Clinical Pharmacology
University of Cagliari, Cagliari
Italy

Contents

List of Committed Contributors

Luigia Cristino Institute of Biomolecular Medicine, CNR of Pozzuoli, Naples, Italy

Liana Fattore Institute of Neuroscience, CNR, Cagliari, Italy

Matthew Hill Departments of Cell Biology and Anatomy and Psychiatry, The Hotchkiss Brain Institute, University of Calgary, Calgary, AB, Canada

Zsolt Lenkei Brain Plasticity Unit, ESPCI-ParisTech, CNRS UMR8249, Paris, France

David Lovinger Laboratory of Integrative Neuroscience, National Institute on Alcohol Abuse and Alcoholism, National Institutes of Health, Rockville, MD, USA

Beat Lutz Institute of Physiological Chemistry, University Medical Center of the Johannes Gutenberg-University Mainz, Mainz, Germany

Olivier Manzoni INMED, INSERM U901, Parc Scientifique de Luminy - BP 13, Marseille, France

Giovanni Marsicano Neurocentre Magendie, U1215 INSERM, Bordeaux, France

Université Bordeaux, Bordeaux, France

Daniela Parolaro Department of Theoretical and Applied Science, Biomedical Research Division, and Neuroscience Center, University of Insubria, Varese, Italy

Loren Parsons The Scripps Research Institute, Committee on the Neurobiology of Addictive Disorders, La Jolla, CA, USA

Marco Pistis Department of Biomedical Sciences, Division of Neuroscience and Clinical Pharmacology, University of Cagliari, Cagliari, Italy

Stephen Skaper Department of Pharmaceutical and Pharmacological Sciences, University of Padua, Padua, Italy

The Endocannabinoid System in Leptin-Driven Changes of Orexinergic Signaling Under Physiological and Pathological Conditions

Luigia Cristino, Roberta Imperatore, Letizia Palomba, and Vincenzo Di Marzo

Abstract In this chapter, the neuroanatomical overlap in the distribution of orexin and endocannabinoid systems, as well as the functional interaction between leptin-driven synaptic rewiring of orexinergic neurons and the orexin receptor-mediated activation of 2-arachidonoylglycerol synthesis, will be presented in the context of their role in the regulation of appetite, reward, sleep/wake, and analgesia. This chapter attempts to piece together what is known about this important cross talk and points out its potential therapeutic implications.

Abbreviations

2-AG	2-arachidonoylglycerol
AA	arachidonic acid
AC	adenylyl cyclase
AEA	anandamide
AgRP	agouti-related peptide
ARC	arcuate nucleus
BBB	blood-brain barrier
cAMP	cyclic adenosine monophosphate
CART	cocaine- and amphetamine-regulated transcript
CB1	cannabinoid receptor 1
CNS	central nervous system
CSF	cerebrospinal fluid
CTB	cholera toxin B subunit
DA	dopaminergic
DAG	diacylglycerol
DAGL	diacylglycerol lipase
DMH	dorsomedial hypothalamic nucleus
DR	dorsal raphe nucleus

L. Cristino, PhD (✉) • R. Imperatore, PhD • L. Palomba, PhD • V. Di Marzo, PhD
Institute of Biomolecular Chemistry, CNR of Pozzuoli, Naples, Italy
e-mail: luigia.cristino@icb.cnr.it

M. Melis (ed.), *Endocannabinoids and Lipid Mediators in Brain Functions*,
DOI 10.1007/978-3-319-57371-7_1

DSE	depolarization-induced suppression of excitation
DSI	depolarization-induced suppression of inhibition
ECs	endocannabinoids
ERK	extracellular signal-regulated kinases
FAAH	fatty acid amide hydrolase
FRET	fluorescence resonance energy transfer
GPCRs	G-protein-coupled receptors
Hcrt	hypocretin
HFD	high-fat diet
IP3	inositol trisphosphate
LC	locus coeruleus
LDT	laterodorsal tegmental nucleus
LH	lateral hypothalamus
LHA	lateral hypothalamic area
LPA	lysophosphatidic acid
MAGL	monoacylglycerol lipase
MAPKKK	mitogen-activated protein kinase kinase kinase
MCH	melanin-concentrating hormone
MSH	melanocyte-stimulating hormone
NAcc	nucleus accumbens
NMDA	*N*-methyl-D-aspartate receptor
NPY	neuropeptide Y
NSCCs	nonselective cation channels
OX-1R	orexin-1 receptor
OX-2R	orexin-2 receptor
OX-A	orexin-A
OX-B	orexin-B
PA	phosphatidic acid
PAG	periaqueductal gray
PAP	phosphatidic acid phosphohydrolase
PC	phosphatidylcholine
PFCx	prefrontal cortex
PI3K	phosphoinositide 3-kinase
PIP2	phosphatidylinositol 4,5-bisphosphate
PKA	protein kinase A
PKC	protein kinase C
PLA1	phospholipase A1
PLA2	phospholipase A2
PLC	phospholipase C
PLD	phospholipase D
POMC	pro-opiomelanocortin
PPO	prepro-orexin
PPT	pedunculopontine nucleus
PVN	paraventricular nucleus

REMS	rapid eye movement sleep
RHT	retinohypothalamic tract
ROCCs	receptor-operated calcium channels
SOCCs	store-operated calcium channels
SON	supraoptic nucleus
THC	tetrahydrocannabinol
TMN	tuberomammillary nucleus
TRP	transient receptor potential channels
TRPV	vanilloid transient receptor potential channels
VMH	ventromedial hypothalamus
VTA	ventral tegmental area

1 Endocannabinoids and Cannabinoid Receptor Type 1: An Overview

The endocannabinoids (ECs) are arachidonic acid-containing messengers generated by esterase and phosphoesterase action, produced on demand at the site of need. The most important ECs are *N*-arachidonylethanolamine (anandamide, AEA) and 2-arachidonoylglycerol (2-AG) (Devane et al. 1992; Sugiura et al. 1995). In the CNS, ECs are often synthesized by postsynaptic neurons in response to a depolarization-induced increase in intracellular Ca^{2+} levels before being rapidly released to act in an autocrine or paracrine manner (Di Marzo 2009). Through the paracrine mode of action, ECs bind CB1 at presynaptic sides and inhibit both inhibitory and excitatory neurotransmission, thereby inducing a retrograde suppression of inhibition (DSI) or suppression of excitation (DSE), respectively, of the cellular target (Kano et al. 2009). After EC synthesis and release, EC signaling is terminated by neuronal reuptake and intracellular hydrolysis of AEA and 2-AG by fatty acid amide hydrolase (FAAH) and monoacylglycerol lipase (MAGL), respectively. A plethora of studies report that the ECs modulate other neurotransmitters like endogenous opioids (Robledo et al. 2008; Parolaro et al. 2010), dopamine and adenosine (Carriba et al. 2007; Ferré et al. 2009; Fernández-Ruiz et al. 2010). Interestingly, strongly emerging evidence points to a cross talk between ECs and orexins in the overall regulation of appetite, nociception, and sleep/wake cycle, thus opening new horizons to potential therapeutic applications of currently existing drugs targeting dysfunctions of these functions. The cannabinoid receptors of type 1 (CB1) are the major cannabinoid receptors expressed in the brain, where they are found both in neurons and astrocytes. CB1 belongs to the superfamily of G-protein-coupled receptors (GPCRs), which inhibit adenylyl cyclase (AC) activity with consequent decrease of cAMP and reduction, among others, of inhibitory c-Raf phosphorylation by PKA activity (Melck et al. 1999; Davis et al. 2003). Through Gi, CB1 regulates ion channels via activation of K^+ channels and inhibition of N-,

P-/Q-, and L-type voltage-gated Ca^{2+} channels (Deadwyler et al. 1995; Hampson et al. 1995). However, CB1 activation is also able to regulate AC via Gs, as well as Ca^{2+} fluxes and phospholipases with subsequent modulation of mitogen-activated protein kinases (p42/p44, p38, and c-Jun N-terminal kinase) regulating nuclear transcription factors (Howlett et al. 2002).

2 Orexin/Hypocretin: Discovery and Characterization

The discovery of orexin/hypocretin neuropeptides and their receptors dates back to 1998 and is attributed to two independent groups. By using subtraction hybridization, de Lecea with colleagues identified the mRNA sequence encoding prepro-hypocretin, the putative precursor of two putative peptide transmitters produced in the mammalian hypothalamus. Simultaneously, while searching for endogenous peptide ligands for multiple orphan G-protein-coupled receptor HFGAN72, Sakurai and Yanagisawa identified the same neuropeptides (Sakurai et al. 1998). De Lecea et al., named these peptides "hypocretins" by combining the suffix "hypo" of hypothalamus, the main source of these peptides in the brain, and "cretin" of "secretin," the gut hormone with high amino acid homology with hypocretins (de Lecea et al. 1998). Sakurai et al., referred to these peptides as orexins from the Greek word "orexis" meaning appetite, in view of their stimulatory action on food intake following central administration (Sakurai et al. 1998). It soon became clear that the peptides isolated by these two groups were identical and both sets of names are still in use. However, the Nomenclature Committee of the International Union of Basic and Clinical Pharmacology (IUPHAR) recommends the use of "orexin" for peptides and receptors and "hypocretin" for the genes encoding them (Gotter et al. 2012). Thus, the peptides are known as prepro-orexin (PPO), orexin-A and orexin-B, and the receptors as OX_1 and OX_2 receptors (OX-1R and OX-2R, respectively), while the rodent genes encoding PPO and the receptors are *Hcrt*, *Hcrtr1*, and *Hcrtr2*, respectively. Mammalian (human, pig, dog, rat, mouse) PPO is composed of 130–131 aa (de Lecea et al. 1998; Sakurai et al. 1998). OX-A is a 33-amino acid peptide of 3562 Da with two intrachain disulfide bonds (Sakurai et al. 1998), which make the peptide particularly prone to misfolding following exposure to free radicals like nitric oxide (Nobunaga et al. 2014). OX-B is a 28-amino acid of 2937 Da with 46% sequence homology with OX-A. The amino acid sequences for OX-A and OX-B are highly conserved in mammals: OX-A is identical in pigs, dogs, rats, mice, and humans, whereas in OX-B there are few changes in one or two amino acids among these species (Kukkonen 2013). Since OX-A and OX-B are very ancient neuromodulators, a strong phylogenetic conservation of their amino acid sequences occurs among nonmammalian vertebrates (Sakurai 2007; Tsujino and Sakurai 2009). Both OX-A and OX-B are cleaved from the precursor PPO (de Lecea et al. 1998; Sakurai et al. 1998), whose expression is restricted to a small number of neuronal cell bodies in the LHA (Peyron et al. 1998), which send widely distributed projections to the rest of the brain (Peyron et al. 1998; Date et al. 1999; Nambu et al. 1999). Indeed, each brain region contains at least one of the two orexin receptors, except for the cerebellum (Trivedi et al. 1998). Orexinergic neurons send widespread projections to other neurons of the brain, including dopaminergic neurons of the reward areas (VTA, NAcc), noradrenergic neurons controlling arousal at LC, serotonergic neurons of the DR, cholinergic neurons of the striatum, and multiple populations of neuropeptide-expressing neurons in the cerebral cortex, mesopontine tegmental nuclei, periaqueductal gray, tuberomammillary nucleus, septal nucleus, and paraventricular thalamus (Peyron et al. 1998; Horvath et al. 1999). Orexinergic neurons also project to regions regulating energy balance in the brain (ARC, VMH, DMH, PVN, LHA) (Lee et al. 2013) and periphery (pancreas, gastrointestinal system, liver, brown and white adipose tissue) (Wu et al. 2004; Grabauskas and Moises 2003; Oldfield et al. 2002; Adler et al. 2012). Notably, the orexins colocalize with many neuropeptides, such as prolactin (Risold et al. 1999), pentraxin (Reti et al. 2002), dynorphin (Chou et al. 2001), and galanin (Håkansson et al. 1999). Many orexin neurons are also glutamatergic as shown by their expression of the vesicular glutamate transporters (Rosin et al. 2003; Torrealba et al. 2003). Of the two peptides, OX-A is the most lipophilic, has been detected in the CSF and plasma, and is able to cross the BBB in both directions (Kastin and Akerstrom 1999).

3 Orexin Receptors: Common Signaling with CB1

Orexins act via OX-1R and OX-2R, which are GPCRs belonging to the rhodopsin family (Sakurai et al. 1998). OX-2R couples to the G_q, G_s, or $G_{i/o}$ subclasses of G proteins and has equal affinity for OX-A and OX-B (reviewed in Kukkonen and Leonard 2014). OX-1R preferentially couples to G_q (Sakurai et al. 1998) and is activated ten times more potently by OX-A than by OX-B. In many regions of the brain, OX-1R and/or OX-2R are distributed complementary to OX-positive projections, especially in the prefrontal and infralimbic cortex, CA of hippocampus, amygdala, BST, paraventricular thalamic nucleus, anterior hypothalamus, DR, VTA, LC, LDT/PPT nucleus (Trivedi et al. 1998; Lu et al. 2000; Marcus et al. 2001), ARC, TMN, DMH, PVN, LH, and medial septal nucleus (Lu et al. 2000; Marcus et al. 2001). OX-1R and OX-2R are also present in peripheral tissues (kidney, adrenal, thyroid, testis, ovaries, lung, pituitary, and jejunum) (Jöhren et al. 2001). Ca^{2+} signaling is among the first functional response triggered by OX-A-mediated stimulation of OX-1R (Sakurai et al. 1998). Indeed, orexin receptors regulate receptor-operated calcium channels (ROCCs), which include nonselective cation channels (NSCCs, such as transient receptor potential (TRP) channels) and Na^+/Ca^{2+} exchanger channels (Louhivuori et al. 2010). In neurons and other excitable cells, unlike CB1, OX-1R also activates N-type voltage-gated Ca^{2+} channels (Uramura et al. 2001). OX-A-mediated elevation of intracellular Ca^{2+} concentration is dependent from many different pathways, including ROCCs, SOCCs, and IP3-mediated Ca^{2+} release (Fig. 1). Unlike CB1, which inhibits AC activity, OX-1R regulates AC activity in both a positive and negative manner (Tang et al. 2008). Both CB1 and OX-1R activate ERK1/2 (Milasta et al. 2005; Ammoun et al. 2006a) and PI3K (Ammoun et al. 2006a; Skrzypski et al. 2011). Upon activation by OX-1R, PKC activates NSCCs (Xia et al. 2009), PLD (Jäntti et al. 2012), and L- and N-type Ca^{2+} channels and inhibits the inwardly rectifying K^{+} channels (Nakajima and Nakajima 2010). Notably, OX-1R modulates the CB1-mediated synaptic plasticity (Borgland et al. 2006; Selbach et al. 2010; Yang et al. 2013) by being able to activate PKC which, in turn, inhibits CB1 activity (Garcia et al. 1998; Uramura et al. 2001). Furthermore, OX-1R activates PLC, PLA2, and PLD (Lund et al. 2000; Johansson et al. 2007, 2008; Turunen et al. 2010, 2012; Jäntti et al. 2012). PLC activation leads to the production of DAG, which can be directly deacylated by DAGL generating a monoacylglycerol (possible 2-AG) or phosphorylated to phosphatidic acid (PA). Both DAG and PA are important lipid mediators with many targets (Fig. 1). PLA2 hydrolyzes glycerophospholipids at the *sn*-2 position by releasing a free fatty acid, mainly arachidonic acid (AA), a precursor of prostaglandins and leukotrienes. AA inhibits the activity of K^{+} and Na^{+} channels and of L- and N-type Ca^{2+} channels, while it can activate TRP channels such as the vanilloid TRPs, TRPV1 and TRPV4 (Kukkonen 2011) (Fig. 1) (reviewed in Meves 2008). Orexin receptors activate PLD, which produces PA and free choline starting from phosphatidylcholine (PC) as a substrate. PA activates many kinases such as phosphatidylinositol-4-phosphate 5-kinase, which generates PIP2, thereby stimulating mitogen-activated protein kinase kinase kinase (MAPKKK) Raf1 and protein kinase Cζ. Furthermore, PA can be converted to DAG by phosphatidic acid phosphohydrolase (PAP) or in lyso-PA (LPA) by PLA1 or PLA2 (Fig. 1). Orexin

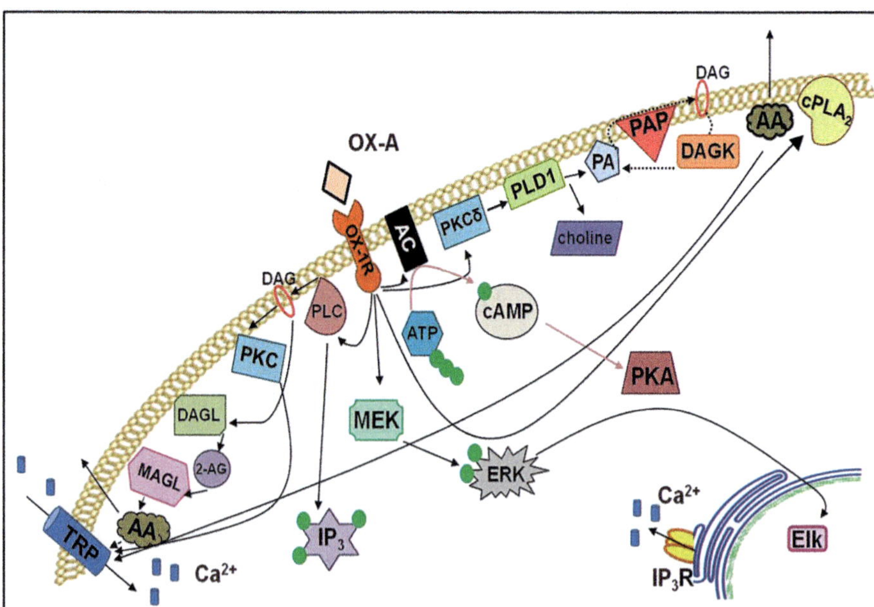

Fig. 1 Simplified overview of orexin receptor signaling

receptor-induced production of DAG and PA may affect CB1 activity through the formation of 2-AG, but the physiological relevance of this remains to be shown.

4 Anatomical Overlap Between Cannabinoid and Orexin Receptor Distribution in the Brain

To the best of our knowledge, no extensive study has been yet performed to describe the anatomical map of cannabinoid and orexin co-distribution in the brain. Evidence from separate studies concerning CB1 or OX-1R or OX-2R suggests an overlap of expression in critical areas involved in the overall regulation of appetite and metabolism, reward, nociception, and sleep/wake cycle. Thus, OX-1R and OX-2R, as well as CB1, are expressed with different densities in all the hypothalamic nuclei, including ARC, DMH, LHA, PVN, SON, and VMH. Prefrontal and infralimbic cortex, CA of hippocampus, amygdala, BST, paraventricular thalamic nucleus, DR, VTA, LC, LDT/PPT nucleus TMN, and medial septal nucleus also express both cannabinoid and orexin receptors (Fig. 2).

5 Molecular Interactions Between Cannabinoid and Orexin Receptors

In 2003, Hilairet et al. provided for the first time, in CHO cells stably co-transfected with CB1 and OX-1R, evidence for a functional cross-talk between these receptors. Indeed, co-expression of both receptors enhanced the ability of OX-A to stimulate

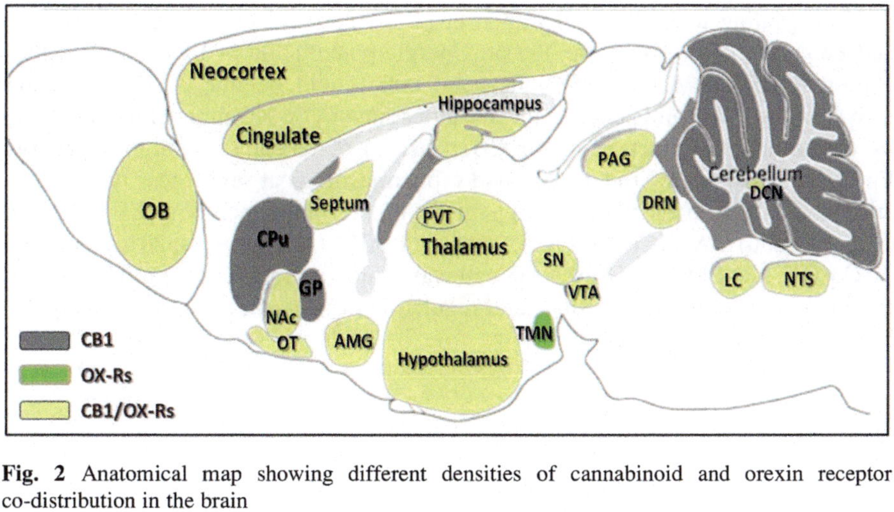

Fig. 2 Anatomical map showing different densities of cannabinoid and orexin receptor co-distribution in the brain

the ERK1/2 pathway ~100-fold more as compared to what is observed when only one of the receptors is expressed in cells and in a manner prevented by SR141716A (a CB1 antagonist/inverse agonist, also known as rimonabant), as well as by pertussis toxin. The authors explained this effect with a direct functional involvement of CB1 in the formation of heteromeric complexes with OX-1R also on the basis of electron microscopy study showing these receptors closely enough to form hetero-oligomers in CHO cells (Hilairet et al. 2003). This possibility was further investigated by Ellis et al. (2006) by using HEK293 cells co-expressing both receptors and displaying spontaneous OX-1R and CB1 internalization following cell exposure to OX-A. In this study, rimonabant reduced the potency of OX-A to activate the MAP kinases ERK1/2 in cells expressing both receptors. Likewise, the OX-1R antagonist SB674042 reduced the potency of the CB1 agonist WIN55,212-2 to phosphorylate ERK1/2. More importantly, orchestrated OX-1R and CB1 trafficking occurred in HEK293 cells following inducible expression of these receptors since (i) treatment with rimonabant resulted in the re-localization of CB1 to the cell surface and (ii) treatment with SB674042 (a selective OX-1R antagonist) induced redistribution of both receptors to the cell surface. Further evidence concerning physical interactions between OX-1R and CB1 was demonstrated by labeling covalently the extracellular domains of CB1 and OX-1R with SNAP-tag® and CLIP-tag™, respectively (Ward et al. 2011). By this approach CB1/OX-1R heteromerization was clearly identified by single-cell fluorescence resonance energy transfer (FRET) assay, which established that CB1 and OX-1R were close enough to produce true heteromers (dimers or oligomers). Besides heteromerization, a further mechanism has been proposed to explain CB1-dependent potentiation of OX-A signaling to ERK1/2. This is based on the enhancement of 2-AG synthesis by OX-A-mediated activation of the PLC-DAGL pathway via OX-1R (Turunen et al. 2012). Since both OX-1R and CB1 activate ERK1/2 (Bouaboula et al. 1995; Ammoun et al. 2006b), the heterologous or constitutive expression of OX-1R and CB1 in cells very strongly potentiates OX-A signaling to ERK1/2 (Turunen et al. 2012; Jäntti et al. 2013; Kukkonen and Leonard 2014). Therefore, the orexinergic potentiation of ERK1/2 phosphorylation in OX-1R-/CB1-CHO-transfected cells could be due to CB1 receptor activation by 2-AG and hence to the amplification of OX-1R intracellular signaling by endocannabinoid activation of CB1, instead of receptor di-/oligomerization. Emerging studies in rodents provide pharmacological evidence that OX-A enhances 2-AG levels in different regions of the brain including LH (Cristino et al. 2013a, b), PAG (Cristino et al. 2016), and ARC (Morello et al. 2016), possibly by Gq(OX-1R)-PLC-DAGLα-2-AG pathway which is crucial in the modulation of autocrine and paracrine functions and regulation of synaptic transmission (Fig. 3).

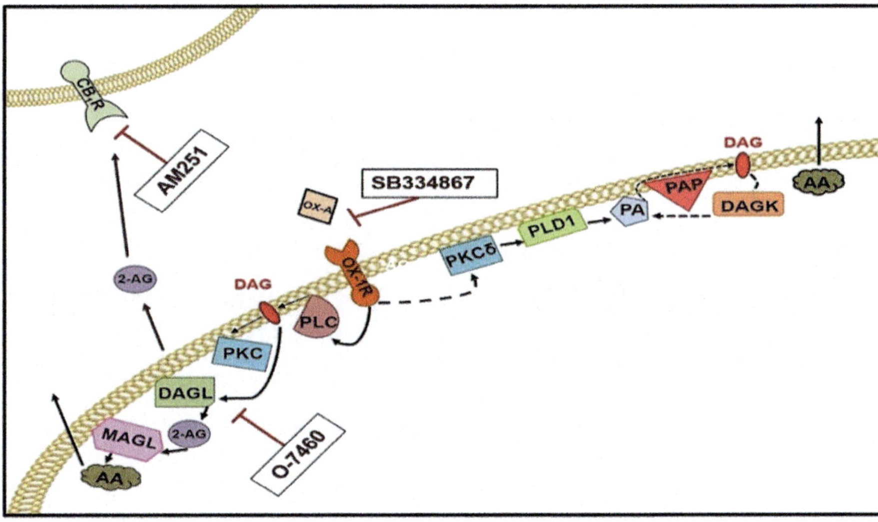

Fig. 3 OX-A signaling cascade regulates 2-AG synthesis and affects CB1 signaling in autocrine and paracrine way

6 Functional Interaction Between Cannabinoid and Orexin Receptors Under Leptin Signal Deficiency: Emerging Studies

Because of their local fast and fine regulation of production, endocannabinoid levels change according to rapid changes in metabolic requirements of the body. Thanks to the pivotal study of Di Marzo and colleagues (Di Marzo et al. 2001), an inverse relationship between leptin and the endocannabinoids (2-AG or AEA) has been established in the hypothalamus of leptin signaling-deficient obese (*ob*/*ob* and *db*/*db*) mice. A previous study found that the CB1 antagonist/inverse agonist rimonabant was able to reduce the food consumption in fasted mice (Colombo et al. 1998). Subsequently, both AEA and 2-AG levels were found to increase or decrease in fasted and sated mice, respectively (Kirkham et al. 2002), whereas CB1-deficient mice were shown to exhibit a leaner phenotype and resistance to diet-induced obesity, also because of lower food intake (Cota et al. 2003). Leptin is a protein of 167 amino acids mainly produced by adipocytes (Fox 2006) and released into the bloodstream. Leptin crosses the BBB by a saturable system (Banks et al. 1996) and signals to the brain to stop eating by inhibiting NPY/AgRP neurons and activating POMC/CART neurons in the arcuate nucleus of the hypothalamus (Schwartz et al. 2000; Sahu 2003) (Fig. 4). Notably, the amount of the circulating leptin is proportional to adiposity and body weight, both in mice and humans (Considine et al. 1996), and leptin administration in rodents causes a profound decrease in food intake and weight loss (Friedman and Halaas 1998). Rodents with genetic defective leptin signaling are obese, as in the case of *db*/*db* mice (knockout

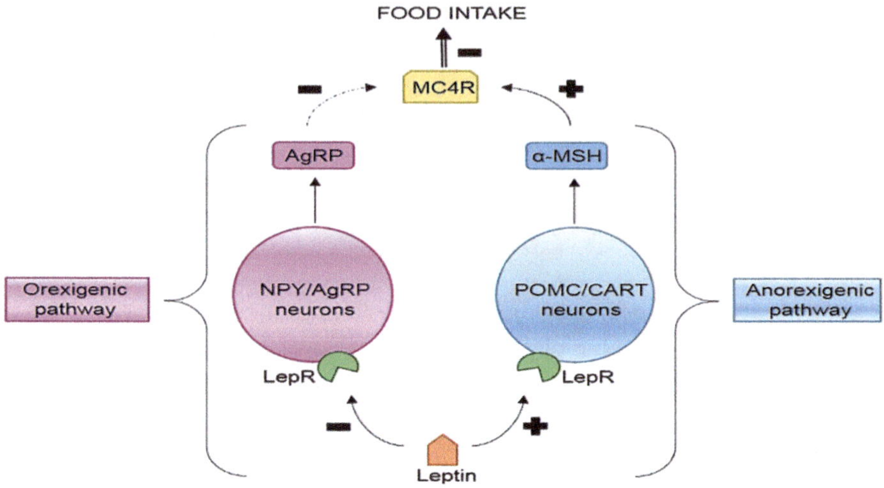

Fig. 4 Schematic representation of leptin signaling at hypothalamic neurons of ARC

for the Ob-Rb gene coding the leptin receptors) or *ob/ob* mice (knockout for the *Ob* gene coding for leptin) (Hill et al. 2010).

6.1 Appetite and Energy Homeostasis

Unlike anatomical studies, growing data are accumulating to unravel the functional cross talk between orexinergic and cannabinoid systems in the regulation of appetite and energy balance. In this regard, the first study concerning the regulation of appetite was by Crespo et al. (2008). By using i.c.v. injection of orexin in pre-fed rats, they reported a dose-dependent increase in the short-term feeding behaviors. The authors also tested the effect of the CB1 inverse agonist SR141716 (rimonabant) when given alone and observed a decrease in feeding behaviors. Finally, the combined effect of orexin and SR141716 was tested. Intraperitoneal injection of SR141716, 10 minutes prior to i.c.v. orexin injection, blocked the orexin-induced feeding already at doses that did not reduce feeding when used alone (Crespo et al. 2008). Similar to a CB1 agonist, also OX-A or OX-B administration stimulates food intake in mice, whereas the OX-1R antagonist, SB334867, reduces feeding (Sakurai et al. 1998; Haynes et al. 2000; Shiraishi et al. 2000). As with endocannabinoid levels, an inverse relationship occurs between circulating leptin and orexin levels. Indeed, fasting results in the upregulation of PPO mRNA (Sakurai et al. 1998), also in obese mice (Yamanaka et al. 2003). However, the regulation of food intake by endocannabinoids at the hypothalamic level results more complex than what was initially believed, because of the occurrence of bimodal orexigenic *vs.* anorexigenic effects of CB1 activation, depending on its expression at glutamatergic or GABAergic inputs. In this scenario, the study of Huang et al. (2007) is clearly understandable, as the authors suggest that fasting-related reduction of leptin levels controls arousal by increasing of orexinergic neuron activity. By using patch-clamp recordings, Huang et al. demonstrated that WIN 55,212-2, a cannabinoid agonist, was able to depolarize MCH neurons, whereas it reduced the spontaneous firing of OX neurons in a manner prevented by AM251 and tetrodotoxin. Using conditions inducing DSI or DSE applied to MCH or OX neurons, the authors revealed that both MCH and OX neurons release endocannabinoids and are innervated by CB1-expressing inhibitory and excitatory inputs, respectively. Notably, the regulation of food intake controlled by endocannabinoids appears further complicated by the observation that hypothalamic circuits are affected by synaptic rewiring regulated by several circulating hormones (leptin, ghrelin, glucocorticoids, etc.). Cristino et al. (2013a, b) investigated the obesity-associated changes in the hypothalamic circuits of orexinergic neurons in the LH. This study employed leptin signaling-deficient mice such as *ob/ob* mice and mice with leptin insensitivity in the ARC caused by high-fat diet (HFD)-induced obesity. It was found that, in these obese mice, in comparison with respective lean control mice, (i) CB1-expressing presynaptic inputs to orexin-A neurons change from predominantly excitatory to inhibitory inputs; (ii) OX-A neurons are able to synthetize 2-AG more than in lean control mice because of elevation of DAGLα expression; and (iii) DSI occurs at OX-A neurons with consequent disinhibition of OX-A neurons and elevation of OX-A trafficking and release to many LH target areas like ARC, PVN, PAG, NAcc, and VTA. The authors also preliminarily investigated the mechanism of the synaptic remodeling at OX-A neurons and found that lack of leptin signaling in the ARC of obese mice, where from most of the fibers innervating orexinergic neurons originate, was the most likely cause. Leptin treatment reversed the synaptic remodeling only in *ob/ob* and not in HFD mice, indicating that this phenomenon is a consequence of leptin deficiency or leptin resistance in the ARC. Interestingly, possibly because *ob/ob* mice lacked endogenous leptin when weaned, but they received it from their heterozygous mothers during lactation, the remodeling of their synapses only occurred after weaning and was reversed by exogenous leptin injection (Cristino et al. 2013a, b) (Fig. 5). Recent findings from Morello et al. (2016) showed a CB1 and OX-1R interaction also in the regulation of POMC neurons, the master subset of the ARC anorexigenic hypothalamic neurons which act to reduce appetite and body weight (Schwartz et al. 2000; Cone 2005). Indeed, ablation of POMC neurons, or loss of α-MSH production, leads to obesity (Yaswen et al. 1999; Balthasar et al. 2004). Manyfold anatomical evidence, including our previous study (Cristino et al. 2013a, b), reveals that POMC neurons send projections to OX-A and regulate OX-A expression (López et al. 2007). OX-A neurons, in turn, send inputs back to POMC neurons (Chemelli et al. 1999) underlying a neuronal circuit perfectly organized to ensure food seeking accompanied by alertness. Notably, CB1 agonism at POMC neurons acutely enhances feeding in a dose-dependent manner (Gómez et al. 2002; Koch et al. 2015) without affecting α-MSH release at low doses (Koch et al. 2015). Furthermore, α-MSH, at doses effective at reducing food intake, does

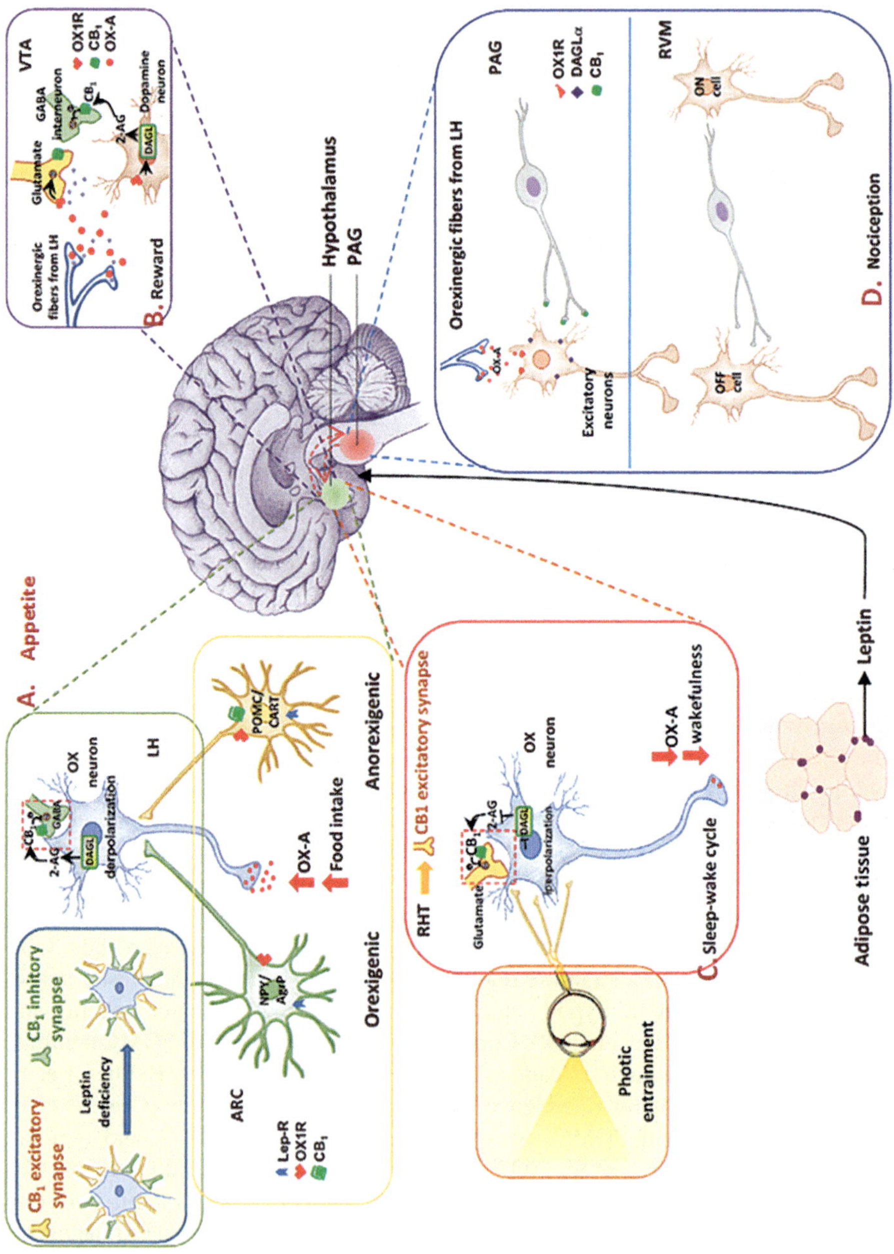

Fig. 5 (continued)

not alter hypothalamic endocannabinoid levels (Matias et al. 2008). In this scenario Morello et al. (2016) found that prolonged OX-1R activation by OX-A promotes appetite by blunting POMC production and α-MSH release as a consequence of OX-A-induced potentiation of 2-AG synthesis and CB1 activation at POMC neurons of mice under satiety state. This pathway relies its mechanism on the potentiation of ERK1/2-mediated Ser-727 phosphorylation/inhibition of STAT3 which affects negatively the expression of the *Pomc* gene (Fig. 6). To further support the occurrence of this autocrine mechanism, the authors also verified its existence in cultured POMC primary neurons. Several preclinical and clinical observations showing an association between obesity and ECS dysregulation in both central and peripheral tissues (Matias and Di Marzo 2007; Bermudez-Silva et al. 2010), as well as the occurrence of a positive correlation between the plasma endocannabinoid levels and markers of obesity or metabolic disorders, have been reported (Engeli et al. 2005; Cota 2007; Di Marzo et al. 2009; Abdulnour et al. 2014). The finding that an endogenously produced peptide such as OX-A can modulate the function of CB1 might shed light on the hyperphagic action of endocannabinoids and on the functional consequences of the inverse relationship between leptin and endocannabinoid levels occurring during obesity also in humans (Monteleone et al. 2005; Nicholson et al. 2015). Indeed, during obesity, the aberrant activation of OX-A-mediated endocannabinoid biosynthesis triggered by deficits in leptin signaling at POMC neurons causes a vicious circle with inhibition of POMC synthesis, hyperphagia, further body weight gain, hypertension, and dysmetabolism, such as hepatosteatosis (Morello et al. 2016; Imperatore et al. 2017). The discovery of this leptin/orexin/endocannabinoid/α-MSH signaling loop and its impact on hyperphagia, obesity, and fatty liver now call for investigations on potential synergies between OX-1R antagonists, CB_1R antagonists, and MC4R agonists for the reduction of body weight. This, in turn, might pave the way

Fig. 5 (continued) Scheme of the leptin-mediated interaction between cannabinoid and orexinergic systems in the control of food intake, sleep/wake cycle, nociception, and reward. Overview of the proposed anatomical pathways, synaptic receptor distribution, and mechanism of endocannabinoid-/orexin-modulated functions. **A**. Appetite. A leptin-driven synaptic rewiring occurs at OX-A neurons. In obese mice, the CB1-positive inputs to OX-A neurons shift from predominantly excitatory to inhibitory drive with consequent disinhibition of OX-A release to LH target areas and increase in food intake. **B**. Reward and seeking behavior. The OX-A-mediated activation of the GqCPR(OX-1R)-PLC-DAGLα-2AG pathway at VTA leads to DSI at CB1-positive inputs to DA neurons and consequent disinhibition of DA release and decrease on reward and seeking behavior. **C**. Wakefulness. OX-A neurons are entrained by a blue light-mediated activation of the retinohypothalamic pathway (RHT). This phenomenon induces a 2-AG synthesis at the OX-A target neurons and consequent DSE-mediated inhibition of OX-A neurons. These effects reduce both OX-A release and wakefulness. **D**. Analgesia. Incoming nociceptive signals are under the control of descending excitatory projections from vlPAG neurons to RVM. These projections undergo disinhibition by glutamate release and 2-AG-mediated DSI, respectively, at OX-A- and CB1-positive inputs to vlPAG output neurons to ON and OFF cells in the RVM. Hypothalamic leptin signal deficiency potentiates OX-A release to PAG and facilitates disinhibition of descending antinociceptive pathway and analgesia

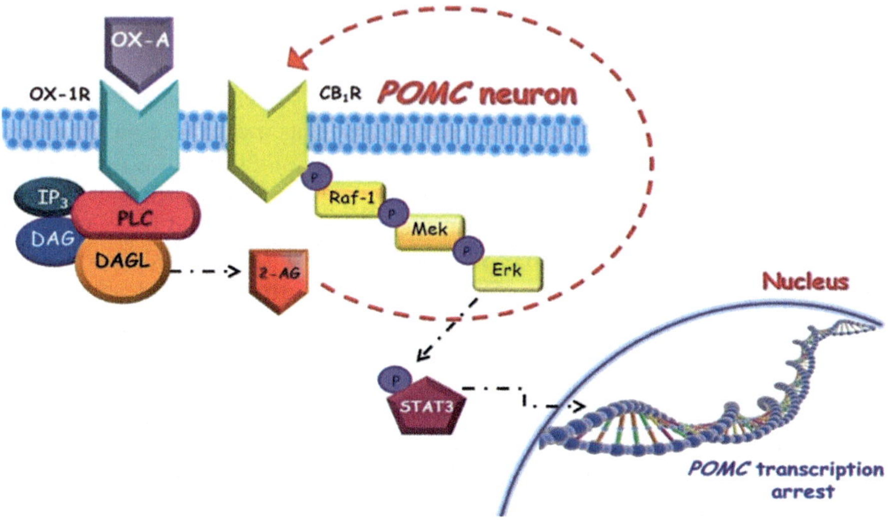

Fig. 6 Scheme of the molecular pathway underlying OX-A-mediated CB1 activation at POMC neurons. OX-A-mediated CB_1R activation via 2-AG production contributes to increase the OX-A potency to phosphorylate ERK1/2 (pERK1/2$^{Thr202/Tyr204}$) in POMC neurons concurrently with enhancement of STAT3 phosphorylation (pSTAT3^{Ser727}). These effects result in the lowering of POMC production and POMC-derived α-MSH levels

to new safer and more efficacious polypharmacological or multi-target treatments against metabolic disorders (Kälin et al. 2015).

6.2 Reward, Arousal, and Seeking Behavior

Orexin and endocannabinoids have been shown to be involved in reward processes, drug-seeking behavior, and addiction in animal models and human (España 2012; Oleson et al. 2012; Baimel et al. 2015; Hernandez and Cheer 2015). Orexinergic neurons project to the VTA and NAcc, the master regions of the brain critical for motivation and reward behavior (Wise and Rompre 1989; España 2012) whose activation is associated with preferences for cues linked with drug and food rewards (Harris et al. 2005). In particular, in the VTA, orexin directly activates dopaminergic neurons through OX-1R (Nakamura et al. 2000) or affects glutamate release onto dopaminergic neurons, inducing excitatory synaptic plasticity (Borgland et al. 2010; Baimel and Borgland 2012) and causing DA release in VTA target regions, such as the NAcc (Vittoz and Berridge 2006; Narita et al. 2006). More recently, orexin was also linked to cannabinoid-induced reward. Chiou and collaborators have proposed another cellular mechanism through which orexin can increase VTA dopaminergic activity by triggering endocannabinoid synthesis (Chiou et al. 2013). Endocannabinoid production and release have been found to occur at VTA-DA neurons (Alger 2002; Melis et al. 2004). Electrophysiological studies demonstrated that activation of OX-1R in VTA-DA neurons initiates a Gq/11-coupled PLC-DAGL pathway leading to the biosynthesis of 2-AG, which, retrogradely, activates CB1 at both inhibitory and excitatory inputs to DA neurons, thus regulating burst firing (French 1997; Wu and French 2000) and neurotransmitter release (Chen et al. 1990; Cheer et al. 2007; Oleson et al. 2012) of VTA-DA neurons. In resting conditions, around 50% of DA neurons are innervated by inhibitory GABAergic inputs (Grace and Bunney 1984). The retrograde activation of CB1 on GABA inputs results in increase of VTA-DA firing and DA release to target areas (Overton and Clark 1997; Zweifel et al. 2009; Mátyás et al. 2008; Chiou et al. 2013). However, although in normal conditions the final endocannabinoid-mediated modulation of DA neuron activity depends on the functional balance between inhibitory and excitatory inputs to these neurons, acute restraint stress activates orexinergic neurons leading to downstream release of OX-A to the VTA and consequent activation of OX-1R of DA neurons. This event triggers the synthesis of 2-AG, which, through a retrograde inhibition of GABA release, induces disinhibition of VTA-DA neurons and initiation of seeking behavior. The opposite scenario occurs in the DR, where the OX-B-mediated inhibition of glutamate release is due to OX-2R-induced enhancement of 2-AG release and activation of CB1 at excitatory inputs and depression of glutamate-mediated synaptic currents of DR serotonergic neurons, as demonstrated by patch-clamp study in male rats (Haj-Dahmane and Shen 2005) (Fig. 5). This effect was mimicked by WIN55,212-2 and abolished by AM-251, confirming the involvement of CB1 receptors. The inhibition of glutamate release was also counteracted by inhibition of G-protein signaling in postsynaptic neurons by GDPβS, a non-hydrolyzable analog of GDP. These results were the first to suggest an orexin/endocannabinoid interaction in the regulation of DR serotonergic neurons by orexin-induced postsynaptic release of ECs. Since serotonergic neurons are activated during wakefulness, the inhibitory action of OX-B, an arousal-increasing neuropeptide, on these neurons therefore seems paradoxical. The authors, however, speculated that this negative feedback induced by endocannabinoid release is necessary for preventing excessive neuronal excitation, therefore ensuring a stable firing (Haj-Dahmane and Shen 2005). Accordingly, the orexin and endocannabinoid interaction is important in the physiological regulation of arousal. In agreement with the role of endocannabinoids and orexins in the regulation of anxiety-like responses, the overlapping distribution of both their receptors has been described in the VTA, NAcc, PFCx, septal nuclei, and amygdaloid nuclei (Maldonado et al. 2006; Aston-Jones et al. 2010; Plaza-Zabala et al. 2012; Flores et al. 2015), besides the DR and LC. Therefore, it can be speculated that orexins exert orexigenic functions by controlling both appetite and reward circuits through stimulation of endocannabinoid release. This is of special relevance to the observation that orexins, via mesolimbic circuits, promote ingestion of highly salient substances (e.g., high-fat diet, drugs of abuse), at least in part, via direct projections onto VTA-DA neurons (Petrovich et al. 2012; Cason and Aston-Jones 2013), which promote NMDA receptor-mediated excitatory postsynaptic potentials at DA neurons and DA release into the NA and PFCx (Borgland et al. 2006; Vittoz and Berridge 2006; España et al. 2010). However, release of orexin and glutamate is required for long-term potentiation of DA signaling that underlies cue-induced reinstatement (seeking) of rewards. Notably, given that both OX-A and OX-B are co-released with dynorphin and glutamate, it has been proposed that both the orexins potentiate the glutamate-mediated long-term modifications that underlie natural reward and addiction to drugs (Mahler et al. 2013). By contrast, lowering the orexins vs. dynorphin ratio suppresses reward responses (Muschamp et al. 2014) by inhibiting DA release to the NAcc and drug, food, or sucrose seeking (Abizaid et al. 2006; España et al. 2010; Sharf et al. 2010; Smith and Aston-Jones 2012; Srinivasan et al. 2012).

6.3 Sleep/Wakefulness

It has been observed that orexinergic neurons are active during wakefulness (Modirrousta et al. 2005), while MCH neurons are active in REMS (Verret et al. 2003). Notably, OX knockout (KO) mice exhibit a narcoleptic-like phenotype, with sudden transitions from wakefulness into REMS (Chemelli et al. 1999). Furthermore, it is known that eCBs have been implicated in sleep regulation because of their strong hypnogenic properties (Pérez-Morales et al. 2012). Indeed, THC, the primary psychoactive agent in marijuana and hashish, tends to generate a distorted sense of time (Tinklenberg et al. 1976). The CB1 agonist CP55940 attenuates light-induced clock-phase advance in hamsters (Sanford et al. 2008), and endocannabinoids, both 2-AG and AEA, show a circadian variation in the brain (Valenti et al. 2004). Interaction between the orexinergic and endocannabinoid systems in the sleep regulation has been described by Pérez-Morales et al. (2013) starting from evidence indicating that WIN55-212-2 depolarizes MCH neurons and hyperpolarizes OX-A neurons in in vitro LH slices (Huang et al. 2007). By injection of 2-AG in the LH of rats, Pérez-Morales and colleagues found that 2-AG increases REMS through a CB1 activation and increases c-Fos expression in MCH neurons, without affecting c-Fos expression in OX-A neurons. Furthermore, Cristino et al. (2013a, b) studied the orexinergic and endocannabinoid interaction in wakefulness. They found the endocannabinoid biosynthesis in the LH of mice induced by a light-modulated excitation of the retinohypothalamic tract (RHT). By electron and confocal microscopy, the authors found CB1 expression at RHT projections to OX-A neurons and identified the retinal ganglion cells/OX-A light-mediated circuit activated by blue light pulse by inducing c-*Fos* expression in the target OX-A neurons (Fig. 5). This activation was paralleled by 2-AG synthesis in a manner prevented by antagonism of mGluR5 receptors which were, in turn, found at OX-A neurons. Collectively, these data suggest that orexinergic and endocannabinoid interaction, in specialized neuronal pathways, regulates different aspects of sleep/wake cycle including the REMS onset and the blue light entrainment of circadian functions.

6.4 Nociception

One common function for both orexin and endocannabinoid systems is the modulation of pain perception at spinal and supraspinal levels. The antinociceptive descending PAG-RVM pathway is crucial in the regulation of pain. PAG contains glutamatergic neurons receiving inhibitory GABAergic inputs from the local interneurons. PAG-mediated control of pain occurs concomitantly with the modulation of pain-responding neurons of the RVM: the ON cells, which are activated, and the OFF cells, which are inhibited, by nociceptive stimuli. In the ventrolateral PAG (vlPAG), activation of excitatory output neurons projecting monosynaptically to OFF cells in the rostral ventromedial medulla (RVM) causes antinociceptive responses via OFF cell stimulation and ON cell inhibition in the RVM, which send inhibitory projections to the dorsal horn of the spinal cord (Behbehani and Zemlan 1990). By morphological approach based on the injection of the retrograde tracing CTB-Alexa488 in the RVM of mice, Cristino and colleagues revealed the occurrence of monosynaptic projections from vlPAG neurons to the OFF cells in the RVM. More importantly, by high-resolution electron microscopy, the authors found that these OX-A neurons targeted by RHT express DAGLα at the cytoplasmic side of the membrane, near to OX-1R at postsynaptic side of excitatory OX-A-positive inputs, and are innervated by CB1-positive inhibitory inputs coming from local GABAergic interneurons (Cristino et al. 2016) (Fig. 5). This study provided the anatomical contribution to understand the OX-1R-mediated PLC-DAGL-2-AG disinhibition of the PAG-RVM descending antinociceptive pathway found by Ho et al. (2011). Indeed, Ho and colleagues were the first to demonstrate that 2-AG can be generated by OX-A-mediated activation of OX-1R in the rat PAG by using O-7460 (a DAGLα inhibitor) during patch-clamp recording of vlPAG slices. They found that, unlike OX-B, OX-A depressed IPSCs in PAG slices in a manner dependent on OX-1R, CB1, PLC, and DAGL activity (Ho et al. 2011) and was mimicked by a MAGL inhibitor. These data suggest that activation of OX-1R at postsynaptic site, as well as activation of mGluR5 and M1/M3 mAChRs, initiates the OX-1R-mediated PLC-DAGL-2-AG disinhibition mechanism in the PAG, thereby contributing to analgesia induced by intra-vlPAG injection of OX-A during the rat hot plate test (Chiou et al. 2013). More recently, Lee et al. found that 30-min restraint stress in mice was able to induce analgesia during hot plate test in a manner prevented by intra-vlPAG injection or intraperitoneal injection of SB334867 or AM251, but not TCS-OX2-29 or naloxone. They found that, during stress, orexins lead to endocannabinoid generation in the vlPAG by engaging the OX-1R-mediated PLC-DAGLα-2-AG pathway, thereby inducing the establishment of stress-induced analgesia (Lee et al. 2016). Accordingly, Cristino et al. found that reduced-pain sensitivity and enhanced OFF-decreased ON cell activity occur in *ob/ob* mice wherein the leptin deficiency strongly enhances OX-A release to vlPAG, among other LH targets. In *ob/ob* mice, these alterations result in (i) increased OX-1R-mediated PLC-DAGL-2-AG disinhibition of vlPAG output neurons by DSI of the neighboring CB1-expressing GABAergic inputs; (ii) subsequent increase of OFF and decrease of ON cell activity in the RVM, as assessed by patch clamp and in vivo electrophysiology; and (iii) analgesia, in both healthy and neuropathic mice. Notably, in HFD obese mice, analgesia was only unmasked following leptin receptor antagonism because of the leptin insensitivity of these mice. All these effects were mimicked by i.c.v. OX-A injection in lean mice and were markedly reduced by treatment with AM251 or SB in *ob/ob* mice at a dose inactive in wt mice (Cristino et al. 2013a, b). According to the orexin/endocannabinoid synergistic effect in the control of analgesia, Kargar et al. found that intra-LC microinjection of OX-A exerted antinociceptive function in the rat formalin test (a model of inflammatory pain), in a manner prevented by intra-LC microinjection of either SB334867 or AM251 (Kargar et al. 2015).

References

Abdulnour J, Yasari S, Rabasa-Lhoret R, Faraj M, Petrosino S, Piscitelli F, Prud' Homme D, Di Marzo V (2014) Circulating endocannabinoids in insulin sensitive vs. insulin resistant obese postmenopausal women. A MONET group study. Obesity 22:211–216

Abizaid A, Gao Q, Horvath TL (2006) Thoughts for food: brain mechanisms and peripheral energy balance. Neuron 51(6):691–702

Adler ES, Hollis JH, Clarke IJ, Grattan DR, Oldfield BJ (2012) Neurochemical characterization and sexual dimorphism of projections from the brain to abdominal and subcutaneous white adipose tissue in the rat. J Neurosci 32(45):15913–15921

Alger BE (2002) Retrograde signaling in the regulation of synaptic transmission: focus on endocannabinoids. Prog Neurobiol 68:247–286

Ammoun S, Johansson L, Ekholm ME, Holmqvist T, Danis AS, Korhonen L, Sergeeva OA, Haas HL, Åkerman KE, Kukkonen JP (2006a) Ox1 orexin receptors activate extracellular signal-regulated kinase (erk) in cho cells via multiple mechanisms: the role of CA2+ influx in OX1 receptor signaling. Mol Endocrinol 20:80–99

Ammoun S, Lindholm D, Wootz H, Åkerman KE, Kukkonen JP (2006b) Gprotein-coupled ox1 orexin/hcrtr-1 hypocretin receptors induce caspase-7 3 dependent and -independent cell death through p38 mitogen-/stress-activated protein kinase. J Biol Chem 281:834–842

Aston-Jones G, Smith RJ, Sartor GC, Moorman DE, Massi L, Tahsili-Fahadan P, Richardson KA (2010) Lateral hypothalamic orexin/hypocretin neurons: a role in reward-seeking and addiction. Brain Res 1314:74–90

Baimel C, Borgland SL (2012) Hypocretin modulation of drug-induced synaptic plasticity. Prog Brain Res 198:123–131

Baimel C, Bartlett SE, Chiou LC, Lawrence AJ, Muschamp JW, Patkar O, Tung LW, Borgland S (2015) Orexin/hypocretin role in reward: implications for opioid and other addictions. Br J Pharmacol 172(2):334–348

Balthasar N, Coppari R, McMinn J, Liu SM, Lee CE, Tang V, Kenny CD, McGovern RA, Chua SC Jr, Elmquist JK, Lowell BB (2004) Leptin receptor signalling in POMC neurons is required for normal body weight homeostasis. Neuron 42:983–991

Banks WA, Kastin AJ, Huang W, Jaspan JB, Maness LM (1996) Leptin enters the brain by a saturable system independent of insulin. Peptides 17(2):305–311

Behbehani MM, Zemlan FP (1990) Bulbospinal and intraspinal thyrotropin releasing hormone systems: modulation of spinal cord pain transmission. Neuropeptides 15(3):161–168

Bermudez-Silva FJ, Vivero MP, McPartland JM, Rodriguez de Fonseca F (2010) The endocannabinoid system, eating behavior and energy homeostasis: the end or a new beginning? Pharmacol Biochem Behav 95:375–382
Borgland SL, Taha SA, Sarti F, Fields HL, Bonci A (2006) Orexin A in the VTA is critical for the induction of synaptic plasticity and behavioral sensitization to cocaine. Neuron 49(4):589–601
Borgland SL, Ungless MA, Bonci A (2010) Convergent actions of orexin/hypocretin and CRF on dopamine neurons: emerging players in addiction. Brain Res 1314:139–144
Bouaboula M, Poinot-Chazel C, Bourrié B, Canat X, Calandra B, Rinaldi-Carmona M, Le Fur G, Casellas P (1995) Activation of mitogen-activated protein kinases by stimulation of the central cannabinoid receptor CB1. Biochem J 312(Pt 2):637–641
Carriba P, Ortiz O, Patkar K, Justinova Z, Stroik J, Themann A, Müller C, Woods AS, Hope BT, Ciruela F, Casadó V, Canela EI, Lluis C, Goldberg SR, Moratalla R, Franco R, Ferré S (2007) Striatal adenosine A2A and cannabinoid CB1 receptors form functional heteromeric complexes that mediate the motor effects of cannabinoids. Neuropsychopharmacology 32 (11):2249–2259
Cason AM, Aston-Jones G (2013) Role of orexin/hypocretin in conditioned sucrose-seeking in rats. Psychopharmacology 226(1):155–165
Cheer JF, Aragona BJ, Heien ML, Seipel AT, Carelli RM, Wightman RM (2007) Coordinated accumbal dopamine release and neural activity drive goal-directed behavior. Neuron 54:237–244
Chemelli RM, Willie JT, Sinton CM, Elmquist JK, Scammell T, Lee C, Richardson JA, Williams SC, Xiong Y, Kisanuki Y, Fitch TE, Nakazato M, Hammer RE, Saper CB, Yanagisawa M (1999) Narcolepsy in orexin knockout mice: molecular genetics of sleep regulation. Cell 98:437–451
Chen J, Paredes W, Lowinson JH, Gardner EL (1990) Delta9-tetrahydrocannabinol enhances presynaptic dopamine efflux in medial prefrontal cortex. Eur J Pharmacol 190:259–262
Chiou LC, Tung LW, Lee YS, Lu GL, Lee HJ, Chagne LY et al. (2013) Orexin-endocannabinoid signaling in stress-induced cocaine relapse. International Narcotic Research Conference, Cairns, Australia
Chou TC, Lee CE, Lu J, Elmquist JK, Hara J, Willie JT, Beuckmann CT, Chemelli RM, Sakurai T, Yanagisawa M, Saper CB, Scammell TE (2001) Orexin (hypocretin) neurons contain dynorphin. J Neurosci 21(19):RC168
Colombo G, Agabio R, Diaz G, Lobina C, Reali R, Gessa GL (1998) Appetite suppression and weight loss after the cannabinoid antagonist SR 141716. Life Sci 63(8):PL113–PL117
Cone RD (2005) Anatomy and regulation of the central melanocortin system. Nat Neurosci 8:571–578
Considine RV, Sinha MK, Heiman ML, Kriauciunas A, Stephens TW, Nyce MR, Ohannesian JP, Marco CC, McKee LJ, Bauer TL et al (1996) Serum immunoreactive-leptin concentrations in normal-weight and obese humans. N Engl J Med 334(5):292–295
Cota D (2007) CB1 receptors: emerging evidence for central and peripheral mechanisms that regulate energy balance, metabolism, and cardiovascular health. Diabetes Metab Res Rev 23:507–517
Cota D, Marsicano G, Lutz B, Vicennati V, Stalla GK, Pasquali R, Pagotto U (2003) Endogenous cannabinoid system as a modulator of food intake. Int J Obes Relat Metab Disord 27 (3):289–301
Crespo I, Gómez de Heras R, Rodríguez de Fonseca F, Navarro M (2008) Pretreatment with subeffective doses of Rimonabant attenuates orexigenic actions of orexin A-hypocretin 1. Neuropharmacology 54(1):219–225
Cristino L, Busetto G, Imperatore R, Ferrandino I, Palomba L, Silvestri C, Petrosino S, Orlando P, Bentivoglio M, Mackie K, Di Marzo V (2013a) Obesity-driven synaptic remodeling affects endocannabinoid control of orexinergic neurons. Proc Natl Acad Sci U S A 110(24):E2229–E2238

Cristino L, Imperatore R, Becker T, Di Spiezio A, Bentivoglio M, Di Marzo V (2013b) The retinohypothalamic patwhay is involved in light-modulated endocannabinoid biosynthesis in the lateral hypothalamus. Abstract B.01 *SfN* 2013

Cristino L, Luongo L, Imperatore R, Boccella S, Becker T, Morello G, Piscitelli F, Busetto G, Maione S, Di Marzo V (2016) Orexin-A and endocannabinoid activation of the descending antinociceptive pathway underlies altered pain perception in leptin signaling deficiency. Neuropsychopharmacology 41(2):508–520

Date Y, Ueta Y, Yamashita H, Yamaguchi H, Matsukura S, Kangawa K, Sakurai T, Yanagisawa M, Nakazato M (1999) Orexins, orexigenic hypothalamic peptides, interact with autonomic, neuroendocrine and neuroregulatory systems. Proc Natl Acad Sci U S A 96 (2):748–753

Davis MI, Ronesi J, Lovinger DM (2003) A predominant role for inhibition of the adenylate cyclase/protein kinase A pathway in ERK activation by cannabinoid receptor 1 in N1E-115 neuroblastoma cells. J Biol Chem 278(49):48973–48980

de Lecea L, Kilduff TS, Peyron C, Gao X, Foye PE, Danielson PE, Fukuhara C, Battenberg EL, Gautvik VT, Bartlett FS 2nd, Frankel WN, van den Pol AN, Bloom FE, Gautvik KM, Sutcliffe JG (1998) The hypocretins: hypothalamus-specific peptides with neuroexcitatory activity. Proc Natl Acad Sci U S A 95:322–327

Deadwyler SA, Hampson RE, Mu J, Whyte A, Childers S (1995) Cannabinoids modulate voltage sensitive potassium A-current in hippocampal neurons via a cAMP-dependent process. J Pharmacol Exp Ther 273(2):734–743

Devane WA, Hanus L, Breuer A, Pertwee RG, Stevenson LA, Griffin G, Gibson D, Mandelbaum A, Etinger A, Mechoulam R (1992) Isolation and structure of a brain constituent that binds to the cannabinoid receptor. Science 258:1946–1949

Di Marzo V (2009) The endocannabinoid system: its general strategy of action, tools for its pharmacological manipulation and potential therapeutic exploitation. Pharmacol Res 60 (2):77–84

Di Marzo V, Goparaju SK, Wang L, Liu J, Bátkai S, Járai Z, Fezza F, Miura GI, Palmiter RD, Sugiura T, Kunos G (2001) Leptin-regulated endocannabinoids are involved in maintaining food intake. Nature 410:822–825

Di Marzo V, Verrijken A, Hakkarainen A, Petrosino S, Mertens I, Lundbom N, Piscitelli F, Westerbacka J, Soro-Paavonen A, Matias I, Van Gaal L, Taskinen MR (2009) Role of insulin as a negative regulator of plasma endocannabinoid levels in obese and non-obese subjects. Eur J Endocrinol 161:715–722

Ellis J, Pediani JD, Canals M, Milasta S, Milligan G (2006) Orexin-1 receptor-cannabinoid CB1 receptor heterodimerization results in both ligand-dependent and -independent coordinated alterations of receptor localization and function. J Biol Chem 281(50):38812–38824

Engeli S, Böhnke J, Feldpausch M, Gorzelniak K, Janke J, Bátkai S, Pacher P, Harvey-White J, Luft FC, Sharma AM, Jordan J (2005) Activation of the peripheral endocannabinoid system in human obesity. Diabetes 54:2838–2843

España RA (2012) Hypocretin/orexin involvement in reward and reinforcement. Vitam Horm 89:185–208

España RA, Oleson EB, Locke JL, Brookshire BR, Roberts DC, Jones SR (2010) The hypocretin-orexin system regulates cocaine self-administration via actions on the mesolimbic dopamine system. Eur J Neurosci 31(2):336–348

Fernández-Ruiz J, Hernández M, Ramos JA (2010) Cannabinoid-dopamine interaction in the pathophysiology and treatment of CNS disorders. CNS Neurosci Ther 16(3):e72–e91

Ferré S, Goldberg SR, Lluis C, Franco R (2009) Looking for the role of cannabinoid receptor heteromers in striatal function. Neuropharmacology 56(Suppl 1):226–234

Flores A, Saravia R, Maldonado R, Berrendero F (2015) Orexins and fear: implications for the treatment of anxiety disorders. Trends Neurosci 38:550–559

Fox EA (2006) A genetic approach for investigating vagal sensory roles in regulation of gastrointestinal function and food intake. Auton Neurosci 12:9–29

French ED (1997) delta9-Tetrahydrocannabinol excites rat VTA dopamine neurons through activation of cannabinoid CB1 but not opioid receptors. Neurosci Lett 226:159–162

Friedman JM, Halaas JL (1998) Leptin and the regulation of body weight in mammals. Nature 395 (6704):763–770

Garcia DE, Brown S, Hille B, Mackie K (1998) Protein kinase C disrupts cannabinoid actions by phosphorylation of the CB1 cannabinoid receptor. J Neurosci 18(8):2834–2841

Gómez R, Navarro M, Ferrer B, Trigo JM, Bilbao A, Del Arco I, Cippitelli A, Nava F, Piomelli D, Rodríguez de Fonseca F (2002) A peripheral mechanism for CB1 cannabinoid receptor-dependent modulation of feeding. J Neurosci 22:9612–9617

Gotter AL, Webber AL, Coleman PJ, Renger JJ, Winrow CJ (2012) International union of basic and clinical pharmacology. Lxxxvi. Orexin receptor function, nomenclature and pharmacology. Pharmacol Rev 64:389–420

Grabauskas G, Moises HC (2003) Gastrointestinal-projecting neurones in the dorsal motor nucleus of the vagus exhibit direct and viscerotopically organized sensitivity to orexin. J Physiol 549 (Pt 1):37–56

Grace AA, Bunney BS (1984) The control of firing pattern in nigral dopamine neurons: burst firing. J Neurosci 4:2877–2890

Haj-Dahmane S, Shen RY (2005) The wake-promoting peptide orexin-B inhibits glutamatergic transmission to dorsal raphe nucleus serotonin neurons through retrograde endocannabinoid signaling. J Neurosci 25(4):896–905

Håkansson M, de Lecea L, Sutcliffe JG, Yanagisawa M, Meister B (1999) Leptin receptor- and STAT3-immunoreactivities in hypocretin/orexin neurones of the lateral hypothalamus. J Neuroendocrinol 11(8):653–663

Hampson RE, Evans GJ, Mu J, Zhuang SY, King VC, Childers SR, Deadwyler SA (1995) Role of cyclic AMP dependent protein kinase in cannabinoid receptor modulation of potassium "A-current" in cultured rat hippocampal neurons. Life Sci 56(23-24):2081–2088

Harris GC, Wimmer M, Aston-Jones G (2005) A role for lateral hypothalamic orexin neurons in reward seeking. Nature 437(7058):556–559

Haynes AC, Jackson B, Chapman H, Tadayyon M, Johns A, Porter RA, Arch JR (2000) A selective orexin-1 receptor antagonist reduces food consumption in male and female rats. Regul Pept 96(1–2):45–51

Hernandez G, Cheer JF (2015) To Act or Not to Act: endocannabinoid/dopamine interactions in decision-making. Front Behav Neurosci 9:336

Hilairet S, Bouaboula M, Carrière D, Le Fur G, Casellas P (2003) Hypersensitization of the Orexin 1 receptor by the CB1 receptor: evidence for cross-talk blocked by the specific CB1 antagonist, SR141716. J Biol Chem 278(26):23731–23737

Hill JW, Elias CF, Fukuda M, Williams KW, Berglund ED, Holland WL, Cho YR, Chuang JC, Xu Y, Choi M, Lauzon D, Lee CE, Coppari R, Richardson JA, Zigman JM, Chua S, Scherer PE, Lowell BB, Brüning JC, Elmquist JK (2010) Direct insulin and leptin action on pro-opiomelanocortin neurons is required for normal glucose homeostasis and fertility. Cell Metab 11:286–297

Ho YC, Lee HJ, Tung LW, Liao YY, Fu SY, Teng SF, Liao HT, Mackie K, Chiou LC (2011) Activation of orexin 1 receptors in the periaqueductal gray of male rats leads to antinociception via retrograde endocannabinoid (2-arachidonoylglycerol)-induced disinhibition. J Neurosci 31:14600–14610

Horvath TL, Diano S, van den Pol AN (1999) Synaptic interaction between hypocretin (orexin) and neuropeptide Y cells in the rodent and primate hypothalamus: a novel circuit implicated in metabolic and endocrine regulations. J Neurosci 19:1072–1087

Howlett AC, Barth F, Bonner TI, Cabral G, Casellas P, Devane WA, Felder CC, Herkenham M, Mackie K, Martin BR, Mechoulam R, Pertwee RG (2002) International Union of Pharmacology. XXVII. Classification of cannabinoid receptors. Pharmacol Rev 54(2):161–202

Huang H, Acuna-Goycolea C, Li Y, Cheng HM, Obrietan K, Van Den Pol AN (2007) Cannabinoids excite hypothalamic melanin-concentrating hormone but inhibit hypocretin/orexin neurons: implications for cannabinoid actions on food intake and cognitive arousal. J Neurosci 27:4870–4881

Imperatore R, Palomba L, Cristino L (2017) The role of orexin-A in hypertension and obesity. Curr Hypertens Rep 19:21–34

Jäntti MH, Putula J, Somerharju P, Frohman MA, Kukkonen JP (2012) OX1 orexin/hypocretin receptor activation of phospholipase D. Br J Pharmacol 165(4b):1109–1123

Jäntti MH, Putula J, Turunen PM, Näsman J, Reijonen S, Lindqvist C, Kukkonen JP (2013) Autocrine endocannabinoid signaling through CB1 receptors potentiates OX1 orexin receptor signaling. Mol Pharmacol 83(3):621–632

Johansson L, Ekholm ME, Kukkonen JP (2007) Regulation of ox(1) orexin/hypocretin receptor-coupling to phospholipase c by ca(2+) influx. Br J Pharmacol 150:97–104

Johansson L, Ekholm ME, Kukkonen JP (2008) Multiple phospholipase activation by ox (1) orexin/hypocretin receptors. Cell Mol Life Sci 65:1948–1956

Jöhren O, Neidert SJ, Kummer M, Dendorfer A, Dominiak P (2001) Preproorexin and orexin receptor mRNAs are differentially expressed in peripheral tissues of male and female rats. Endocrinology 142:3324–3331

Kälin S, Heppner FL, Bechmann I, Prinz M, Tschöp MH, Yi CX (2015) Hypothalamic innate immune reaction in obesity. Nat Rev Endocrinol 11:339–351

Kano M, Ohno-Shosaku T, Hashimotodani Y, Uchigashima M, Watanabe M (2009) Endocannabinoid-mediated control of synaptic transmission. Physiol Rev 89(1):309–380

Kargar HM, Azizi H, Mirnajafi-Zadeh J, Reza MA, Semnanian S (2015) Microinjection of orexin-A into the rat locus coeruleus nucleus induces analgesia via cannabinoid type-1 receptors. Brain Res 1624:424–432

Kastin AJ, Akerstrom V (1999) Orexin A but not orexin B rapidly enters brain from blood by simple diffusion. J Pharmacol Exp Ther 289:219–223

Kirkham TC, Williams CM, Fezza F, Di Marzo V (2002) Endocannabinoid levels in rat limbic forebrain and hypothalamus in relation to fasting, feeding and satiation: stimulation of eating by 2-arachidonoyl glycerol. Br J Pharmacol 136(4):550–557

Koch M, Varela L, Kim JG, Kim JD, Hernández-Nuño F, Simonds SE, Castorena CM, Vianna CR, Elmquist JK, Morozov YM, Rakic P, Bechmann I, Cowley MA, Szigeti-Buck K, Dietrich MO, Gao XB, Diano S, Horvath TL (2015) Hypothalamic POMC neurons promote cannabinoid-induced feeding. Nature 19:45–50

Kukkonen JP (2011) A ménage à trois made in heaven: G-protein-coupled receptors, lipids and TRP channels. Cell Calcium 50(1):9–26

Kukkonen JP (2013) Physiology of the orexinergic/hypocretinergic system: a revisit in 2012. Am J Physiol Cell Physiol 304:C2–32

Kukkonen JP, Leonard CS (2014) Orexin/hypocretin receptor signalling cascades. Br J Pharmacol 171(2):314–331

Lee SJ, Kirigiti M, Lindsley SR, Loche A, Madden CJ, Morrison SF, Smith MS, Grove KL (2013) Efferent projections of neuropeptide Y-expressing neurons of the dorsomedial hypothalamus in chronic hyperphagic models. J Comp Neurol 521(8):1891–1914

Lee HJ, Chang LY, Ho YC, Teng SF, Hwang LL, Mackie K, Chiou LC (2016) Stress induces analgesia via orexin 1 receptor-initiated endocannabinoid/CB1 signaling in the mouse periaqueductal gray. Neuropharmacology 105:577–586

López M, Lage R, Tung YC, Challis BG, Varela L, Virtue S, O'Rahilly S, Vidal-Puig A, Diéguez C, Coll AP (2007) Orexin expression is regulated by alpha-melanocyte-stimulating hormone. J Neuroendocrinol 19:703–707

Louhivuori LM, Jansson L, Nordström T, Bart G, Näsman J, Akerman KE (2010) Selective interference with TRPC3/6 channels disrupts OX1 receptor signalling via NCX and reveals a distinct calcium influx pathway. Cell Calcium 48(2–3):114–123

Lu XY, Bagnol D, Burke S, Akil H, Watson SJ (2000) Differential distribution and regulation of ox1 and ox2 orexin/hypocretin receptor messenger rna in the brain upon fasting. Horm Behav 37:335–344

Lund PE, Shariatmadari R, Uustare A, Detheux M, Parmentier M, Kukkonen JP, Åkerman KEO (2000) The orexin ox1 receptor activates a novel ca2+ influx pathway necessary for coupling to phospholipase c. J Biol Chem 275:30806–30812

Mahler SV, Smith RJ, Aston-Jones G (2013) Interactions between VTA orexin and glutamate in cue-induced reinstatement of cocaine seeking in rats. Psychopharmacology 226(4):687–698

Maldonado R, Valverde O, Berrendero F (2006) Involvement of the endocannabinoid system in drug addiction. Trends Neurosci 29:225–232

Marcus JN, Aschkenasi CJ, Lee CE, Chemelli RM, Saper CB, Yanagisawa M, Elmquist JK (2001) Differential expression of orexin receptors 1 and 2 in the rat brain. J Comp Neurol 435:6–25

Matias I, Di Marzo V (2007) Endocannabinoids and the control of energy balance. Trends Endocrinol Metab 18:27–37

Matias I, Vergoni AV, Petrosino S, Ottani A, Pocai A, Bertolini A, Di Marzo V (2008) Regulation of hypothalamic endocannabinoid levels by neuropeptides and hormones involved in food intake and metabolism: insulin and melanocortins. Neuropharmacology 54:206–212

Mátyás F, Urbán GM, Watanabe M, Mackie K, Zimmer A, Freund TF et al (2008) Identification of the sites of 2-arachidonoylglycerol synthesis and action imply retrograde endocannabinoid signaling at both GABAergic and glutamatergic synapses in the ventral tegmental area. Neuropharmacology 54:95–107

Melck D, Rueda D, Galve-Roperh I, De Petrocellis L, Guzmán M, Di Marzo V (1999) Involvement of the cAMP/protein kinase A pathway and of mitogen-activated protein kinase in the anti-proliferative effects of anandamide in human breast cancer cells. FEBS Lett 463 (3):235–240

Melis M, Pistis M, Perra S, Muntoni AL, Pillolla G, Gessa GL (2004) Endocannabinoids mediate presynaptic inhibition of glutamatergic transmission in rat ventral tegmental area dopamine neurons through activation of CB1 receptors. J Neurosci 24:53–62

Meves H (2008) Arachidonic acid and ion channels: an update. Br J Pharmacol 155(1):4–16

Milasta S, Evans NA, Ormiston L, Wilson S, Lefkowitz RJ, Milligan G (2005) The sustainability of interactions between the orexin-1 receptor and β-arrestin-2 is defined by a single C-terminal cluster of hydroxy amino acids and modulates the kinetics of ERK MAPK regulation. Biochem J 387(Pt 3):573–584

Modirrousta M, Mainville L, Jones BE (2005) Orexin and MCH neurons express c-Fos differently after sleep deprivation vs. recovery and bear different adrenergic receptors. Eur J Neurosci 21 (10):2807–2816

Monteleone P, Matias I, Martiadis V, De Petrocellis L, Maj M, Di Marzo V (2005) Blood levels of the endocannabinoid anandamide are increased in anorexia nervosa and in binge-eating disorder, but not in bulimia nervosa. Neuropsychopharmacology 30:1216–1221

Morello G, Imperatore R, Palomba L, Finelli C, Labruna G, Pasanisi F, Sacchetti L, Buono L, Piscitelli F, Orlando P, Di Marzo V, Cristino L (2016) Orexin-A represses satiety-inducing POMC neurons and contributes to obesity via stimulation of endocannabinoid signalling. Proc Natl Acad Sci U S A 113(17):4759–4764. doi:10.1073/pnas.1521304113

Muschamp JW, Hollander JA, Thompson JL, Voren G, Hassinger LC, Onvani S, Kamenecka TM, Borgland SL, Kenny PJ, Carlezon WA Jr (2014) Hypocretin (orexin) facilitates reward by attenuating the antireward effects of its cotransmitter dynorphin in ventral tegmental area. Proc Natl Acad Sci U S A 111(16):E1648–E1655

Nakajima Y, Nakajima S (2010) Measurement of orexin (hypocretin) and substance p effects on constitutively active inward rectifier k(+) channels in brain neurons. Methods Enzymol 484:613–630

Nakamura T, Uramura K, Nambu T, Yada T, Goto K, Yanagisawa M, Sakurai T (2000) Orexin-induced hyperlocomotion and stereotypy are mediated by the dopaminergic system. Brain Res 873(1):181–187

Nambu T, Sakurai T, Mizukami K, Hosoya Y, Yanagisawa M, Goto K (1999) Distribution of orexin neurons in the adult rat brain. Brain Res 827:243–260

Narita M, Nagumo Y, Hashimoto S, Khotib J, Miyatake M, Sakurai T, Yanagisawa M, Nakamachi T, Shioda S, Suzuki T (2006) Direct involvement of orexinergic systems in the activation of the mesolimbic dopamine pathway and related behaviors induced by morphine. J Neurosci 26:398–405

Nicholson J, Azim S, Rebecchi MJ, Galbavy W, Feng T, Reinsel R, Rizwan S, Fowler CJ, Benveniste H, Kaczocha M (2015) Leptin levels are negatively correlated with 2-arachidonoylglycerol in the cerebrospinal fluid of patients with osteoarthritis. PLoS One 10:e0123132

Nobunaga M, Obukuro K, Kurauchi Y, Hisatsune A, Seki T, Tsutsui M, Katsuki H (2014) High fat diet induces specific pathological changes in hypothalamic orexin neurons in mice. Neurochem Int 78:61–66. doi:10.1016/j.neuint.2014.09.002

Oldfield BJ, Giles ME, Watson A, Anderson C, Colvill LM, McKinley MJ (2002) The neurochemical characterisation of hypothalamic pathways projecting polysynaptically to brown adipose tissue in the rat. Neuroscience 110(3):515–526

Oleson EB, Beckert MV, Morra JT, Lansink CS, Cachope R, Abdullah RA et al (2012) Endocannabinoids shape accumbal encoding of cue-motivated behavior via CB1 receptor activation in the ventral tegmentum. Neuron 73:360–373

Overton PG, Clark D (1997) Burst firing in midbrain dopaminergic neurons. Brain Res Brain Res Rev 25:312–334

Parolaro D, Rubino T, Viganò D, Massi P, Guidali C, Realini N (2010) Cellular mechanisms underlying the interaction between cannabinoid and opioid system. Curr Drug Targets 11 (4):393–405

Pérez-Morales M, Alvarado-Capuleño I, López-Colomé AM, Méndez-Díaz M, Ruiz-Contreras AE, Prospéro-García O (2012) Activation of PAR1 in the lateral hypothalamus of rats enhances food intake and REMS through CB1R. Neuroreport 23(14):814–818

Pérez-Morales M, De La Herrán-Arita AK, Méndez-Díaz M, Ruiz-Contreras AE, Drucker-Colín R, Prospéro-García O (2013) 2-AG into the lateral hypothalamus increases REM sleep and cFos expression in melanin concentrating hormone neurons in rats. Pharmacol Biochem Behav 108:1–7

Petrovich GD, Hobin MP, Reppucci CJ (2012) Selective Fos induction in hypothalamic orexin/ hypocretin, but not melanin-concentrating hormone neurons, by a learned food-cue that stimulates feeding in sated rats. Neuroscience 224:70–80

Peyron C, Tighe DK, Van Den Pol AN, De Lecea L, Heller HC, Sutcliffe JG, Kilduff TS (1998) Neurons containing hypocretin (orexin) project to multiple neuronal systems. J Neurosci 18:9996–10015

Plaza-Zabala A, Maldonado R, Berrendero F (2012) The hypocretin/orexin system: implications for drug reward and relapse. Mol Neurobiol 45(3):424–439

Reti IM, Reddy R, Worley PF, Baraban JM (2002) Selective expression of Narp, a secreted neuronal pentraxin, in orexin neurons. J Neurochem 82(6):1561–1565

Risold PY, Griffond B, Kilduff TS, Sutcliffe JG, Fellmann D (1999) Preprohypocretin (orexin) and prolactin-like immunoreactivity are coexpressed by neurons of the rat lateral hypothalamic area. Neurosci Lett 259(3):153–156

Robledo P, Berrendero F, Ozaita A, Maldonado R (2008) Advances in the field of cannabinoid – opioid cross-talk. Addict Biol 13(2):213–224

Rosin DL, Weston MC, Sevigny CP, Stornetta RL, Guyenet PG (2003) Hypothalamic orexin (hypocretin) neurons express vesicular glutamate transporters VGLUT1 or VGLUT2. J Comp Neurol 465(4):593–603

Sahu A (2003) Leptin signaling in the hypothalamus: emphasis on energy homeostasis and leptin resistance. Front Neuroendocrinol 24(4):225–253

Sakurai T (2007) The neural circuit of orexin (hypocretin): maintaining sleep and wakefulness. Nat Rev Neurosci 8:171–181

Sakurai T, Amemiya A, Ishii M, Matsuzaki I, Chemelli RM, Tanaka H, Williams SC, Richardson JA, Kozlowski GP, Wilson S, Arch JR, Buckingham RE, Haynes AC, Carr SA, Annan RS,

Mcnulty DE, Liu WS, Terrett JA, Elshourbagy NA, Bergsma DJ, Yanagisawa M (1998) Orexins and orexin receptors: a family of hypothalamic neuropeptides and g protein-coupled receptors that regulate feeding behavior. Cell 92:573–585

Sanford AE, Castillo E, Gannon RL (2008) Cannabinoids and hamster circadian activity rhythms. Brain Res 1222:141–148

Schwartz MW, Woods SC, Porte D Jr, Seeley RJ, Baskin DG (2000) Central nervous system control of food intake. Nature 404(6778):661–671

Selbach O, Bohla C, Barbara A, Doreulee N, Eriksson KS, Sergeeva OA, Haas HL (2010) Orexins/hypocretins control bistability of hippocampal long-term synaptic plasticity through co-activation of multiple kinases. Acta Physiol (Oxf) 198(3):277–285

Sharf R, Sarhan M, Dileone RJ (2010) Role of orexin/hypocretin in dependence and addiction. Brain Res. 1314:130–138

Shiraishi T, Oomura Y, Sasaki K, Wayner MJ (2000) Effects of leptin and orexin-A on food intake and feeding related hypothalamic neurons. Physiol Behav 71(3–4):251–261

Skrzypski MT, Le T, Kaczmarek P, Pruszynska-Oszmalek E, Pietrzak P, Szczepankiewicz D, Kolodziejski PA, Sassek M, Arafat A, Wiedenmann B, Nowak KW, Strowski MZ (2011) Orexin A stimulates glucose uptake, lipid accumulation and adiponectin secretion from 3T3-L1 adipocytes and isolated primary rat adipocytes. Diabetologia 54(7):1841–1852

Smith RJ, Aston-Jones G (2012) Orexin/hypocretin 1 receptor antagonist reduces heroin self-administration and cue-induced heroin seeking. Eur J Neurosci 35(5):798–804

Srinivasan S, Simms JA, Nielsen CK, Lieske SP, Bito-Onon JJ, Yi H, Hopf FW, Bonci A, Bartlett SE (2012) The dual orexin/hypocretin receptor antagonist, almorexant, in the ventral tegmental area attenuates ethanol self-administration. PLoS One 7(9):e44726

Sugiura T, Kondo S, Sukagawa A, Nakane S, Shinoda A, Itoh K, Yamashita A, Waku K (1995) 2-Arachidonoylglycerol: a possible endogenous cannabinoid receptor ligand in brain. Biochem Biophys Res Commun 215:89–97

Tang J, Chen J, Ramanjaneya M, Punn A, Conner AC, Randeva HS (2008) The signalling profile of recombinant human orexin-2 receptor. Cell Signal 20:1651–1661

Tinklenberg JR, Roth WT, Kopell BS (1976) Marijuana and ethanol: differential effects on time perception, heart rate, and subjective response. Psychopharmacology 49(3):275–279

Torrealba F, Yanagisawa M, Saper CB (2003) Colocalization of orexin-A and glutamate immunoreactivity in axon terminals in the tuberomammillary nucleus in rats. Neuroscience 119 (4):1033–1044

Trivedi P, Yu H, MacNeil DJ, Van der Ploeg LH, Guan XM (1998) Distribution of orexin receptor mRNA in the rat brain. FEBS Lett 438:71–75

Tsujino N, Sakurai T (2009) Orexin/hypocretin: a neuropeptide at the interface of sleep, energy homeostasis, and reward system. Pharmacol Rev 61:162–176

Turunen PM, Ekholm ME, Somerharju P, Kukkonen JP (2010) Arachidonic acid release mediated by OX1 orexin receptors. Br J Pharmacol 159:212–221

Turunen PM, Jäntti MH, Kukkonen JP (2012) OX1 orexin/hypocretin receptor signaling through arachidonic acid and endocannabinoid release. Mol Pharmacol 82(2):156–167

Uramura K, Funahashi H, Muroya S, Shioda S, Takigawa M, Yada T (2001) Orexin-a activates phospholipase c- and protein kinase c-mediated ca2+ signaling in dopamine neurons of the ventral tegmental area. Neuroreport 12:1885–1889

Valenti M, Viganò D, Casico MG, Rubino T, Steardo L, Parolaro D, Di Marzo V (2004) Differential diurnal variations of anandamide and 2-arachidonoyl-glycerol levels in rat brain. Cell Mol Life Sci 61(7–8):945–950

Verret L, Goutagny R, Fort P, Cagnon L, Salvert D, Léger L, Boissard R, Salin P, Peyron C, Luppi PH (2003) A role of melanin-concentrating hormone producing neurons in the central regulation of paradoxical sleep. BMC Neurosci 4:19

Vittoz NM, Berridge CW (2006) Hypocretin/orexin selectively increases dopamine efflux within the prefrontal cortex: involvement of the ventral tegmental area. Neuropsychopharmacology 31:384–395

Ward RJ, Pediani JD, Milligan G (2011) Ligand-induced internalization of the orexin OX(1) and cannabinoid CB(1) receptors assessed via N-terminal SNAP and CLIP-tagging. Br J Pharmacol 162(6):1439–1452

Wise RA, Rompre PP (1989) Brain dopamine and reward. Annu Rev Psychol 40:191–225

Wu X, French ED (2000) Effects of chronic delta9-tetrahydrocannabinolon rat midbrain dopamine neurons: an electrophysiological assessment. Neuropharmacology 39:391–398

Wu X, Gao J, Yan J, Owyang C, Li Y (2004) Hypothalamus-brain stem circuitry responsible for vagal efferent signaling to the pancreas evoked by hypoglycemia in rat. J Neurophysiol 91:1734–1747

Xia JX, Fan SY, Yan J, Chen F, Li Y, Yu ZP, Hu ZA (2009) Orexin A induced extracellular calcium influx in prefrontal cortex neurons involves I-type calcium channels. J Physiol Biochem 65:125–136

Yamanaka A, Beuckmann CT, Willie JT, Hara J, Tsujino N, Mieda M, Tominaga M, Yagami K, Sugiyama F, Goto K, Yanagisawa M, Sakurai T (2003) Hypothalamic orexin neurons regulate arousal according to energy balance in mice. Neuron 38(5):701–713

Yang L, Zou B, Xiong X, Pascual C, Xie J, Malik A, Xie J, Sakurai T, Xie XS (2013) Hypocretin/orexin neurons contribute to hippocampus-dependent social memory and synaptic plasticity in mice. J Neurosci 33(12):5275–5284

Yaswen L, Diehl N, Brennan MB, Hochgeschwender U (1999) Obesity in the mouse model of pro-opiomelanocortin deficiency responds to peripheral melanocortin. Nat Med 5:1066–1070

Zweifel LS, Parker JG, Lobb CJ, Rainwater A, Wall VZ, Fadok JP et al (2009) Disruption of NMDAR-dependent burst firing by dopamine neurons provides selective assessment of phasic dopamine-dependent behavior. Proc Natl Acad Sci U S A 106:7281–7288

How CB1 Receptor Activity and Distribution Contribute to Make the Male and Female Brain Different Toward Cannabinoid-Induced Effects

Silvia Antinori and Liana Fattore

Abstract Sex-dependent differences have been consistently reported in the prevalence and frequency of use of *Cannabis*, with patterns of use, subjective effects, and progression to dependence being different in male and female smokers. As confirmed by animal studies, cannabinoids exert sex-dependent effects in many physiological and behavioral aspects, including food intake and energy balance (stronger in males) and emotional regulation (stronger in females). Following chronic THC exposure during adolescence, cannabinoid receptors undergo desensitization, which is greater in adolescent female animals than in male animals. Preclinical research is greatly contributing to elucidate the neurobiological bases for sex differences in cannabinoid effects, among which different cannabinoid pharmacodynamic and pharmacokinetic and gonadal hormones play crucial roles. The sexual dimorphism of the brain in cannabinoid subtype 1 receptor (CB1R) distribution and function massively contributes to the variety of differences reported in cannabinoid-induced effects between the two sexes. The distribution and activity of neuronal CB1Rs in the male and female brains and how they can be differently affected by stress and drugs of abuse have been investigated in many brain areas. In this chapter, we will first provide an overview on the brain sexual dimorphism and the sex-dependent effects of cannabinoids. Then we will review both clinical and laboratory-based research evidence revealing important sex-related differences in CB1 receptor level and function in different brain areas. Finally, the influence of gonadal hormones in determining such differences will be discussed.

S. Antinori
Department of Biomedical Sciences, Division of Neuroscience and Clinical Pharmacology, University of Cagliari, Cittadella Universitaria di Monserrato, 09042 Monserrato, CA, Italy

L. Fattore (✉)
Institute of Neuroscience-Cagliari, National Research Council, Cagliari, Italy
e-mail: lfattore@in.cnr.it

M. Melis (ed.), *Endocannabinoids and Lipid Mediators in Brain Functions*,
DOI 10.1007/978-3-319-57371-7_2

1 Brain Sexual Dimorphisms and the Endocannabinoid System

Sex hormones begin to exert their influence on the brain and behavior during the development of the fetus and continue during adulthood (Gore et al. 2014). Various brain regions, like those implicated in sexual and reproductive behaviors (e.g., hypothalamus), display evident differences between males and females. Important differences between men and women have been described in brain circuits associated to stress and emotions as well as to memory and cognition (Cahill 2005). For example, men show larger volumes in the parietal cortex (which is involved in space perception), amygdala (which regulates sexual and social behavior), and hypothalamus (which controls sex drive and food intake), while women show larger volumes in the frontal cortex (responsible for problem-solving and decision-making) and the medial paralimbic cortex (responsible for emotion regulation) (Goldstein et al. 2001). Compelling human neuroimaging in vivo studies have revealed sex-specific differences in brain functions, neural structure, and neurochemistry, thus highlighting the need of considering sex as a modulating factor in brain-related data interpretation (Sacher et al. 2013). Sex-dependent differences also exist at the cellular level, with men and women differing in both cell type and receptor density.

The endogenous cannabinoid system emerges relatively early during brain development. Similar to other brain neurotransmitter systems, the endocannabinoid system is sexually dimorphic, with male and female brains displaying different levels of endocannabinoids (Bradshaw et al. 2006), different endocannabinoid-mediated glial cell genesis in the developing brain (Krebs-Kraft et al. 2010), and different cannabinoid subtype 1 receptor (CB1R) density and affinity (Rodríguez de Fonseca et al. 1994). Parolaro and Viveros' groups, for example, showed that sex affects the endocannabinoid system in many brain areas and that CB1R density and functionality vary in the two sexes in a region-dependent manner (Zamberletti et al. 2012; Llorente-Berzal et al. 2013). Preclinical studies have consistently demonstrated that the density of CB1Rs fluctuates during the estrous cycle and that the affinity of agonist binding is also affected by sex hormones (Rodríguez de Fonseca et al. 1994; Riebe et al. 2010). Moreover, treatment with delta9-tetrahydrocannabinol (delta9-THC) induced greater desensitization of brain CB1Rs in female than in male rats, an effect more evident in adolescent than in adult animals (Burston et al. 2010). These and other more recent findings support the need of considering sex and menstrual phases when analyzing CB1R level and activity in humans (Normandin et al. 2015). The endocannabinoid system is also particularly sensitive to early life stress in a sex- and brain region-dependent manner, with changes evident at gene expression level already in the adolescent brain (Marco et al. 2014).

Human studies revealed that the number of CB1Rs increased with aging in females but not in males (Van Laere et al. 2008). In particular, women show an age-dependent increase of CB1Rs in the basal ganglia, lateral temporal cortex, and hippocampus, while men display an enhanced binding in the limbic system and cortico–striato–thalamic–cortical circuit. Intriguingly, adolescent cannabinoid exposure induces long-term and sex-dependent effects on adult hippocampal neurogenesis and response to stress (Lee et al. 2014). Recently, Woolley's group showed that the compound URB597, a fatty acid amide hydrolase inhibitor which blocks the breakdown of the endogenous cannabinoid anandamide, strongly suppressed inhibitory synapses in female but not male brains, indicating tonic endocannabinoid release in females that is absent in males (Tabatadze et al. 2015). Noteworthy, such a difference is not related to circulating reproductive hormones.

2 Sex-Dependent Effects of Cannabinoids

While there are no racial or ethnic differences for *Cannabis* use worldwide, the prevalence of marijuana smoking, frequency of use, and vulnerability to develop dependence are different between sexes (UNODOC 2015), in both humans and animals (McGregor and Arnold 2007; Fattore et al. 2009; Cooper and Haney 2014). Marijuana is the first illicit drug to be experienced by adolescents, with boys being in general more likely than girls to become "heavy users" of *Cannabis* (Kohn et al. 2004). In all European countries, more male students than females use *Cannabis*, while females are more likely to use sedative drugs (EMCDDA 2006). In Canada, among high school students smoking marijuana, more males than females report poor family relationships and problems at school (Butters 2005). In South America, the gender ratio of *Cannabis* use is lower for the adolescent than the adult population, with a higher proportion of female users in the most developed countries (UNODOC 2015). Gender differences have also been found in self-reported health-related quality of life among marijuana users (Lev-Ran et al. 2012).

Cannabinoids have been shown to influence physiology and behavior in a sex-dependent manner (Craft et al. 2013; Wagner 2016). The effects of cannabinoids on food intake and energy balance, for example, are more manifest in males, while their effects on anxiety-like states and depression are more evident in females (Diaz et al. 2009; Fattore and Fratta 2010). Female marijuana smokers show lower circulating levels of delta9-THC (Jones et al. 2008) and weaker cardiovascular and subjective effects than male smokers (Leatherdale et al. 2007). Notably, male non-marijuana smokers are more sensitive than women to the subjective effects of delta9-THC (Haney 2007). Yet, female marijuana smokers report more frequent dizziness and higher susceptibility to cannabinoid-induced hemodynamic changes and visuospatial memory impairment than male smokers (Mathew et al. 2003). Moreover, women smoking marijuana show less evident withdrawal symptoms than men (Crowley et al. 1998), are more likely to smoke marijuana mostly when under pressure or anxious (Patton et al. 2002), avoid using other drugs (i.e., use *Cannabis* only), and have a lower prevalence of panic attacks and personality disorders (Hasin et al. 2008). The beneficial effects of *Cannabis* use on sexual functioning are consistently reported by women, while men report both facilitatory and inhibiting effects following *Cannabis* use (Gorzalka et al. 2010). In rodents, catalepsy, antinociception, and changes in locomotor activity typically induced by cannabinoids are more evident in females than in males (Tseng and Craft 2004), while cannabinoid-induced effects on exploratory behavior and anxiety are manifest in females only (Biscaia et al. 2003). Differences related to gender (i.e., sociocultural distinctions between men and women) and sex (i.e., anatomical and physiological distinctions between males and females) have also been reported in impulsive actions and compulsive behavior (Fattore and Melis 2016a) as well as in other dysregulated behaviors (Fattore and Melis 2016b).

3 CB1R Density and Function in the Male and Female Brain

Within the brain, neuronal CB1Rs are not distributed homogeneously, being present at higher concentrations in areas regulating cognition and short-term memory (cerebral cortex and hippocampus) and motor functions (basal ganglia and cerebellum) (Herkenham et al. 1990; Tsou et al. 1998; McPartland et al. 2007). CB1Rs are also present in astrocytes and microglia (Stella 2010; Bosier et al. 2013). Similarly, the cannabinoid subtype 2 receptors (CB2Rs) have been detected in microglia, in dendritic cells, in brain endothelial cells, and in subgroups of neurons in various brain regions (Mackie 2005; Maresz et al. 2005; Van Sickle et al. 2005). However, whether sex affects the density and activity of glial CB1Rs or brain CB2Rs is still not known.

This chapter focuses on the distribution and activity of neuronal CB1Rs in brain areas where sex-dependent differences have been explored both at physiological conditions and in response to different events. In particular, sex-dependent differences in CB1R level and function following stressful events (e.g., maternal deprivation) and exposure to drugs of abuse (e.g., cocaine) are reviewed and discussed for each brain area.

3.1 Prefrontal Cortex

The prefrontal cortex is associated with high-order cognitive and memory functions, encodes relevant information in decision-making and goal-directed behavior, and is involved in the storage of information and in control processes, e.g., monitoring and selection (Dembrow and Johnston 2014). Together with the hippocampus and the amygdala, the medial part of the prefrontal cortex regulates fear conditioning and extinction processes likely exerting a top–down control over subcortical structures to induce appropriate behavioral responses. CB1Rs are widely represented in the prefrontal cortex (Glass et al. 1997), which could explain why marijuana use is associated to impairments in working memory, mental flexibility, and selective and sustained attention.

Sex differences in the rat prefrontal cortex have been described for both CB1R density and functionality. One study reported females showing lower basal levels of CB1Rs but higher CB1R activity when compared to male counterparts (Llorente-Berzal et al. 2013), while other studies reported discrepant findings (Zamberletti et al. 2012; Xing et al. 2014). Notably, early maternal separation and adolescent exposure to drugs of abuse were found to affect prefrontal CB1R density and functionality in a sex-dependent manner. That is, after early life stress (i.e., maternal separation) followed by adolescent cocaine exposure, reduced expression and functionality of CB1Rs in the prefrontal cortex were observed in male rats but not in females (Llorente-Berzal et al. 2013). Conversely, following adolescent exposure to nicotine, both male and female rats showed increased CB1R activity (Mateos et al. 2010), while after repeated tail-shock stress, CB1R mRNA expression was decreased in females only (Xing et al. 2014). Also a high-fat diet influences the functionality of CB1Rs in the prefrontal cortex, since it increased CB1R function, but not density, in females (Rojo et al. 2013). Yet, whether the same occurs in males, and to which extent, remains to be determined. Cycling female rats have lower density of CB1Rs than male and ovariectomized female rats in the prefrontal cortex, whereas no sex-dependent differences are present in the functionality of CB1Rs (Castelli et al. 2014). Notably, while ovariectomy increases CB1R expression, replacement with estradiol reduces the amount of CB1Rs to the levels of cycling females, suggesting that ovarian hormones might negatively affect CB1R density in brain areas involved in cognition and emotional processing.

3.2 Thalamus

The thalamus is located in the central portion of the diencephalon and is formed by groups of distinct nuclei, e.g., anterior, posterior, medial, lateral, pulvinar, and reticular nuclei. Extensively connected to virtually all areas of the cerebral cortex, to the brain stem, and to the spinal cord, the thalamus acts as a brain–body *gateway* and represents the most important relay station in the brain (Wolff et al. 2015). By filtering the unnecessary information and allowing conscious mind to register relevant signals only, it is considered to crucially regulate sensory information processing, pain perception, attention, sleep, arousal, learning, cognition, memory, and decision-making (Jones 2007; Baxter 2013; Funahashi 2013; Mitchell and Chakraborty 2013; Mitchell et al. 2014; Mitchell 2015).

This brain area shows a distinctive heterogeneous distribution of CB1Rs, with the highest level located in thalamic nuclei connected with the associational cortical areas (Glass et al. 1997). Curiously, synthetic cannabinoid users show a gray matter density reduction in the right and left thalamus when compared with the healthy control group (Nurmedov et al. 2015). The endocannabinoid system modulates the sleep–wake cycle and the level of arousal, by affecting thalamic rhythms and oscillations (Dasilva et al. 2014), and pain perception, by decreasing nociceptive neurotransmission at the level of the spinal cord and the thalamus (Bradshaw et al. 2006). Altered levels of endocannabinoids in the brain are associated to altered pain sensitivity: activation of thalamic CB1Rs takes part to the antinociceptive efficacy of cannabinoids (Martin et al. 1996), and chronic pain may lead to a specific upregulation of thalamic CB1Rs (Siegling et al. 2001; Knerlich-Lukoschus et al. 2011). Low to moderate levels of estrogen receptor subtypes α- and β-mRNA were detected, respectively, in the subthalamic nucleus and the ventral lateral nucleus of the thalamus (Osterlund and Hurd 2001). It is therefore possible that, being sexually dimorphic, endocannabinoid levels could account for sex differences in pain perception that women reported during different phases of the menstrual cycle (Iacovides et al. 2015).

Unfortunately, to date only few studies examined sex-dependent differences in the expression and function of CB1Rs in thalamic nuclei, most of them being performed on male subjects only. For example, social isolation (Sciolino et al. 2010), but not chronic stress (Hill et al. 2009), was found to affect the density of CB1Rs in the thalamus of male animals, but it is not known whether females are also affected and, if so, to which extent. Conversely, other studies were performed on females only, thus making not possible a direct comparison between the two sexes. A preclinical study that proposed inadequate playful interactions in adolescent animals as a novel model for studying social peer rejection, for example, demonstrated that in play-deprived female rats, the expression levels of CB1Rs were increased in the thalamus (Schneider et al. 2014). But whether the same occurs in male animals, or whether this phenomenon is sexually dimorphic, remains to be demonstrated. Sex-dependent differences in basal CB1R density and function in the thalamus were provided by Zamberletti et al. (2012), who found higher CB1R level and functionality in female rats than in males, and by Llorente-Berzal et al. (2013) that confirmed the higher level of CB1R functionality in females but reported higher CB1R density in males. Importantly, this latter study showed that thalamus CB1R density is significantly affected by early life stress and adolescent exposure to drugs of abuse in a sex-dependent manner. In fact, authors showed that early maternal deprivation caused a reduction in the expression of thalamic CB1Rs in males but not in females, while adolescent cocaine administration induced an increase in thalamic CB1R functionality in females but not in males (Llorente-Berzal et al. 2013).

3.3 Hypothalamus

The hypothalamus is a region comprised of several nuclei that is deputed to the integration of body functions for the maintenance of homeostasis, which is essential for survival and reproduction of the species (Seoane-Collazo et al. 2015). This brain region maintains homeostatic balance in physiology and behavior by integrating massive volumes of information (concerning, among other functions, feeding, satiety, stress, and motivational states) and by assigning appropriate instructions to downstream effector systems (Haas and Lin 2012). The hypothalamus regulates the release of pituitary hormones and the body heat in response to variations in external temperature, influences caloric intake and weight regulation, determines wakefulness/sleep cycles, and regulates fluid intake and sensation of thirst.

The endocannabinoid system plays an important regulatory role on the hypothalamus and broadly interacts with the hypothalamic–pituitary–adrenal axis (Hill and Tasker 2012). The expression of CB1Rs in the hypothalamus is among the lowest in the brain; yet, activation of these receptors is particularly efficient and strongly affects the cross talk between the hypothalamic nuclei and peripheral organs. Among the hypothalamic nuclei, the ventromedial hypothalamic nucleus (involved in homeostatic and behavioral functions including food intake) and the paraventricular nucleus (controlling stress, metabolism, growth, reproduction, immune and autonomic functions) contain CB1Rs (Castelli et al. 2007; Reguero et al. 2011). The endocannabinoid system is intimately involved in appetitive and reward-related behavior. In the hypothalamus, the synthesis of endocannabinoids increases during brief starvation and decreases following food intake (Bellocchio et al. 2006), while expression of CB1Rs increases during fasting (Burdyga et al. 2006).

In the hypothalamus, basal CB1R expression was reported to be higher in female than in male rats (Zamberletti et al. 2012; Xing et al. 2014). Hypothalamic CB1R functionality, but not density, is significantly influenced by neonatal manipulation in a different manner in male and female rats (Llorente-Berzal et al. 2013). Indeed, although females display a higher basal level of CB1R functionality than males (Llorente-Berzal et al. 2013), maternal deprivation significantly diminishes CB1R functionality in males (Llorente-Berzal et al. 2013) and slightly increases CB1R activity in females (Zamberletti et al. 2012). Similar to early life stress, also repeated stress during adolescence (i.e., tail-shock stress) is able to modify CB1R expression in the hypothalamus, by decreasing and increasing CB1R mRNA in stressed females and males, respectively (Xing et al. 2014). Such sex-dependent differences might be explained, at least in part, by the effects of estrogen on CB1R binding. Males and ovariectomized females, in fact, show higher levels of CB1R binding than intact females and estradiol-treated ovariectomized females (Riebe et al. 2010). Findings that ovariectomized females are not different from males in terms of hypothalamic CB1R binding and similar to cycling females when treated with high doses of estradiol suggest that estradiol decreases the expression of CB1Rs in the hypothalamus (Riebe et al. 2010).

3.4 Hippocampus

Located within the brain medial temporal lobe (underneath the cortical surface), the hippocampus is crucial for spatial navigation and storage of long-term memory (Hölscher 2003). It possesses an extraordinary capacity for structural reorganization and controls different types of learning and habit formation processes. In many mammalian species, the total volume of the hippocampus is significantly larger in males than in females (Keeley et al. 2015), a phenomenon that could be due in part to sex-dependent changes in neurogenesis or sex hormone-specific effects.

The hippocampus is among the richest regions of CB1Rs, located presynaptically on the axons of GABAergic interneurons (Mackie 2005). Activation of hippocampal presynaptic CB1Rs has been reported to strongly inhibit GABA release from inhibitory networks in both humans and rats and to reduce the power of hippocampal network oscillations (Hájos et al. 2000). CB1Rs located in the hippocampus contribute to the amnesic-like effects produced in animals by cannabinoid agonists (Akirav 2011) and to impairments in cognition and learning produced in humans by acute or chronic use of marijuana (Paton and Pertwee 1973).

Hippocampal volume was reported to be larger in adult females exposed to delta9-THC during adolescence (Keeley et al. 2015) and to be related to gonadal hormone levels (Galea et al. 1999). Significant sex effects were found in both CB1R density and functionality within the hippocampus, although discrepant findings were also reported. In female rats, for example, Llorente-Berzal et al. (2013) found a slightly lower CB1R expression and a markedly higher CB1R activity than in males, while Zamberletti et al. (2012) reported higher levels of CB1Rs in females and no significant sex differences in CB1R functionality. A third study that used Lister hooded rats instead of Sprague Dawley or Wistar albino strains as in the above studies, found no significant differences between males and females in both CB1R density and function (Castelli et al. 2014). Discrepancies in CB1R density and function in different rat strains have been reported also for other brain areas and are in line with the numerous evidence showing that behavioral (Deiana et al. 2007; Renard et al. 2013; Manduca et al. 2014; Wakeford and Riley 2014), biochemical (Chen et al. 1991; Cadoni et al. 2015), and molecular (Arnold et al. 2010) effects of cannabinoids are strain dependent.

Recently, sex-dependent alterations in hippocampal CB1R expression following adolescent exposure to delta9-THC treatment have been reported in rats (Silva et al. 2015). Specifically, a downregulation of CB1Rs was found in both sexes at 24 h and 2 weeks post chronic treatment with delta9-THC in the CA2 region; however, in females it also persisted in CA1 and CA3 areas. Also Weed et al. (2016) reported sex-dependent long-term alterations in hippocampal CB1R levels after chronic adolescent treatment with delta9-THC. Yet, sex-dependent differences were not found in adolescent rats immediately after the last delta9-THC administration (Weed et al. 2016). Importantly, early life stress induces a significant reduction in CB1R expression in CA1 region of both sexes, but to a greater extent in male than in female rats, and a marked decrease in the CA3 region only in males (Suárez et al. 2009). Conversely, repeated stress during adolescence induces a decreased expression of CB1R mRNA in the hippocampus of female but not male rats (Xing et al. 2014). Notably, the same study found that baseline CB2R mRNA levels were significantly higher in the hippocampus of female animals than in males but were not affected by stress. Another study demonstrated that chronic unpredictable mild stress induces a downregulation of CB1Rs in the hippocampus of male rats (an effect more evident in the dorsal than in the ventral hippocampus), but an upregulation of CB1Rs in the dorsal hippocampus of females (Reich et al. 2009). The above described studies support the notion that the endocannabinoid system is differentially organized in the male and female brains to respond to chronic stress. Finally, Riebe et al. (2010) found no difference in CB1R density between males and females in the hippocampus, although ovariectomized female rats show higher CB1R density compared to cycling female and estradiol-treated ovariectomized female and male rats. Finding that the difference between ovariectomized and cycling female rats is reversed by estradiol replacement not only confirms that CB1R expression is regulated by estradiol in a brain region-dependent manner but also that estradiol has an inhibitory effect on CB1R density in the hippocampus. Similarly, differences in CB1R density between males and ovariectomized females indicate that testosterone may exert an inhibitory effect on the expression of hippocampal CB1Rs.

3.5 Amygdala

The amygdala is a complex area that embraces several interconnected regions including the basolateral amygdala, the central amygdala, and the medial amygdala (Sah et al. 2003). It is a key site for the assignment of emotional salience to external stimuli and the coordination of affective, autonomic, and behavioral responses to such stimuli, i.e., it acts as an interface between sensory inputs and cortical processing (Herman et al. 2003). Activation of the amygdala is associated to the generation of fear and anxiety responses and to the activation of the hypothalamic–pituitary–adrenal axis (Davis 1997; Shin and Liberzon 2010).

Within the cortico–limbic system, and together with the hippocampus and the prefrontal cortex, the basolateral amygdala shows the most prominent expression of CB1Rs, while lower levels of CB1R mRNA expression are present in the central amygdala (McPartland et al. 2007). The endocannabinoid signaling plays an important role in controlling exploratory behavior, motivational states, conditioned fear, and extinction of aversive memories (Marsicano et al. 2002; Chhatwal and Ressler 2007; Ramikie and Patel 2012). A number of evidence revealed a sexually dimorphic amygdala in *Cannabis* users. Female marijuana users, for example, show a larger right amygdala than female controls, while male users have similar volume as male controls (McQueeny et al. 2011). In the amygdala, women show lower baseline cerebral CB1R availability than men; notably, low CB1R availability is related to a more evident novelty-seeking phenotype (Van Laere et al. 2009). An interesting study by Krebs-Kraft et al. (2010) revealed a sexually dimorphic endocannabinoid system within the medial amygdala of developing rats. Females, for example, have higher level of the endocannabinoid degradation enzymes (fatty acid amid hydrolase and monoacylglycerol lipase) than males, lower amounts of endocannabinoids (anandamide and 2-arachidonoylglycerol), and more *ex novo* generated glial cells (but not neurons) than males. Curiously, treatment of neonate rats with cannabinoid agonists reduces glial cell genesis in females but not in males (Krebs-Kraft et al. 2010).

Animal data available so far on sex-dependent differences in CB1R density and function in the amygdala of adult male and female rats are somewhat discrepant. Zamberletti et al. (2012) reported higher density and functionality of CB1Rs in female than in male rats. Yet, Llorente-Berzal et al. (2013) confirmed a higher level of CB1R functionality in females but detected no sex differences in CB1R density in the amygdala. Female animals also exhibited higher basal levels of CB1 mRNA expression in the amygdala (Xing et al. 2014) but significantly reduced CB1R level and function following exposure to delta9-THC (Rubino et al. 2008). Moreover, CB1R density in the amygdala has been reported to be either higher (Riebe et al. 2010) or lower (Castelli et al. 2014) in cycling females than in males and in ovariectomized females. Notably, such a difference appeared to be estradiol dependent, since it was reversed by replacement with estrogen regardless of the direction of effect (Riebe et al. 2010; Castelli et al. 2014). Interestingly, early life stress induced a CB1R downregulation in the basolateral amygdala of male but not female rats (Alteba et al. 2016).

3.6 Periaqueductal Gray

The periaqueductal gray (PAG) matter is a midbrain structure that surrounds the Sylvius aqueduct, receives afferents from nociceptive neurons in the spinal cord, and sends projections to thalamic nuclei that process nociception. It is an important site in ascending pain transmission and a key component of a descending pain inhibitory system. However, it also interacts with the amygdala to regulate fear and anxiety processing and is implicated in modulation of cardiovascular, respiratory, and motor control (Benarroch 2012).

The endocannabinoid system plays an important role in the regulation of aversive responses in the PAG by exerting a "fine-tuning" regulatory control of defensive responses (Moreira et al. 2009, 2012; Fogaça et al. 2012). PAG is also a crucial supraspinal site of action for the CB1R-mediated analgesic effects of cannabinoids (Hu et al. 2014).

Analysis of the density of CB1Rs in the PAG of male and female rats reported discrepant results. Indeed, CB1R density was found to be higher in female rats by Zamberletti et al. (2012), while Llorente-Berzal et al. (2013) reported no differences between the two sexes. However, both studies failed in detecting significant sex-dependent differences in CB1R functionality in the PAG of male and female rats (Zamberletti et al. 2012; Llorente-Berzal et al. 2013). Moreover, adolescent exposure to delta9-THC induces downregulation of CB1Rs in rats, which was greater in female than male adolescent rats (Rubino et al. 2008). Notably, female and male adolescent rats showed higher and lower desensitization in the PAG, respectively, than same-sex adults, while in adults CB1R desensitization was greater in males than in females (Burston et al. 2010). Why CB1R desensitization after adolescent exposure to delta9-THC occurs in the PAG to a greater extent in adolescent females and adult males remains to be elucidated.

3.7 Nucleus Accumbens

The mesolimbic dopamine system that projects from the ventral tegmental area to the nucleus accumbens is crucial for the regulation of the rewarding effects of drugs of abuse. In particular, the nucleus accumbens is required not only for regulating most reward-related behaviors but also for making associations between salient environmental events and rewarding outcomes (Fattore and Diana 2016).

CB1Rs are densely distributed in the human and rat nucleus accumbens, located on both GABAergic and glutamatergic axon terminals (Hoffman and Lupica 2001; Robbe et al. 2001). Noteworthy, although women were reported to be more sensitive to the subjective effects of marijuana (Cooper and Haney 2014), no differences were found between daily marijuana users and nonusers in the morphology (i.e., volume and shape) of the nucleus accumbens (Weiland et al. 2015). During reward anticipation in a monetary reward task, *Cannabis* users displayed attenuated activity in the nucleus accumbens compared to controls (van Hell et al. 2010), but since only male volunteers were employed, whether or not this effect is similar in the two sexes is not known. A more recent study confirmed that marijuana use is associated with decreased nucleus accumbens neural response to anticipation of nondrug rewards (Martz et al. 2016). Unfortunately, although both male and female young adults participated to this study, sex-dependent differences were not evaluated, thus leaving unaddressed the question of potential sex differences in anticipatory reward processing in the nucleus accumbens.

In rats, CB1R density and function are reduced after exposure to cannabinoids during adolescence (Rubino and Parolaro 2008; Marco et al. 2009). Sex-dependent differences have been reported in the density and activity of CB1Rs in the nucleus accumbens, but, as seen for other brain regions, findings are somewhere discrepant among studies. Zamberletti et al. (2012), for example, reported higher CB1R density in females than males with no differences in functionality, while Llorente-Berzal et al. (2013) found similar density of CB1Rs in the two sexes but significantly higher CB1R functionality in females than in males. Conversely, Castelli et al. (2014) did not find any difference between male and female rats as concerns the density and function of CB1Rs in either the shell or the core subregion of the nucleus accumbens.

3.8 Ventral Tegmental Area

The ventral tegmental area receives important dopaminergic inputs from the amygdala and prefrontal cortex and projects relevant dopaminergic efferences back to the amygdala and prefrontal cortex but also to the nucleus accumbens (Fattore and Diana 2016). It is crucially involved in the control of motivated behavior, reward-related learning, salient sensory, and emotional information processing. The tail of the ventral tegmental area, better known as the rostromedial tegmental nucleus, was recently hypothesized to exert a major inhibitory drive on dopaminergic systems (Bourdy and Barrot 2012).

Within the ventral tegmental area, CB1Rs are more numerous on inhibitory terminals rather than on excitatory synapses and disinhibit dopamine neurons in reward-related contexts (Lupica and Riegel 2005; Oleson et al. 2012). CB1R activation results in the inhibition of GABAergic neurotransmission: depression of the GABAergic inhibitory input of dopaminergic neurons increases their firing rate and dopamine release in the projection region of nucleus accumbens (Szabo et al. 2002). Chronic exposure to cannabinoid activates CB1Rs of this area inducing transient depression in locally activated glutamate synapses onto dopamine neurons through activation of glutamate NMDA receptors (Liu et al. 2010).

Both endocannabinoid signaling and CB1Rs within the ventral tegmental area have been reported to be affected by sex by preclinical research (Melis et al. 2013). Yet, no clinical study has been performed so far to specifically assess the existence of sex-dependent differences in the morphology or activation of this area in marijuana smokers. One study, for example, examined whether marijuana smoking was associated with greater cue reactivity in the ventral tegmental area and reported that, when compared to sporadic *Cannabis* users and controls, frequent users show higher neural responses (Cousijn et al. 2013). However, although male and female marijuana users were involved in this study, sex-dependent differences were not evaluated.

In rats, CB1R density and activity in the ventral tegmental area have been found to be higher in females than in males (Zamberletti et al. 2012), although no differences were reported by Castelli et al. (2014). Importantly, delta9-THC exposure during adolescence significantly decreased CB1R level (Rubino et al. 2008; Zamberletti et al. 2012) and function (Zamberletti et al. 2012) in the ventral tegmental area of adult female rats, while no significant alterations were observed in adult male rats (Rubino et al. 2008; Zamberletti et al. 2012). The same cannabinoid treatment significantly decreased CB1R density in both female and male adolescent rats, while CB1R function was significantly reduced in adolescent females only (Rubino et al. 2008).

3.9 *Cerebellum*

The cerebellum has long been acknowledged for its role in motor coordination, but it is also involved in the temporal control of action sequences and in complex cognitive behavior (Schmahmann and Caplan 2006). Along with the basal ganglia, the hippocampus, and the cortex, the cerebellum is a brain area displaying very high expression of CB1Rs, consistent with the profile of motor effects typically induced by cannabinoid agents. The cerebellar cortex is particularly enriched of CB1Rs, located primarily on axons in the molecular layer (Kawamura et al. 2006). Chronic marijuana users show decreased glucose metabolism in the cerebellum compared with controls at baseline, but increased metabolism after acute intravenous delta9-THC administration (Volkow et al. 1996). In the cerebellum, other studies reported decreased regional cerebral blood flow in abstinent marijuana users (Block et al. 2000), but increased regional cerebral blood flow after either intravenous delta9-THC administration or marijuana smoking (Mathew et al. 1998; O'Leary et al. 2002).

Animal studies showed that CB1R levels within the cerebellum are higher in females than in males (Zamberletti et al. 2012) but that CB1R functionality is not significantly different between the two sexes (Zamberletti et al. 2012; Llorente-Berzal et al. 2013). CB1R density and functionality can be differently affected by early life stress. Maternal deprivation, for example, significantly increases the number of CB1Rs exclusively in females and CB1R activity exclusively in males (Llorente-Berzal et al. 2013). Yet, Zamberletti et al. (2012) reported no significant effect of sex on cerebellar CB1R level and function in maternally deprived adult rats. On the other hand, the effect induced on CB1R density and activity by chronic delta9-THC in adolescent animals was similar in the two sexes (Rubino et al. 2008).

3.10 *Basal Ganglia*

CB1Rs are also present in the caudate putamen, substantia nigra, and globus pallidus of the human brain, localized on presynaptic terminals of GABAergic striatonigral and striatopallidal neurons (Hurley et al. 2003; McPartland et al. 2007). In the caudate putamen, the level of CB1R mRNA was significantly lower in marijuana users than in controls (Villares 2007).

In rats, few sex differences were found in the caudate putamen, substantia nigra, and globus pallidus, but, as in other brain regions, data are mostly controversial. Zamberletti et al. (2012), for example, reported a significantly higher CB1R density in females than in males, while the opposite was found by Llorente-Berzal et al. (2013). The two studies reported discrepant data also for CB1R function, but only in the caudate putamen. Rubino et al. (2008) showed a significant reduction of CB1R density following adolescent chronic treatment with delta9-THC in the caudate putamen and substantia nigra of adolescent female rats only. Moreover, the same treatment induced a significant reduction in the function of CB1Rs in the caudate putamen, substantia nigra, and globus pallidus of female rats, but only in the substantia nigra in males (Rubino et al. 2008). As concerns the effect of exposure to drugs of abuse and/or early life stress, it was shown that adolescent exposure to cocaine, but not maternal deprivation, significantly affected CB1R function in all these three regions in male rats while only in the globus pallidus in females (Llorente-Berzal et al. 2013). Notably, CB1R density was not affected by cocaine administrations in both males and females, but it was significantly increased in the substantia nigra of maternally deprived animals of both sexes (Llorente-Berzal et al. 2013).

A third study, however, reported no significant differences in CB1R density and function in the caudate putamen of male and female rats (Castelli et al. 2014). Differences in CB1R binding and function of male and female adult rats among different studies might be due to differences in the methodological approaches (homogenate binding vs. autoradiographic analysis vs. in situ hybridization) and in the strains of rats used (Castelli et al. 2014).

4 Hormonal Influence on CB1Rs

Sex steroids, e.g., estrogens and progestogens, are critical player during brain development since they shape the central nervous system during development and their activating effects strongly influence brain function and behaviors. To further complicate the complex scenario of brain region-dependent differences in the level and activity of CB1Rs in males and females, gonadal hormones have been found to strongly affect CB1R binding and endocannabinoid levels in both humans and animals (Fattore and Fratta 2010; Fattore 2013; Sanchez et al. 2016). Ovarian hormones, for example, mediate the long-lasting behavioral and pharmacodynamic effects induced by chronic exposure to delta9-THC during adolescence in female rats (Winsauer et al. 2011). Estradiol modulates the endocannabinoid tone and rapidly suppresses inhibitory synaptic transmission in hippocampus in a sex-specific manner, i.e., in females but not males (Huang and Woolley 2012). In turn, cannabinoids impinge upon hypothalamic–pituitary–gonadal axis activity, negatively impacting on sexual behavior and sexual motivation (Ferrari et al. 2000; López 2010; López et al. 2010).

In humans, female marijuana smokers appear to be more sensitive to the behavioral and physiological effects of cannabinoids (Craft 2005), with treatment-seeking women presenting more severe withdrawal symptoms than treatment-seeking men (Herrmann et al. 2015). The existence of a reciprocal influence between marijuana smoking and the menstrual cycle is supported by the findings that women who smoke marijuana are at higher risk for delayed ovulation and anovulatory cycles (Jukic et al. 2007) and for reduced fertility due to ovulatory abnormalities (Mueller et al. 1990).

Table 1 Gender- and sex-dependent differences in the regional distribution of CB1 receptors in basal conditions and their modifications in males and/or females after exposure to early life stress or THC

Density	Basal	Early life stress	THC
Prefrontal cortex	F < M (Castelli et al. 2014; Llorente-Berzal et al. 2013)		↓ F (Zamberletti et al. 2012)
	F > M (Zamberletti et al. 2012; Xing et al. 2014)		
Thalamus	F > M (Zamberletti et al. 2012)	↓ M (Llorente-Berzal et al. 2013)	↓ M (Zamberletti et al. 2012)
	F < M (Llorente-Berzal et al. 2013)		
Hypothalamus	F < M (Riebe et al. 2010)		↓ F (Zamberletti et al. 2012)
	F > M (Zamberletti et al. 2012; Xing et al. 2014)		
Hippocampus	F < M (Llorente-Berzal et al. 2013)	↓ F ↓ M (Suárez et al. 2009)	↓ F ↓ M (Silva et al. 2015)
	F > M (Zamberletti et al. 2012; Xing et al. 2014)		
	F ≈ M (Castelli et al. 2014; Riebe et al. 2010)		
Amygdala	F < M (humans) (Van Laere et al. 2009)		↓ F ↓ M (Rubino et al. 2008) ↓ F (Zamberletti et al. 2012)
	F < M (Castelli et al. 2014)		
	F > M (Zamberletti et al. 2012; Xing et al. 2014; Riebe et al. 2010)		
	F ≈ M (Llorente-Berzal et al. 2013)		
Periaqueductal gray	F > M (Zamberletti et al. 2012)		↓ F ↓ M (adolescent rats, Rubino et al. 2008)
	F ≈ M (Llorente-Berzal et al. 2013)		
Nucleus accumbens	F > M (Zamberletti et al. 2012)		↓ F ↓ M (Zamberletti et al. 2012) ↓ F (adult rats); ↓ M (adolescent rats, Rubino et al. 2008)
	F ≈ M (Llorente-Berzal et al. 2013; Castelli et al. 2014)		
Ventral tegmental area	F > M (Zamberletti et al. 2012)		↓ F (adult rats, Rubuno et al. 2008; Zamberletti et al. 2012)
	F ≈ M (Castelli et al. 2014)		

(continued)

Table 1 (continued)

Density	Basal	Early life stress	THC
Cerebellum	$F > M$ (Zamberletti et al. 2012)	$\uparrow$ F (Llorente-Berzal et al. 2013)	$\downarrow$ F $\downarrow$ M (adolescent rats, Rubino et al. 2008)
Basal ganglia (CPu, SN, GB)	$F > M$ (CPu, SN, GP) (Zamberletti et al. 2012)	$\uparrow$ M $\uparrow$ F (SN) (Llorente-Berzal et al. 2013)	$\downarrow$ F (CPu, SN, adolescent rats, Rubino et al. 2008) $\downarrow$ F (CPu, SN, GP); $\uparrow$ M (SN); $\downarrow$ M (CPu) (Zamberletti et al. 2012)
	$F < M$ (CPu, SN, GP) (Llorente-Berzal et al. 2013)		
	$F \approx M$ (CPu) (Castelli et al. 2014)		

In rats, CB1R density and affinity have been reported to be influenced in certain brain areas by the levels of estradiol, progesterone, and testosterone (Rodríguez de Fonseca et al. 1994). Ovarian hormones have been described to significantly influence cannabinoid taking (Fattore et al. 2007) and cannabinoid seeking (Fattore et al. 2010) in rats and to regulate CB1R density and function in the rat prefrontal cortex and amygdala (Castelli et al. 2014). Interestingly, adolescent exposure to cannabinoid agonists caused an immediate (transient) reduction in CB1R expression in the hypothalamus and amygdala of adult rats but did not alter CB1R expression in the nucleus accumbens and globus pallidus at two different time points (Chadwick et al. 2011).

5 Conclusions

Sex-dependent differences in brain CB1R expression and functionality have been reported in almost all brain areas (see Tables 1 and 2). CB1R binding was also found to vary in a sex-specific manner after exposure to both stressful events and drugs of abuse, which suggests a differential activation and functionality of the endocannabinoid system in males and females, and might explain the divergent responses to exogenous cannabinoids often described in the two sexes. The different CB1R distribution and function within the male and female brain and the different gonadal hormonal milieu in the two sexes surely contribute to the different subjective effects induced by exogenous cannabinoids in men and women. Factors determining different CB1R expression in males and females are not well understood; yet, sex steroids have been suggested as potential modulators of CB1R expression both in the brain and periphery. In the anterior pituitary gland, for example, CB1R gene expression was found to be influenced by estrogens

Table 2 Gender- and sex-dependent differences in the functionality of CB1 receptors in basal conditions and their modifications in males and/or females after exposure to early life stress or THC

Functionality	Basal	Early life stress	THC
Prefrontal cortex	F > M (Llorente-Berzal et al. 2013)	↓ M ↓ F (Llorente-Berzal et al. 2013)	↓ M ↓ F (Zamberletti et al. 2012)
	F ≤ M (Zamberletti et al. 2012)		
	F ≈ M (Castelli et al. 2014)		
Thalamus	F > M (Llorente-Berzal et al. 2013, Zamberletti et al. 2012)	↑ M (Zamberletti et al. 2012)	↓ F (Zamberletti et al. 2012)
Hypothalamus	F > M (Llorente-Berzal et al. 2013)	↓ M (Llorente-Berzal et al. 2013) ↑ F (Zamberletti et al. 2012)	↓ F (Zamberletti et al. 2012)
Hippocampus	F > M (Llorente-Berzal et al. 2013)	↑ F ↑ M (Zamberletti et al. 2012)	↓ F ↓ M (Zamberletti et al. 2012) ↓ M (adult rats, Rubino et al. 2008)
	F ≈ M (Castelli et al. 2014; Zamberletti et al. 2012)	↑ F ↑ M (Zamberletti et al. 2012)	↓ F ↓ M (Zamberletti et al. 2012) ↓ M (adult rats, Rubino et al. 2008)
Amygdala	F > M (Zamberletti et al. 2012; Llorente-Berzal et al. 2013)	↓ F ↑ M (Zamberletti et al. 2012)	↓ F (adul rats, Rubino et al. 2008) ↓ M (adolescent rats, Rubino et al. 2008) ↓ F ↓ M (Zamberletti et al. 2012)
Periaqueductal gray	F ≈ M (Zamberletti et al. 2012; Llorente-Berzal et al. 2013)	↑ F (Llorente-Berzal et al. 2013)	↓ F ↓ M (Zamberletti et al. 2012)
Nucleus accumbens	F ≈ M (Castelli et al. 2014, Zamberletti et al. 2012)	↓ M (Llorente-Berzal et al. 2013)	↓ F ↓ M (Zamberletti et al. 2012) ↓ F (Rubino et al. 2008)
	F > M (Llorente-Berzal et al. 2013)		
Ventral tegmental area	F > M (Zamberletti et al. 2012)	↑ M (Zamberlett et al. 2012)	↓ F (adolescent rats, Rubino et al. 2008) ↓ F (Zamberlett et al. 2012)
	F ≈ M (Zamberlett et al. 2014)		
Cerebelium	F ≈ M (Llorente-Berzal et al. 2013; Zamberlett; et al. 2012)	↑ M (Llorente-Berzal et al. 2013)	↓ F ↓ M (Zamberletti et al. 2012) ↓ F ↓ M (adolescent rats, Rubino et al. 2008)

(continued)

Table 2 (continued)

Functionality	Basal	Early life stress	THC
Basal ganglia (CPu, SN, GB	F > M (GP) (Llorente-Berzal et al. 2013; Zamberletti et al. 2012)		↓ F (CPu, SN, GP); ↓ M (SN) (adolescent rats, Rubino et al. 2008) ↓ F ↓ M (CPu, SN) (Zamberletti et al. 2012)
	F < M (SN) (Llorente-Berzal et al. 2013; Zamberletti et al. 2012)		
	F < M (CPu) (Zamberletti et al. 2012)		
	F > M (CPu) (Llorente-Berzal et al. 2013)		
	F ≈ M (CPu) (Castelli et al. 2014)		

(González et al. 2000). Similarly, CB1R density is reduced in parotid glands of castrated male rats and is restored following testosterone replacement (Busch et al. 2006), confirming that sex hormones can modulate CB1R expression. It should be also kept in mind that estradiol differentially modulates cannabinoid regulation of amino acid neurotransmission (Nguyen and Wagner 2006), regulates emotional behavior through an endocannabinoid mechanism (Hill et al. 2007), and negatively modulates cannabinoid-induced effects on appetite, core body temperature, and neurotransmission at proopiomelanocortin (POMC) synapses (Kellert et al. 2009). The numerous evidence of region-dependent fluctuations of brain CB1R density and/or function along the ovarian cycle and following gonadectomy and sex steroid replacement (Bonnin et al. 1993; Rodríguez de Fonseca et al. 1994) strongly support the hypothesis of sex hormone-dependent differences in the sensitivity to cannabinoid treatment (Craft and Leitl 2008).

A conspicuous progress has been made during the recent years in the appreciation of (i) the sexual dimorphic nature of the endocannabinoid system, (ii) the influence of the gonadal hormones on the density and activity of brain CB1Rs, (iii) the necessity to include female subjects in clinical and preclinical research, and (iv) the need for gender-tailored prevention and treatment options for marijuana male and female smokers. Yet, current knowledge on sex-dependent differences in the endocannabinoid system and in response to cannabinoids is only in its infancy, and further rigorous research is needed to understand what makes the female and male brain different to these regards.

References

Akirav I (2011) The role of cannabinoids in modulating emotional and non-emotional memory processes in the hippocampus. Front Behav Neurosci 5:34

Alteba S, Korem N, Akirav I (2016) Cannabinoids reverse the effects of early stress on neurocognitive performance in adulthood. Learn Mem 23:349–358

Arnold JC, Dielenberg RA, McGregor IS (2010) Cannabinoids increase conditioned ultrasonic vocalisations and cat odour avoidance in rats: strain differences in drug-induced anxiety. Life Sci 87:572–578

Baxter MG (2013) Mediodorsal thalamus and cognition in non-human primates. Front Syst Neurosci 7:38

Bellocchio L, Mancini G, Vicennati V et al (2006) Cannabinoid receptors as therapeutic targets for obesity and metabolic diseases. Curr Opin Pharmacol 6:586–591

Benarroch EE (2012) Periaqueductal gray: an interface for behavioral control. Neurology 78:210–217

Biscaia M, Marín S, Fernández B et al (2003) Chronic treatment with CP 55,940 during the peri-adolescent period differentially affects the behavioural responses of male and female rats in adulthood. Psychopharmacology 170:301–308

Block RI, O'Leary DS, Hichwa RD et al (2000) Cerebellar hypoactivity in frequent marijuana users. Neuroreport 11:749–753

Bonnin A, Fernández-Ruiz JJ, Martín M et al (1993) Delta 9-tetrahydrocannabinol affects mesolimbic dopaminergic activity in the female rat brain: interactions with estrogens. J Neural Transm Gen Sect 92:81–95

Bosier B, Bellocchio L, Metna-Laurent M et al (2013) Astroglial CB1 cannabinoid receptors regulate leptin signaling in mouse brain astrocytes. Mol Metab 2:393–404

Bourdy R, Barrot M (2012) A new control center for dopaminergic systems: pulling the VTA by the tail. Trends Neurosci 35:681–690

Bradshaw HB, Rimmerman N, Krey JF, Walker JM (2006) Sex and hormonal cycle differences in rat brain levels of pain-related cannabimimetic lipid mediators. Am J Physiol Regul Integr Comp Physiol 291:349–358

Burdyga G, Varro A, Dimaline R et al (2006) Ghrelin receptors in rat and human nodose ganglia: putative role in regulating CB-1 and MCH receptor abundance. Am J Phys 290:G1289–G1297

Burston JJ, Wiley JL, Craig AA et al (2010) Regional enhancement of cannabinoid CB1 receptor desensitization in female adolescent rats following repeated Δ9-tetrahydrocannabinol exposure. Brit J Pharmacol 161:103–112

Busch L, Sterin-Borda L, Borda E (2006) Effects of castration on cannabinoid cb receptor expression and on the biological actions of cannabinoid in the parotid gland. Clin Exp Pharmacol Physiol 33:258–263

Butters JE (2005) Promoting healthy choices: the importance of differentiating between ordinary and high risk cannabis use among high-school students. Subst Use Misuse 40:845–855

Cadoni C, Simola N, Espa E et al (2015) Strain dependence of adolescent Cannabis influence on heroin reward and mesolimbic dopamine transmission in adult Lewis and Fischer 344 rats. Addict Biol 20:132–142

Cahill L (2005) His brain, her brain. Sci Am 292:40–47

Castelli MP, Piras AP, Melis T et al (2007) Cannabinoid CB1 receptors in the paraventricular nucleus and central control of penile erection: immunocytochemistry, autoradiography and behavioral studies. Neuroscience 147:197–206

Castelli MP, Fadda P, Casu A et al (2014) Male and female rats differ in brain cannabinoid CB1 receptor density and function and in behavioural traits predisposing to drug addiction: effect of ovarian hormones. Curr Pharm Des 20:2100–2113

Chadwick B, Saylor AJ, López HH (2011) Adolescent cannabinoid exposure attenuates adult female sexual motivation but does not alter adulthood CB1R expression or estrous cyclicity. Pharmacol Biochem Behav 100:157–164

Chen JP, Paredes W, Lowinson JH, Gardner EL (1991) Strain-specific facilitation of dopamine efflux by delta 9-tetrahydrocannabinol in the nucleus accumbens of rat: an in vivo microdialysis study. Neurosci Lett 129:136–180

Chhatwal JP, Ressler KJ (2007) Modulation of fear and anxiety by the endogenous cannabinoid system. CNS Spectr 12:211–220

Cooper ZD, Haney M (2014) Investigation of sex-dependent effects of cannabis in daily cannabis smokers. Drug Alcohol Depend 136:85–91

Cousijn J, Goudriaan AE, Ridderinkhof KR et al (2013) Neural responses associated with cue-reactivity in frequent cannabis users. Addict Biol 18:570–580

Craft RM (2005) Sex differences in behavioral effects of cannabinoids. Life Sci 77:2471–2478

Craft RM, Leitl MD (2008) Gonadal hormone modulation of the behavioral effects of delta9-tetrahydrocannabinol in male and female rats. Eur J Pharmacol 578:37–42

Craft RM, Marusich JA, Wiley JL (2013) Sex differences in cannabinoid pharmacology: a reflection of differences in the endocannabinoid system? Life Sci 92:476–481

Crowley TJ, Macdonald MJ, Whitmore EA, Mikulich SK (1998) Cannabis dependence, withdrawal, and reinforcing effects among adolescents with conduct symptoms and substance use disorders. Drug Alcohol Depend 50:27–37

Dasilva M, Grieve KL, Cudeiro J, Rivadulla C (2014) Anandamide activation of CB1 receptors increases spontaneous bursting and oscillatory activity in the thalamus. Neuroscience 265:72–82

Davis M (1997) Neurobiology of fear responses: the role of the amygdala. J Neuropsychiatry Clin Neurosci 9:382–402

Deiana S, Fattore L, Spano MS et al (2007) Strain and schedule-dependent differences in the acquisition, maintenance and extinction of intravenous cannabinoid self-administration in rats. Neuropharmacology 52:646–654

Dembrow N, Johnston D (2014) Subcircuit-specific neuromodulation in the prefrontal cortex. Front Neural Circuits 8:54

Diaz S, Farhang B, Hoien J (2009) Sex differences in the cannabinoid modulation of appetite, body temperature and neurotransmission at POMC synapses. Neuroendocrinology 89:424–440

EMCDDA (2006) A gender perspective on drug use and responding to drug problems. Lisbon 2006

Fattore L (2013) Considering gender in cannabinoid research: a step towards personalized treatment of marijuana addicts. Drug Test Anal 5:57–61

Fattore L, Diana M (2016) Drug addiction: an affective-cognitive disorder in need of a cure. Neurosci Biobehav Rev 65:341–361

Fattore L, Fratta W (2010) How important are sex differences in cannabinoid action? Br J Pharmacol 160:544–548

Fattore L, Melis M (2016a) Sex differences in impulsive and compulsive behaviors: a focus on drug addiction. Addict Biol 21(5):1043–1051. doi:10.1111/adb.12381

Fattore L, Melis M (2016b) Exploring gender and sex differences in behavioral dyscontrol: from drug addiction to impulse control disorders. Front Psych 7:19

Fattore L, Spano MS, Altea S et al (2007) Cannabinoid self-administration in rats: sex differences and the influence of ovarian function. Br J Pharmacol 152:795–804

Fattore L, Fadda P, Fratta W (2009) Sex differences in the self-administration of cannabinoids and other drugs of abuse. Psychoneuroendocrinology 34:S227–S236

Fattore L, Spano MS, Altea S et al (2010) Drug- and cue-induced reinstatement of cannabinoid-seeking behaviour in male and female rats: influence of ovarian hormones. Br J Pharmacol 160:724–735

Ferrari F, Ottani A, Giuliani D (2000) Inhibitory effects of the cannabinoid agonist HU 210 on rat sexual behaviour. Physiol Behav 69:547–554

Fogaça MV, Lisboa SF, Aguiar DC et al (2012) Fine-tuning of defensive behaviors in the dorsal periaqueductal gray by atypical neurotransmitters. Braz J Med Biol Res 45:357–365

Funahashi S (2013) Thalamic mediodorsal nucleus and its participation in spatial working memory processes: comparison with the prefrontal cortex. Front Syst Neurosci 7:36

Galea LA, Perrot-Sinal TS, Kavaliers M, Ossenkopp KP (1999) Relations of hippocampal volume and dentate gyrus width to gonadal hormone levels in male and female meadow voles. Brain Res 821:383–391

Glass M, Dragunow M, Faull RL (1997) Cannabinoid receptors in the human brain: a detailed anatomical and quantitative autoradiographic study in the fetal, neonatal and adult human brain. Neuroscience 77:299–318

Goldstein JM, Seidman LJ, Horton NJ et al (2001) Normal sexual dimorphism of the adult human brain assessed by in vivo magnetic resonance imaging. Cereb Cortex 11:490–497

González S, Bisogno T, Wenger T et al (2000) Sex steroid influence on cannabinoid CB1 receptor mRNA and endocannabinoid levels in the anterior pituitary gland. Biochem Biophys Res Commun 270:260–266

Gore AC, Martien KM, Gagnidze K, Pfaff D (2014) Implications of prenatal steroid perturbations for neurodevelopment, behavior, and autism. Endocr Rev 35:961–991

Gorzalka BB, Hill MN, Chang SC (2010) Male-female differences in the effects of cannabinoids on sexual behavior and gonadal hormone function. Horm Behav 58:91–99

Haas HL, Lin JS (2012) Waking with the hypothalamus. Pflugers Arch 463:31–42

Hájos N, Katona I, Naiem SS et al (2000) Cannabinoids inhibit hippocampal GABAergic transmission and network oscillations. Eur J Neurosci 12:3239–3249

Haney M (2007) Opioid antagonism of cannabinoid effects: differences between marijuana smokers and nonmarijuana smokers. Neuropsychopharmacology 32:1391–1403

Hasin DS, Keyes KM, Alderson D et al (2008) Cannabis withdrawal in the United States: results from NESARC. J Clin Psychiatry 69:1354–1363

Herkenham M, Lynn AB, Little MD et al (1990) Cannabinoid receptor localization in brain. Proc Natl Acad Sci U S A 87:1932–1936

Herman JP, Figueiredo H, Mueller NK et al (2003) Central mechanisms of stress integration: hierarchical circuitry controlling hypothalamo-pituitary-adrenocortical responsiveness. Front Neuroendocrinol 24:151–180

Herrmann ES, Weerts EM, Vandrey R (2015) Sex differences in cannabis withdrawal symptoms among treatment-seeking cannabis users. Exp Clin Psychopharmacol 23:415–421

Hill MN, Tasker JG (2012) Endocannabinoid signaling, glucocorticoid-mediated negative feedback, and regulation of the hypothalamic–pituitary–adrenal axis. Neuroscience 204:5–16

Hill MN, Karacabeyli ES, Gorzalka BB (2007) Estrogen recruits the endocannabinoid system to modulate emotionality. Psychoneuroendocrinology 32:350–357

Hill MN, Hunter RG, McEwen BS (2009) Chronic stress differentially regulates cannabinoid CB1 receptor binding in distinct hippocampal subfields. Eur J Pharmacol 614:66–69

Hoffman AF, Lupica CR (2001) Direct actions of cannabinoids on synaptic transmission in the nucleus accumbens: a comparison with opioids. J Neurophysiol 85:72–83

Hölscher C (2003) Time, space and hippocampal functions. Rev Neurosci 14:253–284

Hu SS, Ho YC, Chiou LC (2014) No more pain upon Gq-protein-coupled receptor activation: role of endocannabinoids. Eur J Neurosci 39:467–484

Huang GZ, Woolley CS (2012) Estradiol acutely suppresses inhibition in the hippocampus through a sex-specific endocannabinoid and mGluR-dependent mechanism. Neuron 74:801–808

Hurley MJ, Mash DC, Jenner P (2003) Expression of cannabinoid CB1 receptor mRNA in basal ganglia of normal and parkinsonian human brain. J Neural Transm 110:1279–1288

Iacovides S, Avidon I, Baker FC (2015) Does pain vary across the menstrual cycle? A review. Eur J Pain 19:1389–1405

Jones EG (2007) The thalamus 2 Cambridge. Cambridge University Press, New York

Jones AW, Holmgren A, Kugelberg FC (2008) Driving under the influence of cannabis: a 10-year study of age and gender differences in the concentrations of tetrahydrocannabinol in blood. Addiction 103:452–461

Jukic AM, Weinberg CR, Baird DD et al (2007) Lifestyle and reproductive factors associated with follicular phase length. J Womens Health (Larchmt) 16:1340–1347
Kawamura Y, Fukaya M, Maejima T et al (2006) The CB1 cannabinoid receptor is the major cannabinoid receptor at excitatory presynaptic sites in the hippocampus and cerebellum. J Neurosci 26:2991–3001
Keeley RJ, Trow J, McDonald RJ (2015) Strain and sex differences in puberty onset and the effects of THC administration on weight gain and brain volumes. Neuroscience 305:328–342
Kellert BA, Nguyen MC, Nguyen C et al (2009) Estrogen rapidly attenuates cannabinoid-induced changes in energy homeostasis. Eur J Pharmacol 622:15–24
Knerlich-Lukoschus F, Noack M, von der Ropp-Brenner B et al (2011) Spinal cord injuries induce changes in cb1 cannabinoid receptor and C–C chemokine expression in brain areas underlying circuitry of chronic pain conditions. J Neurotrauma 28:619–634
Kohn L, Kittel F, Piette D (2004) Peer, family integration and other determinants of cannabis use among teenagers. Int J Adolesc Med Health 16:359–370
Krebs-Kraft DL, Hill MN, Hillard CJ, McCarthy MM (2010) Sex difference in cell proliferation in developing rat amygdala mediated by endocannabinoids has implications for social behavior. Proc Natl Acad Sci U S A 107:20535–20540
Leatherdale ST, Hammond DG, Kaiserman M, Ahmed R (2007) Marijuana and tobacco use among young adults in Canada: are they smoking what we think they are smoking? Cancer Causes Control 18:391–397
Lee TT, Wainwright SR, Hill MN (2014) Sex, drugs, and adult neurogenesis: sex-dependent effects of escalating adolescent cannabinoid exposure on adult hippocampal neurogenesis, stress reactivity, and amphetamine sensitization. Hippocampus 24:280–292
Lev-Ran S, Imtiaz S, Taylor BJ et al (2012) Gender differences in health-related quality of life among cannabis users: results from the National Epidemiologic Survey on Alcohol and Related Conditions. Drug Alcohol Depend 123:190–200
Liu Z, Han J, Jia L et al (2010) Synaptic neurotransmission depression in ventral tegmental dopamine neurons and cannabinoid-associated addictive learning. PLoS One 5:e15634
Llorente-Berzal A, Assis MA, Rubino T et al (2013) Sex-dependent changes in brain CB1R expression and functionality and immune CB2R expression as a consequence of maternal deprivation and adolescent cocaine exposure. Pharmacol Res 74:23–33
López HH (2010) Cannabinoid-hormone interactions in the regulation of motivational processes. Horm Behav 58:100–110
López HH, Zappia K, Cushman CL, Chadwick B (2010) Acute cannabinoid administration attenuates female socio-sexual motivation. Pharmacol Biochem Behav 94:482–487
Lupica CR, Riegel AC (2005) Endocannabinoid release from midbrain dopamine neurons: a potential substrate for cannabinoid receptor antagonist treatment of addiction. Neuropharmacology 48:1105–1116
Mackie K (2005) Distribution of cannabinoid receptors in the central and peripheral nervous system. Handb Exp Pharmacol 168:299–325
Manduca A, Servadio M, Campolongo P et al (2014) Strain- and context-dependent effects of the anandamide hydrolysis inhibitor URB597 on social behavior in rats. Eur Neuropsychopharmacol 24:1337–1348
Marco EM, Rubino T, Adriani W et al (2009) Long-term consequences of URB597 administration during adolescence on cannabinoid CB1 receptor binding in brain areas. Brain Res 125:25–31
Marco EM, Echeverry-Alzate V, López-Moreno JA (2014) Consequences of early life stress on the expression of endocannabinoid-related genes in the rat brain. Behav Pharmacol 25:547–556
Maresz K, Carrier EJ, Ponomarev ED et al (2005) Modulation of the cannabinoid CB2 receptor in microglial cells in response to inflammatory stimuli. J Neurochem 95:437–445
Marsicano G, Wotjak CT, Azad SC et al (2002) The endogenous cannabinoid system controls extinction of aversive memories. Nature 418:530–534

Martin WJ, Hohmann AG, Walker JM (1996) Suppression of noxious stimulus-evoked activity in the ventral posterolateral nucleus of the thalamus by a cannabinoid agonist: correlation between electrophysiological and antinociceptive effects. J Neurosci 16:6601

Martz ME, Trucco EM, Cope LM et al (2016) Association of marijuana use with blunted nucleus accumbens response to reward anticipation. JAMA Psychiat 73(8):838–844. doi:10.1001/jamapsychiatry.2016.1161

Mateos B, Borcel E, Loriga R et al (2010) Adolescent exposure to nicotine and/or the cannabinoid agonist CP 55,940 induces gender-dependent long-lasting memory impairments and changes in brain nicotinic and CB(1) cannabinoid receptors. J Psychopharmacol 25:1676–1690

Mathew RJ, Wilson WH, Turkington TG, Coleman RE (1998) Cerebellar activity and disturbed time sense after THC. Brain Res 797:183–189

Mathew RJ, Wilson WH, Davis R (2003) Postural syncope after marijuana: a transcranial Doppler study of the hemodynamics. Pharmacol Biochem Behav 75:309–318

McGregor IS, Arnold JC (2007) Cannabis reward: biased towards the fairer sex? Br J Pharmacol 152:562–564

McPartland JM, Glass M, Pertwee RG (2007) Meta-analysis of cannabinoid ligand binding affinity and receptor distribution: interspecies differences. Br J Pharmacol 152:583–593

McQueeny T, Padula CB, Price J et al (2011) Gender effects on amygdala morphometry in adolescent marijuana users. Behav Brain Res 224:128–134

Melis M, De Felice M, Lecca S et al (2013) Sex-specific tonic 2-arachidonoylglycerol signaling at inhibitory inputs onto dopamine neurons of Lister Hooded rats. Front Integr Neurosci 7:93

Mitchell AS (2015) The mediodorsal thalamus as a higher order thalamic relay nucleus important for learning and decision-making. Neurosci Biobehav Rev 54:76–88

Mitchell AS, Chakraborty S (2013) What does the mediodorsal thalamus do? Front Syst Neurosci 7:37

Mitchell AS, Sherman SM, Sommer MA et al (2014) Advances in understanding mechanisms of thalamic relays in cognition and behavior. J Neurosci 34:15340–15346

Moreira FA, Aguiar DC, Campos AC et al (2009) Antiaversive effects of cannabinoids: is the periaqueductal gray involved? Neural Plast 2009:625469

Moreira FA, Aguiar DC, Resstel LB et al (2012) Neuroanatomical substrates involved in cannabinoid modulation of defensive responses. J Psychopharmacol 26:40–55

Mueller BA, Daling JR, Weiss NS et al (1990) Recreational drug use and the risk of primary infertility. Epidemiology 1:195–200

Nguyen QH, Wagner EJ (2006) Estrogen differentially modulates the cannabinoid- induced presynaptic inhibition of amino acid neurotransmission in proopiomelanocortin neurons of the arcuate nucleus. Neuroendocrinology 84:123–137

Normandin MD, Zheng MQ, Lin KS et al (2015) Imaging the cannabinoid CB1 receptor in humans with [11C]OMAR: assessment of kinetic analysis methods, test-retest reproducibility, and gender differences. J Cereb Blood Flow Metab 35:1313–1322

Nurmedov S, Metin B, Ekmen S et al (2015) Thalamic and cerebellar gray matter volume reduction in synthetic cannabinoids users. Eur Addict Res 21:315–320

O'Leary D, Block R, Koeppel J et al (2002) Effects of smoking marijuana on brain perfusion and cognition. Neuropsychopharmacology 26:802–816

Oleson EB, Beckert MV, Morra JT et al (2012) Endocannabinoids shape accumbal encoding of cue-motivated behavior via CB1 receptor activation in the ventral tegmentum. Neuron 73:360–373

Osterlund MK, Hurd YL (2001) Estrogen receptors in the human forebrain and the relation to neuropsychiatric disorders. Prog Neurobiol 64:251–267

Paton WDM, Pertwee RG (1973) The actions of cannabis in man. In: Mechoulam R (ed) Marijuana: chemistry, pharmacology, metabolism and clinical effects. Academic Press, New York, pp 287–333

Patton GC, Coffey C, Carlin JB et al (2002) Cannabis use and mental health in young people: cohort study. BMJ 325:1195–1198

Ramikie TS, Patel S (2012) Endocannabinoid signaling in the amygdala: anatomy, synaptic signaling, behavior, and adaptations to stress. Neuroscience 204:38–52
Reguero L, Puente N, Elezgarai I et al (2011) GABAergic and cortical and subcortical glutamatergic axon terminals contain CB1 cannabinoid receptors in the ventromedial nucleus of the hypothalamus. PLoS One 6:e26167
Reich CG, Taylor ME, McCarthy MM (2009) Differential effects of chronic unpredictable stress on hippocampal CB1 receptors in male and female rats. Behav Brain Res 203:264–269
Renard J, Krebs MO, Jay TM, Le Pen G (2013) Long-term cognitive impairments induced by chronic cannabinoid exposure during adolescence in rats: a strain comparison. Psychopharmacology 225:781–790
Riebe CJ, Hill MN, Lee TT et al (2010) Estrogenic regulation of limbic cannabinoid receptor binding. Psychoneuroendocrinology 35:1265–1269
Robbe D, Alonso G, Duchamp F et al (2001) Localization and mechanisms of action of cannabinoid receptors at the glutamatergic synapses of the mouse nucleus accumbens. J Neurosci 21:109–116
Rodríguez de Fonseca F, Cebeira M, Ramos JA et al (1994) Cannabinoid receptors in rat brain areas: sexual differences, fluctuations during estrous cycle and changes after gonadectomy and sex steroid replacement. Life Sci 54:159–170
Rojo ML, Söderström I, Olsson T, Fowler CJ (2013) Changes in cannabinoid CB(1) receptor functionality in the female rat prefrontal cortex following a high fat diet. Life Sci 92:757–762
Rubino T, Parolaro D (2008) Long lasting consequences of cannabis exposure in adolescence. Mol Cell Endocrinol 286:S108–S113
Rubino T, Vigano D, Realini N et al (2008) Chronic delta 9-tetrahydrocannabinol during adolescence provokes sex-dependent changes in the emotional profile in adult rats: behavioral and biochemical correlates. Neuropsychopharmacology 33:1760–1769
Sacher J, Neumann J, Okon-Singer H et al (2013) Sexual dimorphism in the human brain: evidence from neuroimaging. Magn Reson Imaging 31:366–375
Sah P, Faber ES, Lopez De Armentia M, Power J (2003) The amygdaloid complex: anatomy and physiology. Physiol Rev 83:803–834
Sanchez AM, Cioffi R, Viganò P et al (2016) Elevated systemic levels of endocannabinoids and related mediators across the menstrual cycle in women with endometriosis. Reprod Sci 23 (8):1071–1079. doi:10.1177/1933719116630414
Schmahmann JD, Caplan D (2006) Cognition, emotion and the cerebellum. Brain 129:290–292
Schneider P, Hannusch C, Schmahl C et al (2014) Adolescent peer-rejection persistently alters pain perception and CB1 receptor expression in female rats. Eur Neuropsychopharmacol 24:290–301
Sciolino NR, Bortolato M, Eisenstein SA et al (2010) Social isolation and chronic handling alter endocannabinoid signaling and behavioral reactivity to context in adult rats. Neuroscience 168:371–386
Seoane-Collazo P, Fernø J, Gonzalez F, Diéguez C, Leis R, Nogueiras R, López M (2015) Hypothalamic-autonomic control of energy homeostasis. Endocrine 50:276–291
Shin LM, Liberzon I (2010) The neurocircuitry of fear, stress, and anxiety disorders. Neuropsychopharmacology 35:169–191
Siegling A, Hofmann HA, Denzer D et al (2001) Cannabinoid CB1 receptor upregulation in a rat model of chronic neuropathic pain. Eur J Pharmacol 415:R5–R7
Silva L, Harte-Hargrove L, Izenwasser S et al (2015) Sex-specific alterations in hippocampal cannabinoid 1 receptor expression following adolescent delta-9-tetrahydrocannabinol treatment in the rat. Neurosci Lett 602:89–94
Stella N (2010) Cannabinoid and cannabinoid-like receptors in microglia, astrocytes, and astrocytomas. Glia 58:1017–1030
Suárez J, Llorente R, Romero-Zerbo SY et al (2009) Early maternal deprivation induces gender-dependent changes on the expression of hippocampal CB(1) and CB(2) cannabinoid receptors of neonatal rats. Hippocampus 19:623–632

Szabo B, Siemes S, Wallmichrath I (2002) Inhibition of GABAergic neurotransmission in the ventral tegmental area by cannabinoids. Eur J Neurosci 15:2057–2061

Tabatadze N, Huang G, May RM et al (2015) Sex differences in molecular signaling at inhibitory synapses in the hippocampus. J Neurosci 35:11252–11265

Tseng AH, Craft RM (2004) CB(1) receptor mediation of cannabinoid behavioral effects in male and female rats. Psychopharmacology 172:25–30

Tsou K, Brown S, Sanudo-Pena MC et al (1998) Immunohistochemical distribution of cannabinoid CB1 receptors in the rat central nervous system. Neuroscience 83:393–411

United Nations Office on Drugs and Crime (UNODOC), World Drug Report 2015 (United Nations publication, Sales No. E15XI.6)

van Hell HH, Vink M, Ossewaarde L et al (2010) Chronic effects of cannabis use on the human reward system: an fMRI study. Eur Neuropsychopharmacol 20:153–163

Van Laere K, Goffin K, Casteels C et al (2008) Gender-dependent increases with healthy aging of the human cerebral cannabinoid-type 1 receptor binding using [(18)F]MK-9470 PET. NeuroImage 39:1533–1541

Van Laere K, Goffin K, Bormans G et al (2009) Relationship of type 1 cannabinoid receptor availability in the human brain to novelty seeking temperament. Arch Gen Psychiatry 66:196–204

Van Sickle MD, Duncan M, Kingsley PJ et al (2005) Identification and functional characterization of brainstem cannabinoid CB2 receptors. Science 310:329–332

Villares J (2007) Chronic use of marijuana decreases cannabinoid receptor binding and mRNA expression in the human brain. Neuroscience 145:323–334

Volkow N, Gillespie H, Mullani N et al (1996) Brain glucose metabolism in chronic marijuana users at baseline and during marijuana intoxication. Psychiatry Res 67:29–38

Wagner EJ (2016) Sex differences in cannabinoid-regulated biology: a focus on energy homeostasis. Front Neuroendocrinol 40:101–109

Wakeford AG, Riley AL (2014) Conditioned taste avoidance induced by Δ(9)-tetrahydrocannabinol in the Fischer (F344) and Lewis (LEW) rat strains. Pharmacol Biochem Behav 116:39–44

Weed PF, Filipeanu CM, Ketchum MJ, Winsauer PJ (2016) Chronic Δ9-tetrahydrocannabinol during adolescence differentially modulates striatal CB1 receptor expression and the acute and chronic effects on learning in adult rats. J Pharmacol Exp Ther 356:20–31

Weiland BJ, Thayer RE, Depue BE et al (2015) Daily marijuana use is not associated with brain morphometric measures in adolescents or adults. J Neurosci 35:1505–1512

Winsauer PJ, Daniel JM, Filipeanu CM et al (2011) Long-term behavioral and pharmacodynamic effects of delta-9-tetrahydrocannabinol in female rats depend on ovarian hormone status. Addict Biol 16:64–81

Wolff M, Alcaraz F, Marchand AR, Coutureau E (2015) Functional heterogeneity of the limbic thalamus: from hippocampal to cortical functions. Neurosci Biobehav Rev 54:120–130

Xing G, Carlton J, Jiang X et al (2014) Differential expression of brain cannabinoid receptors between repeatedly stressed males and females may play a role in age and gender-related difference in traumatic brain injury: implications from animal studies. Front Neurol 5:161

Zamberletti E, Prini P, Speziali S et al (2012) Gender-dependent behavioral and biochemical effects of adolescent delta-9-tetrahydrocannabinol in adult maternally deprived rats. Neuroscience 204:245–257

Endocannabinoids, Stress, and Negative Affect

Anthony L. Berger, Angela M. Henricks, Matthew N. Hill, and Ryan J. McLaughlin

Abstract The stress response is an evolutionarily conserved mechanism to both allow an organism to cope with a threat and to restore homeostasis following exposure to a stressor. With respect to this response, preclinical research demonstrates that the endogenous cannabinoid (ECB) system constrains the hypothalamic-pituitary-adrenal axis and plays a major role in the habituation to stressors. Specifically, anandamide tonically constrains activation of stress responsive circuits in the brain under basal conditions; however, exposure to stress or glucocorticoids initiates a cascade of events whereby corticotropin-releasing hormone (CRH) rapidly reduces anandamide metabolism through CRHR1-mediated activation of fatty acid amide hydrolase (FAAH) in the basolateral amygdala (BLA), which ultimately facilitates activation of the neuroendocrine axis and emotional response to stress. On the other hand, 2-arachidonoylglycerol (2-AG) provides on-demand synaptic modulation and promotes short-term adaptation to stress via glucocorticoid receptor-dependent mobilization in the hypothalamus and extrahypothalamic inhibitory feedback centers such as the medial prefrontal cortex (mPFC) and hippocampus. Additionally, 2-AG is recruited in the BLA to facilitate habituation to chronic homotypic stress, thus facilitating long-term adaptation as well. Accordingly, impairments in ECB signaling within the hypothalamus, BLA, mPFC, and/or hippocampus may confer maladaptive neuroendocrine and behavioral responses to stress, thereby contributing to the emergence of stress-related disorders such as anxiety, depression, and substance abuse. Moreover, sexual dimorphism and genetic differences in the ECB system may contribute to individual differences in stress coping strategies, which can have profound ramifications

---

A.L. Berger • A.M. Henricks
Department of Psychology, Washington State University, Pullman, WA, USA

M.N. Hill
Departments of Cell Biology and Anatomy & Psychiatry, The Hotchkiss Brain Institute, University of Calgary, Calgary, AB, Canada

R.J. McLaughlin (✉)
Department of Psychology, Washington State University, Pullman, WA, USA

Department of Integrative Physiology & Neuroscience, Washington State University, Pullman, WA, USA
e-mail: ryanmclaughlin@vetmed.wsu.edu

M. Melis (ed.), *Endocannabinoids and Lipid Mediators in Brain Functions*,
DOI 10.1007/978-3-319-57371-7_3

for susceptibility and resilience to stress-related mental disorders. This chapter will review the growing preclinical and clinical evidence demonstrating that the ECB system may be a promising therapeutic target for a host of human affective disorders.

1 Neuroendocrine Response to Stress

The stress response consists of a cascade of events that occur to both prime an organism to deal with a real or perceived threat in the environment and restore bodily homeostasis following resolution of that threat (Cannon 1935; Selye and Fortier 1950). Activation of the stress response initiates a number of behavioral and physiological changes that improve an individual's chance of survival when faced with homeostatic challenges. This occurs through two distinct biological pathways that govern the autonomic and neuroendocrine response to stress. The sympathoadrenal medullary axis coordinates the immediate autonomic response to stress, which is characterized by a release of catecholamines and functions to coordinate a constellation of physiological events, such as increased blood pressure, heart rate, respiratory rate, and pupil dilation, which maximize survival during fight-or-flight-type situations. By contrast, the hypothalamic-pituitary-adrenal (HPA) axis orchestrates the neuroendocrine stress response, which is characterized by the release of glucocorticoids that promote homeostatic recovery. This review will focus more specifically on the neuroendocrine arm of the stress response.

The HPA axis represents the primary branch of the neuroendocrine response to stress (Goel et al. 2014; Handa and Weiser 2014). The HPA response to stress is initiated by activation of corticotropin-releasing hormone (CRH) and arginine-vasopressin (AVP) neurosecretory cells in the parvocellular region of the hypothalamic paraventricular nucleus (PVN) (Dunn and Berridge 1990). CRH and AVP are released into portal circulation of the median eminence, which results in the mobilization of ACTH from the anterior pituitary into the blood stream (Antoni 1986). Finally, ACTH induces the release of glucocorticoids (corticosterone in rodents, cortisol in humans) from the adrenal glands (Axelrod and Reisine 1984; de Kloet 1984), which subsequently act on glucocorticoid receptors (GR) and mineralocorticoid receptors (MR) to mobilize energy stores, thus increasing arousal and focusing attention on the threat at hand (Chrousos and Gold 1992; Tsigos and Chrousos 2002).

Apart from the hypothalamus, GRs are expressed in high levels in the basolateral amygdala (BLA), medial prefrontal cortex (mPFC), and the hippocampus; three brain regions intimately involved in the activation, and subsequent termination, of the neuroendocrine stress response (McEwen 2012; Goldstein et al. 1996). The amygdala primarily exerts a stress-activating role during the processing of environmental stimuli, while the hippocampus and mPFC largely exert inhibitory roles over the stress response. Projections arising from the hippocampus and mPFC serve to dampen HPA axis activity through the activation of a GABAergic relay in the bed nucleus of the stria terminalis (BNST), which inhibits neurosecretory cells in the PVN (Radley et al. 2009, 2013; Radley and Sawchenko 2011, 2015). Acting as an interface between the endocrine stress response and the response of the brain to stress, glucocorticoids acting on GRs in the mPFC (Diorio et al. 1993), hippocampus (Jacobson and Sapolsky 1991), anterior pituitary (Keller-Wood and Dallman 1984), and hypothalamus (Hinz and Hirschelmann 2000) serve to deactivate the HPA axis, thereby promoting stress recovery.

Dysfunction of the HPA axis can lead to overproduction of glucocorticoids, which can produce structural and functional changes that may increase susceptibility to psychiatric disorders, such as anxiety and major depressive disorder (de Kloet et al. 2005; Hammen 2005; Miller et al. 2007; Selye and Fortier 1950; Bodnoff et al. 1995; Checkley 1996; Chrousos and Gold 1992). Chronic stressors result in atrophy and compromised function of the mPFC and hippocampus while increasing dendritic complexity, synaptic connectivity, and functionality of the amygdala (McEwen et al. 2015). The result of this impaired top-down inhibitory control is that amygdala-based mechanisms of information processing predominate, resulting in a sustained state of stress and anxiety and persistent drive onto the HPA axis.

Interestingly, female rats, compared to their male counterparts, have higher basal corticosterone and higher restraint stress-induced corticosterone secretion (Aloisi et al. 1998; Goel and Bale 2008, 2010; Mitsushima et al. 2008; Sterrenburg et al. 2012; Viau 2002). Moreover, corticosterone remains elevated for longer in female rodents following stress offset (Heinsbroek et al. 1991; Iwasaki-Sekino et al. 2009; Weinstock et al. 1998). Human research into sex differences in the HPA axis are less conclusive (Collins and Frankenhaeuser 1978; Earle et al. 1999), but women do tend to exhibit higher cortisol secretion than men in response to social stressors (Stroud et al. 2002). This chapter will review the preclinical evidence supporting a role for the endocannabinoid (ECB) system in regulating the neuroendocrine and behavioral response to acute and chronic stress and also briefly describe potential sex differences in ECB signaling that may underlie discrepancies in the prevalence of stress-related disorders in males and females in the clinical population.

2 Introduction to the Endocannabinoid System

The *Cannabis sativa* plant, or marijuana, has long been used for its mood-enhancing properties (Bonn-Miller et al. 2007; Schofield et al. 2006); however, the biological system that is responsible for its physiological and behavioral effects has only recently been characterized (Matsuda et al. 1990). This system, collectively referred to as the ECB system, is primarily composed of the presynaptically expressed cannabinoid 1 receptor (CB1R), one of the most abundant G-protein-coupled receptor in the central nervous system (Devane et al. 1992; Herkenham et al. 1990; Matsuda et al. 1990) as well as the cannabinoid 2 receptor (CB2R), which is thought to be primarily expressed on glial cells in response to an inflammatory challenge (Munro et al. 1993). Neuroanatomical data has demonstrated that these receptors are widely expressed throughout the forebrain, basal ganglia, and limbic system, which suggests a significant neuromodulatory role for this receptor subtype in humans (Glass et al. 1997) and rodents (Herkenham et al. 1991). The CB2R, on the other hand, is primarily located in the periphery on immune cells (Howlett 2002; Pertwee 2005) and, compared to CB1Rs, is expressed in much lower densities in the central nervous system (Van Sickle et al. 2005). CB1Rs are located on presynaptic axon terminals (Freund et al. 2003; Chevaleyre et al. 2006) of glutamatergic principal neurons as well as on a subpopulation of calbindin-negative and cholecystokinin-positive GABAergic basket cells (Freund et al. 2003; Tsou et al. 1998). Activation of the CB1R suppresses neurotransmitter release through the inhibition of adenylate cyclase, the inhibition of calcium channels, and the activation of potassium channels (Howlett and Fleming 1984; Howlett 1987; Mackie and Hille 1992; Twitchell et al. 1997; Mackie et al. 1995), making them vital in the balance of excitation and inhibition within a given neural circuit. Additionally, CB1R mRNA (Berrendero et al. 1999; Mailleux and Vanderhaeghen 1992; Tsou et al. 1998; Egertová and Elphick 2000; Katona et al. 2001) as well as the CB1R itself (Dalton and Zavitsanou 2010; Glass et al. 1997; Herkenham et al. 1990, 1991) has been localized in areas associated with hypothalamic-pituitary-adrenal (HPA) axis control such as the hypothalamus, BLA, mPFC, and hippocampus in both rodents and humans.

The ECB system is unique in that many endogenous ligands exist for the same receptor. The two primary endogenous ligands that have been characterized to date are N-arachidonylethanolamide (anandamide, AEA) and 2-arachidonoylglycerol (2-AG) (Devane et al. 1992; Sugiura et al. 1995). AEA is primarily synthesized by a specific phospholipase D (NAPE-PLD) (Okamoto et al. 2004), though other redundant biosynthetic pathways exist (Liu et al. 2008). AEA is primarily metabolized by fatty acid amide hydrolase (FAAH) (Cravatt et al. 1996), which is predominantly located in postsynaptic membrane (Tsou et al. 1998). On the other hand, 2-AG is synthesized by two sn-1-selective DAG lipases (DAGL-α and DAGL-β) (Bisogno et al. 2003) and is primarily metabolized by monoacylglycerol lipase (MAGL), which is localized within presynaptic terminals (Dinh et al. 2002). There is also a small but relevant role of alpha-beta hydrolase 6 (ABHD6) in the metabolism of 2-AG (Marrs et al. 2010). Moreover, both AEA and 2-AG are also oxygenated by cyclo-oxygnease-2 (COX-2) to form bioactive prostamides (Hermanson et al. 2014).

ECBs are not stored in vesicles like traditional neurotransmitters but are instead synthesized in postsynaptic cells in an "on-demand" fashion following depolarization of the postsynaptic neuron (Piomelli 2003). They are then transported into the synapse to activate the presynaptically located cannabinoid receptors (Freund et al. 2003). The on-demand nature of ECB release, as well as the expression profile of CB1R on both inhibitory and excitatory terminals (Kano et al. 2009), makes this system ideally positioned to regulate synaptic feedback and is thus a prime target for the treatment of disorders where disruptions in homeostatic function are a hallmark symptom.

Although AEA and 2-AG both share affinity for the same receptor, these two bioactive lipids are functionally distinct. It is unknown why two endogenous ligands exist to compete for the same receptor, but the pharmacokinetic profile of these two ECBs has offered some insight. For instance, AEA has a high affinity for the CB1R but is a low efficacy agonist with respect to inducing intracellular signal transduction. By contrast, 2-AG appears to have a lower affinity for the CB1R but a much higher signaling efficacy than AEA (Hillard 2000). Based on these characteristics, it has been proposed that AEA functions under basal conditions to maintain normal functioning (i.e., a tonic signal), while 2-AG is recruited to modulate activity-induced synaptic plasticity according to changes in the environment (i.e., a phasic signal) (Morena et al. 2016; McLaughlin et al. 2014).

3 Role of the Endocannabinoid System in the Acute Effects of Stress

The anatomical localization and unique functional profile of the ECB system suggests that it could be an important regulator of the acute response to stress. At the level of the hypothalamus, CB1Rs are located on the axon terminals of glutamatergic neurons which activate CRH-expressing neurons in the PVN, making the ECB system ideally positioned to gate excitatory drive of the HPA axis (Di et al. 2003). For instance, in vitro application of corticosterone to hypothalamic slices induces the synthesis and release of ECBs, which then act to inhibit glutamatergic release onto CRH neurosecretory cells in the PVN, thereby dampening activation of the HPA axis (Di et al. 2003; Malcher-Lopes et al. 2006). Notably, CB1R knockout mice show not only increased CRH mRNA in the PVN of the hypothalamus (Cota et al. 2003) but also increased circulating ACTH and corticosterone (Haller et al. 2004a, b; Barna et al. 2004; Steiner et al. 2008). This relationship of glucocorticoids and ECBs is also recapitulated in rats subjected to acute restraint stress, such that CB1R antagonism significantly exacerbates both basal and stress-induced corticosterone secretion (Patel et al. 2004) and increases 2-AG in the hypothalamus (Evanson et al. 2010). Moreover, in vivo administration of corticosterone also increases AEA and 2-AG content in the hypothalamus (Hill et al. 2010; Evanson et al. 2010), while intra-PVN blockade of CB1Rs prevents glucocorticoid-mediated suppression of the HPA axis (Evanson et al. 2010). Thus, CB1Rs appear to play a pivotal role in negatively regulating excitatory inputs to the hypothalamic PVN, inactivation of which increases excitability of the PVN in a manner similar to what is seen following stress exposure.

ECB signaling in extrahypothalamic regions, such as the amygdala, hippocampus, and mPFC, also serves an important regulatory role in stress-induced HPA axis activation and recovery. With respect to activation of the HPA axis, the amygdala has emerged as a critical region governing stress reactivity and anxiety-like behavior, and a role for the ECB system in constraining amygdala activity has recently been demonstrated. For instance, acute restraint stress leads to reductions in AEA in the amygdala via enhanced FAAH activity, and AEA content is negatively correlated with serum corticosterone levels (Hill et al. 2009). Additionally, inhibition of FAAH activity in the BLA partially reverses stress-induced increases in serum corticosterone (Hill et al. 2009). These effects of stress are also recapitulated by central administration of CRH, which was found to significantly reduce AEA content in the amygdala through an increase in FAAH activity and subsequently contribute to increased anxiety-like behavior in rats (Gray et al. 2015). Furthermore, the ability of systemic FAAH inhibition to reduce HPA axis activation is mediated by the local activation of CB1R in the BLA (Bedse et al. 2014). The mechanisms by which AEA signaling modulates BLA excitability are likely via the ability of CB1R signaling to gate excitatory inputs to the BLA, such that under ambient conditions, an AEA tone constrains excitatory inputs to the BLA, while following exposure to stress, when AEA levels decline, disinhibition of these excitatory inputs facilitates stress-induced activation of the BLA (Morena et al. 2016; Ahn et al. 2008; Gorzalka et al. 2008). These findings appear to hold true in humans as well given that individuals possessing a gene variant in FAAH, which results in reduced FAAH activity and elevated AEA levels (Sipe et al. 2002; Dincheva et al. 2015), exhibit reduced activation, and accelerated habituation, of the amygdala in response to stressful threat cues (Hariri et al. 2009; Gunduz-Cinar et al. 2013). Quite interestingly, a recent report has similarly demonstrated that variants in CRH receptor signaling seem to moderate the effects of changes in FAAH activity to regulate amygdala activation (Demers et al. 2016), which is consistent with the relationship between CRH and FAAH detailed above. These data collectively suggest that the actions of CRH in the amygdala serve to enhance FAAH-mediated AEA hydrolysis, which ultimately removes the break on the system and allows for an exaggerated neuroendocrine response to stress.

Moreover, AEA and 2-AG appear to serve distinct functions in the amygdala with respect to the regulation of HPA axis activity in response to acute stress (Hill and Tasker 2012). For instance, while acute stress results in relatively rapid suppression in AEA content in the BLA (Hill et al. 2009; Rademacher et al. 2008), 2-AG content is only increased in the BLA after repeated homotypic stress (Patel et al. 2009), suggesting that changes in amygdalar AEA signaling contribute to initiation of the stress response, whereas 2-AG becomes recruited to promote stress habituation. The importance of 2-AG signaling in the amygdala has further been shown using mice deficient for DAGLα, which is the enzyme primarily responsible for the synthesis of 2-AG in the brain. Male and female DAGLα knockout mice exhibit increased anxiety-like behavior as well as a female-specific depression-like phenotype as measured by decreased sucrose preference in the sucrose preference test (Shonesy et al. 2014). These authors further established that DAGLα deletion impaired retrograde eCB signaling within the amygdala, highlighting the importance of ECB-mediated inhibition of this brain region.

ECBs in the hippocampus and mPFC appear to play a major role in stress recovery. In the hippocampus, acute stress results in reduced AEA and increased 2-AG content in the hippocampus (Wang et al. 2012). These authors also demonstrated that stress-induced suppression of GABAergic transmission in the hippocampus occurs through an ECB mechanism, which is dependent on activation of glucocorticoid receptors. Specifically, these data suggest that glucocorticoids increase 2-AG content, which in turn acts to reduce GABA release and, potentially, increase excitability of the hippocampus (Wang et al. 2012). Thus, glucocorticoids regulate synaptic function in response to stress via the ECB system. In the mPFC, stress results in increased 2-AG content, which is also prevented with a glucocorticoid receptor antagonist (Hill et al. 2011a), thus suggesting that stress-induced changes in ECB signaling are also dependent on glucocorticoid release in the mPFC. Moreover, intra-mPFC CB1R blockade delays termination of the corticosterone response to acute restraint stress, and bath application of corticosterone to mPFC cortical slices suppresses GABA release in a CB1R-dependent manner (Hil et al. 2011a). This suggests that, like in the hippocampus, glucocorticoids recruit 2-AG signaling to suppress GABA release and promote excitability of a circuit. Thus, 2-AG is particularly important for promoting termination of the stress response, presumably by increasing the activation of afferents arising from the mPFC and hippocampus which act in concert to inhibit the HPA axis.

Whereas stress exposure leads to a delayed increase in 2-AG in the mPFC following the termination of stress (Hill et al. 2011a), stress also elicits a rapid reduction in AEA content in the mPFC (McLaughlin et al. 2012). As AEA and 2-AG exhibit a "yin-yang" relationship, where AEA is depressed quickly after stress followed by a delayed increase in 2-AG, it has been proposed that AEA functions under basal conditions to maintain normal functioning, while 2-AG is recruited to shut down the stress response once the environment is deemed no longer threatening (Hill and Tasker 2012). Consistent with this hypothesis, intra-mPFC administration of a FAAH inhibitor can blunt the HPA axis response to stress (McLaughlin et al. 2014).

While the majority of preclinical data has been collected in male animals, significant sex differences exist in how men and women respond to an acute stressor. For example, men and women use different coping strategies to deal with stress (Schmied et al. 2015), and the psychological impact of stress may be greater in women (Schmied et al. 2015; Taylor 2014), which may indicate sex differences in HPA axis reactivity. Interestingly, ECBs and gonadal hormones interact in a bidirectional manner to modulate behavior (Hill et al. 2007a; Carroll and Anker 2010; Gorzalka and Dang 2012), so sexual dimorphisms in the stress response may arise, at least partially, from sex differences in the ECB system. Acute stress increases circulating ECBs in men and women, but the effect depends on hormone stage in women (Dlugos et al. 2012). Specifically, ECBs are highest in women during the luteal phase, when levels of gonadal hormones are also high. In ovariectomized female rats, estradiol reduces anxiety, but this effect is blocked by a CB1R antagonist and recapitulated by the administration of an FAAH inhibitor (Hill et al. 2007a), suggesting that gonadal hormones such as estradiol may alleviate stress-induced negative affective symptoms through an ECB mechanism. Additionally, a series of elegant studies from Catharine Woolley's group have recently demonstrated that estradiol increases AEA levels in the hippocampus via mGluR1-dependent stimulation of phospholipase C and IP3R activation, which ultimately serves to dampen GABAergic inhibition of hippocampal principal outputs (Huang and Woolley 2012; Tabatadze et al. 2015). Notably, this estradiol-mediated stimulation of hippocampal AEA synthesis only occurs in females, thus providing a mechanism by which estradiol may modulate hippocampal-dependent stress recovery processes in a sex-specific manner, which could have important implications for the sexually dimorphic response to stress.

4 Role of the ECB System in the Effects of Chronic Stress

As one might expect, exposure to chronic stress also results in alterations to multiple components of the ECB system in both hypothalamic and extrahypothalamic structures. For instance, 7 days of social defeat stress reduces AEA and increases 2-AG content in the hypothalamus of defeated mice (Dubreucq et al. 2012). Chronic stress has also been shown to increase basal and stress-induced corticosterone secretion in parallel with these decreased AEA levels in the hypothalamus (Hill et al. 2010). Furthermore, CB1Rs are downregulated in the hypothalamus following both postweaning social isolation (Sciolino et al. 2010) and repeated restraint stress through the actions of corticosterone (Wamsteeker et al. 2010). Similar to acute stress, these findings indicate that chronic stress causes functional adaptations in the hypothalamic ECB system that serve to reduce the threshold for HPA axis activation and render the system less sensitive to heightened glucocorticoid-mediated negative feedback.

Changes in the ECB system following chronic stress seem to both contribute to alterations in amygdalar function and also function to maintain homeostasis in the face of repeated stress. Following chronic restraint stress, pyramidal neurons of the BLA exhibit an increase in dendritic arborization (Hill et al. 2013a, b), indicative of enhanced BLA output. These changes were mediated by reductions in AEA via elevated FAAH, as this phenotype was reversed by genetic and pharmacological ablation of FAAH (Hill et al. 2013a, b). Repeated restraint stress also causes an increase in basal circulating corticosterone, while decreasing AEA content throughout the corticolimbic circuit, specifically in the PFC and amygdala (Hill et al. 2010; Patel et al. 2005; Rademacher et al. 2008). Inhibition of FAAH does not modulate HPA dynamics during habituation to repeated stress; however, it does prevent basal hypersecretion of corticosterone, suggesting that the decrease in AEA signaling contributes to changes in basal HPA axis function. Exposure to repeated restraint (Patel et al. 2005; Hill et al. 2010), or chronic treatment with corticosterone, results in increased 2-AG content in the amygdala (Hill et al. 2005), implying that the increase in 2-AG levels could also be a byproduct of chronically elevated corticosterone. A putative mechanism underlying this enhancement of 2-AG signaling is the downregulation of MAGL, the 2-AG hydrolase, in the BLA following chronic restraint stress (Sumislawski et al. 2011). This increase in 2-AG content within the amygdala appears to contribute to the adaptive process of stress habituation as local blockade of CB1R within the BLA can dishabituate the HPA axis to repeated stress (Hill et al. 2010). Together, these studies highlight a key, albeit functionally distinct, role for AEA and 2-AG in regulating the amygdalar response to chronic stress by preventing basal corticosterone hypersecretion and promoting habituation to stress, respectively.

In the hippocampus and mPFC, exposure to chronic stress or corticosterone similarly produces an increase in FAAH activity, a decrease in AEA content, and an increase in 2-AG (Bowles et al. 2012; Patel et al. 2005; Rademacher et al. 2008; Hill et al. 2006, 2007b; Gray et al. 2016). However, the CB1R response to chronic stress in the PFC is unique with respect to the rest of the corticolimbic circuit. Whereas chronic stress unanimously decreases CB1R binding in subcortical areas, CB1R binding in the PFC becomes paradoxically upregulated (Hill et al. 2008a, b; McLaughlin et al. 2013). Moreover, this increase in CB1R binding and functionality has been recapitulated in postmortem tissue samples from depressed (Choi et al. 2012; Hungund et al. 2004) and alcoholic (Vinod et al. 2010) suicide victims. When the initial postmortem studies were published, the prevailing notion was that the increase in CB1R functionality was a detrimental consequence of prolonged exposure to stress that ultimately contributed to the depressive-like phenotype (Hungund et al. 2004). Other research has arrived at the conclusion that despite higher receptor number, receptor functionality may be reduced (see Chapter "Endocannabinoid-Dependent Synaptic Plasticity in the Striatum"). However, an additional explanation could be that this region-specific increase in the number of CB1R binding sites may serve an adaptive purpose that is recruited in an attempt to dampen the negative effects of chronic stress. In support of this notion, we have shown that following chronic unpredictable stress exposure, intra-mPFC CB1R blockade significantly exacerbates behavioral despair in the forced swim test as operationalized by immobility behavior during the session (McLaughlin et al. 2013). These data suggest that negating the chronic stress-induced increase in CB1R binding in the mPFC promotes despair-like coping responses beyond that which would be expected from chronic stress alone. In line with this protective role for CB1Rs in the mPFC, glial cells have been shown to release an increased amount of ECBs and over-express CB1Rs in the PFC under neuropathological conditions, which may constitute an endogenous defense mechanism that prevents additional cell damage (Massi et al. 2008). Moreover, CB1R knockout mice, which exhibit pronounced neuroinflammatory responses in the PFC (Zoppi et al. 2011), also show significant dendritic retraction in the PFC and expansion in the BLA, thus mirroring the effects of chronic stress exposure (Hill et al. 2011b). These data collectively suggest that endocannabinoid signaling protects against the detrimental effects of chronic stress in a region-specific manner, with breakdown of this system likely contributing to the development of stress-related psychiatric disorders.

Little is known regarding how ECB signaling is altered in females under conditions of chronic stress, although the consensus is that the response is unique from that of males. For instance, chronic unpredictable stress has been shown to actually increase CB1R binding in the dorsal hippocampus of intact and ovariectomized female rats (Reich et al. 2009). Similarly, hippocampal CB1R immunoreactivity is decreased in neonatal male rats, but not female rats that experienced maternal deprivation (Suárez et al. 2009). Thus, sex differences in the ECB response to chronic stress could contribute to the wide gap in prevalence rates for stress-related illnesses among men and women in the clinical population.

5 Stress-Related Negative Affect and Drug Withdrawal

5.1 Stress-Related Negative Affective Behavior

Considering that ECB signaling becomes depressed after chronic stress resulting in hyperactivity of the HPA axis, as well as the fact that high levels of stress are a risk factor for many psychiatric disorders such as anxiety, depression, and PTSD (Smoller 2015), it is not surprising that perturbation of the ECB system results in affective symptoms of stress-related disorders (Gray et al. 2015; Gorzalka and Hill 2011; Neumeister et al. 2015). For instance, CB1R knockout mice exhibit increased depressive-like passive coping responses (i.e., immobility) in the forced swim test and tail suspension test (Steiner et al. 2008; Aso et al. 2008), two commonly implemented paradigms used to assess the antidepressant potential of novel pharmacotherapeutic compounds. Similarly, these mice are particularly susceptible to the anhedonic effects of chronic stress (Martin et al. 2002) and exhibit reduced responsiveness to rewarding stimuli such as alcohol (Poncelet et al. 2003) and sucrose (Sanchis-Segura et al. 2004; Cota et al. 2003), in addition to reductions in food intake and weight gain (Haller et al. 2002). CB1R-deficient mice also display an increase in anxiogenic-like traits in tests such as the light-dark box (Martin et al. 2002) and elevated plus maze (Haller et al. 2002, 2004a, b) and increased aggression in the resident-intruder test (Martin et al. 2002). Moreover, these animals exhibit strongly impaired short-term and long-term extinction in fear-conditioning tests (Marsicano et al. 2002) but show no differences during extinction of appetitive motivated tasks (Hölter et al. 2005). Lastly, these mice exhibit impaired hippocampal neurogenesis (Jin et al. 2004) and reduced brain-derived neurotrophic factor (BDNF) release in response to neurotoxic insults (Khaspekov et al. 2004). Therefore, CB1R-deficient mice exhibit a phenotype that is strikingly reminiscent of the symptomatic profile of both depression and anxiety (Hill and Gorzalka 2005), suggesting that disruptions in ECB signaling fundamentally contribute to stress-related affective disorders.

In line with these data, chronic CB1R antagonist administration mimics the effects of chronic unpredictable stress, including increased behavioral despair in the forced swim test, reduced sucrose consumption, decreased serotonin levels in the frontal cortex, reduced hippocampal cell proliferation, survival, and BDNF levels, and increased concentrations of pro-inflammatory cytokines (Beyer et al. 2010). Thus, much like the emotional phenotype displayed by CB1R-deficient mice, chronic CB1R blockade is capable of recapitulating the symptom profile of major depression in preclinical models (Hill and Gorzalka 2005; Valverde and Torrens 2012). Evidence for the detrimental effects of chronic CB1R blockade has since been demonstrated in human clinical trials for the CB1R antagonist rimonabant as a prospective treatment for obesity. A significant proportion of individuals taking the drug spontaneously developed anxiety, adverse depressive-like symptoms, and suicidal ideations that inevitably led to the suspension of clinical trials in both North America and Europe (Christensen et al. 2007; Hill and Gorzalka 2009). Moreover, these effects were also observed in clinical trials for rimonabant in the treatment of atherosclerosis (the STRADIVARIUS, or Strategy to Reduce Atherosclerosis Development Involving Administration of Rimonabant: The Intravascular Ultrasound Study) (Nissen et al. 2008). In these trials, 43% of individuals taking rimonabant developed adverse mood and anxiety responses, compared to 28% in the placebo condition (Nissen et al. 2008). In fact, the emergence of these symptoms was sufficient for one in 13 individuals to discontinue use of rimonabant, compared to one in 47 individuals who discontinued placebo treatment for these same reasons (Nissen et al. 2008). A likely explanation for the strikingly high incidence of adverse depressive-like symptoms observed in these trials is the non-exclusion of patients with prior psychiatric disorders. This inevitably resulted in a less selected study population that more closely reflects the risks of depression and anxiety with rimonabant treatment in routine clinical practice (Rumsfeld and Nallamothu 2008). Thus, chronic CB1R blockade confers symptoms of stress-related affective disorders in both preclinical models and clinical populations, underscoring the importance of this system for proper physiological, behavioral, and emotional responses to environmental adversity.

5.2 *Drug Withdrawal*

While drug and alcohol dependence is associated with altered motivational states, tolerance, and escalated consumption, withdrawal is associated with a profound negative affective state, the alleviation of which serves as the driving factor in relapse (Boden and Fergusson 2011; Sánchez-Peña et al. 2012). Specifically, the self-medication hypothesis predicts that drug use is perpetuated by these negative affective symptoms present during withdrawal (Markou et al. 1998). Given that chronic substance abuse alters the neuroendocrine stress response and that stress hormones recruit the ECB system to modulate emotional behavior, it is unsurprising that the ECB system also fundamentally contributes to both drug dependence and withdrawal symptomology.

Most evidence demonstrates a role for the ECB system in the adverse effects of drug dependence, and withdrawal comes from studies using alcohol-dependent rodents. Rats exposed to 52 days of forced alcohol consumption show reduced CB1R gene expression in the caudate putamen, hypothalamus, and hippocampus (Ortiz et al. 2004), indicating that the downregulation of CB1Rs may be the result of a compensatory mechanism in response to dynamic changes in ECB mobilization.

Moreover, after two cycles of withdrawal, chronic alcohol exposure leads to reductions in CB1R mRNA expression in the BLA (Serrano et al. 2012), which likely contributes to the anxiety-like phenotype observed during withdrawal. Accordingly, a similar phenomenon has also been observed in human alcoholics abstaining from alcohol who show reduced CB1R occupancy and lower AEA levels throughout the brain (Hirvonen et al. 2013; Ceccarini et al. 2014).

In preclinical models, CB1R antagonists prevent acquisition of alcohol drinking in alcohol-preferring rats (Serra et al. 2001) and decrease drinking, as well as sucrose intake, in outbred mice and rats (Arnone et al. 1997; Freedland et al. 2001). Furthermore, some studies have demonstrated that CB1R antagonists decrease escalated alcohol self-administration in alcohol-dependent rats (Rodriguez de Fonseca et al. 1999). Enhanced AEA in mice with a human FAAH polymorphism exhibiting decreased enzymatic activity has also recently been shown to display enhanced binge alcohol intake and preference (Zhou et al. 2016), suggesting that the ECB system is not only important for the affective symptoms of withdrawal but also for the induction of alcohol dependence.

An interaction between the opioid and ECB systems has recently been expanded on (Scavone et al. 2013; Coutts and Pertwee 1998), with preclinical evidence demonstrating that treatment with Δ-9-tetrahydrocannabinol (THC), AEA, or 2-AG can alleviate the symptoms of morphine withdrawal (Cichewicz and Welch 2003; Vela et al. 1995; Yamaguchi et al. 2001; Ramesh et al. 2011, 2013). Additionally, a CB1R antagonist reduces heroin self-administration in rats and morphine-conditioned place preference in mice (Navarro et al. 2001), indicating that reducing ECB signaling can alleviate the rewarding and motivating properties of opioids. This has led researchers to propose the investigation of cannabinoids in the treatment of opioid dependence (Scavone et al. 2013).

Furthermore, after chronic nicotine exposure, AEA and 2-AG is increased in the brainstem and decreased in the hippocampus and cortex (Gonzalez et al. 2002). This same study showed that AEA is also decreased in the striatum after chronic nicotine exposure, with no changes in 2-AG in this region. Thus, the symptomatology of nicotine withdrawal may be driven by changes in ECB signaling. Conversely, FAAH inhibition enhances nicotine-conditioned place preference in mice (Merritt et al. 2008) while also decreasing nicotine withdrawal-induced anxiety without affecting the physical signs of withdrawal (Cippitelli et al. 2011). Therefore, increasing AEA signaling is a potential therapeutic target for specifically alleviating mood disruptions during nicotine withdrawal.

Finally, there is some evidence to indicate that ECB system changes contribute to cocaine withdrawal. Animals chronically treated with cocaine show a decrease in 2-AG content in the limbic forebrain (Gonzalez et al. 2002). Interestingly, this study also evaluated alcohol and nicotine, and cocaine withdrawal resulted in the least amount of changes in ECBs. Furthermore, CB1R antagonists do not reduce cocaine self-administration but do block reinstatement of cocaine seeking and cue- or stress-induced relapse (Filip et al. 2006; Chauvet et al. 2014; De Vries et al. 2001; McReynolds et al. 2016; Vaughn et al. 2012). There has been some speculation as to why CB1R antagonists do not block the reinforcing effects of cocaine, as they do for alcohol, nicotine, and opioids. It has been suggested that the difference may result from how cocaine modulates dopamine, by directly increasing dopamine in the nucleus accumbens as opposed to acting through the VTA (Parolaro et al. 2007). For example, CB1R deletion does not block cocaine-induced increases in nucleus accumbens dopamine (Soria et al. 2005), suggesting that ECBs are not directly involved in the rewarding effects of cocaine, but play a role in mechanisms driving relapse. On the whole, however, drug withdrawal depresses ECB signaling likely via the CRH-mediated mechanism described above, which ultimately serves a permissive role in the induction of withdrawal-induced negative affective symptoms.

5.3 *Sex Differences in Negative Affect and Drug Withdrawal*

Women report mood disorders approximately twice as often as men, and sexual dimorphism of the ECB system may contribute to these differences. Multiple lines of evidence suggest that gonadal hormones, such as estradiol, modulate the ECB system (Gorzalka and Dang 2012). For instance, long-term ovariectomy induces depressive-like behaviors in rats (de Chaves et al. 2009). As mentioned previously, the anxiolytic effect of estradiol is blocked by a CB1R antagonist (Hill et al. 2007a), indicating that estrogens may provide anxiolytic-like effects through an ECB mechanism. Clinical data has also demonstrated that major depressive disorder is correlated with reductions in circulating ECBs in women (Hill et al. 2008a, b). Additionally, circulating 2-AG is also depressed in men and women diagnosed with PTSD (Hill et al. 2013a, b), perhaps suggesting an impaired ability to terminate the stress response. Overall, sex differences seen in psychiatric disorders, especially those related to disruptions in mood, may arise from sexual dimorphisms in the ECB system.

Sex differences also exist in the way men and women develop and express the symptoms of alcohol dependence. Alcohol dependence may develop more rapidly in women, and adolescent girls show greater cognitive deficits after excessive alcohol use than adolescent boys (for review see Sharrett-Field et al. 2013). Women also report significantly more psychiatric symptoms in response to chronic alcohol use (Hernandez-Avila et al. 2004), as well as fertility issues and menstrual cycle disruptions (Wilsnack et al. 1984; Emanuele and Emanuele 2001). Consequently, gonadal hormones, and thus changes in ECB signaling, may contribute to sex differences in alcoholism. For example, female wild-type mice drink significantly more than male wild-type mice, but this sex difference is abolished by CB1R deletion (Hungund et al. 2003). Additionally, estradiol protects against the motoric symptoms of withdrawal, as well as neuronal damage to the cerebellum observed following chronic alcohol exposure (Jung et al. 2002). As estradiol has been shown to stimulate AEA biosynthesis in the brain (Huang and Woolley 2012; Tabatadze et al. 2015) and downregulate FAAH expression, thereby increasing AEA levels

(Maccarrone et al. 2000), estradiol may protect against the affective symptoms of alcohol withdrawal through an ECB mechanism.

6 Implications for the Development of ECB-Based Therapeutics

There is some evidence that exogenous cannabinoid use can alleviate mood disorders. For example, a recent controlled trial examining the effects of cannabis consumption on chronic neuropathic pain revealed significant improvements in measures of anxiety and depression, in addition to reduced pain and better quality of sleep (Ware et al. 2010). Unfortunately, a major obstacle in the implementation of cannabis-derived drugs for mood disorders is that long-term heavy cannabis use, particularly during adolescence, is associated with increased risk of experiencing depressive-like symptoms that persist into adulthood (Bovasso 2001; Patton 2002). However, considerable evidence suggests that enhancement of AEA or 2-AG signaling specifically may attenuate the symptoms of stress-related affective disorders.

Inhibition of FAAH reduces anxiety behavior in mice and rats, which is coupled with an increase in AEA in the frontal cortex, striatum, and hippocampus (Marco et al. 2015; Kathuria et al. 2003; Hill et al. 2007a). FAAH inhibition has also been shown to reverse the anxiety and depressive phenotype expressed in rats exposed to chronic stress (Hill et al. 2013a, b; Bortolato et al. 2007; Rossi et al. 2010; Lomazzo et al. 2015). For example, FAAH knockout mice do not show the same anxiety behavior, decreases in AEA, and structural remodeling of the amygdala seen in wild-type mice exposed to chronic stress (Hill et al. 2013a, b). Systemic FAAH inhibition also prevents the negative affective phenotype induced by social defeat stress in rodents (Rossi et al. 2010), suggesting that FAAH inhibition may alleviate symptoms of affective disorders.

In fact, in humans, anxiety is negatively correlated with circulating AEA (Dlugos et al. 2012), suggesting that increasing AEA signaling could be therapeutic for mood disorders. Additionally, a polymorphism of the FAAH gene (C385A) in humans is correlated with both decreased anxiety and increased resting state functional connectivity between the mPFC and BLA (Dincheva et al. 2015; Hariri et al. 2009; Gunduz-Cinar et al. 2013). These authors recapitulated this polymorphism in the mouse and observed an anxiolytic phenotype as well as increased projections from the mPFC to the BLA. Thus, AEA signaling in the mPFC may be constraining BLA activity, perhaps reducing stress-induced negative affect.

In terms of 2-AG, DAGLα-deficient mice show increased anxiety- and depressive-like behavior, which is reversible with a MAGL inhibitor (Shonesy et al. 2014). However, it is important to note that the negative affective behaviors observed were sex specific, with only female DAGLα-deficient mice showing depressive-like behavior. Conversely, systemic MAGL inhibition reduces the anxiety- and depressive-like behavior seen in mice following chronic unpredictable stress exposure (Zhang et al. 2015), indicating that increasing 2-AG signaling during stress may also be efficacious for treating anxiety and depression. A reduction in circulating 2-AG levels has been reported in humans who have experienced chronic stress of simulated space travel (Yi et al. 2016), supporting the notion that 2-AG tone may play a role in stress-related pathology. Overall, considering the preclinical and clinical data, impairments in ECB signaling may well contribute to stress-related affective disorders, making the use of enzyme inhibitors an attractive option for pharmacotherapeutic development as they avoid the pitfalls of global agonism seen in THC or synthetic CB1R agonists by specifically augmenting AEA or 2-AG mobilization (Gunduz-Cinar et al. 2013).

7 Summary

The stress response is an evolutionarily conserved mechanism to both allow an organism to cope with a threat and to restore homeostasis following the occurrence of a stressor. In addition, preclinical research demonstrates that the ECB system constrains the HPA axis and plays a major role in the habituation to stressors. AEA serves to tonically activate the ECB system, and the reduction of AEA, via increased FAAH activity, in the BLA serves a permissive role in the initiation of the stress response. 2-AG signaling, on the other hand, is recruited to activate CB1Rs and promote termination of the stress response in the short term and to facilitate habituation to stressors in the long term.

The impairment of the ECB system following stress may contribute to stress-related disorders such as anxiety, depression, and substance abuse. Chronic stress downregulates the ECB system, resulting in reduced inhibitory control to the amygdala and a hyperactive stress response. Sex-dependent and genetic differences in the ECB system speak both to the detriment of having reduced ECB signaling and the conferred resiliency to stressors when ECB signaling is augmented.

The growing body of preclinical and clinical evidence demonstrates that the ECB system represents a promising therapeutic avenue for a host of affective human disorders. Pharmacological inhibition of ECB hydrolases has emerged as an effective strategy for managing behavioral and neuroendocrine responses to acute and chronic stress. Moreover, genetic polymorphism in genes that encode these enzymes has offered considerable insight into the stress-resilient phenotype conferred by increased circulating AEA in humans. This piece of information, in addition to extensive animal research, points to selectively augmenting ECB activity via FAAH inhibition as an attractive therapeutic strategy for treating human stress-related affective illnesses.

References

Ahn K, McKinney MK, Cravatt BF (2008) Enzymatic pathways that regulat endocannabinoid signaling in the nervous system. Chem Rev 108(5):1687–1707

Aloisi AM, Ceccarelli I, Lupo C (1998) Behavioural and hormonal effects of restraint stress and formalin test in male and female rats. Brain Res Bull 47(1):57–62. Available at: http://www.ncbi.nlm.nih.gov/entrez/query.fcgi?cmd=Retrieve&db=PubMed&dopt=Citation&list_uids=9766390

Antoni FA (1986) Hypothalamic control of adrenocorticotropin secretion: advances since the discovery of 41-residue corticotropin-releasing factor. Endocr Rev 7(4):351–378

Arnone M et al (1997) Selective inhibition of sucrose and ethanol intake by SR 141716, an antagonist of central cannabinoid (CB1) receptors. Psychopharmacology 132(1):104–106

Aso E et al (2008) BDNF impairment in the hippocampus is related to enhanced despair behavior in CB1 knockout mice. J Neurochem 105(2):565–572

Axelrod J, Reisine TD (1984) Stress hormones: their interaction and regulation (Review). Science 224(4648):452–459

Barna I et al (2004) The role of endogenous cannabinoids in the hypothalamo-pituitary-adrenal axis regulation: in vivo and in vitro studies in CB1 receptor knockout mice. Life Sci 75 (24):2959–2970

Bedse G et al (2014) Role of the basolateral amygdala in mediating the effects of the fatty acid amide hydrolase inhibitor URB597 on HPA axis response to stress. Eur Neuropsychopharmacol 24(9):1511–1523

Berrendero F et al (1999) Analysis of cannabinoid receptor binding and mRNA expression and endogenous cannabinoid contents in the developing rat brain during late gestation and early postnatal period. Synapse 33(3):181–191. Available at: http://www.ncbi.nlm.nih.gov/entrez/query.fcgi?cmd=Retrieve&db=PubMed&dopt=Citation&list_uids=10420166

Beyer CE et al (2010) Depression-like phenotype following chronic CB1 receptor antagonism. Neurobiol Dis 39(2):148–155

Bisogno T et al (2003) Cloning of the first sn1-DAG lipases points to the spatial and temporal regulation of endocannabinoid signaling in the brain. J Cell Biol 163(3):463–468

Boden JM, Fergusson DM (2011) Alcohol and depression. Addiction 106(5):906–914

Bodnoff SR et al (1995) Enduring effects of chronic corticosterone treatment on spatial learning, synaptic plasticity, and hippocampal neuropathology in young and mid-aged rats. J Neurosci 15(1 Pt 1):61–69

Bonn-Miller MO, Zvolensky MJ, Bernstein A (2007) Marijuana use motives: concurrent relations to frequency of past 30-day use and anxiety sensitivity among young adult marijuana smokers. Addict Behav 32(1):49–62

Bortolato M et al (2007) Antidepressant-like activity of the fatty acid amide hydrolase inhibitor URB597 in a rat model of chronic mild stress. Biol Psychiatry 62(10):1103–1110

Bovasso GB (2001) Cannabis abuse as a risk factor for depressive symptoms. Am J Psychiatry 158 (12):2033–2037. Available at: http://www.ncbi.nlm.nih.gov/pubmed/11729021

Bowles NP et al (2012) Chronic, noninvasive glucocorticoid administration suppresses limbic endocannabinoid signaling in mice. Neuroscience 204:83–89

Cannon W (1935) Stress and strains of homeostasis. Am J Med Sci 189:1–14

Carroll ME, Anker JJ (2010) Sex differences and ovarian hormones in animal models of drug dependence. Horm Behav 58(1):44–56

Ceccarini J et al (2014) Changes in cerebral CB1 receptor availability after acute and chronic alcohol abuse and monitored abstinence. J Neurosci 34(8):2822–2831. Available at: http://www.ncbi.nlm.nih.gov/pubmed/24553924

Chauvet C et al (2014) Chronic stimulation of the tone of endogenous anandamide reduces cue- and stress-induced relapse in rats. Int J Neuropsychopharmacol 18:1–5. Available at: http://www.ncbi.nlm.nih.gov/pubmed/25522382

Checkley S (1996) The neuroendocrinology of depression and chronic stress. Br Med Bull 52 (3):597–617

Chevaleyre V, Takahashi KA, Castillo PE (2006) Endocannabinoid-mediated synaptic plasticity in the CNS. Annu Rev Neurosci 29:37–76. Available at: http://www.ncbi.nlm.nih.gov/pubmed/16776579
Choi K et al (2012) Expression pattern of the cannabinoid receptor genes in the frontal cortex of mood disorder patients and mice selectively bred for high and low fear. J Psychiatr Res 46 (7):882–889
Christensen R et al (2007) Efficacy and safety of the weight-loss drug rimonabant: a meta-analysis of randomised trials. Lancet 370(9600):1706–1713
Chrousos GP, Gold PW (1992) The concepts of stress and stress system disorders. JAMA 267 (9):1244–1252
Cichewicz DL, Welch SP (2003) Modulation of oral morphine antinociceptive tolerance and naloxone-precipitated withdrawal signs by oral Delta 9-tetrahydrocannabinol. J Pharmacol Exp Ther 305(3):812–817
Cippitelli A et al (2011) Endocannabinoid regulation of acute and protracted nicotine withdrawal: effect of FAAH inhibition. PLoS One 6(11)
Collins A, Frankenhaeuser M (1978) Stress responses in male and female engineering students. J Hum Stress 4(2):43–48
Cota D et al (2003) The endogenous cannabinoid system affects energy balance via central orexigenic drive and peripheral lipogenesis. J Clin Invest 112(3):423–431
Coutts A, Pertwee R (1998) Evidence that cannabinoid-induced inhibition of electrically evoked contractions of the myenteric plexus–longitudinal muscle preparation of guinea-pig small intestine can be modulated by Ca2+ and cAMP. Can J Physiol Pharmacol 76(3):340–346
Cravatt BF et al (1996) Molecular characterization of an enzyme that degrades neuromodulatory fatty-acid amides. Nature 384:83–87
Dalton VS, Zavitsanou K (2010) Cannabinoid effects on CB1 receptor density in the adolescent brain: an autoradiographic study using the synthetic cannabinoid HU210. Synapse (New York, NY) 64(11):845–54. Available at: http://www.ncbi.nlm.nih.gov/pubmed/20842718 [Accessed January 6, 2015]
de Chaves G et al (2009) Effects of long-term ovariectomy on anxiety and behavioral despair in rats. Physiol Behav 97(3–4):420–425
de Kloet ER (1984) Adrenal steroids as modulators of nerve cell function. J Steroid Biochem 20 (1):175–181. Available at: http://www.ncbi.nlm.nih.gov/pubmed/6200696
de Kloet ER, Joëls M, Holsboer F (2005) Stress and the brain: from adaptation to disease. Nature reviews. Neuroscience 6(6):463–475
De Vries TJ et al (2001) A cannabinoid mechanism in relapse to cocaine seeking. Nat Med 7 (10):1151–1154. Available at: http://www.ncbi.nlm.nih.gov/entrez/query.fcgi?cmd=Retrieve&db=PubMed&dopt=Citation&list_uids=11590440\nhttp://www.nature.com/nm/journal/v7/n10/pdf/nm1001-1151.pdf
Demers C et al (2016) Interactions between anandamide & corticotropin-releasing hormone signaling modulate human amygdala function & risk for anxiety disorders: An imaging genetics strategy for modeling molecular interactions. Biol Psychiatry 80(5):356–362
Devane WA et al (1992) Isolation and structure of a brain constituent that binds to the cannabinoid receptor. Science 258(5090):1946–1949
Di S et al (2003) Nongenomic glucocorticoid inhibition via endocannabinoid release in the hypothalamus: a fast feedback mechanism. J Neurosci 23(12):4850–4857
Dincheva I et al (2015) FAAH genetic variation enhances fronto-amygdala function in mouse and human. Nat Commun 6:6395. Available at: http://www.nature.com/doifinder/10.1038/ncomms7395
Dinh TP et al (2002) Brain monoglyceride lipase participating in endocannabinoid inactivation. Proc Natl Acad Sci U S A 99(16):10819–10824. Available at: http://www.pnas.org/cgi/content/long/99/16/10819
Diorio D, Viau V, Meaney MJ (1993) The role of the medial prefrontal cortex (cingulate gyrus) in the regulation of hypothalamic-pituitary-adrenal responses to stress. J Neurosci 13

(9):3839–3847. Available at: http://www.scopus.com/inward/record.url?eid=2-s2.0-0027200692&partnerID=40
Dlugos A et al (2012) Acute stress increases circulating anandamide and other N-acylethanolamines in healthy humans. Neuropsychopharmacology 37(11):2416–2427. Available at: http://www.ncbi.nlm.nih.gov/pubmed/22763622\nhttp://www.nature.com/npp/journal/v37/n11/pdf/npp2012100a.pdf
Dubreucq S et al (2012) Genetic dissection of the role of cannabinoid type-1 receptors in the emotional consequences of repeated social stress in mice. Neuropsychopharmacology 37 (8):1885–1900. Available at: http://www.nature.com/doifinder/10.1038/npp.2012.36
Dunn AJ, Berridge CW (1990) Physiological and behavioral responses to corticotropin-releasing factor administration: is CRF a mediator of anxiety or stress responses? Brain Res Rev 15 (2):71–100. Available at: http://www.sciencedirect.com/science/article/pii/016501739090012D
Earle TL, Linden W, Weinberg J (1999) Differential effects of harassment on cardiovascular and salivary cortisol stress reactivity and recovery in women and men. J Psychosom Res 46 (2):125–141
Egertová M, Elphick MR (2000) Localisation of cannabinoid receptors in the rat brain using antibodies to the intracellular C-terminal tail of CB. J Comp Neurol 422(2):159–171
Emanuele MA, Emanuele N (2001) Alcohol and the male reproductive system. Alcohol Res Health 25(4):282–287
Evanson NK et al (2010) Fast feedback inhibition of the HPA axis by glucocorticoids is mediated by endocannabinoid signaling. Endocrinology 151(10):4811–4819
Filip M et al (2006) Involvement of cannabinoid CB1 receptors in drug addiction: effects of rimonabant on behavioral responses induced by cocaine. Pharmacol Rep 58(6):806–819
Freedland CS et al (2001) Effects of SR141716A on ethanol and sucrose self-administration. Alcohol Clin Exp Res 25(2):277–282. Available at: http://www.ncbi.nlm.nih.gov/htbin-post/Entrez/query?db=m&form=6&dopt=r&uid=11236843\nhttp://onlinelibrary.wiley.com/store/10.1111/j.1530-0277.2001.tb02209.x/asset/j.1530-0277.2001.tb02209.x.pdf?v=1&t=hlovu4cy&s=ca6288e841c7e96ea4e05e9a90910711cc6ff7f0
Freund TF, Katona I, Piomelli D et al (2003) Role of endogenous cannabinoids in synaptic signaling. Physiol Rev 83:1017–1066
Glass M, Faull RL, Dragunow M (1997) Cannabinoid receptors in the human brain: a detailed anatomical and quantitative autoradiographic study in the fetal, neonatal and adult human brain. Neuroscience 77(2):299–318. Available at: http://linkinghub.elsevier.com/retrieve/pii/S0306452296004289
Goel N, Bale TL (2008) Organizational and activational effects of testosterone on masculinization of female physiological and behavioral stress responses. Endocrinology 149(12):6399–6405
Goel N, Bale TL (2010) Sex differences in the serotonergic influence on the hypothalamic-pituitary-adrenal stress axis. Endocrinology 151(4):1784–1794
Goel N et al (2014) Sex differences in the HPA axis. Compr Physiol 4(3):1121–1155. Available at: http://www.ncbi.nlm.nih.gov/pubmed/24944032 [Accessed January 6, 2015]
Goldstein LE et al (1996) Role of the amygdala in the coordination of behavioral, neuroendocrine, and prefrontal cortical monoamine responses to psychological stress in the rat. J Neurosci 16 (15):4787–4798
Gonzalez S et al (2002) Changes in endocannabinoid contents in the brain of rats chronically exposed to nicotine, ethanol or cocaine. Brain Res 954(1):73–81
Gorzalka BB, Dang SS (2012) Minireview: endocannabinoids and gonadal hormones: bidirectional interactions in physiology and behavior. Endocrinology 153(3):1016–1024
Gorzalka BB, Hill MN (2011) Putative role of endocannabinoid signaling in the etiology of depression and actions of antidepressants. Prog Neuro-Psychopharmacol Biol Psychiatry 35 (7):1575–1585. Available at: http://dx.doi.org/10.1016/j.pnpbp.2010.11.021
Gorzalka BB, Hill MN, Hillard CJ (2008) Regulation of endocannabinoid signaling by stress: implications for stress-related affective disorders. Neurosci Biobehav Rev 32(6):1152–1160

Gray JM et al (2015) Corticotropin-releasing hormone drives anandamide hydrolysis in the amygdala to promote anxiety. J Neurosci 35(9):3879–3892. Available at: http://www.ncbi.nlm.nih.gov/pubmed/25740517
Gray J et al (2016) Sustained glucocorticoid exposure recruits cortico-limbic CRH signaling to modulate endocannabinoid function. Psychoneuroendocrinology 66(1):151–158
Gunduz-Cinar O et al (2013) Amygdala FAAH and anandamide: mediating protection and recovery from stress. Trends Pharmacol Sci 34(11):637–644. Available at: http://dx.doi.org/10.1016/j.tips.2013.08.008
Haller J et al (2002) The effects of genetic and pharmacological blockade of the CB1 cannabinoid receptor on anxiety. Eur J Neurosci 16(7):1395–1398
Haller J et al (2004a) CB1 cannabinoid receptors mediate anxiolytic effects: convergent genetic and pharmacological evidence with CB1-specific agents. Behav Pharmacol 15(4):299–304. Available at: http://www.ncbi.nlm.nih.gov/pubmed/15252281\nhttp://graphics.tx.ovid.com/ovftpdfs/FPDDNCOBABIDMP00/fs047/ovft/live/gv024/00008877/00008877-200407000-00007.pdf
Haller J et al (2004b) Context-dependent effects of CB1 cannabinoid gene disruption on anxiety-like and social behaviour in mice. Eur J Neurosci 19(7):1906–1912
Hammen C (2005) Stress and depression. Annu Rev Clin Psychol 1:293–319
Handa RJ, Weiser MJ (2014) Gonadal steroid hormones and the hypothalamo-pituitary-adrenal axis. Front Neuroendocrinol 35:197–220
Hariri AR et al (2009) Divergent effects of genetic variation in endocannabinoid signaling on human threat- and reward-related brain function. Biol Psychiatry 66(1):9–16
Heinsbroek RPW et al (1991) Sex- and time-dependent changes in neurochemical and hormonal variables induced by predictable and unpredictable footshock. Physiol Behav 49(6):1251–1256
Herkenham M et al (1990) Cannabinoid receptor localization in brain. Proc Natl Acad Sci U S A 87(5):1932–1936
Herkenham M et al (1991) Characterization and localization of cannabinoid receptors in rat brain: a quantitative in vitro autoradiographic study. J Neurosci 11:563–583
Hermanson DJ et al (2014) Substrate-selective COX-2 inhibition as a novel strategy for therapeutic endocannabinoid augmentation. Trends Pharmacol Sci 35(7):358–367
Hernandez-Avila CA, Rounsaville BJ, Kranzler HR (2004) Opioid-, cannabis- and alcohol-dependent women show more rapid progression to substance abuse treatment. Drug Alcohol Depend 74(3):265–272
Hill MN, Gorzalka BB (2005) Pharmacological enhancement of cannabinoid CB1 receptor activity elicits an antidepressant-like response in the rat forced swim test. Eur Neuropsychopharmacol 15(6):593–599
Hill M, Gorzalka B (2009) Impairments in endocannabinoid signaling and depressive illness. JAMA 301(11):1165–1166
Hill MNN, Tasker JGG (2012) Endocannabinoid signaling, glucocorticoid-mediated negative feedback, and regulation of the hypothalamic-pituitary-adrenal axis. Neuroscience 204:5–16. Available at: http://linkinghub.elsevier.com/retrieve/pii/S0306452211014278\nhttp://www.sciencedirect.com/science/article/pii/S0306452211014278
Hill MN et al (2005) Downregulation of endocannabinoid signaling in the hippocampus following chronic unpredictable stress. Neuropsychopharmacology 30(3):508–515. Available at: http://www.ncbi.nlm.nih.gov/entrez/query.fcgi?cmd=Retrieve&db=PubMed&dopt=Citation&list_uids=15525997\nhttp://www.nature.com/npp/journal/v30/n3/pdf/1300601a.pdf
Hill MN et al (2006) Involvement of the endocannabinoid system in the ability of long-term tricyclic antidepressant treatment to suppress stress-induced activation of the hypothalamic-pituitary-adrenal axis. Neuropsychopharmacology 31(12):2591–2599. Available at: http://www.ncbi.nlm.nih.gov/entrez/query.fcgi?cmd=Retrieve&db=PubMed&dopt=Citation&list_uids=16710317\nhttp://www.nature.com/npp/journal/v31/n12/pdf/1301092a.pdf
Hill MN, Karacabeyli ES, Gorzalka BB (2007a) Estrogen recruits the endocannabinoid system to modulate emotionality. Psychoneuroendocrinology 32(4):350–357

Hill MN, Barr AM et al (2007b) Electroconvulsive shock treatment differentially modulates cortical and subcortical endocannabinoid activity. J Neurochem 103(1):47–56

Hill M et al (2008a) Regional alterations in the endocannabinoid system in an animal model of depression: effects of concurrent antidepressant treatment. J Neurochem 106(6):2322–2336. Available at: http://www.pubmedcentral.nih.gov/articlerender.fcgi?artid=2606621&tool=pmcentrez&rendertype=abstract [Accessed December 16, 2014]

Hill MN et al (2008b) Serum endocannabinoid content is altered in females with depressive disorders: a preliminary report. Pharmacopsychiatry 41:48–53

Hill MN et al (2009) Suppression of amygdalar endocannabinoid signaling by stress contributes to activation of the hypothalamic-pituitary-adrenal axis. Neuropsychopharmacology 34 (13):2733–2745. Available at: http://www.pubmedcentral.nih.gov/articlerender.fcgi?artid=3197779&tool=pmcentrez&rendertype=abstract [Accessed December 16, 2014]

Hill MN et al (2010) Endogenous cannabinoid signaling is essential for stress adaptation. Proc Natl Acad Sci U S A 107(20):9406–9411. Available at: http://www.pubmedcentral.nih.gov/articlerender.fcgi?artid=2889099&tool=pmcentrez&rendertype=abstract [Accessed December 18, 2014]

Hill MN, McLaughlin RJ et al (2011a) Recruitment of prefrontal cortical endocannabinoid signaling by glucocorticoids contributes to termination of the stress response. J Neurosci 31 (29):10506–10515. Available at: http://www.pubmedcentral.nih.gov/articlerender.fcgi?artid=3179266&tool=pmcentrez&rendertype=abstract [Accessed December 16, 2014]

Hill MN, Hillard CJ, McEwen BS (2011b) Alterations in corticolimbic dendritic morphology and emotional behavior in cannabinoid CB1 receptor-deficient mice parallel the effects of chronic stress. Cereb Cortex (New York, NY: 1991) 21(9): 2056–64. Available at: http://www.pubmedcentral.nih.gov/articlerender.fcgi?artid=3155602&tool=pmcentrez&rendertype=abstract

Hill MN et al (2013a) Disruption of fatty acid amide hydrolase activity prevents the effects of chronic stress on anxiety and amygdalar microstructure. Mol Psychiatry 18(10):1125–1135. Available at: http://www.pubmedcentral.nih.gov/articlerender.fcgi?artid=4148304&tool=pmcentrez&rendertype=abstract

Hill MN et al (2013b) Reductions in circulating endocannabinoid levels in individuals with post-traumatic stress disorder following exposure to the world trade center attacks. Psychoneuroendocrinology 38(12):2952–2961

Hillard CJ (2000) Biochemistry and pharmacology of the endocannabinoids arachidonylethanolamide and 2-arachidonylglycerol. Prostaglandins Other Lipid Mediat 61 (1-2):3–18

Hinz B, Hirschelmann R (2000) Rapid non-genomic feedback effects of glucocorticoids on CRF-induced ACTH secretion in rats. Pharm Res 17(10):1273–1277

Hirvonen J et al (2013) Reduced cannabinoid CB1 receptor binding in alcohol dependence measured with positron emission tomography. Mol Psychiatry 18(8):916–921. Available at: http://www.ncbi.nlm.nih.gov/pubmed/22776901\nhttp://www.nature.com/mp/journal/v18/n8/pdf/mp2012100a.pdf

Hölter SM et al (2005) Cannabinoid CB1 receptor is dispensable for memory extinction in an appetitively-motivated learning task. Eur J Pharmacol 510(1–2):69–74

Howlett AC (1987) Cannabinoid inhibition of adenylate cyclase: relative activity of constituents and metabolites of marihuana. Neuropharmacology 26(5):507–512

Howlett AC (2002) The cannabinoid receptors. Prostaglandins Other Lipid Mediat 68–69:619–631. Available at: http://www.sciencedirect.com/science/article/pii/S0090698002000606

Howlett AC, Fleming RM (1984) Cannabinoid inhibition of adenylate cyclase. Pharmacology of the response in neuroblastoma cell membranes. Mol Pharmacol 26(3):532–538. Available at: http://molpharm.aspetjournals.org/content/26/3/532.abstract

Huang GZ, Woolley CS (2012) Estradiol acutely suppresses inhibition in the hippocampus through a sex-specific endocannabinoid and mGluR-dependent mechanism. Neuron 74 (5):801–808

Hungund BL et al (2003) Cannabinoid CB1 receptor knockout mice exhibit markedly reduced voluntary alcohol consumption and lack alcohol-induced dopamine release in the nucleus accumbens. J Neurochem 84(4):698–704

Hungund BL et al (2004) Upregulation of CB1 receptors and agonist-stimulated [35S]GTPgammaS binding in the prefrontal cortex of depressed suicide victims. Mol Psychiatry 9(2):184–190. Available at: http://www.ncbi.nlm.nih.gov/entrez/query.fcgi?cmd=Retrieve&db=PubMed&dopt=Citation&list_uids=14966476\nhttp://www.nature.com/mp/journal/v9/n2/pdf/4001376a.pdf

Iwasaki-Sekino A et al (2009) Gender differences in corticotropin and corticosterone secretion and corticotropin-releasing factor mRNA expression in the paraventricular nucleus of the hypothalamus and the central nucleus of the amygdala in response to footshock stress or psychological. Psychoneuroendocrinology 34(2):226–237

Jacobson L, Sapolsky R (1991) The role of the hippocampus in feedback regulation of the hypothalamic-pituitary-adrenocortical axis. Endocr Rev 12(2):118–134

Jin K et al (2004) Defective adult neurogenesis in CB1 cannabinoid receptor knockout mice. Mol Pharmacol 66(2):204–208

Jung ME et al (2002) Estradiol protects against cerebellar damage and motor deficit in ethanol-withdrawn rats. Alcohol 26(2):83–93

Kano M et al (2009) Endocannabinoid-mediated control of synaptic transmission. Physiol Rev 89 (1):309–380. Available at: http://www.ncbi.nlm.nih.gov/entrez/query.fcgi?cmd=Retrieve&db=PubMed&dopt=Citation&list_uids=19126760\nhttp://physrev.physiology.org/content/89/1/309.full.pdf

Kathuria S et al (2003) Modulation of anxiety through blockade of anandamide hydrolysis. Nat Med 9(1):76–81

Katona I et al (2001) Distribution of CB1 cannabinoid receptors in the amygdala and their role in the control of GABAergic transmission. J Neurosci 21(23):9506–9518

Keller-Wood ME, Dallman MF (1984) Corticosteroid inhibition of ACTH secretion. Endocr Rev 5 (1):1–24

Khaspekov LG et al (2004) Involvement of brain-derived neurotrophic factor in cannabinoid receptor-dependent protection against excitotoxicity. Eur J Neurosci 19(7):1691–1698

Liu J et al (2008) Multiple pathways involved in the biosynthesis of anandamide. Neuropharmacology 54(1):1–7

Lomazzo E et al (2015) Therapeutic potential of inhibitors of endocannabinoid degradation for the treatment of stress-related hyperalgesia in an animal model of chronic pain. Neuropsychopharmacology 40(2):48–501

Maccarrone M et al (2000) Down-regulation of anandamide hydrolase in mouse uterus by sex hormones. Eur J Biochem 267(10):2991–2997

Mackie K, Hille B (1992) Cannabinoids inhibit N-type calcium channels in neuroblastoma-glioma cells. Proc Natl Acad Sci U S A 89(9):3825–3829

Mackie K et al (1995) Cannabinoids activate an inwardly rectifying potassium conductance and inhibit Q-type calcium currents in AtT20 cells transfected with rat brain cannabinoid receptor. J Neurosci 15(10):6552–6561. Available at: http://www.ncbi.nlm.nih.gov/entrez/query.fcgi?cmd=Retrieve&db=PubMed&dopt=Citation&list_uids=7472417\nhttp://www.jneurosci.org/content/15/10/6552.full.pdf

Mailleux P, Vanderhaeghen JJ (1992) Localization of cannabinoid receptor in the human developing and adult basal ganglia. Higher levels in the striatonigral neurons. Neurosci Lett 148:173–176

Malcher-Lopes R et al (2006) Opposing crosstalk between leptin and glucocorticoids rapidly modulates synaptic excitation via endocannabinoid release. J Neurosci 26(24):6643–6650.

Available at: http://www.ncbi.nlm.nih.gov/entrez/query.fcgi?cmd=Retrieve&db=PubMed&dopt=Citation&list_uids=16775153

Marco EM et al (2015) Potential therapeutic value of a novel FAAH inhibitor for the treatment of anxiety. PLoS One 10(9)

Markou A, Kosten TR, Koob GF (1998) Neurobiological similarities in depression and drug dependence: a self-medication hypothesis. Neuropsychopharmacology 18:135–174. Available at: PM:9471114

Marrs WR et al (2010) The serine hydrolase ABHD6 controls the accumulation and efficacy of 2-AG at cannabinoid receptors. Nat Neurosci 13(8):951–957. Available at: http://dx.doi.org/10.1038/nn.2601

Marsicano G et al (2002) The endogenous cannabinoid system controls extinction of aversive memories. Nature 418(6897):530–534

Martin M et al (2002) Involvement of CB1 cannabinoid receptors in emotional behaviour. Psychopharmacology 159(4):379–387

Massi P et al (2008) Expression and function of the endocannabinoid system in glial cells. Curr Pharm Des 14(23):2289–2298. Available at: http://www.ncbi.nlm.nih.gov/pubmed/18781979

Matsuda LA et al (1990) Structure of a cannabinoid receptor and functional expression of the cloned cDNA. Nature 346(6284):561–564. Available at: http://www.ncbi.nlm.nih.gov/pubmed/2165569

McEwen BS (2012) Brain on stress: How the social environment gets under the skin. Proc Natl Acad Sci U S A 109(Supplement_2):17180–17185

McEwen BS et al (2015) Mechanisms of stress in the brain. Nat Neurosci 18(10):1353–1363. Available at https://www.ncbi.nlm.nih.gov/pmc/articles/PMC4933289/

McLaughlin RJ et al (2012) Prefrontal cortical anandamide signaling coordinates coping responses to stress through a serotonergic pathway. Eur Neuropsychopharmacol 22(9):664–671. Available at: http://www.pubmedcentral.nih.gov/articlerender.fcgi?artid=3366159&tool=pmcentrez&rendertype=abstract [Accessed January 6, 2015]

McLaughlin RJ et al (2013) Upregulation of CB_1 receptor binding in the ventromedial prefrontal cortex promotes proactive stress-coping strategies following chronic stress exposure. Behav Brain Res 237:333–337. Available at: http://www.pubmedcentral.nih.gov/articlerender.fcgi?artid=3501995&tool=pmcentrez&rendertype=abstract [Accessed January 6, 2015]

McLaughlin RJ, Hill MN, Gorzalka BB (2014) A critical role for prefrontocortical endocannabinoid signaling in the regulation of stress and emotional behavior. Neurosci Biobehav Rev 42:116–131. Available at: http://www.ncbi.nlm.nih.gov/pubmed/24582908 [Accessed December 4, 2014]

McReynolds JR et al (2016) CB1 receptor antagonism blocks stress-potentiated reinstatement of cocaine seeking in rats. Psychopharmacology 233(1):99–109

Merritt LL et al (2008) The endogenous cannabinoid system modulates nicotine reward and dependence. J Pharmacol Exp Ther 326(2):483–492. Available at: http://www.pubmedcentral.nih.gov/articlerender.fcgi?artid=2746999&tool=pmcentrez&rendertype=abstract

Miller GE, Chen E, Zhou ES (2007) If it goes up, must it come down? Chronic stress and the hypothalamic-pituitary-adrenocortical axis in humans. Psychol Bull 133(1):25–45

Mitsushima D et al (2008) Gonadal steroid hormones maintain the stress-induced acetylcholine release in the hippocampus: simultaneous measurements of the extracellular acetylcholine and serum corticosterone levels in the same subjects. Endocrinology 149(2):802–811

Morena M et al (2016) Neurobiological interactions between stress and the endocannabinoid system. Neuropsychopharmacology 41(1):80–102. Available at: http://dx.doi.org/10.1038/npp.2015.166

Munro S, Thomas KL, Abu-Shaar M (1993) Molecular characterization of a peripheral receptor for cannabinoids. Nature 365(6441):61–65

Navarro M et al (2001) Functional interaction between opioid and cannabinoid receptors in drug self-administration. J Neurosci 21(14):5344–5350. Available at: http://www.ncbi.nlm.nih.gov/htbin-post/Entrez/query?db=m&form=6&dopt=r&uid=11438610\nhttp://www.jneurosci.org/cgi/

content/full/21/14/5344\nhttp://www.jneurosci.org/cgi/content/abstract/21/14/5344\nhttp://www.jneurosci.org/content/21/14/5344.full.pdf

Neumeister A et al (2015) Translational evidence for a role of endocannabinoids in the etiology and treatment of posttraumatic stress disorder. Psychoneuroendocrinology 51:577–584

Nissen SE et al (2008) Effect of rimonabant on progression of atherosclerosis in patients with abdominal obesity and coronary artery disease: the STRADIVARIUS randomized controlled trial. JAMA 299(13):1547–1560

Okamoto Y et al (2004) Molecular characterization of a phospholipase D generating anandamide and its congeners. J Biol Chem 279(7):5298–5305

Ortiz S et al (2004) Differences in basal cannabinoid CB1 receptor function in selective brain areas and vulnerability to voluntary alcohol consumption in fawn hooded and wistar rats. Alcohol Alcohol 39(4):297–302

Parolaro D et al (2007) Role of endocannabinoids in regulating drug dependence. Neuropsychiatr Dis Treat 3(6):711–721

Patel S et al (2004) Endocannabinoid signaling negatively modulates stress-induced activation of the hypothalamic-pituitary-adrenal axis. Endocrinology 145(12):5431–5438

Patel S et al (2005) Inhibition of restraint stress-induced neural and behavioural activation by endogenous cannabinoid signalling. Eur J Neurosci 21(4):1057–1069

Patel S et al (2009) Repeated homotypic stress elevates 2-arachidonoylglycerol levels and enhances short-term endocannabinoid signaling at inhibitory synapses in basolateral amygdala. Neuropsychopharmacology 34(13):2699–2709. Available at: http://dx.doi.org/10.1038/npp.2009.101

Patton GC (2002) Cannabis use and mental health in young people: cohort study. BMJ 325 (7374):1195–1198. Available at: http://www.bmj.com/content/325/7374/1195.1?variant=full-text&rss=1&ssource=mfr

Pertwee RG (2005) Pharmacological actions of cannabinoids. Handb Exp Pharmacol 168:1–51

Piomelli D (2003) The molecular logic of endocannabinoid signalling. Nat Rev Neurosci 4 (11):873–884

Poncelet M et al (2003) Overeating, alcohol and sucrose consumption decrease in CB1 receptor deleted mice. Neurosci Lett 343(3):216–218

Rademacher DJ et al (2008) Effects of acute and repeated restraint stress on endocannabinoid content in the amygdala, ventral striatum, and medial prefrontal cortex in mice. Neuropharmacology 54(1):108–116

Radley JJ, Sawchenko PE (2011) A common substrate for prefrontal and hippocampal inhibition of the neuroendocrine stress response. J Neurosci 31(26):9683–9695. Available at: http://www.pubmedcentral.nih.gov/articlerender.fcgi?artid=3197245&tool=pmcentrez&rendertype=abstract

Radley J, Sawchenko P (2015) Evidence for involvement of a limbic paraventricular hypothalamic inhibitory network in hypothalamic-pituitary-adrenal axis adaptations to repeated stress. J Comp Neurol 523(18):2769–2787

Radley JJ, Gosselink KL, Sawchenko PE (2009) A discrete GABAergic relay mediates medial prefrontal cortical inhibition of the neuroendocrine stress response. J Neurosci 29 (22):7330–7340

Radley JJ et al (2013) Chronic stress-induced alterations of dendritic spine subtypes predict functional decrements in an hypothalamo-pituitary-adrenal-inhibitory prefrontal circuit. J Neurosci 33(36):14379–14391. Available at: http://www.ncbi.nlm.nih.gov/pubmed/24005291

Ramesh D et al (2011) Blockade of endocannabinoid hydrolytic enzymes attenuates precipitated opioid withdrawal symptoms in mice. J Pharmacol Exp Ther 339(1):173–185. Available at: http://www.pubmedcentral.nih.gov/articlerender.fcgi?artid=3186294&tool=pmcentrez&rendertype=abstract

Ramesh D et al (2013) Dual inhibition of endocannabinoid catabolic enzymes produces enhanced antiwithdrawal effects in morphine-dependent mice. Neuropsychopharmacology 38 (6):1039–1049. Available at: http://www.pubmedcentral.nih.gov/articlerender.fcgi?artid=3629394&tool=pmcentrez&rendertype=abstract

Reich CG, Taylor ME, McCarthy MM (2009) Differential effects of chronic unpredictable stress on hippocampal CB1 receptors in male and female rats. Behav Brain Res 203(2):264–269
Rodriguez de Fonseca F et al (1999) Cannabinoid receptor antagonist SR141716A decreases operant ethanol self administration in rats exposed to ethanol-vapor chambers. Zhongguo Yao Li Xue Bao 20(12):1109–1114
Rossi S et al (2010) Preservation of striatal cannabinoid CB1 receptor function correlates with the antianxiety effects of fatty acid amide hydrolase inhibition. Mol Pharmacol 78(2):260–268
Rumsfeld JS, Nallamothu BK (2008) The hope and fear of rimonabant. JAMA 299 (13):1601–1602. Available at: http://www.ncbi.nlm.nih.gov/pubmed/18387935
Sánchez-Peña JF, Alvarez-Cotoli P, Rodríguez-Solano JJ (2012) Psychiatric disorders associated with alcoholism: 2 year follow-up of treatment. Actas Esp Psiquiatr 40(3):129–135. Available at: http://www.ncbi.nlm.nih.gov/pubmed/22723131
Sanchis-Segura C et al (2004) Reduced sensitivity to reward in CB1 knockout mice. Psychopharmacology 176(2):223–232
Scavone JL, Sterling RC, Van Bockstaele EJ (2013) Cannabinoid and opioid interactions: implications for opiate dependence and withdrawal. Neuroscience 248:637–654. Available at: http://www.pubmedcentral.nih.gov/articlerender.fcgi?artid=3742578&tool=pmcentrez&rendertype=abstract [Accessed January 6, 2015]
Schmied EA et al (2015) Sex differences in coping strategies in military survival school. J Anxiety Disord 29:7–13
Schofield D et al (2006) Reasons for cannabis use in psychosis. Aust N Z J Psychiatry 40 (6–7):570–574
Sciolino NR et al (2010) Social isolation and chronic handling alter endocannabinoid signaling and behavioral reactivity to context in adult rats. Neuroscience 168(2):371–386
Selye H, Fortier C (1950) Adaptive reaction to stress. Psychosom Med 12:149–157
Serra S et al (2001) The cannabinoid receptor antagonist SR 141716 prevents acquisition of drinking behavior in alcohol-preferring rats. Eur J Pharmacol 430(2-3):369–371
Serrano A et al (2012) Differential effects of single versus repeated alcohol withdrawal on the expression of endocannabinoid system-related genes in the rat amygdala. Alcohol Clin Exp Res 36(6):984–994
Sharrett-Field L et al (2013) Sex differences in neuroadaptation to alcohol and withdrawal neurotoxicity. Pflugers Arch Eur J Physiol 465(5):643–654
Shonesy BC et al (2014) Genetic disruption of 2-arachidonoylglycerol synthesis reveals a key role for endocannabinoid signaling in anxiety modulation. Cell Rep 9(5):1644–1654
Sipe JC et al (2002) A missense mutation in human fatty acid amide hydrolase associated with problem drug use. Proc Natl Acad Sci U S A 99(12):8394–8399
Smoller JW (2015) The genetics of stress-related disorders: PTSD, depression, and anxiety disorders. Neuropsychopharmacology 41(1):297–319. Available at: http://www.nature.com/doifinder/10.1038/npp.2015.266
Soria G et al (2005) Lack of CB1 cannabinoid receptor impairs cocaine self-administration. Neuropsychopharmacology 30(9):1670–1680
Steiner MA et al (2008) Antidepressant-like behavioral effects of impaired cannabinoid receptor type 1 signaling coincide with exaggerated corticosterone secretion in mice. Psychoneuroendocrinology 33(1):54–67
Sterrenburg L et al (2012) Sex-dependent and differential responses to acute restraint stress of corticotropin-releasing factor-producing neurons in the rat paraventricular nucleus, central amygdala, and bed nucleus of the stria terminalis. J Neurosci Res 90(1):179–192
Stroud LR, Salovey P, Epel ES (2002) Sex differences in stress responses: social rejection versus achievement stress. Biol Psychiatry 52(4):318–327
Suárez J et al (2009) Early maternal deprivation induces gender-dependent changes on the expression of hippocampal CB1 and CB2 cannabinoid receptors of neonatal rats. Hippocampus 19(7):623–632

Sugiura T et al (1995) 2-Arachidonoylglycerol a possible endogenous cannabinoid receptor ligand in brain. Biochem Biophys Res Commun 215(1):89–97

Sumislawski JJ, Ramikie TS, Patel S (2011) Reversible gating of endocannabinoid plasticity in the amygdala by chronic stress: a potential role for monoacylglycerol lipase inhibition in the prevention of stress-induced behavioral adaptation. Neuropsychopharmacology 36 (13):2750–2761. Available at: http://dx.doi.org/10.1038/npp.2011.166

Tabatadze N et al (2015) Sex differences in molecular signaling at inhibitory synapses in the hippocampus. J Neurosci 35(32):11252–11265. Available at: http://www.ncbi.nlm.nih.gov/pubmed/26269634

Taylor CJ (2014) Physiological stress response to loss of social influence and threats to masculinity. Soc Sci Med 103:51–59

Tsigos C, Chrousos GP (2002) Hypothalamic-pituitary-adrenal axis, neuroendocrine factors and stress. J Psychosom Res:865–871

Tsou K et al (1998) Immunohistochemical distribution of cannabinoid CB1 receptors in the rat central nervous system. Neuroscience 83(2):393–411. Available at: http://www.ncbi.nlm.nih.gov/entrez/query.fcgi?cmd=Retrieve&db=PubMed&dopt=Citation&list_uids=9460749

Twitchell W, Brown S, Mackie K (1997) Cannabinoids inhibit N- and P/Q-type calcium channels in cultured rat hippocampal neurons. J Neurophysiol 78(1):43–50. Available at: http://www.ncbi.nlm.nih.gov/entrez/query.fcgi?cmd=Retrieve&db=PubMed&dopt=Citation&list_uids=9242259 \nhttp://jn.physiology.org/content/78/1/43.full.pdf

Valverde O, Torrens M (2012) CB1 receptor-deficient mice as a model for depression. Neuroscience 204:193–206

Van Sickle MD et al (2005) Identification and functional characterization of brainstem cannabinoid CB2 receptors. Science (New York, NY) 310(5746):329–332

Vaughn L et al (2012) Cannabinoid receptor involvement in stress-induced cocaine reinstatement: potential interaction with noradrenergic pathways. Neuroscience 204:117–124

Vela G, Ruiz-gayo M, Fuentes JA (1995) Anandamide decreases naloxone-precipitated withdrawal signs in mice chronically treated with morphine. Neuropharmacology 34(6):665–668

Viau V (2002) Functional cross-talk between the hypothalamic-pituitary-gonadal and -adrenal axes. J Neuroendocrinol 14(6):506–513

Vinod KY et al (2010) Selective alterations of the CB1 receptors and the fatty acid amide hydrolase in the ventral striatum of alcoholics and suicides. J Psychiatr Res 44(9):591–597

Wamsteeker JI, Kuzmiski JB, Bains JS (2010) Repeated stress impairs endocannabinoid signaling in the paraventricular nucleus of the hypothalamus. J Neurosci 30(33):11188–11196

Wang M et al (2012) Acute restraint stress enhances hippocampal endocannabinoid function via glucocorticoid receptor activation. J Psychopharmacol (Oxford, England) 26(1):56–70. Available at: http://jop.sagepub.com/content/26/1/56.long

Ware M a et al (2010) Smoked cannabis for chronic neuropathic pain: a randomized controlled trial. CMAJ 182(14):E694–E701. Available at: http://www.pubmedcentral.nih.gov/articlerender.fcgi?artid=2950205&tool=pmcentrez&rendertype=abstract

Weinstock M et al (1998) Gender differences in sympathoadrenal activity in rats at rest and in response to footshock stress. Int J Dev Neurosci 16(3-4):289–295

Wilsnack SC, Klassen AD, Wilsnack RW (1984) Drinking and reproductive dysfunction among women in a 1981 national survey. Alcohol Clin Exp Res 8(5):451–458. Available at: http://www.ncbi.nlm.nih.gov/pubmed/6391255

Yamaguchi T et al (2001) Endogenous cannabinoid, 2-arachidonoylglycerol, attenuates naloxone-precipitated withdrawal signs in morphine-dependent mice. Brain Res 909(1–2):121–126

Yi B et al (2016) Reductions in circulating endocannabinoid 2-arachidonoylglycerol levels in healthy human subjects exposed to chronic stressors. Prog Neuro-Psychopharmacol Biol Psychiatry 67:92–97

Zhang J et al (2015) Inhibition of monoacylglycerol lipase prevents chronic traumatic encephalopathy-like neuropathology in a mouse model of repetitive mild closed head injury.

J Cereb Blood Flow Metab 35(3):443–453. Available at: http://www.pubmedcentral.nih.gov/articlerender.fcgi?artid=4348384&tool=pmcentrez&rendertype=abstract
Zhou Y et al (2016) Involvement of endocannabinoids in alcohol "binge" drinking: studies of mice with human fatty acid amide hydrolase genetic variation and after CB1 receptor antagonists. Alcohol Clin Exp Res 40(3):467–473
Zoppi S et al (2011) Regulatory role of cannabinoid receptor 1 in stress-induced excitotoxicity and neuroinflammation. Neuropsychopharmacology 36(4):805–818

Cell-Autonomous Endocannabinoid Production Shapes Polarized and Dynamic Distribution and Signaling Patterns of Cannabinoid CB_1 Receptors in Neurons

Delphine Ladarre and Zsolt Lenkei

Abstract Interaction with the highly regulated local lipid environment is emerging as key dynamic component of cellular function through the control of the structure, conformation, and function of cell-membrane-embedded proteins, such as G-protein-coupled receptors (GPCRs). The type-1 cannabinoid receptor CB_1, because of a relatively unstable GPCR structure and specific entry sites for lipids diffusing from the plasma membrane, may be particularly sensitive to such effects. In this chapter, we will discuss the first level of this lipid–protein interaction: the cell-autonomous scale, the foundation on which other important layers, such as paracrine or transsynaptic signaling systems, are built in vivo. Recent studies reveal an intricate balance between the polarized production of endocannabinoid (eCB) lipids and the polarized targeting and signaling of CB_1. The endocannabinoid 2-arachidonoylglycerol (2-AG), which is specifically produced in the somatodendritic plasma membrane, exerts cell-autonomous tonic activation on somatodendritic CB_1 receptors. This activation, in addition to important local signaling effects, also regulates CB_1 responses to other cannabinoids and provides the driving force for important basal endocytosis, which is followed by transcytotic CB_1 delivery to the axonal plasma membrane, where the large majority of CB1Rs accumulate at steady state. This cell-autonomous tonic CB_1 activation is based on two important properties of the endocannabinoid system: the elevated basal production of eCBs in specific regions of the plasma membrane (i.e., basal activation) and the structural instability of the CB_1 protein (i.e., constitutive activity). Key elements of this unusually dynamic functional model are valuable to better understand activation mechanisms of presynaptic CB_1 receptors and may also explain the high diversity of reported CB_1 ligands, ranging from peptide and lipid allo- and orthosteric regulators to the phytocannabinoid Δ^9-THC, the psychoactive component of marijuana.

D. Ladarre • Z. Lenkei (✉)
Brain Plasticity Unit, CNRS and ESPCI Paris, PSL Research University, Paris, France
e-mail: zsolt.lenkei@inserm.fr

M. Melis (ed.), *Endocannabinoids and Lipid Mediators in Brain Functions*,
DOI 10.1007/978-3-319-57371-7_4

1 Polarized Distribution of Neuronal Plasma Membrane Components

Most differentiated cells organize their plasma membrane into distinct domains, and this compartmentalization of membrane proteins and lipids is critical for proper cell functioning. According to an emerging view, membrane proteins and lipids are organized in constitutive dynamic molecular assemblies, regulated by a combination of protein–protein and protein–lipid interactions (Saka et al. 2013). The highly regulated lipid environment controls the structure, conformation, and function of embedded proteins (Phillips et al. 2009), and the dynamics of the underlying actin cytoskeleton regulate the mobility of many of the membrane constituents, playing a key role in controlling receptor signaling through the regulation of their interactions with functional protein partners (Mattila et al. 2016). As compared to other cell types, archetypal neurons possess a vast plasma membrane surface, which, similarly to the underlying neuronal cytoplasm, is highly compartmentalized and heterogeneous both in physical structure and chemical composition. The polarized neuronal architecture, crucial to establish the directionality of information flow through neuronal networks, is maintained by a nonuniform distribution of organelles and proteins (Craig and Banker 1994). Accordingly, the protein and lipid composition of the plasma membrane also differs greatly between axons and somatodendritic compartments (Horton and Ehlers 2003). This polarization is established by specific targeting mechanisms as discussed below and is maintained by the axon initial segment (AIS), which acts as a surface diffusion barrier and an intracellular traffic filter for both lipids and proteins (for review see Leterrier and Dargent (2014)).

Importantly, neuronal proteins are constantly renewed, and the polarity of their distribution is maintained due to specific targeting processes following their synthesis in the Golgi apparatus, typically localized in the soma (reviewed in Lasiecka and Winckler (2011)). Most dendritic proteins are targeted to the somatodendritic plasma membrane through direct *selective delivery*. For example, cargoes transporting the dendritic transferrin receptor from the Golgi apparatus to the plasma membrane are excluded from the axon, leading to an exclusively dendritic delivery (Burack et al. 2000). For axonal targeting, three main processes have been described, *selective delivery*, *selective fusion*, and *selective retention/elimination* (Horton and Ehlers 2003). While direct targeting through *selective delivery* is predominant in dendrites, it seems that only a few axonal proteins reach the axon through this mechanism (Lasiecka and Winckler 2011). As microtubule orientation-based carriers cannot traffic selectively to the axon due to the mixed dendritic microtubule orientation, either *selective fusion* or *selective retention/elimination* may be used to achieve specific axonal targeting. The latter process assumes that carriers fuse to the dendritic membrane, but proteins are rapidly removed by endocytosis and delivered to the axon in endosomes through *transcytosis* (Lasiecka and Winckler 2011). Experimentally, the difference between both mechanisms is hard to detect but recent studies tend to demonstrate that transcytosis is the predominant axonal targeting mechanism, as first shown for the neuronal sodium channel $Na_V1.2$, where a specific C-terminus sequence is recognized by the clathrin-dependent endocytic pathway in the somatodendritic compartment, leading to transcytosis and axonal targeting (Garrido et al. 2001). Subsequently, similar targeting processes were found for various other axonal proteins, such as vesicle-associated membrane protein 2 (VAMP2), involved in synaptic vesicle fusion with the presynaptic membrane (Sampo et al. 2003); tropomyosin-related kinase A (TrkA) receptors, involved in growth and survival of developing neurons (Ascano et al. 2009); contactin-associated protein 2, a cell adhesion molecule (Bel et al. 2009); and β-site APP-cleaving enzyme 1 (BACE1) of amyloid precursor protein (Buggia-Prévot et al. 2014). Recent results, discussed below, also indicate that several important members of the G-protein-coupled receptor (GPCR) family, a large group of sensory and signaling membrane proteins with a vast array of critical functions in neurons, are addressed to axons through a similar mechanism, but with the notable exception that their targeting depends on their pharmacological activation state (Carrel et al. 2011; Simon et al. 2013).

2 GPCR Activation State Depends on the Plasma Membrane Environment

GPCRs are dynamic sensory molecules, composed of rigid alpha-helix transmembrane domains (TMs) that are connected by supple loops, leading to a highly flexible three-dimensional structure (Deupi and Kobilka 2010; Venkatakrishnan et al. 2013). Evidence from both functional and biophysical studies suggests that GPCRs are not simple bimodal switches but permanently sample multiple conformations (Deupi and Kobilka 2010; Venkatakrishnan et al. 2013). They are mostly held in a basal conformational equilibrium at steady state by intervening loops and non-covalent intramolecular interactions such as the proposed "ionic lock" between the third and sixth TMs (Ballesteros et al. 2001). These interactions, as well as others, may collectively form a central hydrophobic core in most Family A GPCRs (Tehan et al. 2007).

At the basal state, inactive conformations typically display a lower energy level than active conformations (Kobilka and Deupi 2007; Deupi and Kobilka 2010). This free-energy difference between active and inactive states, called *activation energy barrier*, leads to a lower probability for active states to appear spontaneously. Agonist-binding level modifies the energy landscape resulting in increased number of receptors in active conformations, i.e., showing rearrangement of the cytoplasmic side of the receptor, which leads to mobilization of intracellular signaling pathways such as the characteristic activation of cognate heterotrimeric G-proteins (Deupi and Kobilka 2010; Venkatakrishnan et al. 2013). Signaling by stabilized activated states is typically terminated in seconds by phosphorylation of specific intracellular serine and threonine residues, which leads to decoupling of the receptor from G-proteins (desensitization) and to the recruitment of scaffolding/effector proteins such as ß-arrestins, ultimately resulting in GPCR endocytosis in the timescale of minutes, typically through the clathrin-mediated endocytic pathway (Hanyaloglu and Zastrow 2008). After elimination of bound ligands and dephosphorylation in endosomes, GPCRs are either recycled back to the plasma membrane or become degraded in lysosomes, depending on the receptor subtype, its level of activation, and the cell type (Hanyaloglu and Zastrow 2008). While the above-described agonist-induced GPCR endocytosis was initially associated with rapid desensitization of G-protein-mediated cellular responses, accumulating evidence now indicates both G-protein-independent signaling (Reiter et al. 2012) and G-protein-mediated signaling (Irannejad and Zastrow 2014) also occur from endosomal compartments.

Theoretical energy landscapes and experimental results both suggest that the multitude of possible GPCR conformational states leads to an energy continuum (Deupi and Kobilka 2010). Thus, at the basal state, even if active conformations are not favored due to a high activation energy barrier, some receptors may still display active conformations. This ligand-free basal activity is called *constitutive activity*. This concept leads to the characterization of antagonist ligands by their ability to promote inactive states: an *inverse agonist* ligand stabilizes inactive receptor conformations, decreasing the basal biological response, while a *neutral antagonist* binds the receptor without inducing changes in the conformational state. Finally, *allosteric modulators* modify either the active (positive allosteric modulators or PAMs) or the inactive (negative allosteric modulators or NAMs) state of the receptor by binding to other parts of the GPCR than the major ligand-binding (orthosteric) site.

The above model shows that constitutive GPCR activity, which could be perceived as counterintuitive in the now outdated bimodal switch model of GPCR function, is rather a logical result of the inherent characteristics of molecular GPCR mechanisms. Notably, the level of basal activity of a given GPCR depends not only of the level of the activation energy barrier but also of the presence of active conformations at low energy levels, since high conformation dynamics may lead to smaller energetic differences between inactive, intermediate, and active states (Manglik and Kobilka 2014). Depending on the GPCR, one or both of these mechanisms can explain constitutive activity. Importantly, the level of basal activity in ligand-free conditions depends both on factors intrinsic to the receptor protein—mostly the ensemble of inactive-state conserving non-covalent interactions between its transmembrane domains, which determine *constitutive activity*—and on extrinsic local factors of *basal activation*, such as cytosolic signaling and regulatory proteins, plasma membrane lipids, pH, ions, transmembrane voltage gradients (reviewed in Manglik and Kobilka (2014)), and even the curvature of the plasma membrane (Brown 2012). Notably, direct and indirect interactions with membrane lipids are now thought to control various essential aspects of GPCR function (reviewed in Oates and Watts (2011)). For example, it was recently speculated that cholesterol-induced structural changes might induce important changes in GPCR ligand binding (Zocher et al. 2012) and, very recently, differences in the phospholipid composition of the plasma membrane—a function of the cell type, subcellular domains, or nanoscale lipid domains—were shown to play a key role in GPCR activation through genuine allosteric mechanisms (Dawaliby et al. 2016). In conclusion, the combination of intrinsic *constitutive activity* and extrinsic *basal activation* yields the level of *tonic activation*, which is typically low for many but not all GPCRs (Seifert and Wenzel-Seifert 2002).

The above-described lipid-induced interactions may be particularly relevant for GPCRs with a relatively unstable structure (i.e., high conformational flexibility), such as the type-1 cannabinoid receptor CB_1.

3 The CB_1 Cannabinoid Receptor: A Structurally Instable GPCR

The type-1 cannabinoid receptor (CB_1), a class A (or rhodopsin-like) GPCR, is the major neuronal target of endocannabinoid lipids (eCBs) and of exogenous cannabinoids, such as Δ^9-tetrahydrocannabinol (THC), the main psychoactive constituent of marijuana (Pertwee et al. 2010). CB_1, like numerous other GPCRs, displays a high level of constitutive activity both when heterogeneously expressed in non-neuronal cells and in neurons where CB_1 receptors are endogenous (Pertwee 2005). However, in contrast to other constitutively active GPCRs, where constitutive activity is characteristically a result of structural instability (Gether et al. 1997; Alewijnse et al. 2000; Zhang 2004; Holst et al. 2004; Engelstoft et al. 2014; Gilliland et al. 2015), for CB_1, the lipid composition of the surrounding plasma membrane contributes significantly to tonic activation, blurring the distinction between classical constitutive activity and tonic activation, as discussed below.

The cannabinoid receptor family shares several highly conserved stretches of residues, such as the aspartic acid-arginine-tyrosine (DRY) amino acid motif as well as several neighboring residues with the GPCR consensus sequence. However, Debra Kendall's group has observed (D'Antona et al. 2006) that all cannabinoid receptors identified to date differ from the consensus GPCR sequence by containing Thr at position 3.46 according to the Ballesteros–Weinstein numbering scheme (Ballesteros and Weinstein 1995), named T210 in the original article (D'Antona et al. 2006), one helical turn above Arg 3.50, which is part of the basal-state stabilizing, relatively conserved ionic lock in other GPCRs (Ballesteros et al. 2001). Strikingly, mutations of this single residue lead to dramatic effects in CB_1 activity: substitution of Thr 3.46 with Ala (T3.46A or T210A), a residue with a higher helical packing moment, results in a receptor that shows diminished constitutive activity, while mutation to isoleucine (T3.46I or T210I) yields a hyperactive receptor (D'Antona et al. 2006). These mutations were proposed to result in changes in the conformational state of CB_1, yielding receptor forms that display significantly different levels of constitutive activity but preserve responsiveness to agonist and inverse agonist ligands (D'Antona et al. 2006). Indeed, we found that treatment of T210I-CB_1 expressing neurons with the antagonist/inverse agonist AM281 or treatment of T210A-CB_1 expressing neurons with the agonist WIN22,512-2 resulted in a significant redistribution toward the wild-type (WT) phenotype, confirming that T210-CB1R mutants are true conformational CB_1 isoforms (Simon et al. 2013).

Further studies that used modeling and experimental confirmation with double mutant CB1Rs have significantly extended our comprehension of the key conformational states of CB1Rs and probably also of other GPCRs (Scott et al. 2012; Ahn et al. 2013). These results, yet to be directly confirmed by crystallography, suggest a simple and elegant mechanistic explanation for the graded levels of constitutive activation of CB_1: inactive mutant receptor forms have two salt bridges between transmembrane domains 2 and 6 (TM2–TM6) and TM3–TM6. These interactions between opposed TMs keep the receptor in a tightly packed form with low conformational flexibility and high thermal stability and presumably preclude efficient activation of intracellular effectors in the absence of an agonist. Fully active receptor forms have no salt bridges between opposing TMs but do between neighboring TMs, leading to a loosely packed instable form that efficiently activates intracellular signaling. Finally, partially active receptor forms, such as the wild-type CB_1, contain only one salt bridge between the opposing TM3 and TM6, accompanied by a salt bridge between neighboring TMs (Scott et al. 2012; Ahn et al. 2013). This model indicates that wild-type CB_1 receptors are partially destabilized GPCRs at the basal state.

Due to this structural instability, the activation energy barrier of wild-type CB_1 is lower than in typical GPCRs, providing a mechanistic explanation for constitutive CB_1 activity and possibly also for the elevated sensitivity of the receptor to ortho- and allosteric ligands. Indeed, to date, in vitro evidence suggests that there are at least 15 endogenous lipidic compounds that can target cannabinoid receptors either orthosterically or allosterically (Pertwee 2015; see more details below). Other ligands with reported CB_1-activating effects include sodium, peptides such as pepcan-12, a 12-amino-acid neuropeptide or hemopressin (for a recent review, see Morales et al. (2016)), and pregnenolone (Vallee et al. 2014) which has traditionally been viewed as an inactive precursor of neurosteroids. Some of these compounds also target non-cannabinoid receptors or ion channels, at concentrations at which they seem to interact with cannabinoid receptors (Pertwee 2015). Notably, plasma membrane lipids that are either considered to be the orthosteric ligands of CB_1, such as anandamide (N-arachidonoyl ethanolamide or AEA) and 2-arachidonoylglycerol (2-AG), or endogenous allosteric modulators, such as oleamide (an endogenous fatty acid amide) or the arachidonic acid derivative lipoxin A4, may also nonspecifically modulate GPCR conformations, as suggested by their allo- or orthosteric modulator role for several other membrane proteins (for recent reviews, see van der Westhuizen et al. (2015)). For example, AEA, first described as the bona fide CB_1 ligand, now is recognized as an activator of at least 19 other plasma membrane proteins at micromolar concentrations (Pertwee 2005), and 2-AG was recently shown to potentiate GABA-mediated activation of the $GABA_A$ ligand-gated ion channel, through positive allosteric modulation of its β2-subunit (Sigel et al. 2011; Baur et al. 2013). Moreover, in pancreatic β-cells, 2-AG blocks K_{ATP} potassium channels through a CB_1-independent mechanism, given that the CB_1-specific antagonist/inverse agonist AM251 does not remove this effect (Spivak et al. 2012). Finally, recent reports have failed to confirm CB1R-related allosteric modulator effects of pregnenolone and lipoxin A4 in a standard model cell line (Khajehali et al. 2015) and, together with hemopressin, also by using physiologically relevant cultured neurons and the endocannabinoid 2-AG ligand (Straiker et al. 2015). These diverging results underline the delicate pharmacology of CB_1 and indicate that standard experimental approaches developed for typical GPCRs, which are more stable at steady state and have well-known specific endogenous agonists, should be used under enhanced scrutiny for the naturally destabilized CB_1.

4 Ubiquitous Neuronal Membrane Component Endocannabinoids Produce Tonic CB_1 Activation

The lipophilic eCBs were proposed to be synthesized on demand from ubiquitous plasma membrane components through various biosynthetic pathways, either in response to intracellular calcium elevations or combined with activation of several $G_{q/11}$-protein-coupled receptors (for recent detailed reviews, see Alger and Kim (2011), Pertwee (2015)). The first eCBs identified were AEA and 2-AG, both belonging to the eicosanoid family of polyunsaturated fatty acids. AEA is synthesized mainly by hydrolysis of the corresponding N-acyl-phosphatidylethanolamines (NAPEs), catalyzed by a phospholipase D selective for NAPEs (NAPE-PLD), and is degraded by hydrolysis by fatty acid amine hydrolase (FAAH). The other major endocannabinoid, 2-AG, is released by hydrolysis of the cell membrane diacylglycerols (DAGs) by two isoforms of DAG lipases (DAGLα and DAGLβ). The degradation of 2-AG is mainly due to monoacylglycerol lipases (MAGLs). Importantly, DAGLα, the major DAGL in the postnatal brain, is segregated to axonal tracts during embryonic development but was shown to accumulate after birth in the somatodendritic plasma membrane in several brain areas, such as the cerebellum (Bisogno et al. 2003), striatum (Uchigashima et al. 2007), hippocampus (Yoshida et al. 2006; Katona et al. 2006), and amygdala (Yoshida et al. 2011). Similarly, we found somatodendritic segregation of DAGLα in fully polarized mature cultured hippocampal neurons (Ladarre et al. 2015), indicating local production of 2-AG in the plasma membrane of the somatodendritic compartment but not in its axonal counterpart.

Neuronal eCBs are generally considered to be retrograde transsynaptic signals, produced in the postsynaptic cell, released and traveling “backwards” across the synaptic cleft, through still undetermined mechanisms, to activate presynaptic CB_1, as shown by a large body of indirect experimental evidence, gathered using electrophysiological, anatomical, and genetic tools (reviewed in Freund et al.

(2003), Kano et al. (2009)). Studies aimed at understanding eCB transport—by focusing mostly on AEA—have proposed protein carriers (Chicca et al. 2012), such as fatty acid-binding proteins, heat shock proteins, or albumin (Maccarrone et al. 2009), as potential transporters to move eCBs intracellularly and across the plasma membrane (Chicca et al. 2012; Fowler 2013). However, the actual role of such a transporter, the FAAH-like AEA transporter (FLAT) (Fu et al. 2012), has also been questioned (Leung et al. 2013). Neo-synthesized AEA was also suggested to be released into the extracellular fluid and to reach its targets bound to circulating proteins, such as lipocalins or albumin (Piomelli 2003). Interestingly, a recent report suggests vesicle-based extracellular transport for anandamide by showing its enriched presence in exosomes and ectosomes (i.e., shed microvesicles), produced by microglia (Gabrielli et al. 2015). Such lipidic-vesicle-based transport may well explain efficient distribution of lipid eCBs in aqueous environments and opens up an entire new field in neurotransmitter research aimed at studying the liberation and metabolism of these extracellular vesicle transporters.

Importantly, however, in addition to this established retrograde transsynaptic signaling role, it was also reported that significant amounts of 2-AG are present in resting neurons, that only a fraction of stimulation-induced 2-AG is released from cells, and that basal levels of AEA are sufficient to constitutively induce CB_1 signaling (for review refer to Alger and Kim (2011)). Notably, both AEA and 2-AG are directly derived from membrane phospholipids and synthesized in the lipid bilayer of the plasma membrane, and no effective intramembrane sequestration of the important basal pool of eCBs has yet been identified. Consequently, these endocannabinoids may reach CB_1 through two-dimensional diffusion in—and through direct entry from—the lipid bilayer, a long-held assumption in the field recently confirmed by experimental and computational results (Reggio 2010). Indeed, analogously to the now structurally resolved S1PR1 lipidergic GPCR, 2-AG was shown to access the binding pocket of the type-2 cannabinoid receptor CB_2 through entry between TM6 and TM7 (Hurst et al. 2013). As the CB_1 amino acid sequence is highly similar to that of CB_2, we may assume that their structure, binding sites, and agonist entry mode are analogous. Structure–activity relationship studies investigating the interaction of cannabinoid receptors with eCBs have also indicated in AEA, 2-AG, and congeners high "dynamic plasticity," indicating that they are able to extend as well as form tightly curved conformations (Reggio 2010). Strikingly, only analogs that are capable of these two extremes in conformation are recognized by CB_1, suggesting that one of these conformations is important for getting to the receptor in the membrane while the other is important for getting into the receptor (Reggio 2010). Due to these unusual ligand characteristics, the basal activation stemming from the steady-state endocannabinoid tone in the plasma membrane may be a major contributor to tonic CB_1 activation (reviewed in Howlett et al. (2011)). Indeed, blocking the synthesis of 2-AG, the major endocannabinoid in the mammalian brain (Stella et al. 1997), abolishes (or strongly decreases) both constitutive CB_1 signaling and CB_1 internalization in heterologous systems (Turu et al. 2007). These results were recently further refined by our laboratory in cultured hippocampal neurons by showing that the antagonist/inverse agonist AM281 produces a marked rapid increase of basal PKA activation in somata and dendrites, but not in axons, and that DAGL inhibition abolishes this effect, suggesting that tonic activation of somatodendritic CB_1 is generated by local somatodendritic production of the endocannabinoid 2-AG (Ladarre et al. 2015). Interestingly, the 2-AG content of the somatodendritic plasma membrane also modulates CB_1 responses to exogenous cannabinoids, as we have observed that the presence of 2-AG is necessary for WIN55-212,2 and THC-mediated activation of somatodendritic CB_1, but, on the contrary, CP55,940-mediated decrease of PKA activity in the somatodendritic compartment is higher in the absence of 2-AG (Ladarre et al. 2015). These results suggest that 2-AG, which is specifically produced in the somatodendritic plasma membrane, has important cell-autonomous local actions on somatodendritic CB_1 receptors, both by achieving significant basal activation and by regulating CB_1 activation by other cannabinoids.

Taken together, several lines of experimental evidence show that eCBs acting on CB_1 may be, in many instances, of cell-autonomous origin. This mode of action is often referred to in the endocannabinoid literature as "autocrine" (from Greek *krīnō*, "to separate" or "to secrete"), a term usually applied to hormones or other chemical messengers that are secreted to the extracellular milieu before binding to receptors on the same cell. However, as seen above, eCBs do not need to leave the neuronal plasma membrane to gain access to the ligand-binding site of CB_1 receptors on the same neuron. We have also shown previously that tonic CB_1 activation does not depend on eCBs in the extracellular milieu (Leterrier et al. 2004). Consequently, "cell-autonomous" appears to be a more adapted term for this important element of eCB action.

In conclusion, the above-detailed results suggest that important CB1R-activating lipids may modify relatively nonspecifically the conformation of plasma-membrane-embedded proteins. Indeed, the lack of clearly understood storage and release mechanisms for eCBs, the structural instability of CB_1, the high level of cellular eCB content in membranes of non-stimulated neurons, the multiple reports of cell-autonomous CB_1 activation mechanisms, and high level of CB_1 receptors outside of presynaptic specializations (Mátyás et al. 2006; Leterrier et al. 2006) suggest that in addition to the classical modus operandi of presynaptic GPCR function (Starke 1981), novel functional paradigms should also be envisioned for CB_1, both in the adult and developing brain. An interesting possibility is that CB_1 may be particularly sensitive to the lipid composition of the plasma membrane, due to its structural instability and the presence of specific entry sites that are accessible for lipids present in the lipid bilayer. As a consequence, CB_1 may be able to continuously translate the plasma membrane concentration of highly regulated lipid eCBs into the activation of specific intracellular signaling pathways (and be the subject of steady-state endocytosis, as discussed below). However, this relative nonspecificity of CB_1 activation mechanisms does not exclude their physiological pertinence: cannabinoid receptors may have been evolutionary optimized to become highly sensitive but relatively nonspecific sensory molecules, able to report changes in the level of a broad range of lipidic and non-lipidic components, several of them being synthesis intermediates of other important molecules such as arachidonic acid. Importantly, recent insights into GPCR function have started to reveal that the very efficient and rather on–off mode of rhodopsin activation, traditionally considered as the gold standard of GPCR action, may correspond only to one extreme of the GPCR function spectrum, being optimized for the highly efficient (i.e., very low signal-to-noise ratio) response to light (Manglik and Kobilka 2014). On the contrary, the relatively inefficient coupling of the ß2AR receptor may be important for its more complex functional repertoire and its diversity of responses to different ligands (Manglik and Kobilka 2014). In this context, the instable and promiscuous CB_1 would rather be located at the opposite side of the GPCR function spectrum, as compared to rhodopsin. Finally, the structural characteristics of CB_1 may also explain its role as a major drug target: the phytocannabinoid Δ^9-THC, which was likely not optimized by evolution in plants to act on a human brain receptor, is able to efficiently activate CB_1, because of its high sensitivity that results from structural instability.

5 Subcellular Distribution of Neuronal CB_1 Receptors

CB_1 is one of the most abundant GPCRs in the brain. Its distribution has been studied with various techniques such as autoradiography, in situ hybridization, and immunohistochemistry, all reporting high receptor densities in the cerebral cortex, hippocampus, amygdala, basal ganglia, and cerebellum, in accordance with behavioral effects of cannabis (Freund et al. 2003). CB_1 receptors are highly expressed in GABAergic neurons and display a lower but significant expression in glutamatergic neurons (Marsicano and Lutz 1999; Kawamura et al. 2006; Katona et al. 2006). Interestingly, while most CB_1 receptors are localized to the plasma membrane in axons—typically outside of synaptic terminals as discussed below—of mature neurons both in vitro (Coutts et al. 2001; Leterrier et al. 2006; McDonald et al. 2007) and in vivo (Katona et al. 1999, 2001; Pickel et al. 2004; Nyiri et al. 2005; Bodor 2005; Kawamura et al. 2006) (Mátyás et al. 2006; Thibault et al. 2013), high levels of intracellular CB_1 localization were reported in non-polarized cells, both when endogenously expressed (McIntosh et al. 1998; Hsieh et al. 1999; Graham et al. 2006; Ellis et al. 2006; Tappe-Theodor et al. 2007) or transfected (Wu et al. 2008; Rozenfeld and Devi 2008; Grimsey et al. 2010). Similar prominent intracellular CB_1 localization was reported in the somatodendritic domain of mature neurons (Fig. 1) (Coutts et al. 2001; Katona et al. 2001; Pickel et al. 2004; Bodor 2005; Leterrier et al. 2006; McDonald et al. 2007; Scavone et al. 2010; Thibault et al. 2013) as well as in the somatodendritic and axonal domain of embryonic neurons (Vitalis et al. 2008).

The exact origin and role of this intracellular CB_1 population is a hotly debated subject in the cannabinoid field. Since eCBs are lipid ligands that are thought to easily cross the plasma membrane, several teams have proposed "stationary" intracellular localization of functional CB_1 receptors on different intracellular organelles. First, a stationary and functional intracellular localization guided by

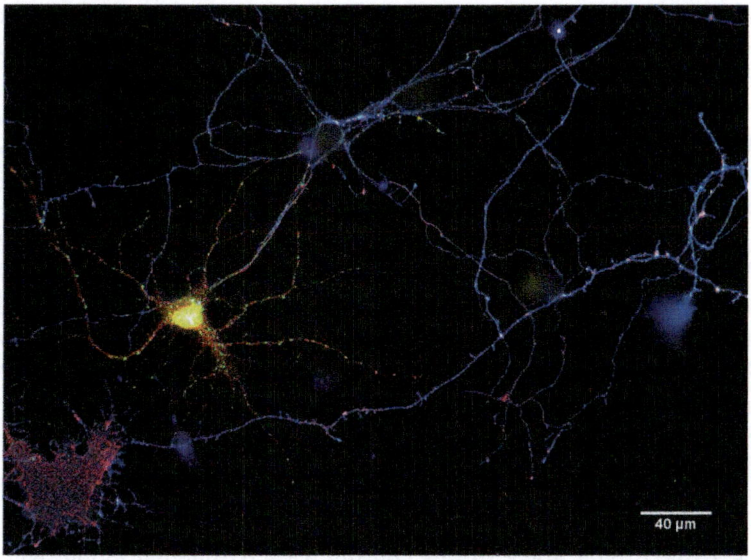

Fig. 1 CB_1 receptors are constantly endocytosed in the somatodendritic region but not on the axonal plasma membrane. Living neurons co-expressing *Flag-CB₁-eGFP* (*green*) and a *clathrin light chain-DsRed* (*red*) fusion proteins at *day* in vitro 8 were incubated for 1 h with an anti-Flag M1 antibody to bind to the extracellular Flag epitope (i.e., antibody feeding), revealed after fixation and permeabilization in blue to label CB_1 that transited at the neuronal surface in the last 1 h. This figure shows that in the somatodendritic compartment (at the left side of the image), CB_1 receptors are predominantly internalized and often co-localize with clathrin, indicating their internalization through a clathrin-dependent mechanism. On the contrary, in axons (network of thin neurites at the right side of the image), CB_1 accumulates on the plasma membrane. In the lower left corner, a glial cell expresses *clathrin light chain-DsRed* but not *Flag-CB1-eGFP*. For experimental details and a higher magnification example of a similar preparation, showing endocytic vesicles in the soma and proximal dendrites co-labeled for *Flag-*CB_1*-eGFP*, *clathrin-DsRed*, and M1 antibody, please refer to Leterrier et al. (2006). Endogenous CB1Rs show similar activation-dependent constitutive endocytosis as shown by live labeling with a N-terminus-targeted antibody (Simon et al. 2013)

the adaptor-protein AP-3 to late endosomal/lysosomal compartments was proposed (Rozenfeld and Devi 2008). Lysosomal CB_1 was subsequently suggested to release calcium from the endoplasmic reticulum and lysosomal calcium stores following anandamide stimulation (Brailoiu et al. 2011). Next, functional CB_1 receptors, found by immunohistochemistry to be expressed on the mitochondrial outer membrane, were shown to inhibit mitochondrial respiration (Benard et al. 2012) reducing it by 20% at saturating ligand concentrations (Hebert-Chatelain et al. 2014). However, others have also reported that both agonist and inverse agonist cannabinoid ligands may inhibit mitochondrial respiration through off-target (i.e., CB_1-independent) mechanisms (Fišar et al. 2014; Schirris et al. 2015; Singh et al. 2015) and that anti-CB_1 antibodies show off-target labeling of another important mitochondrial protein, which is upregulated by neuronal anoxia (Morozov et al. 2013, 2015). In addition, important aspects of general GPCR function, such as targeting, signaling, desensitization by phosphorylation, and resensitization/downregulation through endocytic pathways, remain poorly understood for mitochondrial CB_1 receptors, warranting further studies to assess their physiological relevance.

Contrarily to this "static" view of intracellular CB_1 localization, several studies, including reports from our laboratory, have indicated a different and dynamic source for intracellular CB_1 receptors: constitutive endocytosis from the plasma membrane. First, live imaging of GFP-tagged CB_1 expressing HEK293 cells shows highly dynamic endosomal labeling with traffic from and to the plasma membrane (see Fig. 3 and Supplementary Movie 1 in Leterrier et al. 2004). Second, the endosomal CB_1 population is readily labeled by incubation of non-permeabilized living cells with antibodies directed toward the extracellular N-terminus, both for transfected (Fig. 1 and Leterrier et al. 2004, 2006; Koch et al. 2006; Ellis et al. 2006; Wu et al. 2008; Grimsey et al. 2010; Ward et al. 2011; Simon et al. 2013) and endogenous CB_1 (Leterrier et al. (2006) and Supplementary Figure S2 in Simon et al. (2013)). Third, pharmacological or genetic inhibition of the endocytic machinery upregulates the proportion of plasma-membrane-located CB_1 receptors and diminishes the intracellular CB_1 population (Leterrier et al. 2004, 2006).

What is the driving force of this constitutive CB_1 internalization, which was reported both in a wide range of immortalized cell lines and in neurons? Results obtained by either pharmacological (Leterrier et al. 2004, 2006; Ellis et al. 2006; Ward et al. 2011; Simon et al. 2013) or genetic (D'Antona et al. 2006; Simon et al. 2013) tools show that constitutive CB_1 endocytosis is a consequence of tonic CB_1 activation at steady state and indicate that the presence of both structural CB_1 determinants (D'Antona et al. 2006; Simon et al. 2013) and basal activation through steady-state cell-autonomous production of endocannabinoids such as 2-AG (Turu et al. 2007) are necessary. These results are in line with the view that the important intracellular pool of CB_1 receptors, a distribution pattern characteristic also for several other constitutively active GPCRs (Morisset et al. 2000; Marion et al. 2004; Morris 2004; Jacquier et al. 2006; Mohammad et al. 2007; Holliday et al. 2007; Chanrion et al. 2008), is a dynamic phenomenon and is related to tonic CB_1 activity. Finally, similarly to other GPCRs, exposure to agonists induces additional internalization of the plasma-membrane-located CB_1 population in non-polarized cell lines (Hsieh et al. 1999; Leterrier et al. 2004; Tappe-Theodor et al. 2007; Wu et al. 2008; Grimsey et al. 2010).

Poststimulation endocytosis is important both for resensitization and downregulation of CB_1 receptors, either through recycling to the plasma membrane or through lysosomal destruction mediated by association of post-endocytic sorting proteins such as GASP1, respectively (Whistler et al. 2002; Martini et al. 2007, 2010). The choice between recycling and degradation may depend both on the concentration of the agonist ligand and on the treatment duration (Hsieh et al. 1999; Whistler et al. 2002; Leterrier et al. 2004; Martini et al. 2007, 2010; Tappe-Theodor et al. 2007; Wu et al. 2008; Grimsey et al. 2010). However, GPCR recycling is difficult to measure directly, putatively contributing to a debate about this specific point of the CB_1 life cycle (Rozenfeld and Devi 2008; Grimsey et al. 2010).

What is the in vivo relevance of the above results, which, by using highly resolved in vitro experimental approaches, indicate elevated activation-dependent plasticity of subcellular CB_1 distribution phenotypes? The first indication of activation-dependent plasticity in brain CB_1 distribution came from reports showing that chronic cannabis use leads to downregulation of CB_1 in humans and that these effects subside within days to several weeks of sustained abstinence (for review see Curran et al. (2016)). Similarly, chronic cannabinoid treatment in rodents leads to a reduction in CB_1 function throughout the brain (Breivogel and Childers (1999) and reviewed in Sim-Selley (2003), González et al. (2005)) that persists for days to weeks before functional recovery that varies between brain regions (Sim-Selley et al. 2006). These results suggested the occurrence of agonist-induced CB_1 internalization in vivo, and we have indeed reported that acute systemic cannabinoid treatment of rats induces marked CB_1 internalization into somatodendritic endosomes (Thibault et al. 2013). On the contrary, antagonist/inverse agonist treatment decreased the basal level of endosomal CB_1 receptors in the somatodendritic compartment, strikingly demonstrating that at steady state CB_1 is already significantly internalized through basal activation (Thibault et al. 2013). Notably, at 16h after a single cannabinoid treatment, the number of perikarya displaying CB_1-containing endosomes returned near to control levels in the neocortex without change in brain CB_1 protein levels, suggesting that CB_1 are efficiently recycled after moderate activation (Thibault et al. 2013). However, more sustained activation led to enhanced endosomal labeling and a parallel decrease in brain CB_1 protein levels (Thibault et al. 2013), suggesting redirection of CB_1 to degradation pathways, leading to the characteristic downregulation reported to parallel the development of tolerance (Sim-Selley et al. 2006).

Interestingly, near-maximal increase of the somatodendritic endosomal labeling after a single agonist treatment was not accompanied by a significant decrease in axonal labeling, which required repeated daily agonist treatments even in the most responsive regions, such as the neocortex and hippocampus (Thibault et al. 2013). These results were recently confirmed by showing slow activation-dependent elimination of CB_1 from synaptic boutons following chronic CB_1 activation in rats (Dudok et al. 2015). The relatively slow axonal translocation kinetics likely results from restrained endocytosis of axonal CB_1 receptors. Indeed, maximal agonist-induced CB_1 internalization in axons of cultured neurons is obtained only after 16 hours of continuous agonist treatment (Coutts et al. 2001; Leterrier et al. 2006), while immortalized cell lines display maximal levels of CB_1 endocytosis already at 15–30 minutes following activation (Rinaldi-Carmona et al. 1998; Hsieh et al. 1999; Leterrier et al. 2004), indicating the relatively low efficacy of the axonal endocytotic machinery. Notably, endocytic capacity at the neuronal surface is also highly polarized and compartmentalized, plasma membrane endocytosis at the axonal cell membrane is restricted to nerve terminals and varicosities, whereas dendrites and immature axons are capable of internalization along their whole length, as shown in cultured hippocampal neurons (Sinclair et al. 1988; Parton et al. 1992; Parton and Dotti 1993). Thus, CB_1 receptors, located mainly at preterminal axons and axon shafts (Katona et al. 1999, 2001; Pickel et al. 2004;

Nyiri et al. 2005; Bodor 2005; Kawamura et al. 2006; Mátyás et al. 2006; Thibault et al. 2013), may need to be transported for considerable distances following activation in order to reach synapses and varicosities and access the endocytic machinery. In certain pathways of the basal ganglia that express functional CB_1 receptors on thin, unmyelinated fibers devoid of synaptic specializations, such as the pallido-nigral and striato-pallidal projections (Mátyás et al. 2006), this distance may reach several hundreds of micrometers. Indeed, we have found a striking lack of CB_1 redistribution in pallido-nigral axons after prolonged agonist treatment (Thibault et al. 2013), indicating their pronounced inability to internalize. Interestingly, basal ganglia display also very moderate downregulation of CB_1 following chronic exposure to Δ^9-THC or synthetic agonists, as compared to other brain regions (Sim-Selley 2003; González et al. 2005). Since GPCR endocytosis is the first step of receptor downregulation, a simple mechanistic model may explain the marked region-specific differences in downregulation of CB_1 receptors by suggesting that endocytosis and downregulation of axonal CB_1 receptors are inversely proportional to their distance from potential internalization sites, such as terminals and axonal varicosities (Thibault et al. 2013). Finally, the reduced internalization capacity of the basal ganglia may also explain the reported lack of difference in desensitization in this region after treatment with agonists having high and low capacity to induce endocytosis (Wu et al. 2008): in the neocortex and hippocampus, efficient internalization, dephosphorylation, and recycling allow partial CB_1 resensitization in animals treated with high, but not low, endocytotic agonists (Wu et al. 2008), while the basal ganglia mostly lack this internalization-dependent resensitizing mechanism (Sim-Selley and Martin 2002).

In conclusion, CB_1 displays polarized distribution in neurons: in the somatodendritic compartment, CB_1 is mainly internalized due to a combined effect of structural instability and basal activation stemming from the constitutive endocannabinoid tone in the plasma membrane, while in the axon CB_1 accumulates on the plasma membrane due to a lack of endocannabinoids and the lower endocytic capacity of this neuronal compartment.

6 Polarized CB_1 Targeting in Neurons

Since polarized distribution of GPCRs to one of the main neuronal sub-domains—soma, dendrites, or axons—has critical functional consequences, considerable effort has been deployed in the last decades toward both classifying GPCRs as pre- or postsynaptic and identifying protein motifs ensuring targeting and anchoring of receptors either to presynaptic (axonal) or postsynaptic (somatodendritic) domains. However, this enterprise has only yielded the identification of a limited number of targeting motifs (reviewed in Lasiecka and Winckler (2011), Winckler (2012)), so specific targeting and anchoring signals have not yet been identified for the vast majority of neuronal GPCRs. In addition, detailed ultrastructural analysis revealed that most GPCRs are located outside of pre- and postsynaptic specializations (for references see Simon et al. (2013)). Finally, by demonstrating an often surprising and yet unexplained diversity of distribution phenotypes, GCPRs may display different polarized distribution patterns depending on the neuronal subtype or on the neurochemical environment (for references see Simon et al. (2013)). Such differences in polarized distribution of neuronal GPCRs may have important but presently mostly unexplored consequences on intraneuronal mobilization of signaling pathways.

Recent studies have started to reveal an intricate dynamic relationship between neuronal polarization and GPCR activation, reporting previously unexpected polarized and dynamic patterns of CB_1 targeting and signaling. The first approach to study CB_1 trafficking in neurons was the comparison of steady-state (i.e., non-stimulated) endocytosis of endogenous and transfected CB_1, by using "antibody feeding" of live neurons, which showed significantly higher CB_1 endocytosis from the somatodendritic surface as compared to the axonal plasma membrane (Leterrier et al. 2006; Simon et al. 2013), yielding two consistently offset sigmoid curves of CB_1 endocytosis over a large scale of CB_1 activation levels (Simon et al. 2013). This finding is notable since, as described above, in neurons surface-localized CB_1 receptors are mostly found in axons, so one could expect higher levels of axonal CB_1 endocytosis assuming similar endocytic efficiency from the somatodendritic and axonal plasma membranes. Therefore, this result directly confirms the above-proposed difference in CB1R internalization capacity between dendrites and axons. We also established that, similarly to the above-reported findings in non-polarized cells, the driving force of this constitutive somatodendritic endocytosis is tonic CB_1 activation, as shown by using either pharmacological (Leterrier et al. 2006; Simon et al. 2013) or genetic (Simon et al. 2013) tools. This tonic activation depends both on structural CB_1 determinants (Simon et al. 2013) and steady-state cell-autonomous production of the eCB 2-AG (Turu et al. 2007). The substantial difference between the somatodendritic and axonal endocytosis rates, present at a broad range of CB_1 activation levels (Simon et al. 2013), suggests that polarized CB_1 distribution may be a result of selective elimination of CB_1 from the somatodendritic plasma membrane, as previously described for several axonal proteins which are also first delivered to the dendritic plasma membrane but rapidly removed by endocytosis and translocated to the axon through transcytosis (Lasiecka and Winckler 2011). This hypothesis was confirmed by using microfluidic devices that allow separate incubation of the somatodendritic and axonal plasma membrane, allowing us to demonstrate that CB_1 receptors transiently present on the somatodendritic plasma membrane are constitutively endocytosed and transported to the axonal surface through transcytosis (Simon et al. 2013). Direct axonal export of CB_1 is also present but does not lead to efficient GPCR delivery to distal axonal portions. Conversely, the transcytotic route originating from somatodendritic recycling endosomes delivers CB_1 in vesicles specialized for long-range axonal delivery directly to distal portions of axons (Simon et al. 2013). Interestingly, once delivered to the axonal plasma membrane, CB_1 receptors appear to diffuse relatively freely, as shown both by diffusion analysis using single-quantum dot imaging (Mikasova et al. 2008) and by our

FRAP analysis (Simon et al. 2013), suggesting the lack of anchoring to scaffolding proteins. Interestingly, a recent study showed, by using STORM super-resolution microscopy, a uniform distribution of CB_1 in presynaptic terminals of GABAergic neurons in the mouse hippocampus (Dudok et al. 2015) without specific enrichment in nanoscale domains in synaptic terminals. This is in line with our model of passive accumulation of CB_1 on the axonal plasma membrane, which suggests that CB_1 receptors are "retained" in the axonal plasma membrane simply through a reduced basal endocytosis rate. Based on the ensemble of the above results, we suggest that the major driving force of polarized CB_1 targeting to axons is a polarized difference in steady-state endocytosis, which is higher in the somatodendritic region (Leterrier et al. 2006; Simon et al. 2013). This model was confirmed by an independent report (McDonald et al. 2007), and we have tested its general validity for other axonal GPCRs by showing that the 5-HT_{1B} serotonin receptor is addressed to axons by similar mechanisms (Carrel et al. 2011).

The pharmacological activation state of CB_1 as a driving force of this transcytotic targeting was debated initially (McDonald et al. 2007), possibly due to the important natural variability of polarized distribution patterns of CB_1 in cultured hippocampal neurons and to a yet unexplained mixed phenotype of CB_1 expression following inverse agonist treatment (Simon et al. 2013). We have overcome this variability in our reports (Leterrier et al. 2006; Simon et al. 2013) by using a Brefeldin A (BFA) release protocol, a standard experimental approach in protein targeting studies, which enhances experimental sensitivity by first accumulating newly synthetized proteins in the Golgi apparatus before synchronized release through washout of BFA (Cid-Arregui et al. 1995; Wisco et al. 2003; Fache et al. 2004). By using this tool, antagonist/inverse agonist AM281 treatment led to the accumulation of newly released CB_1 on the somatodendritic plasma membrane, showing that the pharmacological conformational state of CB_1 is a major regulator of steady-state somatodendritic endocytosis and transcytotic axonal targeting (Leterrier et al. 2006; Simon et al. 2013), a finding also confirmed by our subsequent studies, which used additional pharmacological and genetic tools, the T210I-CB_1 and T210A-CB_1 mutants (Simon et al. 2013). Notably, we have shown that tonic activation-dependent constitutive endocytosis is also crucial for axonal targeting of at least one other GPCR, the 5-HT_{1B} serotonin receptor, since inverse agonist treatment, which prevents tonic activation, leads to atypical accumulation of newly synthesized 5-HT_{1B}s on the somatodendritic plasma membrane (Carrel et al. 2011). Strikingly, kinetic modeling confirmed by our experimental results indicated that gradual changes of basal activation lead to gradual shifts of polarized CB_1 localization and that changes in basal activation can completely switch distribution phenotypes from somatodendritic (i.e., postsynaptic) to axonal (i.e., presynaptic) and vice versa, for several GPCRs (Simon et al. 2013). Finally, as discussed above, in vivo inhibition of basal CB_1 activation by inverse agonist treatment decreases the number of cell bodies containing CB_1-immunoreactive endosomes, confirming that a substantial amount of CB_1 receptors is permanently activated and internalized in the brain (Thibault et al. 2013).

Taken together, our results (summarized in Fig. 2A) demonstrate an unexpectedly high activation-dependent plasticity of polarized CB_1 distribution. We have

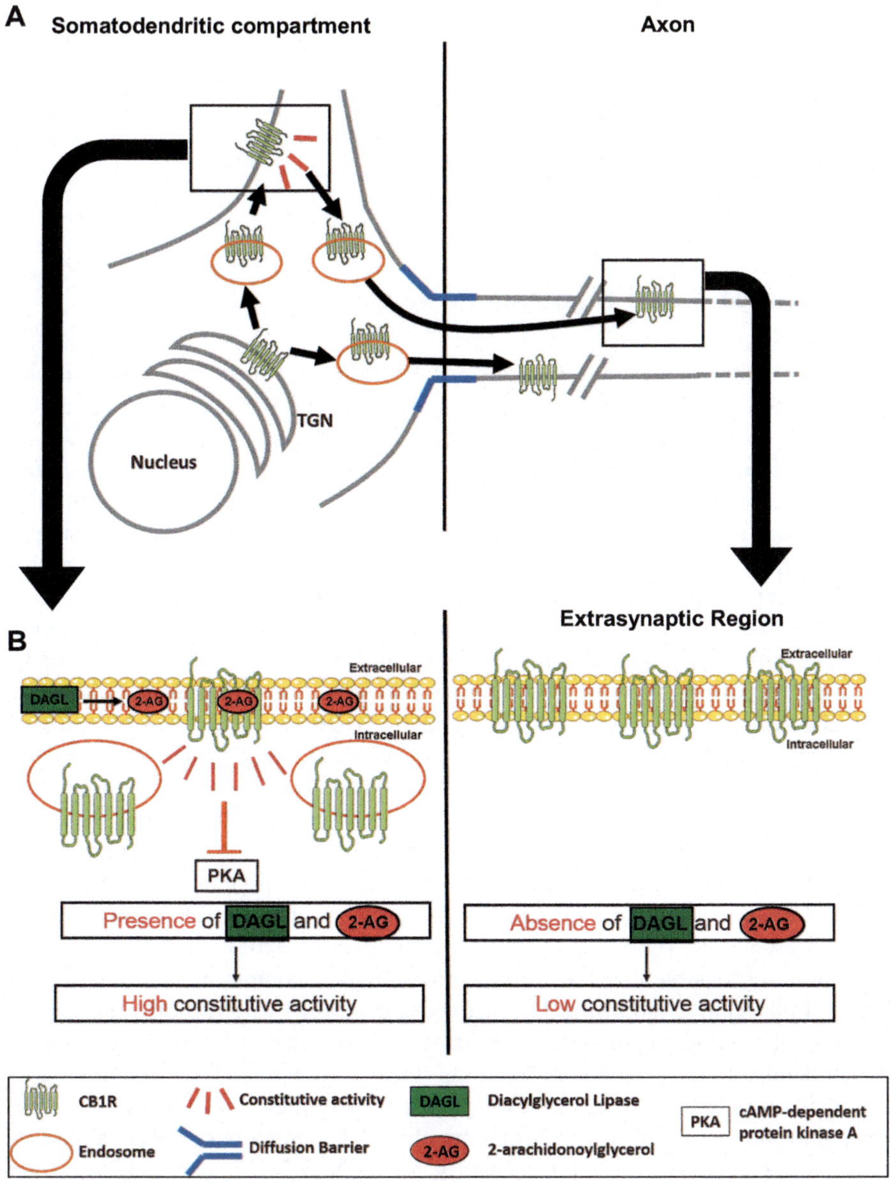

Fig. 2 Polarized patterns of CB_1 targeting and signaling in neurons. (**a**) Schematic representation of CB_1 targeting in neurons. After their synthesis, CB_1 is targeted both to the axonal plasma membrane and to the somatodendritic plasma membrane. Somatodendritic CB_1 receptors are rapidly internalized by constitutive endocytosis and transferred by transcytosis to the axonal plasma membrane where they accumulate. Importantly, transcytosis enables CB1R transport to distal axonal regions, as compared to direct targeting to proximal axonal regions. (**b**) Schematic representation of basal CB_1 activation in the somatodendritic compartment and extrasynaptic axonal regions, where the majority of axonal CB_1 locate. In the somatodendritic compartment, CB_1 receptors constitutively inhibit cAMP/PKA activity and are constitutively endocytosed because of basal activation, both due to elevated local levels of the endocannabinoid 2-AG,

characterized several key points of neuronal GPCR targeting by first showing how prolonged pharmacological treatments modify neuronal GPCR distribution (Simon et al. 2013). We have also shown that the change of basal activation levels may lead to a striking reversion of polarized GPCR distribution and that the natural constitutive activation of CB_1 is near to the optimal moderate level to ensure polarized distribution to the axonal plasma membrane (Simon et al. 2013).

7 Polarized CB1R Signaling in Neurons

The above studies show that trafficking of CB_1 differs between the somatodendritic compartment, where it is mostly internalized by constitutive activation, and axons, where CB_1 receptors are predominantly localized on the axonal plasma membrane. We have recently asked whether these polarized differences in traffic and localization are accompanied by polarized differences in signaling. By using cultured hippocampal neurons of rat embryos, we measured the PKA activity downstream of endogenous CB_1 receptors with FRET (Förster resonance energy transfer) probes in individual axons, dendrites, and neuronal somata (Ladarre et al. 2015). In both somata and dendrites, CB_1 activation with WIN55-212,2 led to a decrease of PKA activity while CB_1 blockade with antagonist/inverse agonist AM281 produced a rapid and significant increase of PKA activity. Thus, somatodendritic CB_1 receptors, in transient passage at the plasma membrane (Leterrier et al. 2006; Simon et al. 2013), are available to exogenous cannabinoids, potentially explaining important local effects such as endocannabinoid-mediated somatodendritic slow self-inhibition (SSI) (Bacci et al. 2004; Marinelli et al. 2008) and the tonic 2-AG-mediated enhancement of the h-current through the hyperpolarization-activated cyclic nucleotide-gated (HCN) channels in superficial CA1 pyramidal cells (Maroso et al. 2016). In addition, somatodendritic CB_1 also constitutively inhibit PKA activity (Ladarre et al. 2015), which is consistent with their constitutive activation responsible for their endocytosis in this compartment, as described above. Importantly, inhibiting DAGLα before AM281 application removes the effect of the antagonist/inverse agonist, suggesting that the tonic activation of somatodendritic CB_1 receptors requires local cell-autonomous production of 2-AG (Ladarre et al. 2015). Interestingly, as already mentioned above, the 2-AG content of the somatodendritic plasma membrane also modulates CB_1 responses to exogenous cannabinoids, being necessary for WIN55-212,2- and THC-mediated CB1R activation but dispensable (and competitively antagonist) for CP55,940

Fig. 2 (continued) synthetized by DAGLα. On the contrary, in axons, CB_1 receptors accumulate on plasma membrane of axonal shafts (i.e., preterminal axons) and are not tonically activated due to a lack of DAGLα action on CB_1 (Ladarre et al. 2015). Therefore, 2-AG, which is specifically produced in the somatodendritic plasma membrane (Bisogno et al. 2003), has important cell-autonomous local actions on somatodendritic CB_1 receptors, by regulating, in addition to the significant basal activation of downstream signaling pathways, their activation by other cannabinoids and by providing the driving force for significant basal endocytosis followed by transcytotic delivery to the axonal plasma membrane.

Interestingly, in axons the rapid decrease of PKA activity following CB_1 activation is significantly stronger than in dendrites (Ladarre et al. 2015). This result is consistent with the higher level of CB_1 receptors on axonal plasma membrane as compared to somatodendritic plasma membrane (Coutts et al. 2001; Pickel et al. 2004; Leterrier et al. 2006). Moreover, inverse agonist application does not lead to an increase of PKA activity, confirming the absence of cell-autonomous tonic activation for axonal CB_1 receptors (Ladarre et al. 2015). Notably, DAGLα is absent in mature axons (Bisogno et al. 2003). Thus, absence of 2-AG in this compartment may explain the lack of cell-autonomous tonic CB_1 activation (Fig. 2B), further proving that tonic CB_1 activation, reported in a wide range of cell lines and in the somatodendritic domain, requires cell-autonomous production of eCBs in addition to the structural instability of CB_1.

8 A Special Case: CB_1 at the Presynaptic Terminal

CB_1 was characterized originally as a major presynaptic GPCR implicated in retrograde synaptic signaling (recent review in Katona and Freund (2012)), and this discovery is registered among the most important advances of the last decades in neuroscience (Südhof and Malenka 2008). Intriguingly, this key physiological role is apparently assured by only a small percentage of CB1Rs, since subcellular localization studies have consistently found a significantly higher density of CB1Rs in axon shafts (also called preterminal axons) than in terminals, indicating a predominantly extrasynaptic cell membrane localization of axonal CB1Rs both in vitro (Leterrier et al. 2006) and in vivo (Bodor et al. 2005; Nyiri et al. 2005; Mátyás et al. 2006). The physiological role of the large extrasynaptic CB_1 population, often located at great distances from synapses (Thibault et al. 2013), is still enigmatic. Nevertheless, endocannabinoids induce critical changes in synaptic strength both in a tonic and in a phasic, activity-dependent manner (Katona and Freund 2012; Soltesz et al. 2015). While early studies focused more on phasic synaptic modulation, in which eCBs transiently depress neurotransmission after being produced "on demand" in response to specific postsynaptic signals, such as an increase in Ca^{2+} concentration or activation of $G_{q/11}$-coupled GPCRs, tonic eCB effect on the probability of GABA release is increasingly recognized as a major feature of brain eCB function (Katona and Freund 2012; Soltesz et al. 2015).

What is the probable relevance of the results, reviewed above in the chapter, to the presynaptic terminal, a small and highly specialized subcellular compartment?

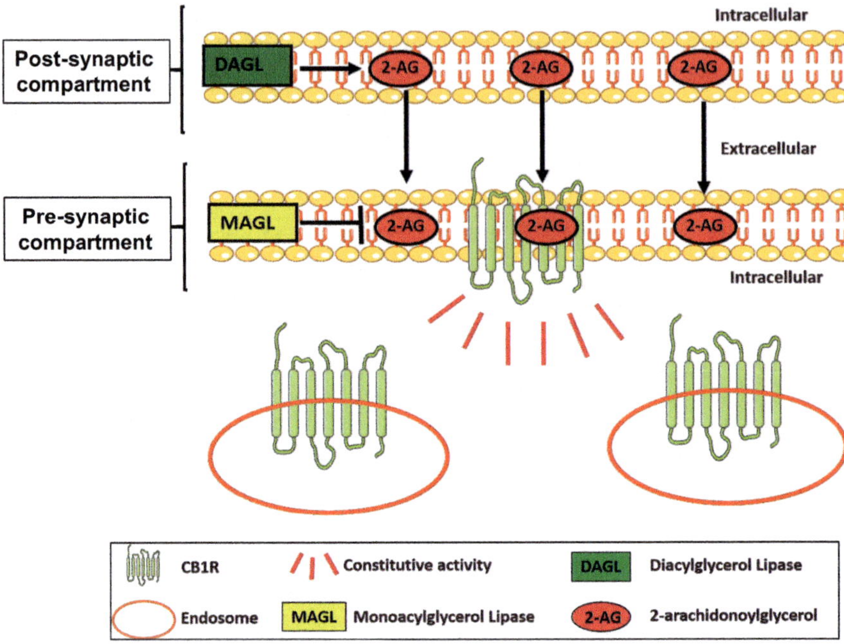

Fig. 3 Schematic representation of CB_1 receptor activation in synapses. CB_1 in the presynaptic compartment is tonically activated by paracrine 2-AG secretion and still not well-understood extracellular 2-AG transport from the postsynaptic membrane. In the presynaptic compartment, 2-AG levels are tightly controlled by MAGL-mediated degradation, and activated CB_1 is likely to endocytose into endosomes

These results (summarized in Fig. 2) were typically obtained in isolated neurons and are useful to understand the cell-autonomous components of tonic CB_1 activation, localization, and signaling. They indicate that tonic activation of CB_1 receptors in axons, where presynaptic terminals are located, is not of cell-autonomous origin. Accordingly, a key role for constitutive release of endocannabinoids from the postsynaptic neuron was demonstrated by showing that postsynaptic BAPTA chelation of intracellular Ca^{2+} signals abolished tonic eCB signaling (Hentges et al. 2005; Neu et al. 2007). Indeed, in the brain, tonic paracrine liberation of eCBs from neighboring neurons or glial cells adds a major additional layer of regulation for tonic CB_1 activation (Losonczy et al. 2004; Neu et al. 2007; Szabó et al. 2014; Lenkey et al. 2015). As a representative example, constitutive presynaptic CB_1 activity in the hippocampus, having a pivotal function in the tonic control of GABA release, is generated by continuous postsynaptic production of 2-arachidonoylglycerol (2-AG) and is strictly regulated by presynaptic monoacylglycerol lipase activity (Fig. 3) (Lee et al. 2015).

Unfortunately, the difficulty in reconstructing physiologically relevant synapses in vitro and the small size of the presynaptic terminal—close to the resolution limit of optical microscopy—has hindered us in the past from finding direct answers to several important questions. First, while it has been shown that the presence of eCBs is needed for tonic activation, we do not know whether the structural instability of CB_1 has also a key role in the regulation of presynaptic plasticity. Recently, the contribution of structural determinants (i.e., constitutive activity) and endocannabinoid tone (i.e., basal activation) in tonic CB_1 activation was started to be evaluated by measuring ex vivo effects of a recently developed neutral antagonist NESS0327 on the release of GABA from perisomatic GABAergic axon terminals in the mouse hippocampus (Lee et al. 2015). Interestingly, NESS0327 inhibited CB_1 activation by both exogenous and endogenous cannabinoids, but a residual tonic CB_1 activation was still measured. However, it is not clear if this remaining constitutive activation is due to a genuine intrinsic constitutive activity or to the activation of CB_1 receptors bound to 2-AG before NESS0327 application. This question could be investigated in the future by genetically introducing stabilizing mutations to create, for example, mice expressing T3.46A (or T210A) mutant CB_1 receptors. Second, we do not know whether presynaptic CB_1 receptors show the activation-dependent subcellular targeting patterns described in the somatodendritic region, which would lead to a relatively low level of surface-located CB_1 in synapses presenting high levels of tonic CB_1 activation. However, several lines of indirect evidence suggest that this may be the case. CB_1 containing endosomal structures were reported in presynaptic terminals not only in agonist-treated (Thibault et al. 2013) but also in untreated animals (Bodor 2005). Chronic cannabinoid treatment leads to a measurable downregulation of CB_1 from terminals and axons, suggesting that CB_1 activation leads to local endocytosis at the presynaptic terminal (Thibault et al. 2013; Dudok et al. 2015). Finally, a recent study reported a striking lack of correlation between CB_1 levels at the presynaptic plasma membrane and the level of tonic endocannabinoid-mediated modulation of GABA release at single synapses, a result that may probably be explained by tonic activation-induced endocytic elimination of CB_1 receptors from the presynaptic plasma membrane (Lenkey et al. 2015). Undoubtedly, further technical development of highly resolved dynamic experimental approaches is needed to achieve a better understanding of the dynamic behavior of CB_1 receptors in the presynaptic terminal.

9 Conclusion

Taken together, the results presented in this chapter suggest that basal activation of the structurally instable CB_1 through cell-autonomous somatodendritic 2-AG production, which is polarized in neurons, is essential in obtaining tonic activation of somatodendritic CB_1. This tonic activation, besides other important local effects, drives the correct, i.e., axonally polarized, CB_1 distribution. Thus, the polarized production of a neuronal membrane lipid, 2-AG, leads to dynamic and polarized distribution and signaling patterns of CB_1, a major neurotransmitter receptor. Key elements of this functional model are also valuable to better understand activation of presynaptic CB_1 receptors. The ensemble of the reviewed data strongly differentiates CB_1 from other, more classical GPCRs, since the combined result of ubiquitous eCB lipids and the instable CB_1 structure leads to an unusually dynamic model of receptor function, providing a novel conceptual framework for the comprehension of the brain endocannabinoid system.

Acknowledgments We wish to thank Christophe Leterrier for the photography of Fig. 1 and Maureen McFadden for the help with the English syntax.

References

Ahn KH, Scott CE, Abrol R et al (2013) Computationally-predicted CB1 cannabinoid receptor mutants show distinct patterns of salt-bridges that correlate with their level of constitutive activity reflected in G protein coupling levels, thermal stability, and ligand binding. Proteins 81:1304–1317. doi:10.1002/prot.24264

Alewijnse AE, Timmerman H, Jacobs EH et al (2000) The effect of mutations in the DRY motif on the constitutive activity and structural instability of the histamine H(2) receptor. Mol Pharmacol 57:890–898

Alger BE, Kim J (2011) Supply and demand for endocannabinoids. Trends Neurosci:1–12. doi:10.1016/j.tins.2011.03.003

Ascano M, Richmond A, Borden P, Kuruvilla R (2009) Axonal targeting of Trk receptors via transcytosis regulates sensitivity to neurotrophin responses. J Neurosci 29:11674–11685. doi:10.1523/JNEUROSCI.1542-09.2009

Bacci A, Huguenard JR, Prince DA (2004) Long-lasting self-inhibition of neocortical interneurons mediated by endocannabinoids. Nature 431:312–316. doi:10.1038/nature02913

Ballesteros JA, Weinstein H (1995) [19] Integrated methods for the construction of three-dimensional models and computational probing of structure-function relations in G protein-coupled receptors. In: Receptor molecular biology. Elsevier, pp 366–428

Ballesteros JA, Jensen AD, Liapakis G (2001) Activation of the β2-adrenergic receptor involves disruption of an ionic lock between the cytoplasmic ends of transmembrane segments 3 and 6. J Biol Chem 276:29171–29177

Baur R, Kielar M, Richter L et al (2013) Molecular analysis of the site for 2-arachidonylglycerol (2-AG) on the β2 subunit of GABAA receptors. J Neurochem 126:29–36. doi:10.1111/jnc.12270

Bel C, Oguievetskaia K, Pitaval C et al (2009) Axonal targeting of Caspr2 in hippocampal neurons via selective somatodendritic endocytosis. J Cell Sci 122:3403–3413. doi:10.1242/jcs.050526

Benard G, Massa F, Puente N et al (2012) Mitochondrial CB1 receptors regulate neuronal energy metabolism. Nat Neurosci 15:558–564. doi:10.1038/nn.3053

Bisogno T, Howell F, Williams G et al (2003) Cloning of the first sn1-DAG lipases points to the spatial and temporal regulation of endocannabinoid signaling in the brain. J Cell Biol 163:463–468. doi:10.1083/jcb.200305129

Bodor AL (2005) Endocannabinoid signaling in rat somatosensory cortex: laminar differences and involvement of specific interneuron types. J Neurosci 25:6845–6856. doi:10.1523/JNEUROSCI.0442-05.2005

Brailoiu GC, Oprea TI, Zhao P et al (2011) Intracellular Cannabinoid Type 1 (CB1) receptors are activated by anandamide. J Biol Chem 286:29166–29174. doi:10.1074/jbc.M110.217463

Breivogel CS, Childers SR (1999) Chronic□ 9-tetrahydrocannabinol treatment produces a time-dependent loss of cannabinoid receptors and cannabinoid receptor-activated G proteins .... J Neurochem 73:2447–2459

Brown MF (2012) Curvature forces in membrane lipid–protein interactions. Biochemistry 51:9782–9795. doi:10.1021/bi301332v

Buggia-Prévot V, Fernandez CG, Riordan S et al (2014) Axonal BACE1 dynamics and targeting in hippocampal neurons: a role for Rab11 GTPase. Mol Neurodegener 9(1). doi:10.1186/1750-1326-9-1

Burack MA, Silverman MA, Banker G (2000) The role of selective transport in neuronal protein sorting. Neuron 26:465–472

Carrel D, Simon A, Emerit MB et al (2011) Axonal targeting of the 5-HT1B serotonin receptor relies on structure-specific constitutive activation. Traffic 12:1501–1520. doi:10.1111/j.1600-0854.2011.01260.x

Chanrion B, Mannoury la Cour C, Gavarini S et al (2008) Inverse agonist and neutral antagonist actions of antidepressants at recombinant and native 5-hydroxytryptamine2C receptors: differential modulation of cell surface expression and signal transduction. Mol Pharmacol 73:748–757. doi:10.1124/mol.107.041574

Chicca A, Marazzi J, Nicolussi S, Gertsch J (2012) Evidence for bidirectional endocannabinoid transport across cell membranes. J Biol Chem 287:34660–34682. doi:10.1074/jbc.M112.373241

Cid-Arregui A, Parton RG, Simons K, Dotti CG (1995) Nocodazole-dependent transport, and brefeldin A – sensitive processing and sorting, of newly synthesized membrane proteins in cultured neurons. J Neurosci 15:4259–4269

Coutts AA, Anavi-Goffer S, Ross RA et al (2001) Agonist-induced internalization and trafficking of cannabinoid CB1 receptors in hippocampal neurons. J Neurosci 21:2425–2433

Craig AM, Banker G (1994) Neuronal polarity. Annu Rev Neurosci 17:267–310. doi:10.1146/annurev.ne.17.030194.001411

Curran HV, Freeman TP, Mokrysz C et al (2016) Keep off the grass? Cannabis, cognition and addiction. Nat Rev Neurosci 17:293–306. doi:10.1038/nrn.2016.28

D'Antona AM, Ahn KH, Kendall DA (2006) Mutations of CB1 T210 produce active and inactive receptor forms: correlations with ligand affinity, receptor stability, and cellular localization. Biochemistry 45:5606–5617. doi:10.1021/bi060067k

Dawaliby R, Trubbia C, Delporte C et al (2016) Allosteric regulation of G protein-coupled receptor activity by phospholipids. Nat Chem Biol 12:35–39. doi:10.1038/nchembio.1960

Deupi X, Kobilka BK (2010) Energy landscapes as a tool to integrate GPCR structure, dynamics, and function. Physiology 25:293–303. doi:10.1152/physiol.00002.2010

Dudok B, Barna L, Ledri M et al (2015) Cell-specific STORM super-resolution imaging reveals nanoscale organization of cannabinoid signaling. Nat Neurosci. doi:10.1038/nn.3892

Ellis J, Pediani JD, Canals M et al (2006) Orexin-1 receptor-cannabinoid CB1 receptor heterodimerization results in both ligand-dependent and -independent coordinated alterations of receptor localization and function. J Biol Chem 281:38812–38824. doi:10.1074/jbc.M602494200

Engelstoft MS, Norn C, Hauge M et al (2014) Structural basis for constitutive activity and agonist-induced activation of the enteroendocrine fat sensor GPR119. Br J Pharmacol 171:5774–5789. doi:10.1111/bph.12877

Fache M-P, Moussif A, Fernandes F et al (2004) Endocytotic elimination and domain-selective tethering constitute a potential mechanism of protein segregation at the axonal initial segment. J Cell Biol 166:571–578. doi:10.1083/jcb.200312155

Fišar Z, Singh N, Hroudová J (2014) Cannabinoid-induced changes in respiration of brain mitochondria. Toxicol Lett 231:62–71. doi:10.1016/j.toxlet.2014.09.002

Fowler CJ (2013) Transport of endocannabinoids across the plasma membrane and within the cell. FEBS J 280:1895–1904. doi:10.1111/febs.12212

Freund TF, Katona I, Piomelli D (2003) Role of endogenous cannabinoids in synaptic signaling. Physiol Rev 83:1017–1066. doi:10.1152/physrev.00004.2003

Fu J, Bottegoni G, Sasso O et al (2012) A catalytically silent FAAH-1 variant drives anandamide transport in neurons. Nat Neurosci 15:64–69. doi:10.1038/nn.2986

Gabrielli M, Battista N, Riganti L et al (2015) Active endocannabinoids are secreted on extracellular membrane vesicles. EMBO Rep 16:213–220. doi:10.15252/embr.201439668

Garrido JJ, Fernandes F, Giraud P et al (2001) Identification of an axonal determinant in the C-terminus of the sodium channel Na(v)1.2. EMBO J 20:5950–5961. doi:10.1093/emboj/20.21.5950

Gether U, Ballesteros JA, Seifert R et al (1997) Structural instability of a constitutively active G protein-coupled receptor. Agonist-independent activation due to conformational flexibility. J Biol Chem 272:2587–2590

Gilliland CT, Kufareva I, Handel T (2015) Structural analysis of the constitutive activity of the chemokine receptor CCR1. FASEB J 29:893.5. doi:10.1096/fj.1530-6860

González S, Cebeira M, Fernández-Ruiz J (2005) Cannabinoid tolerance and dependence: a review of studies in laboratory animals. Pharmacol Biochem Behav 81:300–318. doi:10.1016/j.pbb.2005.01.028

Graham ES, Ball N, Scotter EL et al (2006) Induction of Krox-24 by endogenous cannabinoid type 1 receptors in Neuro2A cells is mediated by the MEK-ERK MAPK pathway and is suppressed by the phosphatidylinositol 3-kinase pathway. J Biol Chem 281:29085–29095. doi:10.1074/jbc.M602516200

Grimsey NL, Graham ES, Dragunow M, Glass M (2010) Cannabinoid receptor 1 trafficking and the role of the intracellular pool: implications for therapeutics. Biochem Pharmacol:1–13. doi:10.1016/j.bcp.2010.06.007

Hanyaloglu AC, Zastrow MV (2008) Regulation of GPCRs by endocytic membrane trafficking and its potential implications. Annu Rev Pharmacol Toxicol 48:537–568. doi:10.1146/annurev.pharmtox.48.113006.094830

Hebert-Chatelain E, Reguero L, Puente N et al (2014) Cannabinoid control of brain bioenergetics: exploring the subcellular localization of the CB1 receptor. Mol Metab 3:495–504. doi:10.1016/j.molmet.2014.03.007

Hentges ST, Low MJ, Williams JT (2005) Differential regulation of synaptic inputs by constitutively released endocannabinoids and exogenous cannabinoids. J Neurosci 25:9746–9751. doi:10.1523/JNEUROSCI.2769-05.2005

Holliday ND, Holst B, Rodionova EA et al (2007) Importance of constitutive activity and arrestin-independent mechanisms for intracellular trafficking of the ghrelin receptor. Mol Endocrinol 21:3100–3112. doi:10.1210/me.2007-0254

Holst B, Holliday ND, Bach A et al (2004) Common structural basis for constitutive activity of the ghrelin receptor family. J Biol Chem 279:53806–53817. doi:10.1074/jbc.M407676200

Horton AC, Ehlers MD (2003) Neuronal polarity and trafficking. Neuron 40:277–295

Howlett AC, Reggio PH, Childers SR et al (2011) Endocannabinoid tone versus constitutive activity of cannabinoid receptors. Br J Pharmacol 163:1329–1343. doi:10.1111/j.1476-5381.2011.01364.x

Hsieh C, Brown S, Derleth C, Mackie K (1999) Internalization and recycling of the CB1 cannabinoid receptor. J Neurochem 73:493–501

Hurst DP, Schmeisser M, Reggio PH (2013) Endogenous lipid activated G protein-coupled receptors: emerging structural features from crystallography and molecular dynamics simulations. Chem Phys Lipids 169:46–56. doi:10.1016/j.chemphyslip.2013.01.009

Irannejad R, Zastrow von M (2014) GPCR signaling along the endocytic pathway. Curr Opin Cell Biol 27:109–116. doi:10.1016/j.ceb.2013.10.003

Jacquier V, Prummer M, Segura J-M et al (2006) Visualizing odorant receptor trafficking in living cells down to the single-molecule level. Proc Natl Acad Sci U S A 103:14325–14330. doi:10.1073/pnas.0603942103

Katona I, Freund TF (2012) Multiple functions of endocannabinoid signaling in the brain. Annu Rev Neurosci 35:529–558. doi:10.1146/annurev-neuro-062111-150420

Katona I, Sperlágh B, Sík A et al (1999) Presynaptically located CB1 cannabinoid receptors regulate GABA release from axon terminals of specific hippocampal interneurons. J Neurosci 19:4544–4558

Katona I, Rancz EA, Acsady L et al (2001) Distribution of CB1 cannabinoid receptors in the amygdala and their role in the control of GABAergic transmission. J Neurosci 21:9506–9518
Katona I, Urbán GM, Wallace M et al (2006) Molecular composition of the endocannabinoid system at glutamatergic synapses. J Neurosci 26:5628–5637. doi:10.1523/JNEUROSCI.0309-06.2006
Kawamura Y, Fukaya M, Maejima T et al (2006) The CB1 cannabinoid receptor is the major cannabinoid receptor at excitatory presynaptic sites in the hippocampus and cerebellum. J Neurosci 26:2991–3001. doi:10.1523/JNEUROSCI.4872-05.2006
Khajehali E, Malone DT, Glass M et al (2015) Biased agonism and biased allosteric modulation at the CB1 cannabinoid receptor. Mol Pharmacol 88:368–379. doi:10.1124/mol.115.099192
Kobilka BK, Deupi X (2007) Conformational complexity of G-protein-coupled receptors. Trends Pharmacol Sci 28:397–406. doi:10.1016/j.tips.2007.06.003
Koch T, Wu D-F, Yang L-Q et al (2006) Role of phospholipase D2 in the agonist-induced and constitutive endocytosis of G-protein coupled receptors. J Neurochem 97:365–372. doi:10.1111/j.1471-4159.2006.03736.x
Ladarre D, Roland AB, Biedzinski S et al (2015) Polarized cellular patterns of endocannabinoid production and detection shape cannabinoid signaling in neurons. Front Cell Neurosci 8:426–426. doi:10.3389/fncel.2014.00426
Lasiecka ZM, Winckler B (2011) Mechanisms of polarized membrane trafficking in neurons – Focusing in on endosomes. Mol Cell Neurosci 48:278–287. doi:10.1016/j.mcn.2011.06.013
Lee SH, Ledri M, Toth B et al (2015) Multiple forms of endocannabinoid and endovanilloid signaling regulate the tonic control of GABA release. J Neurosci 35:10039–10057. doi:10.1523/JNEUROSCI.4112-14.2015
Lenkey N, Kirizs T, Holderith N et al (2015) Tonic endocannabinoid-mediated modulation of GABA release is independent of the CB1 content of axon terminals. Nat Commun 6:6557. doi:10.1038/ncomms7557
Leterrier C, Dargent B (2014) No Pasaran! Role of the axon initial segment in the regulation of protein transport and the maintenance of axonal identity. Semin Cell Dev Biol 27:44–51. doi:10.1016/j.semcdb.2013.11.001
Leterrier C, Bonnard D, Carrel D et al (2004) Constitutive endocytic cycle of the CB1 cannabinoid receptor. J Biol Chem 279:36013–36021
Leterrier C, Lainé J, Darmon M et al (2006) Constitutive activation drives compartment-selective endocytosis and axonal targeting of type 1 cannabinoid receptors. J Neurosci 26:3141–3153. doi:10.1523/JNEUROSCI.5437-05.2006
Leung K, Elmes MW, Glaser ST et al (2013) Role of FAAH-like anandamide transporter in anandamide inactivation. PLoS One 8:e79355. doi:10.1371/journal.pone.0079355
Losonczy A, Biró ÁA, Nusser Z (2004) Persistently active cannabinoid receptors mute a subpopulation of hippocampal interneurons. Proc Natl Acad Sci U S A 101:1362–1367. doi:10.1073/pnas.0304752101
Maccarrone M, Dainese E, Oddi S (2009) Intracellular trafficking of anandamide: new concepts for signaling. Trends Biochem Sci 35:601–608. doi:10.1016/j.tibs.2010.05.008
Manglik A, Kobilka B (2014) The role of protein dynamics in GPCR function: insights from the β2AR and rhodopsin. Curr Opin Cell Biol 27:136–143. doi:10.1016/j.ceb.2014.01.008
Marinelli S, Pacioni S, Bisogno T et al (2008) The endocannabinoid 2-arachidonoylglycerol is responsible for the slow self-inhibition in neocortical interneurons. J Neurosci 28:13532–13541
Marion S, Weiner DM, Caron MG (2004) RNA editing induces variation in desensitization and trafficking of 5-hydroxytryptamine 2c receptor isoforms. J Biol Chem 279:2945–2954. doi:10.1074/jbc.M308742200
Maroso M, Szabó GG, Kim HK et al (2016) Cannabinoid control of learning and memory through HCN channels. Neuron 89:1059–1073. doi:10.1016/j.neuron.2016.01.023
Marsicano G, Lutz B (1999) Expression of the cannabinoid receptor CB1 in distinct neuronal subpopulations in the adult mouse forebrain. Eur J Neurosci 11:4213–4225

Martini L, Waldhoer M, Pusch M et al (2007) Ligand-induced down-regulation of the cannabinoid 1 receptor is mediated by the G-protein-coupled receptor-associated sorting protein GASP1. FASEB J 21:802–811. doi:10.1096/fj.06-7132com

Martini L, Thompson D, Kharazia V, Whistler JL (2010) Differential regulation of behavioral tolerance to WIN55,212-2 by GASP1. Neuropsychopharmacology 35:1363–1373. doi:10.1038/npp.2010.6

Mattila PK, Batista FD, Treanor B (2016) Dynamics of the actin cytoskeleton mediates receptor cross talk: an emerging concept in tuning receptor signaling. J Cell Biol 212:267–280. doi:10.1083/jcb.201504137

Mátyás F, Yanovsky Y, Mackie K et al (2006) Subcellular localization of type 1 cannabinoid receptors in the rat basal ganglia. NSC 137:337–361. doi:10.1016/j.neuroscience.2005.09.005

McDonald NA, Henstridge CM, Connolly CN, Irving AJ (2007) An essential role for constitutive endocytosis, but not activity, in the axonal targeting of the CB1 cannabinoid receptor. Mol Pharmacol 71:976–984. doi:10.1124/mol.106.029348

McIntosh HH, Song C, Howlett AC (1998) CB1 cannabinoid receptor: cellular regulation and distribution in N18TG2 neuroblastoma cells. Mol Brain Res 53:163–173. doi:10.1016/S0169-328X(97)00294-5

Mikasova L, Groc L, Choquet D, Manzoni OJ (2008) Altered surface trafficking of presynaptic cannabinoid type 1 receptor in and out synaptic terminals parallels receptor desensitization. Proc Natl Acad Sci U S A 105:18596–18601. doi:10.1073/pnas.0805959105

Mohammad S, Baldini G, Granell S et al (2007) Constitutive traffic of melanocortin-4 receptor in Neuro2A cells and immortalized hypothalamic neurons. J Biol Chem 282:4963–4974. doi:10.1074/jbc.M608283200

Morales P, Goya P, Jagerovic N, Hernandez-Folgado L (2016) Allosteric modulators of the CB1 cannabinoid receptor: a structural update review. Cannabis Cannabinoid Res 1:22–30. doi:10.1089/can.2015.0005

Morisset S, Rouleau A, Ligneau X et al (2000) High constitutive activity of native H3 receptors regulates histamine neurons in brain. Nature 408:860–864. doi:10.1038/35048583

Morozov YM, Dominguez MH, Varela L et al (2013) Antibodies to cannabinoid type 1 receptor co-react with stomatin-like protein 2 in mouse brain mitochondria. Eur J Neurosci 38:2341–2348. doi:10.1111/ejn.12237

Morozov YM, Sun Y-Y, Kuan C-Y, Rakic P (2015) Alteration of SLP2-like immunolabeling in mitochondria signifies early cellular damage in developing and adult mouse brain. Eur J Neurosci 43:245–257. doi:10.1111/ejn.13124

Morris DP (2004) Cellular trafficking of human 1a-adrenergic receptors is continuous and primarily agonist-independent. Mol Pharmacol 66:843–854. doi:10.1124/mol.104.000430

Neu A, Földy C, Soltesz I (2007) Postsynaptic origin of CB1-dependent tonic inhibition of GABA release at cholecystokinin-positive basket cell to pyramidal cell synapses in the CA1 region of the rat hippocampus. J Physiol 578:233–247. doi:10.1113/jphysiol.2006.115691

Nyiri G, Cserep C, Szabadits E et al (2005) CB1 cannabinoid receptors are enriched in the perisynaptic annulus and on preterminal segments of hippocampal GABAergic axons. Neuroscience 136:811–822. doi:10.1016/j.neuroscience.2005.01.026

Oates J, Watts A (2011) Uncovering the intimate relationship between lipids, cholesterol and GPCR activation. Curr Opin Struct Biol 21:802–807. doi:10.1016/j.sbi.2011.09.007

Parton RG, Dotti CG (1993) Cell biology of neuronal endocytosis. J Neurosci Res 36:1–9. doi:10.1002/jnr.490360102

Parton RG, Simons K, Dotti CG (1992) Axonal and dendritic endocytic pathways in cultured neurons. J Cell Biol 119:123–137. doi:10.2307/1615258

Pertwee RG (2005) Inverse agonism and neutral antagonism at cannabinoid CB1 receptors. Life Sci 76:1307–1324. doi:10.1016/j.lfs.2004.10.025

Pertwee RG (2015) Endocannabinoids and their pharmacological actions. Handb Exp Pharmacol 231:1–37. doi:10.1007/978-3-319-20825-1_1

Pertwee RG, Howlett AC, Abood ME et al (2010) International union of basic and clinical pharmacology. LXXIX. Cannabinoid receptors and their ligands: beyond CB1 and CB2. Pharmacol Rev 62:588–631. doi:10.1124/pr.110.003004
Phillips R, Ursell T, Wiggins P, Sens P (2009) Emerging roles for lipids in shaping membrane-protein function. Nature 459:379–385. doi:10.1038/nature08147
Pickel VM, Chan J, Kash TL et al (2004) Compartment-specific localization of cannabinoid 1 (CB1) and mu-opioid receptors in rat nucleus accumbens. NSC 127:101–112. doi:10.1016/j.neuroscience.2004.05.015
Piomelli D (2003) The molecular logic of endocannabinoid signalling. Nat Rev Neurosci 4:873–884. doi:10.1038/nrn1247
Reggio PH (2010) Endocannabinoid binding to the cannabinoid receptors: what is known and what remains unknown. Curr Med Chem 17:1468–1486
Reiter E, Ahn S, Shukla AK, Lefkowitz RJ (2012) Molecular mechanism of β-arrestin-biased agonism at seven-transmembrane receptors. Annu Rev Pharmacol Toxicol 52:179–197
Rinaldi-Carmona M, Le Duigou A, Oustric D et al (1998) Modulation of CB1 cannabinoid receptor functions after a long-term exposure to agonist or inverse agonist in the Chinese hamster ovary cell expression system. J Pharmacol Exp Ther 287:1038–1047
Rozenfeld R, Devi LA (2008) Regulation of CB1 cannabinoid receptor trafficking by the adaptor protein AP-3. FASEB J 22:2311–2322. doi:10.1096/fj.07-102731
Saka SK, Honigmann A, Eggeling C et al (2013) Multi-protein assemblies underlie the mesoscale organization of the plasma membrane. Nat Commun 5:4509–4509. doi:10.1038/ncomms5509
Sampo B, Kaech S, Kunz S, Banker G (2003) Two distinct mechanisms target membrane proteins to the axonal surface. Neuron 37:611–624
Scavone JL, Mackie K, Van Bockstaele EJ (2010) Characterization of cannabinoid-1 receptors in the locus coeruleus: relationship with mu-opioid receptors. Brain Res 1312:18–31. doi:10.1016/j.brainres.2009.11.023
Schirris TJJ, Ritschel T, Herma Renkema G et al (2015) Mitochondrial ADP/ATP exchange inhibition: a novel off-target mechanism underlying ibipinabant-induced myotoxicity. Sci Rep 5:14533. doi:10.1038/srep14533
Scott CE, Abrol R, Ahn KH et al (2012) Molecular basis for dramatic changes in cannabinoid CB1 G protein-coupled receptor activation upon single and double point mutations. Protein Sci 22:101–113. doi:10.1002/pro.2192
Seifert R, Wenzel-Seifert K (2002) Constitutive activity of G-protein-coupled receptors: cause of disease and common property of wild-type receptors. Naunyn Schmiedeberg's Arch Pharmacol 366:381–416. doi:10.1007/s00210-002-0588-0
Sigel E, Baur R, Rácz I et al (2011) The major central endocannabinoid directly acts at GABA (A) receptors. Proc Natl Acad Sci U S A 108:18150–18155. doi:10.1073/pnas.1113444108
Simon AC, Loverdo C, Gaffuri A-L et al (2013) Activation-dependent plasticity of polarized GPCR distribution on the neuronal surface. J Mol Cell Biol 5:250–265. doi:10.1093/jmcb/mjt014
Sim-Selley LJ (2003) Regulation of cannabinoid CB1 receptors in the central nervous system by chronic cannabinoids. Crit Rev Neurobiol 15:91–119
Sim-Selley LJ, Martin BR (2002) Effect of chronic administration of R-(+)-[2,3-Dihydro-5-methyl-3-[(morpholinyl)methyl]pyrrolo[1,2,3-de]-1,4-benzoxazinyl]-(1-naphthalenyl) methanone mesylate (WIN55,212-2) or delta(9)-tetrahydrocannabinol on cannabinoid receptor adaptation in mice. J Pharmacol Exp Ther 303:36–44. doi:10.1124/jpet.102.035618
Sim-Selley LJ, Schechter NS, Rorrer WK et al (2006) Prolonged recovery rate of CB1 receptor adaptation after cessation of long-term cannabinoid administration. Mol Pharmacol 70:986–996. doi:10.1124/mol.105.019612
Sinclair GI, Baas PW, Heidemann SR (1988) Role of microtubules in the cytoplasmic compartmentation of neurons. II. Endocytosis in the growth cone and neurite shaft. Brain Res 450:60–68

Singh N, Hroudová J, Fišar Z (2015) Cannabinoid-induced changes in the activity of electron transport chain complexes of brain mitochondria. J Mol Neurosci 56:926–931. doi:10.1007/s12031-015-0545-2

Soltesz I, Alger BE, Kano M et al (2015) Weeding out bad waves: towards selective cannabinoid circuit control in epilepsy. Nat Rev Neurosci 16:264–277. doi:10.1038/nrn3937

Spivak CE, Kim W, Liu Q-R et al (2012) Blockade of β-cell KATP channels by the endocannabinoid, 2-arachidonoylglycerol. Biochem Biophys Res Commun 423:13–18. doi:10.1016/j.bbrc.2012.05.042

Starke K (1981) Presynaptic receptors. Annu Rev Pharmacol Toxicol 21:7–30

Stella N, Schweitzer P, Piomelli D (1997) A second endogenous cannabinoid that modulates long-term potentiation. Nature 388:773–778. doi:10.1038/42015

Straiker A, Mitjavila J, Yin D et al (2015) Aiming for allosterism: evaluation of allosteric modulators of CB1 in a neuronal model. Pharmacol Res 99:370–376. doi:10.1016/j.phrs.2015.07.017

Südhof TC, Malenka RC (2008) Understanding synapses: past, present, and future. Neuron 60:469–476. doi:10.1016/j.neuron.2008.10.011

Szabó GG, Lenkey N, Holderith N et al (2014) Presynaptic calcium channel inhibition underlies CB1 cannabinoid receptor-mediated suppression of GABA release. J Neurosci 34:7958–7963. doi:10.1523/JNEUROSCI.0247-14.2014

Tappe-Theodor A, Agarwal N, Katona I et al (2007) A molecular basis of analgesic tolerance to cannabinoids. J Neurosci 27:4165–4177. doi:10.1523/JNEUROSCI.5648-06.2007

Tehan BG, Bortolato A, Blaney FE et al (2007) Unifying family A GPCR theories of activation. Pharmacol Ther 143:51–60. doi:10.1016/j.pharmthera.2014.02.004

Thibault K, Carrel D, Bonnard D et al (2013) Activation-dependent subcellular distribution patterns of CB1 cannabinoid receptors in the rat forebrain. Cereb Cortex 23:2581–2591. doi:10.1093/cercor/bhs240

Turu G, Simon A, Gyombolai P et al (2007) The role of diacylglycerol lipase in constitutive and angiotensin AT1 receptor-stimulated cannabinoid CB1 receptor activity. J Biol Chem 282:7753–7757. doi:10.1074/jbc.C600318200

Uchigashima M, Narushima M, Fukaya M et al (2007) Subcellular arrangement of molecules for 2-arachidonoyl-glycerol-mediated retrograde signaling and its physiological contribution to synaptic modulation in the striatum. J Neurosci 27:3663–3676. doi:10.1523/JNEUROSCI.0448-07.2007

Vallee M, Vitiello S, Bellocchio L et al (2014) Pregnenolone can protect the brain from cannabis intoxication. Science 343:94–98. doi:10.1126/science.1243985

van der Westhuizen ET, Valant C, Sexton PM, Christopoulos A (2015) Endogenous allosteric modulators of G protein-coupled receptors. J Pharmacol Exp Ther 353:246–260

Venkatakrishnan AJ, Deupi X, Lebon G et al (2013) Molecular signatures of G-protein-coupled receptors. Nature 494:185–194. doi:10.1038/nature11896

Vitalis T, Lainé J, Simon A et al (2008) The type 1 cannabinoid receptor is highly expressed in embryonic cortical projection neurons and negatively regulates neurite growth in vitro. Eur J Neurosci 28:1705–1718. doi:10.1111/j.1460-9568.2008.06484.x

Ward RJ, Pediani JD, Milligan G (2011) Ligand-induced internalization of the orexin OX(1) and cannabinoid CB(1) receptors assessed via N-terminal SNAP and CLIP-tagging. Br J Pharmacol 162:1439–1452. doi:10.1111/j.1476-5381.2010.01156.x

Whistler JL, Enquist J, Marley A et al (2002) Modulation of postendocytic sorting of G protein-coupled receptors. Science 297:615–620. doi:10.1126/science.1073308

Winckler B (2012) Scientiae forum/Models and speculations Pathways for axonal targeting of membrane proteins. Biol Cell 96:669–674. doi:10.1016/j.biolcel.2004.05.005

Wisco D, Anderson ED, Chang MC et al (2003) Uncovering multiple axonal targeting pathways in hippocampal neurons. J Cell Biol 162:1317–1328. doi:10.1083/jcb.200307069

Wu D-F, Yang L-Q, Goschke A et al (2008) Role of receptor internalization in the agonist-induced desensitization of cannabinoid type 1 receptors. J Neurochem 104:1132–1143. doi:10.1111/j.1471-4159.2007.05063.x

Yoshida T, Fukaya M, Uchigashima M et al (2006) Localization of diacylglycerol lipase-alpha around postsynaptic spine suggests close proximity between production site of an endocannabinoid, 2-arachidonoyl-glycerol, and presynaptic cannabinoid CB1 receptor. J Neurosci 26:4740–4751. doi:10.1523/JNEUROSCI.0054-06.2006

Yoshida T, Uchigashima M, Yamasaki M et al (2011) Unique inhibitory synapse with particularly rich endocannabinoid signaling machinery on pyramidal neurons in basal amygdaloid nucleus. Proc Natl Acad Sci U S A 108:3059–3064. doi:10.1073/pnas.1012875108

Zhang M (2004) Constitutively active G protein-coupled receptor mutants block dictyostelium development. Mol Biol Cell 16:562–572. doi:10.1091/mbc.E04-06-0456

Zocher M, Zhang C, Rasmussen SGF et al (2012) Cholesterol increases kinetic, energetic, and mechanical stability of the human β2-adrenergic receptor. Proc Natl Acad Sci U S A 109:E3463–E3472. doi:10.1073/pnas.1210373109

Endocannabinoid-Dependent Synaptic Plasticity in the Striatum

Brady K. Atwood and David M. Lovinger

Abstract The striatum plays a critical role in mediating the goal-directed and habitual behaviors involved in the development of drug abuse and addiction. Cannabinoid receptor-dependent synaptic plasticity is a prominent regulator of striatal circuit function. It is increasingly clear that striatal endocannabinoid signaling impacts the development of drug abuse. The expression of endocannabinoid-mediated plasticity in the striatum is affected by disparate types of drugs of abuse, such as opiates and alcohol. Understanding the mechanism of endocannabinoid-mediated plasticity and its effects on striatal circuit function may elucidate how endocannabinoids shape addiction-related behavior and perhaps offer new therapeutic avenues for treating drug abuse.

1 Introduction

The striatum is a set of large subcortical nuclei with key roles in the cortico-basal ganglia circuitry that control goal-directed and habitual actions (Gerfen and Bolam 2010; Yin and Knowlton 2006). As the entryway for information flow from the cortex to the basal ganglia, the striatum processes information relevant to all sensory, motor, and emotional functions and communicates with downstream basal ganglia nuclei. The striatum also receives the strongest dopaminergic innervation of any part of the brain, and these inputs arise from midbrain neurons (Gerfen and Bolam 2010). Ultimately, the basal ganglia feedback to the cortex controls production and structure of nonreflexive actions.

The striatum can be roughly separated into at least three large subregions that are part of distinct cortico-basal ganglia circuits (although more extensive divisions have been suggested) (Joel and Weiner 2000; Yin and Knowlton 2006). The

B.K. Atwood
Department of Psychiatry, Department of Pharmacology & Toxicology, Indiana University School of Medicine, Indianapolis, IN, USA

D.M. Lovinger (✉)
Laboratory for Integrative Neuroscience, National Institute on Alcohol Abuse and Alcoholism, National Institutes of Health, Bethesda, MD, USA
e-mail: lovindav@mail.nih.gov

M. Melis (ed.), *Endocannabinoids and Lipid Mediators in Brain Functions*,
DOI 10.1007/978-3-319-57371-7_5

dorsomedial striatum (DMS, in rodent, caudate nucleus in primates) is part of an "associative" circuit in which cortical regions such as the prefrontal cortex and other associative cortices innervate the DMS. This circuit has a number of functions, including integration of spatial information and control of behaviors that are driven by outcomes (i.e., goal-directed behaviors) (Yin and Knowlton 2006). The "sensorimotor" circuit includes projections from primary sensory and motor cortices to the dorsolateral striatum (DLS in rodents, putamen nucleus in primates). This circuit controls the production of well-learned behaviors, especially those that are driven by stimuli and context and have become habitual (Yin and Knowlton 2006; Gerdeman et al. 2003; Everitt and Robbins 2016). The limbic circuit includes connections from both allocortical and neocortical regions to the ventral striatum (also known as nucleus accumbens, NAc). In the mammalian brain, this cortico-basal ganglia circuit is involved in affective control of actions, and also has prominent roles in learning and memory driven by conditioned cues, including learning related to the actions of drugs of abuse (Yin and Knowlton 2006). The different striatal subregions can be further subdivided into smaller areas including the striosome/matrix subcompartments of the dorsal striatum, and the core/shell subcompartments of the NAc. These subcompartments are distinguished by differences in afferent inputs and efferent outputs (Eblen and Graybiel 1995; Kincaid and Wilson 1996; Langer and Graybiel 1995).

As described later in this chapter, the striatum has a rich mixture of afferent inputs and microcircuits that ultimately determine output to downstream BG regions. A variety of neurotransmitters, neuromodulators, and neuropeptides act within these circuits, including glutamate, GABA, dopamine, acetylcholine, serotonin, dynorphin, enkephalin, somatostatin, and substance P, among others (Emson et al. 2010). The role of endocannabinoids and CB1 receptors within these circuits is to modulate the actions of the other neurotransmitters, predominantly through presynaptic inhibition of neurotransmitter release (Lovinger et al. 2010). These modulatory roles are the subject of this review.

As extensively reviewed in this volume, endocannabinoids (eCBs) are modified fatty acids produced by all cells in the body, as they are metabolites of arachidonoyl-containing membrane lipids (Pertwee 2015). The endocannabinoids are synthesized and released from cells where they have juxtacrine and paracrine actions on other cells and cellular elements. A few subtypes of cell surface receptors, described later in this review, mediate eCB actions (Pertwee 2015). Within the nervous system, the cannabinoid 1 (CB1) receptor is the predominant mediator of eCB effects. This is a G protein-coupled receptor (GPCR) that generally modulates cells in a way that inhibits neuronal activity and neurotransmitter release. The CB1 receptor is one of the most strongly expressed GPCRs in the central nervous system, with widespread expression across the brain (Hu and Mackie 2015; Allen Brain Atlas, http://mouse.brain-map.org). The eCB signaling CB receptor system has a huge variety of actions within the body and brain, including influences on cognitive functions, action control, emotional regulation, cardiovascular function, and regulation of the immune system (Pertwee 2015). Given the ubiquity of this signaling system in humans, understanding eCB-CB functions has the potential to provide new avenues for therapeutic intervention in a variety of disorders.

The name of the eCBs is derived from the fact that the CB receptors are the major target for delta9-tetrahydrocannabinol (THC), the major psychoactive "phytocannabinoid" compound in *Cannabis sativa* (and hence in marijuana, hashish, and other cannabis drugs) (Pertwee 2015). Cannabis derivatives are among the most widely used and abused drugs in the world, and thus it is important to determine how THC acts through its receptors to produce intoxication, cognitive impairment, habit formation, and abuse liability. This review will focus on phytocannabinoid, synthetic cannabinoid, eCB and CB receptor actions in the striatum, and the ways in which these actions contribute to behavior and alterations in behavior induced by drugs of abuse.

1.1 The Endocannabinoid System in the Striatum

1.1.1 Striatal Neurons

Over 90% of neurons within the striatum are the GABAergic medium spiny neurons (MSNs) and these are the only projection neurons (Tepper et al. 2007). Figure 1 depicts these MSNs and their synaptic inputs. MSNs of the direct pathway express D1 dopamine receptors and primarily project to the substantia nigra pars reticulata, whereas MSNs of the indirect pathway express D2 dopamine receptors and project to the external globus pallidus (Gerfen 1992). MSN activity is driven by cortical and intralaminar thalamic glutamatergic inputs. These inputs mostly synapse onto dendritic spine heads of the MSNs that are found on the outer two thirds of the dendrites, though some intralaminar thalamic inputs synapse onto dendritic shafts (Gerfen and Bolam 2010; Tepper and Bolam 2004). The striatum and nucleus accumbens are also home to a sparse population of GABAergic interneurons (Tepper et al. 2007, 2010) along with one class of large, aspiny, tonically active cholinergic interneurons (CINs) (Bennett and Wilson 1999; Kawaguchi 1992; Zhou et al. 2002). The population of GABAergic interneurons can be subdivided in a number of different ways, one of the simplest is based on the expression of specific proteins and their distinct firing patterns. Fast-spiking interneurons (FSIs) express the calcium-binding protein parvalbumin (Kawaguchi 1993; Kubota et al. 1993; Tepper and Bolam 2004; Tepper et al. 2010). FSIs are electrotonically coupled via gap junctions (Koos and Tepper 1999) and their activity is driven mostly by cortical inputs (Kita 1993), but FSIs also receive input from dopaminergic and cholinergic neurons and other striatal GABAergic neurons. FSIs make strong synaptic connections with MSNs of both pathways and act as mediators of feed-forward inhibition of MSN activity (Tepper et al. 2010). Low-threshold-spiking interneurons (LTSNs) express neuropeptide Y, somatostatin, and nitric oxide synthase (NOS). They also receive input from cortical, dopaminergic, and cholinergic neurons. These LTSNs also form synapses with MSNs, although not as abundantly as with FSIs (Kubota et al. 1993; Partridge et al. 2009; Tepper and Bolam 2004; Tepper et al. 2010). A third class of GABAergic interneurons are the calretinin-positive interneurons (Tepper et al. 2010). Although little is known of their role in striatal function,

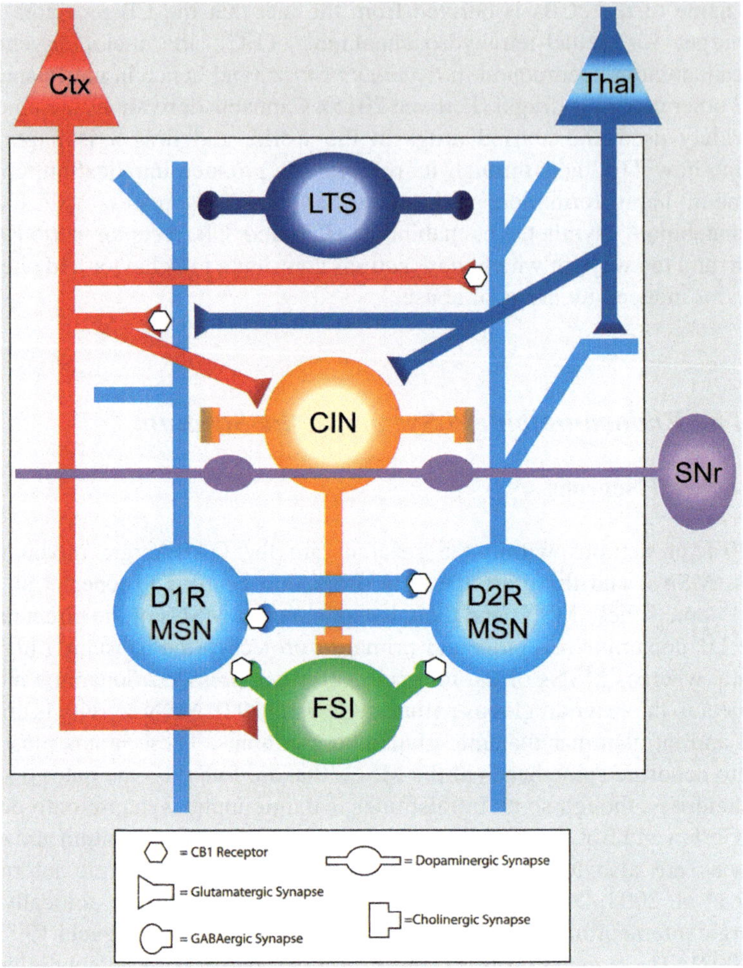

Fig. 1 CB1 cannabinoid receptors at striatal synapses. The striatum is home to a variety of synapse types. Excitatory synapses in the dorsal striatum arise from the neocortex and thalamus. The nucleus accumbens has additional prominent excitatory inputs that have their sources in the allocortex (including the hippocampus and amygdala). The GABAergic medium spiny neurons (MSNs) are the projection neurons of the striatum. MSNs of the direct pathway primarily express D1 dopamine receptors, whereas MSNs of the indirect pathway predominantly express D2 dopamine receptors. Fast-spiking interneurons (FSI) and low-threshold spiking interneurons (LTS) are GABAergic interneurons that modulate striatal output (in addition to other GABAergic interneuron types not pictured). Tonically active cholinergic interneurons (CIN) and dopaminergic input from the midbrain (substantia nigra pars compacta in the dorsal striatum and ventral tegmental area in the nucleus accumbens) further modulate striatal function. CB1 receptors are found at many types of glutamatergic and GABAergic synapses and are predominantly presynaptically expressed

they do not express D1 or D2 dopamine receptors (Petryszyn et al. 2014). The striatum can also be subdivided into intercalated striosome and matrix regions differentiated on their inputs and outputs and enrichment of specific proteins in one region over the other (Bolam et al. 1988; Davis and Puhl 2011; Desban et al. 1993; Eblen and Graybiel 1995; Kincaid and Wilson 1996; Langer and Graybiel 1989). Striosome MSNs receive more prominent input from associative and limbic brain regions and project to the substantia nigra pars compacta (Canales 2005; Gerfen 1992; Joel and Weiner 2000). Matrix MSNs receive input from sensorimotor regions and project predominantly to external globus pallidus (Berretta et al. 1997; Donoghue and Herkenham 1986; Ebrahimi et al. 1992; Gerfen 1992; Malach and Graybiel 1986).

1.2 Endocannabinoid System Expression in the Striatum

The striatum is home to some of the most abundant CNS expression of the CB1R cannabinoid receptor (CB1R) (Herkenham et al. 1991a, b; Tsou et al. 1998), best known as the receptor responsible for mediating the psychoactive properties of Δ9-THC (Matsuda et al. 1990; Monory et al. 2007). Anatomical techniques such as in situ hybridization, receptor autoradiography, and immunohistochemistry as well as functional studies have identified CB1R expression throughout the striatum. Figure 1 outlines the synapses at which CB1R has been detected thus far in the striatum. CB1R mRNA is highly expressed in MSNs (Julian et al. 2003; Matsuda et al. 1993). CB1R is found in both direct and indirect MSNs (Hohmann and Herkenham 2000; Oude Ophuis et al. 2014). Some of the densest expression of CB1R in the CNS is in the terminal fields of MSNs in the substantia nigra and globus pallidus (Herkenham et al. 1991a, b) and this expression is not due to its expression in intrinsic neurons in these brain regions, but is in the MSNs originating from the dorsal striatum (Egertova and Elphick 2000; Julian et al. 2003; Matsuda et al. 1993; Tsou et al. 1998). The highest levels of CB1R expression within the striatum are found in the dorsolateral striatum with a decreasing gradient of expression ventromedially (Egertova and Elphick 2000; Marsicano and Lutz 1999; Oude Ophuis et al. 2014; Tsou et al. 1998; Van Waes et al. 2012). CB1R receptors are found in both striosome and matrix subdivisions of the striatum where they may be more abundantly expressed within matrix MSNs (Fusco et al. 2004; Rodriguez et al. 2001).

CB1Rs are also found on cortical inputs to the dorsal striatum (Gerdeman and Lovinger 2001; Huang et al. 2001; Rodriguez et al. 2001; Wu et al. 2015) and nucleus accumbens (Robbe et al. 2001). CB1R mRNA is not found in the intralaminar nuclei of the thalamus, suggesting an absence of CB1R expression in thalamic inputs to the dorsal striatum, and functional data support this conclusion (Wu et al. 2015). Anatomical and functional studies demonstrate that FSIs abundantly express CB1R (Fusco et al. 2004; Hohmann and Herkenham 2000; Mathur et al. 2013; Uchigashima et al. 2007; Winters et al. 2012). There are conflicting reports regarding the degree to which LTSNs and CINs express CB1R (Azad et al.

2001; Fusco et al. 2004; Hohmann and Herkenham 2000). Calretinin interneurons are not reported to express CB1R (Fusco et al. 2004). One recent study found that FSIs were the only neurons within the nucleus accumbens to express CB1R (Winters et al. 2012). Dopaminergic inputs to the dorsal striatum from the substantia nigra pars compacta do not express CB1R (Herkenham et al. 1991a, b; Mailleux and Vanderhaeghen 1992b; Matsuda et al. 1993; Romero et al. 1997), although there is the possibility that they do transiently during development as CB1R expresses in cultured mesencephalic TH-positive neurons (Hernandez et al. 2000). VTA dopaminergic inputs to nucleus accumbens also do not express CB1R (Mailleux and Vanderhaeghen 1992b; Matsuda et al. 1993; Romero et al. 1997). CB1R is localized predominantly on presynaptic elements in the striatum and is most abundant at inhibitory synapses, such as MSN and FSI terminals, although it is found on glutamatergic terminals as well (Narushima et al. 2006a, b; Rodriguez et al. 2001; Uchigashima et al. 2007). CB1R is also found on glutamatergic terminals in the nucleus accumbens (Matyas et al. 2007; Pickel et al. 2004). In addition to its prevalence on presynaptic elements, CB1R may be found postsynaptically to some degree as well (Kofalvi et al. 2005; Pickel et al. 2006).

One study found that CB1R mRNA levels in the striatum peak at postnatal day 25 and then decrease into adulthood (Van Waes et al. 2012). The highest CB1R mRNA levels in the striatum were found in areas receiving more sensorimotor input, whereas levels were lower in associative and limbic areas. Interestingly, the striatal areas with most CB1R mRNA expression received inputs from cortical regions with lower CB1R expression and vice versa. The authors of this study suggest that CB1Rs in sensorimotor areas are mostly involved in modulating MSN collaterals, whereas in limbic and associative regions, they are more involved in regulating glutamatergic input and interneuron neurotransmission. However, it is important to know if protein expression patterns align with this mRNA expression pattern. In addition to CB1R, other cannabinoid receptors include the type 2 cannabinoid receptor CB2R as well as the ionotropic TRPV1 receptor. CB2R expression in neurons in the striatum is controversial (Atwood and Mackie 2010), and reports of its presence may be due to expression in microglia (Concannon et al. 2015; Palazuelos et al. 2009; Sagredo et al. 2009) or blood vessels (Dowie et al. 2014). TRPV1 is expressed in both the dorsal striatum and nucleus accumbens (Grueter et al. 2010; Kauer and Gibson 2009; Koles et al. 2013; Roberts et al. 2004; Starowicz et al. 2008). TRPV1 expression on inputs to the striatum may be restricted to juvenile stages of development, suggesting that studies of TRPV1 signaling in older animals may be of postsynaptic TRPV1 (Koles et al. 2013).

The endocannabinoid system consists of not just cannabinoid receptors but also the enzymes that are responsible for the production and degradation of endocannabinoids, the endogenous ligands for the cannabinoid receptors and these enzymes are also present in the striatum. Endocannabinoids are produced on demand by biosynthetic enzymes that cleave plasma membrane lipid precursors to form them (Alger 2002; Wilson and Nicoll 2002). There are two primary endocannabinoids. Anandamide (AEA) is an endogenous ligand for both CB1R as well as TRPV1 with little efficacy at CB2R (Devane et al. 1992; Howlett et al.

2002; Zygmunt et al. 1999). 2-arachidonylglycerol (2-AG) is the second major endocannabinoid and it appears to be the primary mediator of endocannabinoid retrograde signaling in the CNS (Gao et al. 2010; Gonsiorek et al. 2000; Howlett et al. 2002; Mechoulam et al. 1995; Stella et al. 1997; Sugiura et al. 1995; Tanimura et al. 2010). 2-AG acts at CB1R and CB2R (Gonsiorek et al. 2000; Howlett et al. 2002). N-acyl-phosphatidylethanolamine-hydrolyzing phospholipase D (NAPE-PLD) is a calcium-sensitive enzyme that can produce the endocannabinoid AEA and is thought to be primarily responsible for its production (Ligresti et al. 2005; Okamoto et al. 2004, 2007; Piomelli et al. 2000). Other biosynthetic pathways do exist to produce AEA, such as through α/β-hydrolase4, or phospholipase A2 or phospholipase C (PLC) pathways (Ligresti et al. 2005; Liu et al. 2006; Simon and Cravatt 2006, 2008, 2010). There are little data as to the subcellular expression of NAPE-PLD in neurons of the striatum, but some evidence indicates that it has at least has some level of expression (Castelli et al. 2007; Ferrer et al. 2007; Marco et al. 2014). In other brain regions, NAPE-PLD is predominantly a presynaptically localized enzyme (Nyilas et al. 2008; Pickel et al. 2012). AEA may be released by stimulating striatal neurons in culture and in vivo (Di Marzo et al. 1994; Giuffrida et al. 1999). 2-AG on the other hand is formed by sn-1-specific diacylglycerol lipases (DGL-α and DGL-β) (Bisogno et al. 2003). DGLα is primarily responsible for producing the 2-AG that is involved in modulation of synaptic transmission, whereas DGLβ may have little role (Gao et al. 2010; Tanimura et al. 2010). Striatal DGL-α expression is almost exclusively expressed postsynaptically and is found most abundantly expressed on dendrites and especially dendritic spines in the striatum, consistent with findings in other brain regions (Hashimotodani et al. 2013; Katona 2009; Matyas et al. 2007; Shonesy et al. 2013, 2015; Uchigashima et al. 2007). Some functional evidence exists suggesting that endocannabinoid release following production occurs via an endocannabinoid membrane transporter (AMT) in the striatum (Adermark and Lovinger 2007b; Ronesi et al. 2004; Seif et al. 2011). Termination of endocannabinoid signaling is accomplished via uptake by the AMT (Beltramo et al. 1997; Fegley et al. 2004a, b; Gerdeman et al. 2002; Hillard et al. 1997; Robbe et al. 2002b) and subsequent hydrolysis by intracellular enzymes. Fatty-acid amide hydrolase (FAAH) degrades AEA (Cravatt et al. 1996; Ueda et al. 1995), whereas monoacylglycerol lipase (MGL) degrades 2-AG (Dinh et al. 2002a, 2002b). Numerous studies demonstrate the anatomical and functional expression of these enzymes in the striatum (Castelli et al. 2007; Gubellini et al. 2002; Maccarrone et al. 2001, 2003; Malone et al. 2008; Marco et al. 2014; Matyas et al. 2007; Micale et al. 2009; Seif et al. 2011; Soria-Gomez et al. 2007). In other brain regions it has been established that FAAH is predominantly a postsynaptic enzyme, whereas MGL is presynaptically localized (Gulyas et al. 2004). Although there are little data to suggest their subcellular distribution in the striatum, it is likely that these enzymes are expressed in a similar manner.

2 Cannabinoid Receptor-Mediated Regulation of Synaptic Transmission

2.1 *Effects of Cannabinoids on GABAergic and Glutamatergic Synaptic Transmission*

CB1R is a G protein-coupled receptor (GPCR) that primarily couples to the Gi/o class of G proteins, and as such its activation leads to an inhibition of adenylyl cyclase and subsequent decrease in cellular cAMP levels (Howlett et al. 2002). CB1R activation also inhibits voltage-gated calcium channels and activates G protein-coupled inwardly rectifying potassium channels (Caulfield and Brown 1992; Mackie et al. 1993, 1995; Mackie and Hille 1992; Shen and Thayer 1998; Twitchell et al. 1997). Cumulatively, these effects on intracellular signaling result in reduced cellular excitability and, due to its proximity to synaptic terminals (Nyiri et al. 2005), a reduction in the probability of neurotransmitter release (Shen et al. 1996). This ability to inhibit neurotransmission allows both exogenous cannabinoid agonists (such as Δ9-THC) and endogenous cannabinoids to have a profound impact on neuronal communication. Since CB1R is found on both glutamatergic and GABAergic terminals in the striatum and other brain regions, its activation can suppress both inhibitory and excitatory synaptic transmission.

Despite earlier hypotheses that CB1R receptor activation in the striatum would increase GABAergic transmission in the striatum via reduced GABA uptake (Glass et al. 1997), the first evidence that cannabinoids could influence striatal fast synaptic transmission demonstrated that synthetic cannabinoid agonists, such as WIN55,212-2, actually reduce the magnitude of inhibitory postsynaptic currents (IPSCs) recorded in MSNs (Szabo et al. 1998). The inhibitory effect of WIN55,212-2 was prevented by the CB1R receptor antagonist SR141617A but WIN55,212-2 failed to inhibit calcium currents in MSNs, suggesting that presynaptic CB1R receptors mediated the inhibition of GABAergic transmission. Subsequently others demonstrated that CB1R agonists had similar effects on inhibitory transmission in nucleus accumbens MSNs (Hoffman et al. 2003; Manzoni and Bockaert 2001). Here it was shown that CB1R receptor activation reduces evoked IPSC amplitudes and changes the frequency, but not the amplitude of miniature IPSCs, also suggesting presynaptic CB1R receptors as the mediators of the cannabinoid ligands' effects. The first studies of the effects of cannabinoids on glutamatergic transmission demonstrated that exogenously applied AEA and synthetic cannabinoids also decrease excitatory postsynaptic current (EPSC) magnitudes in MSNs of the dorsal striatum via CB1R receptor activation (Gerdeman and Lovinger 2001; Gerdeman et al. 2002; Huang et al. 2001). CB1R receptor activation increases paired-pulse ratios and coefficients of variation and reduces the frequency, but not amplitude, of spontaneous EPSCs and evoked asynchronous EPSCs all indicative of a presynaptic locus for inhibition of neurotransmission (Gerdeman and Lovinger 2001; Huang et al. 2001). Once again, it was also demonstrated that CB1R receptor activation via synthetic cannabinoids has similar inhibitory effects on excitatory transmission in nucleus accumbens MSNs (Robbe et al. 2001, 2002b).

Following these studies demonstrating that exogenous application of cannabinoid ligands could inhibit synaptic transmission in both the dorsal striatum and in nucleus accumbens, two studies published very close together demonstrated that endocannabinoids produced via specific stimulation paradigms could also inhibit neurotransmission as well. In the dorsal striatum a pairing of high-frequency stimulations (four 1 s 100 Hz stimulations) with postsynaptic depolarizations induces presynaptically expressed synaptic depression (Choi and Lovinger 1997a, b) that was later demonstrated to be prevented by a CB1R receptor antagonist and absent in CB1R null mutant mice (Gerdeman et al. 2002). In the nucleus accumbens, moderate-frequency stimulation (10 min, 13 Hz) also induces presynaptically expressed synaptic depression that is also blocked by a CB1R receptor antagonist and is absent in CB1R null mice. Furthermore the authors of this study demonstrated that a CB1R agonist occluded the inhibitory effects of the stimulation (Robbe et al. 2002b). In addition both studies demonstrated that the AMT blocker AM404 could induce or enhance synaptic depression, further supporting the evidence for endocannabinoid-mediated inhibition of neurotransmission (Gerdeman et al. 2002; Robbe et al. 2002b). Furthermore it was realized that the synaptic depression elicited by high-frequency stimulation in field recordings that was discovered in earlier studies was actually cannabinoid receptor-mediated (Calabresi et al. 1993, 1994; Lovinger et al. 1993; Walsh 1993).

2.2 *Types of Synaptic Depression Mediated by Cannabinoid Receptors*

There are different forms of synaptic depression that occur. Short-term depression (STD) in response to specific postsynaptic activation patterns, stimulation paradigms, or receptor ligands lasts for seconds to minutes and then returns to pre-stimulation or pre-ligand neurotransmission levels. On the other hand, long-term depression (LTD) lasts from minutes to hours or possibly longer. LTD can be further subdivided into two different forms of LTD. Static LTD is resistant to reversal by a receptor antagonist and is indicative of long-lasting changes in synaptic signaling elements independent of ongoing receptor activation (Atwood et al. 2014b). Labile LTD is long-lasting synaptic depression that is readily reversed by a receptor antagonist and thus is dependent on either persistent receptor activation by a ligand or is mediated by sustained receptor activation independent of the continuous presence of a receptor ligand (Atwood et al. 2014b). The differences between STD and labile LTD are nuanced, but it is clear that for certain receptors such as the Group II mGluRs, at certain synapses, LTD mediated by these receptors can be reversed only so long as the antagonist is present in the preparation (Lodge et al. 2013). Removal of the antagonist restores synaptic depression back to its pre-antagonist application levels in what has been referred to as "reversibly reversible LTD" (Lodge et al. 2013), but can be classified as an example of labile LTD.

It is clear that cannabinoid receptors can induce both STD and LTD (both reversible and irreversible) in multiple brain areas. The studies identifying endocannabinoid-mediated synaptic depression described above determined that endocannabinoids induce LTD of excitatory transmission (eCB-LTD). CB1R receptors also can induce LTD of inhibitory transmission (eCB-iLTD) as first evidenced by Adermark and coworkers (Adermark and Lovinger 2007b, 2009; Adermark et al. 2009) in the dorsal striatum; however, the same has not been demonstrated in the nucleus accumbens. Dorsal striatal eCB-iLTD occurs at both FSI-MSN and MSN-MSN synapses, although different mechanisms induce LTD at each type of synapse (Mathur et al. 2013). Nonetheless, it is now well established that CB1R receptor activation can induce LTD in the striatum. On the other hand, other studies demonstrated that CB1R receptor activation can also induce STD. Cannabinoid receptor-mediated depolarization-induced suppression of inhibition (DSI) and excitation (DSE) were originally described in the hippocampus and cerebellum (Kreitzer and Regehr 2001; Wilson and Nicoll 2001). Cannabinoid receptor-mediated DSE may occur at glutamatergic inputs onto MSNs (Narushima et al. 2006a; Uchigashima et al. 2007), although others have failed to observe this (Gerdeman et al. 2002; Kreitzer and Malenka 2005; Yin and Lovinger 2006). DSI weakly occurs at MSN-MSN synapses in the dorsal striatum (Freiman et al. 2006). DSI at FSI-MSN synapses has been identified by some, but not others in the dorsal striatum (Freiman et al. 2006; Narushima et al. 2006b, 2007; Uchigashima et al. 2007). CB1R-mediated DSI does occur as FSI-MSN synapses in nucleus accumbens (Winters et al. 2012). DSE and DSI utilize differing mechanisms (discussed below). It is clear though that the CB1R receptor is not restricted to mediating one form of synaptic depression. Even at the same synapses, different forms of synaptic depression can exist. Both LTD and STD (DSE) occur at excitatory inputs onto dorsal striatal MSNs (Gerdeman et al. 2002; Narushima et al. 2006a, Yin et al. 2006), and LTD is only reversible (labile LTD) if the antagonist is applied shortly after the LTD-inducing stimulation after which it becomes static LTD (Ronesi et al. 2004). It is also likely that CB1R receptor activation alone is insufficient to induce LTD at these synapses, but requires other signaling components (Ronesi et al. 2004). Excitatory synapses appear to be less sensitive to cannabinoids than inhibitory synapses. Low-frequency stimulation (LFS) induces LTD of only inhibitory transmission onto dorsal striatal MSNs, and inhibitory synapses respond more to lower concentrations of CB1R agonist (Adermark and Lovinger 2009). CB1R-mediated LTD of excitatory inputs onto dorsal striatal and nucleus accumbens MSNs is sometimes irreversible and sometimes reversible by a CB1R receptor antagonist depending on the experimental conditions (Adermark and Lovinger 2007b, 2009; Adermark et al. 2009; Huang et al. 2001; Kreitzer and Malenka 2005, 2007; Robbe et al. 2001; Yin et al. 2006, 2008). CB1R-mediated synaptic depression at inhibitory synapses is also irreversible in some experimental conditions and reversible in others (Adermark and Lovinger 2007b, 2009; Adermark et al. 2009; Freiman et al. 2006).

Whether or not LTD is reversible (labile LTD) or not (static LTD) may depend upon a number of factors. CB1R activation alone is usually insufficient to induce eCB-LTD (Huang et al. 2001; Ronesi et al. 2004; Singla et al. 2007; Yin and Lovinger 2006). One determination of whether CB1R activation induces LTD or not may be the level of presynaptic activity. For example, in stimulating two pathway inputs to a single MSN, the pathway that received 5 min 1 Hz stimulation during application of WIN55,212-2 displayed LTD, whereas the pathway with no stimulation during agonist application did not display LTD. Similar results were found when one pathway received constant 0.05 Hz stim rather than 1 Hz for 5 min. HFS-induced eCB-LTD occurred in a pathway that received 0.05 Hz stimulation, but did not occur in a pathway that received no stimulation for 10 min after the induction protocol (Singla et al. 2007). Indeed postsynaptic loading with AEA induces LTD at excitatory synapses onto dorsal striatal MSNs that also requires presynaptic activity, as does eCB-LTD induced by pharmacological activation of postsynaptic L-type VGCCs (Adermark and Lovinger 2007a, b, 2009; Adermark et al. 2009). On the other hand, AEA loading-induced eCB-iLTD onto dorsal striatal MSNs does not require presynaptic activity, but it is enhanced by it (Adermark and Lovinger 2007b). eCB-iLTD induced by an L-type VGCC activator also occurs independently of presynaptic activity (Adermark et al. 2009). Cannabinoid modulation of net striatal output is the product of the balance of inhibitory and excitatory cannabinoid receptor-mediated synaptic plasticity. For instance, low- to moderate-frequency stimulation induces an LTP-like increase in neuronal output due to endocannabinoid-LTD of inhibitory synapses, whereas HFS induces net LTD by inhibiting excitatory synapses (Adermark 2011; Adermark and Lovinger 2009). The type of depression CB1R mediates is likely a result of differing synaptic environments. The composition and activation state of the presynaptic intracellular signaling pathways that couple CB1R to its mediators of neurotransmitter release inhibition are likely candidates for determining the type of synaptic depression that occurs. This likely explains why LTD is inducible at inhibitory synapses without needing presynaptic activity, whereas the opposite is true for excitatory synapses (Adermark and Lovinger 2007a; Adermark et al. 2009; Singla et al. 2007). Additionally a certain length of CB1R activation may be necessary for the receptor to engage the signaling pathways that are responsible for LTD expression rather than the typical STD. The composition of the machinery responsible for producing endocannabinoids (the type and levels of production) may differ from synapse to synapse and therefore may play some role in determining whether LTD or STD is induced. This could also explain why different synapses require different stimulation patterns to induce CB1R-mediated plasticity. This may also determine to some degree the role of endocannabinoid signaling in filtering striatal transmission. Some data suggest that endocannabinoids only inhibit neurotransmitter release from less active terminals in the dorsal striatum and nucleus accumbens core (Wang et al. 2012; Wong et al. 2015), thus acting as a high-pass filtering mechanism, though in dopamine-depleted mice, this filtering capacity is lost (Wong et al. 2015).

Multiple types of stimulation paradigms can induce endocannabinoid-mediated LTD. As indicated above 100 Hz, HFS paired with postsynaptic depolarization induces LTD of excitatory synapses in the dorsal striatum (Atwood et al. 2014a, b; Choi and Lovinger 1997a, b; Gerdeman et al. 2002; Kreitzer and Malenka 2007; Ronesi et al. 2004; Wang et al. 2006; Yin et al. 2006), and 10 min of 13 Hz MFS induces LTD of excitatory synapses in the nucleus accumbens (Mato et al. 2005, 2008; Robbe et al. 2002b). A multitude of studies have demonstrated that in field recordings HFS is sufficient to induce LTD that is cannabinoid-mediated, similar to the HFS pairing with postsynaptic depolarization utilized in whole-cell recordings (e.g., see Gerdeman et al. 2002). In addition 60–80 s of 1 Hz LFS induces eCB-iLTD in MSNs in the dorsal striatum (Adermark and Lovinger 2009; Mathur et al. 2013). HFS can also induce iLTD when MSNs are held in a slightly depolarized "up-state" (Mathur et al. 2013). 20 Hz stimulation paired with postsynaptic depolarization induces LTD in dorsal striatal indirect pathway MSNs (Lerner et al. 2010; Lerner and Kreitzer 2012). 20 min of LFS 4 Hz stimulation in nucleus accumbens induces CB1R-dependent LTD as well (Burattini et al. 2014). In addition, 5 min of MFS 10 Hz stimulation induces CB1-mediated LTD in the nucleus accumbens (Hoffman et al. 2003; Mato et al. 2005) and in the dorsal striatum where it is presynaptically expressed static LTD and is not occlusive with HFS-mediated eCB-LTD (Ronesi and Lovinger 2005). Another study that used this same LFS paradigm in the nucleus accumbens core found LTD only in indirect pathway MSNs and found that a CB1R receptor antagonist only partially blocked this LTD (Grueter et al. 2010). Instead a complete block was achieved when the CB1R antagonist was paired with a TRPV1 antagonist. TRPV1 is one of the other targets of AEA (Smart and Jerman 2000; Zygmunt et al. 1999). LFS-LTD was reduced in the TRPV1 null mutant mouse (Grueter et al. 2010). Unlike the CB1R receptor-mediated LTD component, the TRPV1 component was postsynaptically expressed. This is consistent with a recent study that found that presynaptic TRPV1 receptors stimulate glutamate release, but this effect is absent after postnatal day 27 (Koles et al. 2013). The presynaptic TRPV1 receptors do not cross talk with presynaptic CB1R signaling as well.

Spike-timing protocols have also been utilized to induce cannabinoid receptor-mediated plasticity in the striatum. An electrical and an optogenetic negative-timing spike-timing-dependent plasticity protocol (postsynaptic depolarization prior to presynaptic stimulation) can induce presynaptic LTD of glutamatergic inputs onto MSNs that requires CB1R receptor activation (Shen et al. 2008; Wu et al. 2015). The conditions that determine whether or not LTD is induced using these protocols are discussed below. The direction of the spike timing can change depending on whether GABAergic transmission is blocked. In the absence of $GABA_A$ receptor antagonists a positive-timing paradigm (presynaptic stimulation precedes postsynaptic depolarization) leads to CB1R-mediated LTD (Fino et al. 2005, 2010). A recent study found that altering the number of stimulation pairings using STDP could produce both LTD at pairings greater than 100, but at low pairing numbers (5–10 pairings), the stimulation actually induced long-term potentiation (LTP) (Cui et al. 2015). The LTP required CB1R and TRPV1 activation as well as postsynaptic endocannabinoid production. eCB-LTD and eCB-LTP could be sequentially induced in the same MSN.

2.3 Local Effects of Cannabinoids on Striatal Dopamine Release

It is well established that CB1Rs in the midbrain can alter synaptic transmission onto dopaminergic neurons that project to the striatum, thus modulating dopamine release in the striatum (French et al. 1997; Gardner 2005; Lupica et al. 2004; Parsons and Hurd 2015). In addition to their effects on glutamate and GABAergic transmission, there is some limited evidence that cannabinoids can also modulate striatal dopaminergic transmission locally within the striatum and not just indirectly via midbrain CB1Rs. CB1R agonists can inhibit electrically evoked dopamine release in striatal slices, and their effects were blocked by CB1R antagonists (Cadogan et al. 1997). CB1R activation had no effect on dopamine uptake or acetylcholine release. Similar findings were obtained when dopamine release was induced via NMDAR activation (Kathmann et al. 1999). However, Szabo and colleagues (Szabo et al. 1999) found conflicting results, showing that CB1R agonists and a CB1R antagonist had no effect on electrically evoked dopamine release in the dorsal striatum and nucleus accumbens. These findings were later reconciled to some degree by findings that a CB1R agonist indeed does not inhibit dopamine release in response to single electrical stimuli, but dopamine release elicited by trains of stimulation (30 pulses 10 Hz) was inhibited by a CB1R agonist (Sidlo et al. 2008). This effect was blocked by a CB1R antagonist but persists in the presence of a dopamine transporter inhibitor and occurs equally in guinea pig, mouse, and rat. These data are all curious given the absence of CB1Rs on striatal dopaminergic inputs (Herkenham et al. 1991a, b; Mailleux and Vanderhaeghen 1992b; Matsuda et al. 1993; Romero et al. 1997). Therefore the effects of cannabinoids are likely indirect. In support of this, additional experiments suggested a role of $GABA_AR$ signaling as well as hydrogen peroxide and KATP channels (Sidlo et al. 2008). The authors of this study hypothesized that CB1R activation reduces GABA release, which decreases GABAAR activity to increase H_2O_2 production. This would cause more opening of presynaptic KATP channels on dopaminergic terminals, which reduces dopamine release. Additional work will be needed to conclusively determine if this is indeed the mechanism. One additional minor note is that TRPV1 activation enhances dopamine release in the striatum, although its ability to do so decreases during development. By postnatal day 27, it has no impact (Koles et al. 2013).

3 Mechanisms of Cannabinoid Receptor-Mediated Synaptic Plasticity

3.1 Postsynaptic Production of Endocannabinoids

Endocannabinoid signaling as indicated above involves postsynaptic production of endocannabinoids that transmit in a retrograde fashion. A number of different mechanisms exist in the striatum that initiate the formation of endocannabinoids by their biosynthetic enzymes (Fig. 2). It is probable that different types of synapses utilize distinct pathways leading to the production and release of endocannabinoids. Ultimately endocannabinoids are formed by enzymatic cleavage of membrane phospholipids by specific phospholipases (such as NAPE-PLD), although the identity of these enzymes and the routes by which they are activated vary. It is well established, however, that increases in postsynaptic calcium levels and activation of specific GPCRs are primarily involved in initiating the endocannabinoid production process, although other molecular players may contribute.

Increases in postsynaptic calcium levels are important for the expression of some forms of eCB-induced plasticity, such as some forms of eCB-LTD. Postsynaptic calcium chelation prevents eCB-LTD in the dorsal striatum and in the nucleus accumbens (Choi and Lovinger 1997a, b; Robbe et al. 2002b; Yin et al. 2008; Yin and Lovinger 2006). In the striatum the predominant source of postsynaptic calcium are the L-type VGCCs that are localized postsynaptically at synapses onto MSNs (Olson et al. 2005) and have no impact on presynaptic function (Lovinger et al. 1994; Kupferschmidt et al. 2015). L-type VGCC blockade prevents eCB-LTD (Calabresi et al. 1992b, 1993; Choi and Lovinger 1997a, b; Kreitzer and Malenka 2005; Wang et al. 2006). The most convincing data for a role of L-type VGCCs in eCB-LTD though is that this LTD does not occur in a mutant mouse that is null for the low-threshold, CaV1.3 form of the L-type channel VGCC (Wang et al. 2006). Furthermore, direct pharmacological activation of L-type VGCCs can induce eCB-LTD as long as it is paired with presynaptic activity (Adermark and Lovinger 2007a). A recent study in the nucleus accumbens demonstrated that 5HT2R activation induces LTD that requires L-type VGCCs. 20 min 4 Hz stimulation induces 5-HT release, which activates postsynaptic 5HT2Rs to increase L-type VGCC activity to induce 2-AG release (Burattini et al. 2014). eCB-mediated long-lasting disinhibition (DLL) of striatal output is also prevented by an L-type channel blocker, suggesting that L-type VGCCs are important for eCB-iLTD as well (Adermark 2011). Indeed these channels are critical for the "up-state" form of eCB-iLTD, but interestingly not the "down-state" form of eCB-iLTD that occurs when MSNs are held at more hyperpolarized potentials (Mathur et al. 2013). Blocking L-type VGCCs converted the "up-state" eCB-iLTD to become like the "down-state" eCB-iLTD. Additional experiments (discussed below) determined that the "up-state" L-type VGCC-dependent eCB-iLTD is mediated by 2-AG, whereas the "down-state" eCB-iLTD is mediated by AEA and is calcium independent. Another source of postsynaptic calcium are the intracellular stores (Yin and

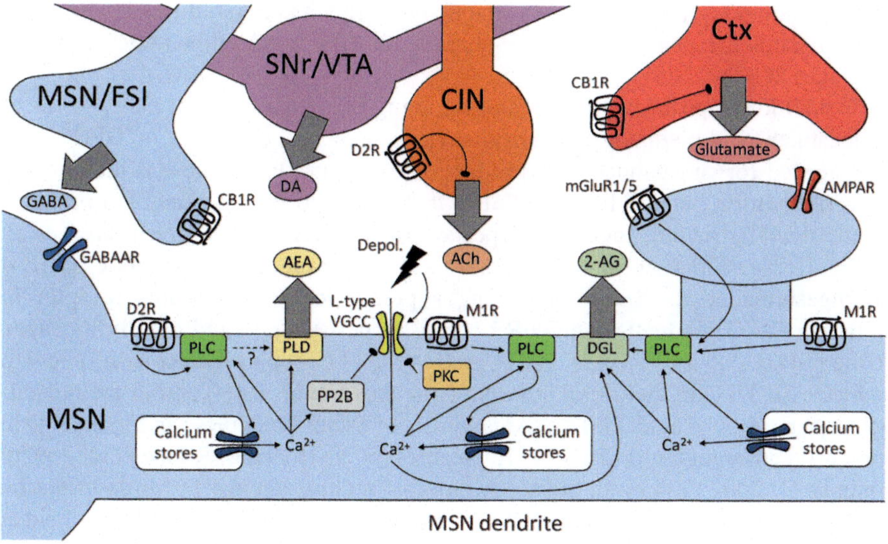

Fig. 2 Endocannabinoid production and signaling in the striatum. Endocannabinoids are produced from postsynaptic neurons through a number of parallel and intersecting pathways. 2-arachidonoyl glycerol (2-AG) can be produced via activation of Gq-coupled G protein-coupled receptors, such as Group I metabotropic glutamate receptors (mGluR1/mGluR5) or M1 muscarinic receptors (M1Rs). Activation of phospholipase C (PLC) and diacylglycerol lipase (DGL) along with increased intracellular calcium, via IP_3 receptor-regulated intracellular stores or through L-type voltage-gated calcium channels (VGCC), can result in 2-AG production. Activation of postsynaptic D2 dopamine receptors may also result in anandamide (AEA) production via PLC and phospholipase D (PLD) activation coupled with increased intracellular calcium levels. M1R may exert an inhibitory influence on endocannabinoid production via indirect inhibition of VGCCs. D2Rs may relieve this inhibition through activation of calcineurin (PP2B) or by inhibiting acetylcholine release from CINs. Presynaptic CB1 cannabinoid receptors are targeted by the postsynaptically released endocannabinoids to inhibit GABA and glutamate release

Lovinger 2006). HFS-induced eCB-LTD in indirect pathway MSNs requires some calcium from L-type VGCCs and RyR stores (Lerner and Kreitzer 2012). On the other hand, STDP LTD requires postsynaptic Ca^{2+} arising from L-type VGCCs and IP3R-regulated stores, but not RyR stores (Fino et al. 2010). Blockade of Ca^{2+} stores with thapsigargin and a RyR antagonist, but not an IP3R antagonist, prevents MFS (13 Hz 10 min) eCB-LTD in nucleus accumbens (Robbe et al. 2002b).

Like the "down-state" eCB-iLTD described above, other forms of eCB-mediated plasticity in the striatum appear to operate independently of L-type VGCCs. L-type VGCC blockade does not prevent MFS (13 Hz 10 min) eCB-LTD in nucleus accumbens (Mato et al. 2008), although as indicated above, this form of LTD does still require postsynaptic calcium from RyR stores (Robbe et al. 2002b). Other forms of LTD are even completely independent of postsynaptic calcium. LFS (20 Hz)-LTD in indirect MSNs does not require postsynaptic Ca^{2+} (Lerner and Kreitzer 2012). The MFS-induced (5 min 10 Hz stimulation) eCB-LTD in the dorsal striatum is not prevented by L-type VGCC antagonists or intracellular calcium chelation (Ronesi and Lovinger 2005). AEA loading LTD of inhibitory transmission does not require postsynaptic calcium signaling, but this is likely due to bypassing the need for calcium-stimulated AEA production (Adermark and Lovinger 2007b, 2009).

In addition to postsynaptic calcium signaling being requisite for many forms of endocannabinoid-mediated plasticity, activation of specific types of GPCRs are also needed for endocannabinoid production to occur. One of these is the Group I class of mGluRs (mGluR1 and mGluR5) that couple to the Gq class of G proteins. In the dorsal striatum, mGluR5 is postsynaptically expressed (Narushima et al. 2007; Uchigashima et al. 2007) and is localized near postsynaptic DGL-α (Uchigashima et al. 2007), suggesting a predominant role of this receptor in initiating 2-AG production. mGluR5 is mostly perisynaptic and closely localized to the PSD (Uchigashima et al. 2007). Activation of Group I mGluRs is sufficient to induce eCB-LTD in the dorsal striatum (Chepkova et al. 2009) and in the nucleus accumbens (Robbe et al. 2002b). Antagonists of group I mGluRs can block LTD in the dorsal striatum (Adermark 2011; Calabresi et al. 1999a, b; Robbe et al. 2002b; Sung et al. 2001; Yin et al. 2008). A Group I mGluR agonist has little effect on transmission in neurons held at −70 mV, but produces eCB-LTD in neurons held at −50 mV, suggesting that there might be state dependence to mGluR-mediated eCB-LTD (Kreitzer and Malenka 2005). Indeed, blocking L-type VGCCs prevents mGluR1/5-induced eCB-LTD in the dorsal striatum (Kreitzer and Malenka 2005). This suggests that, for at least this form of eCB-LTD, a convergence of G protein and calcium signaling is necessary. In addition, via their Gq coupling, mGluRs can stimulate release of calcium from internal stores, linking mGluR activation with calcium signaling in eCB production. mGluR1/5 activation in nucleus accumbens occludes LFS-induced eCB-LTD (Grueter et al. 2010). eCB-LTD can often be blocked by antagonists of Group I mGluRs. mGluR1/5 antagonists block eCB-LTD of excitatory transmission induced by MFS (13 Hz 10 min) in nucleus accumbens (Robbe et al. 2002b). HFS-induced eCB-LTD at excitatory inputs to indirect pathway MSNs in the dorsal striatum is mGluR1/5-dependent (Kreitzer and Malenka 2007; Lerner and Kreitzer 2012; Sung et al. 2001) as is MFS/LFS (20 Hz)-induced eCB-LTD onto the same neurons (Lerner and Kreitzer 2012). Neurotensin-induced eCB-LTD in the dorsal striatum is blocked by a Group I mGluR antagonist (Yin et al. 2008). On the other hand, eCB-LTD induced by L-type VGCC activators occurs independently of mGluRs, suggesting that with sufficient postsynaptic calcium levels, the need for mGluR to stimulate endocannabinoid release is bypassed (Adermark and Lovinger 2007a). mGluR1/5 activation could induce eCB-LTD by stimulating calcium release from intracellular stores or also by more directly activating endocannabinoid-producing enzymes. eCB-LTD induced by mGluR I activation occurs at corticostriatal but not thalamostriatal inputs (Wu et al. 2015). eCB-iLTD in DLS also requires Group I mGluRs (Adermark and Lovinger 2009), and the long-lasting disinhibition (DLL) of striatal output that results from this eCB-iLTD also requires mGluR1/5 signaling (Adermark 2011). One recent study found that mGluR I agonist-induced eCB-LTD occurs in the nucleus accumbens in adolescent, but not adult mice, and there is no LTD in either age in DLS, suggesting that there may be an age dependence to eCB-LTD (Zhang et al. 2015). However, the dorsal striatal data could explain an age dependence of DLL (Adermark and Lovinger 2009; Clarke and Adermark 2010; Zhang et al. 2015). An age dependence of eCB-LTD has been observed by others as well, but contrasts to some degree with others' findings (Ade and Lovinger 2007; Choi and Lovinger 1997a, b). These age-dependent differences may be explained by reduced CB1R and mGluR1 expression in nucleus accumbens and DLS in adults (Ade and Lovinger 2007; Mailleux and Vanderhaeghen 1992a; Van Waes et al. 2012; Zhang et al. 2015) and increased AEA tone in DLS of adults (Zhang et al. 2015). Group I mGluR activation induces CB1R-mediated depression of FSI-MSN and MSN-MSN synapses (Freiman et al. 2006). Interestingly, Group I mGluR activation is reported to preferentially induce 2-AG production (Jung et al. 2005), fitting with the data that suggest that the 2-AG-mediated "up-state" eCB-iLTD may be induced via mGluR1/5 activation (Mathur et al. 2013). eCB STD also involves Group I mGluRs. DSE in MSNs requires mGluR1/5 activation (Narushima et al. 2006a; Uchigashima et al. 2007), and DSI that occurs at FSI inputs to MSNs is enhanced by mGluR1/5 activation as is DSI at MSN-MSN synapses (Freiman et al. 2006; Narushima et al. 2006b).

mGluR1 and mGluR5 are the two members of the Group I class of mGluRs. Which one is responsible for mediating endocannabinoid production in the striatum and nucleus accumbens? STDP induces presynaptic cannabinoid LTD that requires mGluR5 receptor activation (Fino et al. 2010; Shen et al. 2008). mGluR5 is involved in Group I mGluR-driven eCB-LTD in the dorsal striatum of adolescent and adult mice, whereas mGluR1 is only involved in older mice (Chepkova et al. 2009). This may explain other data from older animals that HFS-induced eCB-LTD requires mGluR1, but not mGluR5 (Gubellini et al. 2001). Disruption of mGluR5 regulation by knocking out the PSD protein SAPAP allows for eCB-LTD in the dorsal striatum to be more readily inducible (Chen et al. 2011b). In the nucleus accumbens, an mGluR5-selective antagonist blocks eCB-LTD induced by multiple types of stimulation protocols (Grueter et al. 2010; Robbe et al. 2002b). Disruption of mGluR5 surface expression (via an in vivo cocaine exposure) in nucleus accumbens disrupts eCB-LTD (Fourgeaud et al. 2004). mGluR5 is likely the mGluR involved in DSE in the striatum (Uchigashima et al. 2007). There are also some forms of eCB-LTD that appear to be independent of mGluR activation. Clearly, loading a postsynaptic cell with AEA bypasses the need to form endocannabinoids and as such AEA loading-induced LTD does not require mGluR signaling (Adermark and Lovinger 2007b, 2009). The MFS-induced (5 min 10 Hz stimulation) LTD in the dorsal striatum that is CB1-dependent but not occlusive with eCB HFS LTD is not dependent on mGluRs (Ronesi and Lovinger 2005).

Cholinergic signaling also influences eCB-mediated synaptic plasticity. The Gq-coupled mAChRs (e.g., M1R), for instance, can play similar roles in the induction of eCB production as mGluRs (Kano et al. 2009; Martin et al. 2015) albeit the evidence is less abundant for such in the striatum and nucleus accumbens. In the dorsal striatum, M1Rs are postsynaptically expressed (Narushima et al. 2007; Uchigashima et al. 2007) and are localized near postsynaptic DGL-α, although not as close to the PSD as mGluR5 (Uchigashima et al. 2007). STDP-induced eCB-LTD is blocked by an M1R antagonist (Fino et al. 2010). eCB-LTD in the dorsal striatum is blocked by an M1R antagonist, but does not occlude a CB1R agonist's inhibitory effect, suggesting that M1R activation is upstream of CB1R activation (Tozzi et al. 2011). An mAChR agonist enhances eCB-LTD in the striatum (Calabresi et al. 1992b). eCB-mediated long-lasting disinhibition (DLL) of striatal output is not prevented by an mAChR antagonist, but is occluded by an mAChR agonist (Adermark 2011). DSE in MSNs requires mGluR1/5 activation but not mAChR activation (Narushima et al. 2006a), but a combination of mGluR5 and mAChR activation enhances DSE (Uchigashima et al. 2007). On the other hand, DSI that occurs at FSI inputs to MSNs (Narushima et al. 2006b) is dependent on postsynaptic M1R signaling (Narushima et al. 2007) and is enhanced by mGluR1/5 activation (Uchigashima et al. 2007). Activation of M1R or inhibiting AChE enhances DSI, whereas an M1R antagonist blocks DSI, and DSI is absent in M1 null mutant mice. Reducing the firing of CINs decreases DSI as well. On the other hand, M1R activation might actually impair CB1-induced LTD. M1R activation reduces L-type VGCC opening in MSNs via PLC-mediated induction of calcium release from IP3R-sensitive intracellular calcium stores and subsequent activation of PKC (Howe and Surmeier 1995; Perez-Burgos et al. 2008). Thus activation of CINs could prevent LTD induction by activating M1Rs and thus decrease Ca^{2+} influx from L-type VGCCs (Wang et al. 2006).

D2 dopamine receptors (D2Rs) are another critical component for eCB-mediated synaptic plasticity, at least in the dorsal striatum. In this region D2R activation induces synaptic depression of excitatory and inhibitory synapses (Bamford et al. 2004a, b; Calabresi et al. 1992a, b; Centonze et al. 2004b; Cepeda et al. 1993; Flores-Hernandez et al. 1997; Hsu et al. 1995; Tang et al. 2001). Some of these studies identified D2R-mediated synaptic depression as presynaptically expressed, and this may indeed be the case to some degree. Presynaptic D2R activation can reduce MFS but not LFS-induced synaptic depression of corticostriatal transmission, which may allow them to act as a low-pass filter, although D2R activation alone in these experiments was insufficient to induce LTD (Bamford et al. 2004b). However, D2Rs are only sparsely expressed at presynaptic sites in DMS and absent in DLS (Wang and Pickel 2002), but are abundantly expressed postsynaptically in the striatum (Ambrose et al. 2004; Fuxe et al. 2007; Wang and Pickel 2002). This suggests then that D2R activation induces retrograde messenger-induced presynaptic depression, which suggests endocannabinoids may be involved. Indeed in vivo D2 activation stimulates AEA release but not 2-AG release in the dorsal striatum, whereas a D1R dopamine receptor agonist has no effect (Centonze et al. 2004a; Giuffrida et al. 1999). This is via stimulation of NAPE-PLD and inhibition of FAAH (Centonze et al. 2004a). Indeed, D2R activation in MSNs can result in increases in postsynaptic calcium levels that arise from a PLC-IP3R-mediated calcium release from intracellular stores (Hernandez-Lopez et al. 2000). Increased intracellular calcium could stimulate NAPE-PLD and other calcium-sensitive enzymes that play a role in AEA production (Okamoto et al. 2004, 2007). The age-dependent shift from LTP to LTD in response to HFS in the dorsal striatum is paralleled by an increase in AEA levels. AEA induces LTD in young rats demonstrating that CB1R receptors are present and functional at these synapses by the early stages of postnatal development. Modulating 2-AG levels using THL has no effect on the HFS-induced expression of eCB-LTD, supporting the evidence that AEA mediates HFS-induced LTD (Ade and Lovinger 2007). Cocaine has the same effect as the D2R agonist on AEA levels. In addition cocaine-mediated inhibition of GABAergic transmission in the striatum is via D2R and endocannabinoids (Centonze et al. 2004a). D2R-mediated STD is occluded by a CB1R agonist, blocked by a CB1R antagonist, and absent in a CB1R null mutant mouse (Yin and Lovinger 2006). Interestingly this effect is also blocked by an mGluR1/5 antagonist suggesting a convergence of dopaminergic and glutamatergic signaling. Indeed D2R-mediated synaptic depression is enhanced when paired with stimulation of glutamatergic striatal inputs (Bamford et al. 2004b; Yin and Lovinger 2006). In addition, D2R agonist-induced inhibition of corticostriatal neurotransmitter release is blocked by CB1 antagonists in control and dopamine-depleted mouse dorsal striatum and mGluR5 antagonists in dopamine-depleted mice (Wong et al. 2015). The role of mGluR5 in control mice was not addressed in this study. In the nucleus accumbens core, D2R-mediated inhibition of neurotransmitter release from cortical inputs onto MSNs is also dependent on endocannabinoid signaling as well as mGluR1 and mGluR5 (Wang et al. 2012). These studies also indicate that CB1R activation is downstream of D2R and mGluR activation. In the dorsal striatum, D2Rs are often as requisite as mGluRs in determining whether eCB-LTD takes place. HFS-induced eCB-LTD is blocked by D2R antagonists or occluded by D2R activation (Kreitzer and Malenka 2005, 2007; Lerner and Kreitzer 2012; Yin and Lovinger 2006). This LTD is also lost in dopamine-depleted states (Calabresi et al. 1993). In D2R null mice, HFS fails to induce LTD (Tang et al. 2001) or may induce NMDAR-dependent LTP instead of eCB-LTD (Calabresi et al. 1997). It has also been shown that eCB-LTD at excitatory inputs to indirect pathway MSNs is D2R-dependent (Kreitzer and Malenka 2007). Neurotensin-induced CB1-mediated inhibition of corticostriatal transmission is blocked by a D2 antagonist (Yin et al. 2008). eCB-mediated long-lasting disinhibition (DLL) of striatal output is prevented by a D2R antagonist in addition to the previously mentioned block by an mGluR antagonist (Adermark 2011). The MFS-induced (5 min 10 Hz stimulation) eCB-LTD in the dorsal striatum is blocked by a D2R antagonist (Ronesi and Lovinger 2005). D2R antagonists block LFS (20 Hz) LTD (Lerner and Kreitzer 2012). As is the case with mGluR1/5 in L-type VGCC activator-induced LTD, D2Rs are also not required for this form of eCB-LTD, suggesting that increasing intracellular calcium also bypasses the cellular consequences of activating D2Rs (Adermark and Lovinger 2007a). Negative-timing STDP induces presynaptic cannabinoid LTD that requires both D2 and mGluR5 receptor activation in indirect MSNs. The same protocol fails to induce LTD in direct MSNs unless D1 dopamine receptors are blocked after which eCB-LTD is induced (Shen et al. 2008). The case for D1R having an inhibitory role in eCB-LTD is curious considering early reports that D1-like receptors block eCB-LTD (Calabresi et al. 1992a, b) Later work determined, however, that a D1R antagonist inhibited LTP, whereas an antagonist for the D1R-like D5R prevented LTD induction (Centonze et al. 2003a, b). The D5Rs involved were possibly localized in NOS+ interneurons (discussed in further detail below).

Also of interest is whether D2R-mediated eCB-LTD is pathway dependent. eCB-LTD is induced in the dorsal striatum in both direct and indirect pathway MSNs and is inhibited in both by a D2R antagonist (Tozzi et al. 2011; Wang et al. 2006). This is comparable to what has been found by others (Bagetta et al. 2011; Picconi et al. 2011; Shen et al. 2008; Wu et al. 2015), but contrasts with others who only find LTD in indirect pathway MSNs (Kreitzer and Malenka 2007; Nazzaro et al. 2012). This may be due to the transgenic mice used (Bagetta et al. 2011; Kramer et al. 2011; Kreitzer and Malenka 2007) or the form of stimulation (Kreitzer and Malenka 2007, 2008). It bears mentioning here that up-state eCB-iLTD occurs equally in both pathways, whereas down-state eCB-iLTD only occurs in direct pathway MSNs (Mathur et al. 2013). Although there is no clear role for dopamine receptors in these forms of iLTD, the pathway dependence may implicate some differential involvement of dopamine receptors. STDP-induced eCB-tLTP in the dorsal striatum is also dopamine dependent and occurs in both direct and indirect pathways (Cui et al. 2015). On the other hand, in nucleus accumbens, dopamine receptor antagonists do not appear to block eCB-LTD (Robbe et al. 2002b).

Postsynaptic depolarization is not needed for D2R-mediated CB1R-dependent STD in the dorsal striatum (Yin and Lovinger 2006). For eCB-LTD, depolarization is not sufficient to induce LTD (Calabresi et al. 1999a; Kreitzer and Malenka 2005); however, the state-dependent mGluR-induced LTD that requires modest depolarization in conjunction with 25 Hz stimulation requires D2Rs and is enhanced by a D2R agonist (Kreitzer and Malenka 2005). This LTD requires L-type VGCCs, indicating that the depolarization is permissive for L-type VGCC opening. Despite not requiring depolarization though, the D2R-induced STD still requires calcium release from postsynaptic stores (Yin and Lovinger 2006). D2R antagonists block both HFS and LFS (20 Hz) LTD by altering cAMP/PKA signaling, and this requires RGS4 functionality (Lerner and Kreitzer 2012). D2-mediated eCB-LTD may involve AC5 in the striatum. AC5 is expressed in MSNs and AC5 null mutant mice have disrupted LTD, suggesting that postsynaptic cAMP signaling is important for D2-mediated (but not mGluR-mediated) eCB-LTD (Kheirbek et al. 2009). An interesting study in adult rat nucleus accumbens core demonstrates that a combination of D1R and D2R agonists increase neuron firing, but neither alone does (Seif et al. 2011). This effect is blocked by CB1R and PLC antagonists. The agonist effects are independent of GABAergic, glutamatergic, and cholinergic transmission. MGL, but not FAAH, inhibition allows subthreshold concentration of the agonists to increase neuron firing. mGluR5 inhibition prevents agonist effects, but not when the postsynaptic cell is filled with 2-AG and an AMT blocker prevents the effect of 2-AG. These agonist effects are mediated by reduction of A-type K+ currents. The authors of this study suggest that postsynaptic CB1Rs mediate the dopamine receptor effects (Kofalvi et al. 2005; Pickel et al. 2006; Seif et al. 2011).

Additional studies implicate other signaling systems in D2R-mediated CB1-dependent synaptic depression. One of these is the striatal adenosine system. A2A adenosine receptors (A2ARs) are mostly postsynaptic receptors and co-localize with D2Rs, but not D1Rs (Fuxe et al. 2007; Quiroz et al. 2009). A2AR activation is necessary for induction of STDP-induced striatal LTP in striatal indirect pathway MSNs (Shen et al. 2008). On the other hand, antagonism of A2ARs coupled with a D2R agonist is permissive for presynaptically expressed synaptic depression (Tozzi et al. 2007, 2011). In addition inhibition of A2AR activation enhances D2R-mediated synaptic depression (Ferre et al. 1997; Kim and Palmiter 2008; Stromberg et al. 2000; Tozzi et al. 2007, 2011, 2012). A2AR inhibition coupled with D2R activation induces LTD that is at least in part dependent on CB1R (Tozzi et al. 2011, 2012). This LTD is presynaptically expressed, requires postsynaptic calcium, and occurs in both direct and indirect pathways (Tozzi et al. 2011). Some glutamatergic terminals that contact direct pathway MSNs contain A2ARs, and activation of these receptors increases glutamatergic transmission and counters the effects of CB1R activation (Martire et al. 2011; Quiroz et al. 2009). Thus A2ARs counter the effects of D2R activation in indirect pathway MSNs via postsynaptic signaling mechanisms and antagonize the effects of CB1R activation presynaptically in direct pathway MSNs. An additional site of action of D2Rs and A2ARs may be on CINs in the striatum. There is some evidence that D2R and A2ARs are co-expressed in CINs (Brown et al. 1990; James and Richardson 1993; Song et al. 2000; Tozzi et al. 2011), but other studies suggest that CINs might not express A2ARs (Svenningsson et al. 1997). A2AR activation stimulates ACh release in the striatum (Brown et al. 1990). Activation of D2Rs and inhibition of A2ARs reduces firing of CINs (Tozzi et al. 2011, 2012). Reduced firing of CINs decreases ACh release. Additionally CINs also express the inhibitory A1 adenosine receptor which reduces ACh release, offering another avenue for adenosine to reduce ACh release (Brown et al. 1990; Song et al. 2000). One hypothesis that connects D2R activation with cholinergic transmission is that D2R-mediated inhibition of acetylcholine release from CINs relieves M1R-mediated tonic inhibition of L-type VGCCs on MSNs as discussed above, thus allowing LTD to occur (Wang et al. 2006). Additional data demonstrate that an M1R antagonist reduces EPSC amplitude, as does application of AChE. The effect of the M1R antagonist requires presynaptic activity and a postsynaptic potential more depolarized than −90 mV. The M1R and AChE effects are preserved in the presence of a D2R antagonist. In the presence of a D2R antagonist, HFS paired with depolarization usually does not induce eCB-LTD, but does when a M1R antagonist or AChE is applied prior to the induction protocol (Wang et al. 2006). Interestingly, postsynaptic D2Rs exert an inhibitory effect on L-type VGCCs via calcineurin/PP-2B, suggesting that these receptors actually could provide some antagonistic effects against the relief of M1R signaling (Hernandez-Lopez et al. 2000; Perez-Burgos et al. 2008). Much work remains to understand how D2Rs, Group I mGluRs, and cholinergic signaling interact to differentially produce specific endocannabinoids and subsequent forms synaptic plasticity.

Another element to D2R-, cholinergic-, and mGluR-dependent CB1R-mediated plasticity in the striatum is nitric oxide (NO) signaling (Centonze et al. 2003a, b). NO arises from the NOS-positive interneurons. In addition to membrane depolarization, mGluR activation, and dopamine, NO appears to be required for HFS-induced eCB-LTD (Calabresi et al. 1999a). This LTD is mutually occlusive with LTD induced by cGMP-dependent phosphodiesterase inhibitor (Calabresi et al. 1999b) and is prevented by NO signaling blockade. Activating intracellular PKG induces LTD as do NO donors. mGluR1/5 activation-induced eCB-LTD can be blocked but not reversed by nNOS inhibitors, and this LTD is absent in the eNOS null mutant mouse. The NO donor SNAP mimics the effect of mGluR-driven LTD. The NO effect is likely downstream of CB1R activation as a NOS inhibitor can block the effect of a CB1R agonist (Sergeeva et al. 2007). NO is critical for the expression of mGluR1/5-driven eCB-LTD (Chepkova et al. 2009). Other data implicate the D1-like D5R inducing NO release from interneurons (Centonze et al. 2003a, b). While other findings indicate that D2R-dependent LTD in the striatum is reduced in eNOS KO mice. NOS inhibitor enhances LTD expression, although this LTD is NMDAR-dependent, suggesting it may be a different entity than CB1R-mediated LTD (Doreulee et al. 2003).

Some combination of receptor activation (mGluR1/5, mAChR, D2R) and calcium signaling ultimately leads to the activation of enzymes that cleave membrane lipids to produce anandamide (AEA) and 2-arachidonylglycerol (2-AG). These GPCRs may signal through a number of pathways to induce eCB production. One such pathway is through activation of phospholipases such as PLC, most likely PLCbeta (Hernandez-Lopez et al. 2000; Perez-Burgos et al. 2008; Uchigashima et al. 2007). STDP LTD is blocked by a PLC inhibitor (Fino et al. 2010). Postsynaptic PLC is critical for MFS/LFS (20Hz)/Depolarization eCB-LTD on indirect MSNs, but interestingly does not appear to be required for HFS LTD onto the same neurons (Lerner and Kreitzer 2012). D2R activation-induced eCB signaling in the dorsal striatum is dependent on PLC (Yin and Lovinger 2006). Neurotensin-induced CB1-mediated plasticity is blocked by a PLC inhibitor (Yin et al. 2008). L-type VGCC activator-induced eCB-LTD occurs independently of PLC (Adermark and Lovinger 2007a). PLD may be another phospholipase necessary for the expression of HFS-induced eCB-LTD (Lerner and Kreitzer 2012), but it is not clear that this is the NAPE-PLD implicated in AEA production.

HFS in young rats (<P14) induces long-term potentiation (LTP), but induces LTD in older rats. HFS induces LTP in older rats in the presence of a CB1R receptor antagonist (Ade and Lovinger 2007). This shift from LTP to LTD is paralleled by an increase in AEA levels. AEA induces LTD in young rats demonstrating that CB1R receptors are present and functional at these synapses by the early stages of postnatal development. Modulating 2-AG levels using THL has no effect on the HFS-induced expression of CB1R receptor-mediated LTD, supporting the evidence that AEA mediates HFS-induced LTD (Ade and Lovinger 2007). Down-state iLTD was enhanced by FAAH inhibitor, but this treatment had no effect on up-state iLTD. Subthreshold stimulation induced down-state iLTD only in FAAH null mutant mouse. It had no effect in wild-type mice. Subthreshold

stimulation also had no effect on up-state iLTD in either wild-type or FAAH null mice (Mathur et al. 2013). A FAAH inhibitor enhances subthreshold stimulation to allow induction of LTD (Kreitzer and Malenka 2007). On the other hand, LFS (20 Hz)-induced eCB-LTD in indirect MSNs is 2-AG mediated (Lerner and Kreitzer 2012). An A2A antagonist blocks both this form of eCB-LTD and HFS-induced eCB-LTD, but A2A inhibition can also increase MFS-induced eCB-LTD in indirect MSNs, and this is blocked by THL, also suggesting 2-AG dependence (Lerner et al. 2010; Lerner and Kreitzer 2012). Up-state iLTD is mediated by 2-AG and mGluR activation. DGL inhibition blocked up-state iLTD, demonstrating it is mediated by 2-AG. MGL inhibition enhanced subthreshold LFS to induce iLTD in the up-state, but had no effect in the down-state (Mathur et al. 2013). Furthermore, inhibiting 2-AG production blocks the expression of DSE and DSI in striatal MSNs as well as mGluR1/5- and mAChR-induced inhibition of excitatory and inhibitory transmission (Uchigashima et al. 2007).

Following their production, eCBs are released from a neuron presumably via a membrane transporter, such as the as yet molecularly unidentified AMT. In support of this, intracellular, but not extracellular, application of an AMT blocker prevents eCB-LTD, and AEA and 2-AG loading-induced plasticity require AMT (Ronesi et al. 2004). L-type VGCC activator eCB-LTD and eCB-iLTD are prevented by an AMT blocker as well (Adermark et al. 2009). However, the AMT also can serve to limit the spread of endocannabinoids by cellular uptake (Day et al. 2001; Deutsch et al. 2001; Hillard et al. 1997; Hillard and Jarrahian 2000). AMT inhibitors therefore can also induce eCB-LTD on their own (Gerdeman et al. 2002; Robbe et al. 2002b). Since the AMT operates by facilitated diffusion, pharmacological manipulations of the AMT must be considered in terms of the concentration gradient of endocannabinoids. MFS-induced (5 min 10 Hz stimulation) LTD in the dorsal striatum appears to be independent of the AMT (Ronesi and Lovinger 2005).

3.2 Presynaptic Expression of Cannabinoid Receptor-Mediated Plasticity

It is clear that as a presynaptically localized Gi/o-coupled GPCR, CB1R is in a prime location to inhibit neurotransmitter release (Fig. 3). For example, CB1R inhibits neurotransmitter release through inhibition of presynaptic VGCCs. CB1R can inhibit N-type and P/Q-type VGCCs, the types that are typically found presynaptically (Caulfield and Brown 1992; Kushmerick et al. 2004; Mackie et al. 1993, 1995; Mackie and Hille 1992; Shen and Thayer 1998; Twitchell et al. 1997). Indeed, in the dorsal striatum, the N-type VGCC blocker conotoxin prevents the inhibitory effect of WIN55,212-2 (Huang et al. 2001). CB1-mediated LTD in the amygdala is mediated via P/Q-type VGCC inhibition (Huang et al. 2003) as is mGluR2/3-mediated LTD in the nucleus accumbens (Robbe et al. 2002a); thus it is

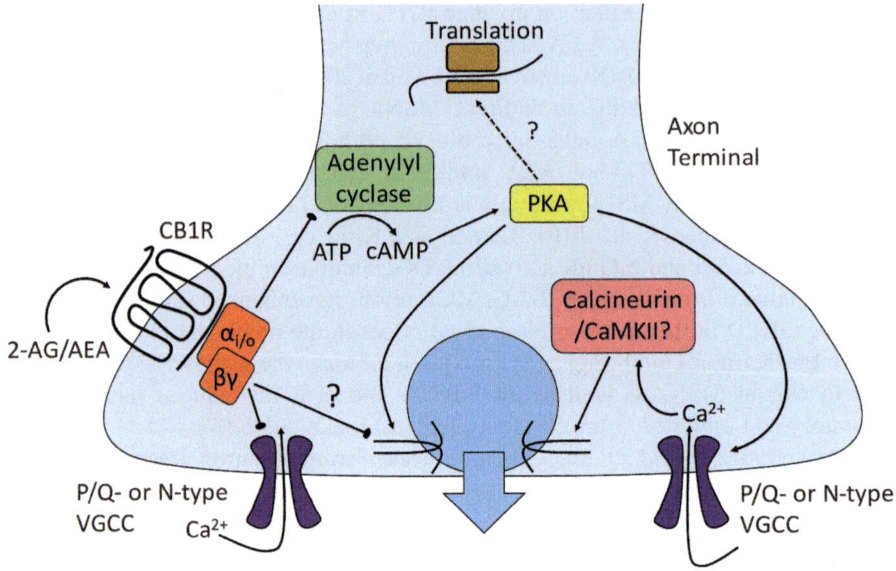

Fig. 3 Mechanisms of presynaptic CB1 cannabinoid receptor-mediated inhibition of neurotransmitter release. CB1 activation by the endocannabinoids 2-AG and AEA results in activation of Gi/o-type G proteins. G protein beta-gamma subunits have an inhibitory impact on presynaptic voltage-gated calcium channels (VGCCs) and release machinery, reducing neurotransmitter release. Signaling through the Gαi subunit results in inhibition of adenylyl cyclase and subsequent reduction in cAMP levels and reduced protein kinase A (PKA) signaling. Decreased PKA signaling impacts release machinery and VGCC function, altering neurotransmitter release probabilities. Decreases in calcium entry likely alter the function of calcium-sensitive kinases or phosphatases that may further impact neurotransmission. These changes may also underlie CB1-mediated long-term synaptic depression. Presynaptic protein translation may also play a role in this long-term depression

conceivable that CB1R could do likewise in the nucleus accumbens and possibly in the dorsal striatum. However Robbe and colleagues (Robbe et al. 2001) provided evidence that presynaptically localized, CB1R agonist-mediated synaptic depression occurred via K+ channel modulation and not via L-, N-, or P/Q-type VGCCs in nucleus accumbens. However, these findings must be interpreted with caution, as manipulations of presynaptic potassium channels usually alter presynaptic calcium levels. There is currently no evidence in the dorsal striatum that CB1R affects neurotransmitter release via coupling to K+ channels. CB1R coupling to VGCCs and/or GIRKs is the likely mechanism for CB1-mediated STD, such as DSE and DSI (Narushima et al. 2006a, b; Uchigashima et al. 2007). There are multiple possible routes that connect CB1R activation with LTD of synaptic transmission (Atwood et al. 2014b). Presynaptic calcium appears to play some role in this process as decreasing extracellular Ca^{2+} blocks direct CB1R-agonist-induced LTD, whereas chelating postsynaptic intracellular Ca^{2+} has no effect (Singla et al. 2007). Reducing EPSC amplitude by 95% using ionotropic glutamate receptor antagonists also does not prevent this form of LTD. Thus postsynaptic Ca^{2+} is not needed for LTD mediated strictly by CB1R activation. What is required is sustained presynaptic activity (Adermark and Lovinger 2007b, 2009; Adermark et al. 2009; Singla et al. 2007). As indicated above, CB1R-mediated inhibition of excitatory transmission by WIN55,212-2 in NAc can be prevented by K+ channel blockers, but not VGCC blockers (Robbe et al. 2001). However, MFS-induced eCB-LTD in nucleus accumbens is dependent on P/Q VGCCs, but interestingly not N-type VGCCs or even K+ channels (Mato et al. 2008). Another source of presynaptic calcium that could determine the expression of LTD is NMDARs. These receptors are important for some forms of presynaptically expressed LTD (Rodriguez-Moreno et al. 2010). However, they do not appear necessary for any form of definitive eCB-LTD described thus far (Calabresi et al. 1992b; Chepkova et al. 2009; Fino et al. 2010; Grueter et al. 2010; Kreitzer and Malenka 2005; Lovinger et al. 1993; Robbe et al. 2002b; Ronesi and Lovinger 2005; Walsh 1993). eCB-LTD induced by mGluR1/5 activation can be reduced by NMDAR antagonists, but it is not dependent on them (Sergeeva et al. 2007). Interestingly in juvenile mice, mGluR1/5-driven LTD is NMDAR-dependent and CB1R-independent and becomes NMDAR-independent and CB1-dependent in adolescence (Chepkova et al. 2009). The change is likely due to an increased CB1R expression in the striatum during development. Alterations in presynaptic calcium levels can alter the function of presynaptic calcium-sensitive proteins such as calcineurin, calmodulin, and calmodulin-dependent protein kinase II, proteins that may be involved in Gi/o-coupled GPCR-mediated LTD in other brain regions (Atwood et al. 2014b; Heifets et al. 2008; Kobayashi et al. 1999; Li et al. 2002; Stanton and Gage 1996). It is unknown whether these proteins play a role in striatal CB1R-mediated LTD.

Another clear pathway downstream from CB1R activation is the cAMP/PKA pathway. Increasing cAMP levels with forskolin or inhibiting PKA blocks the effect of a CB1R agonist in the dorsal striatum (Huang et al. 2002). In addition, Beta2-adrenergic receptor activation that increases adenylyl cyclase activity reduces the inhibitory effect of a CB1R agonist (Huang et al. 2002). The authors of this study also found that a likely target of PKA itself is CB1R as forskolin increased CB1R phosphorylation, although the link with this and synaptic depression is not clear. CB1R-mediated STD of excitatory transmission in the nucleus accumbens occurs independently from cAMP signaling; however eCB-LTD and the mutually occlusive mGluR2/3 LTD are mediated by cAMP/PKA signaling in addition to the P/Q VGCC discussed above (Mato et al. 2005, 2008; Robbe et al. 2001). It is conceivable therefore that eCB-LTD is expressed as an alteration of PKA-mediated phosphorylation of P/Q type of VGCCs. 5-HT1bR-mediated LTD in the dorsal striatum is cAMP/PKA pathway dependent, and this form of LTD is also mutually occlusive with eCB-LTD suggesting that this signaling is also important for cannabinoid-mediated LTD. Early work on corticostriatal LTD found that protein kinase inhibitors prevent eCB-LTD in the striatum, although it is not clear if these inhibitors affected presynaptic kinases (Calabresi et al. 1994). Another target of presynaptic cAMP/PKA signaling is the vesicular release machinery itself (Atwood et al. 2014b). One potential member of this machinery that may be involved in eCB-LTD is RIM1. RIM1 has been shown to be involved in eCB-LTD in other brain regions. A RIM1 null mutant mouse has reduced LFS and mGluR1/5-induced LTD in nucleus accumbens in indirect MSNs (Grueter et al. 2010). It is also possible that cAMP/PKA signaling may alter local presynaptic protein synthesis and this could determine the expression of CB1R-mediated LTD. Extracellular application of protein translation inhibitors blocks eCB-LTD, but intracellular application of translation and transcription inhibitors does not. Translation inhibitors also prevent mGluR1/5-induced LTD. Protein synthesis inhibitors do not prevent inhibition mediated by direct CB1R activation. Translation occurring in the presynaptic terminal is required for the maintenance, but not the induction of CB1-mediated LTD (Yin et al. 2006). AEA loading-mediated static LTD of inhibitory synapses in the dorsal striatum requires protein synthesis as does L-type VGCC activator-induced eCB-LTD (Adermark and Lovinger 2009; Adermark et al. 2009). It is abundantly clear that further work is needed to solidify a link between CB1R activation and the molecular mechanisms involved in presynaptic expression of LTD.

4 In Vivo Functions of Striatal Cannabinoid Signaling

4.1 Role of the Striatal Endocannabinoid System in Behavior

The striatum participates in a variety of behaviors involving action control, reward- and reinforcement-based learning, and addiction. Investigators have just begun to explore the roles in behavioral control of eCBs and CB1 receptors on the different striatal cellular elements.

It has long been known that activation of striatal CB1 receptors via local THC injection can produce akinesia and catalepsy in rodents (Gough and Olley 1978). The extent of this effect depends on the drug dose. Thus, some subpopulation of striatal CB1 receptors can exert strong control over movement initiation. Severe locomotor reductions are observed at doses that produce ~60% occupancy of striatal receptors (Dhawan et al. 2006). Catalepsy is one of the "tetrad" behaviors used to characterize in vivo effects of CB1 receptor activation. A recent study indicates that the CB1 receptors expressed by MSNs are not the subpopulation involved in catalepsy. Naydenov et al. (2014) have re-expressed CB1 receptors in knockout mice, using a GPR88 Cre mouse to obtain expression in MSNs. This strategy did not rescue the loss of agonist-induced catalepsy. It is possible that developmental effects or agonist effects on parallel circuits or other parts of the circuitry (e.g., globus pallidus, Pertwee and Wickens 1991) can produce catalepsy independent of direct actions within the striatum.

Effects of cannabinoid drugs, as well as the role of eCBs and CB1 in striatal-based learning, have been examined in some studies (for an excellent review, see Goodman and Packard 2015). Stimulus-response learning in the T-maze and similar paradigms is dependent on striatal circuitry (Packard and McGaugh 1996). Jarbe and Henriksson showed that delta9-THC slows task acquisition and reversal of T-maze learning when given prior to training (Henriksson and Järbe 1972; Jarbe and Henriksson 1973). However, the use of the striatal-dependent response strategy was not assessed in this study, and thus the effects of peripherally delivered drug may involve brain regions other than the striatum. In a more specific assay of response learning, Goodman and Packard (2014) found that post-training injection of the CB1 agonist WIN 55,212-2 impaired memory for the correct turn in a water plus-maze task. Thus, activation of CB1 receptors impairs response learning and could involve the striatum and associated basal ganglia circuitry. Additional studies are needed to assess the anatomical locus of the receptors that mediate this impairment.

Endocannabinoid actions on CB1 receptors are implicated in the flexibility of stimulus-response learning in a reversal learning phase of the T-maze paradigm. When the CB1 antagonist AM251 was injected into DLS following reversal training, it impaired consolidation of the reversal. Injection of the antagonist into the dorsal hippocampus actually facilitated reversal consolidation, consistent with long-standing evidence that the hippocampus mediates spatial learning in this task that can compete with DLS-mediated S-R learning (Goodman and Packard 2014).

The Morris water maze has also been used to assess the effect of cannabinoids on striatal-based S-R learning. While the "hidden platform" variant of this task is well known to depend on circuitry that includes the hippocampus, this circuitry is not required for performance of the "cued" water maze version in which the platform location is marked by a visible flag or other cue. Instead, this cued version requires an intact striatum (Packard and McGaugh 1996). Consolidation of memory in this water maze task variant is impaired by either peripheral or intra-DLS post-training injection of WIN 55,212-2 (Goodman and Packard 2014). This finding supports the T-maze data in suggesting that strong activation of CB1 receptors in DLS has S-R memory impairing actions. This is certainly relevant to our understanding of cognitive impairment during cannabinoid intoxication.

Endocannabinoids appear to play a role in extinction of striatal-based learning. Instrumental learning, the acquisition of new actions initially driven by reward, involves two prominent corticostriatal circuits. The associative circuit that includes prefrontal cortices and the dorsomedial striatum (caudate nucleus in primates) is involved early in learning and controls behavior when it is goal-directed (i.e., sensitive to reward). The sensorimotor circuit that contains motor cortices and the dorsolateral striatum (putamen nucleus in primates) controls behavior when it becomes habitual (i.e., insensitive to reward). Activation of CB1 receptors appears to be necessary for habit learning and action control (Hilario et al. 2007). Synapses between the orbitofrontal cortex and the ventral portion of the dorsomedial striatum are one location where eCBs and CB1 have important roles in habit learning. Knocking out CB1 receptors at these synapses prevents mice from developing habitual instrumental learning in a lever pressing task (Gremel et al. 2016). This effect is probably due to suppression of information about outcome value that is important for supporting goal-directed actions. The idea that presynaptic suppression of OFC-DMS inputs is important for establishment and maintenance of habitual behaviors is supported by experiments in which a Gi/o-coupled designer receptor activated by a designer drug (DREADD) construct is expressed in OFC. Activation of this receptor suppresses OFC output and promotes habitual behavior even under conditions where it is normally goal-directed (Gremel and Costa 2013; Gremel et al. 2016). These findings highlight the important role of eCBs and CB1 receptors in maintaining the proper balance of circuit function in the constant competition for goal-directed versus habitual control of behavior.

There is also evidence that altering eCB signaling affects learning-related neuronal activity in the nucleus accumbens/ventral striatum. Treatment with a CB1 antagonist suppresses food-rewarded operant responding, and firing of putative MSNs encoding predictive cue presentation and reward delivery was altered by this antagonist treatment (Hernandez and Cheer 2012). Activation of CB1 receptors by eCBs has also been implicated in dendritic spine changes produced by operant responding for highly palatable food (Guegan et al. 2013).

4.2 Striatal Endocannabinoids in Drug Abuse and Addiction

There is growing evidence of roles for eCBs and CB1 receptors in response to drugs of abuse, drug use disorders, and addiction. The role of CB1 in cannabis drug intoxication is clear, as the major psychoactive component of *Cannabis sativa*, THC, is a CB1 partial agonist. Activation of CB1 induces a behavioral "tetrad" that is associated with the cannabis high (Martin et al. 1991). As mentioned above, catalepsy is one of these behaviors and may well involve CB1 receptors in the striatum (Naydenov et al. 2014). Single (Mato et al. 2004), and prolonged exposure to THC reduces actions of other cannabinoid agonists and endocannabinoids on striatal receptors (Hoffman et al. 2004; Mato et al. 2005; Nazzaro et al. 2012), as mentioned above. While there has been some debate as to whether THC is "addictive," there are clear signs of dependence following chronic exposure, and cannabis can be abused with clear signs of craving (e.g., Hart 2005; Haughey et al. 2008; Marshall et al. 2014). In animal models, THC appears to have rewarding effects, as it lowers the threshold for rewarding brain stimulation (reviewed in Gardner 2002), is self-administered intravenously (Tanda et al. 2000), and can induce a place preference (Valjent and Maldonado 2000). The role of striatal CB1 receptors in cannabis dependence is still being determined, but animals will operantly respond for direct injection of THC into the NAC shell region, indicating a role for this part of the ventral striatum in the rewarding or reinforcing effects of the drug (Zangen et al. 2006). Rats will also self-administer 2-AG, and this is associated with increased DA in the NAc shell (De Luca et al. 2014). In the striatum of human chronic marijuana users, positron emission tomography (PET) reveals a blunting of drug-induced dopamine release (Volkow et al. 2014). Decreases in CB1 receptor availability are also observed in the ventral striatum of chronic cannabis-using humans, as measured with PET (Ceccarini et al. 2015). Thus, there is increasing evidence that chronic THC exposure produces signs of addiction in both experimental animals and humans, which can be accompanied by changes in striatal neurochemistry.

There is also evidence that drugs of abuse alter striatal endocannabinoid levels and striatal-based reward and habit learning. Single (Mato et al. 2004) and chronic exposure to THC is known to impair CB1 agonist-induced and endocannabinoid-mediated synaptic depression in a variety of brain regions including the dorsal and ventral striatum (Hoffman et al. 2004; Mato et al. 2005; Nazzaro et al. 2012). Habit learning in an instrumental nose-poke task is enhanced following 7 days of THC exposure, with concomitant loss of eCB-dependent corticostriatal LTD (Nazzaro et al. 2012). Interestingly, the deficit in LTD and the behavioral changes could be restored by boosting MSN excitability with a blocker of small-conductance potassium channels. This finding indicates that loss of eCB signaling is the result of deficient eCB production by MSNs.

Endocannabinoids in the ventral striatum have also been implicated in the rewarding and/or motivational effects of abused drugs. Blocking CB1 receptors reduces nicotine self-administration, and this is associated with decreased nicotine-induced increase in DA release in the NAc (Cohen et al. 2002). CB1 blockade in the NAc also reduces intravenous heroin self-administration (Caillé and Parsons 2006) and cue-induced heroin-seeking behavior (Alvarez-Jaimes et al. 2008); reduces drug-primed reinstatement of cocaine seeking (Xi et al. 2006), the "breakpoint" for cocaine self-administration in a progressive ratio procedure after prolonged drug exposure (Orio et al. 2009); and reduces ethanol self-administration in alcohol-preferring rats (Malinen and Hyytiä 2008). Cocaine effects on mGluR activation of eCB signaling also play a role in synaptic transmission changes in the NAc that may contribute to cocaine self-administration. After a withdrawal period following long-term cocaine self-administration, Group I mGluR signaling is changed from mGluR5-stimulated eCB production to mGluR1-stimulated AMPA receptor subunit alteration (McCutcheon et al. 2011). Cocaine effects on NAc dopamine and impulsivity also appear to involve CB1 (Hernandez et al. 2014). More recent evidence also suggests a role in cocaine self-administration of accumbens CB2 receptors (Xi et al. 2011), but the localization of these receptors on cellular elements in this region remains to be determined. Acute exposure to ethanol inhibits the firing of NAc neurons when that firing is driven by activation of glutamatergic afferents from the basolateral amygdala (Perra et al. 2005). These alcohol effects are greatly reduced by inhibition of FAAH or application of a CB1 agonist.

Using a chronic intermittent ethanol (CIE) exposure procedure, DePoy et al. (2013) found increased 2-AG levels in the DLS. This increase was accompanied by a loss of eCB-dependent LTD at glutamatergic synapses. Similar results were obtained using chronic ethanol exposure being administered via drinking water (Adermark et al. 2011; Xia et al. 2006). CIE enhanced DLS-dependent learning on touchscreen discrimination and reversal and conditioned approach tasks (DePoy et al. 2013). Hypertrophy of MSN dendritic spines was also observed following the CIE exposure regimen. While it is not yet clear that all of these behavioral and morphological changes are the result of alterations in endocannabinoid signaling, this is certainly a reasonable mechanism to explain some of these ethanol effects. There is growing evidence that the DLS is involved in ethanol promotion of habitual behavior and habitual alcohol seeking (Chen et al. 2011a; Corbit et al. 2012; Cuzon Carlson et al. 2011). It will be important to determine how endocannabinoids contribute to these behavioral changes.

Activation of striatal CB1 receptors has been implicated in psychostimulant-induced increases in locomotion and sensitization of these locomotor effects (Corbille et al. 2007; Thiemann et al. 2008). In addition to the alcohol effects on striatal eCB-LTD mentioned above, other drugs of abuse also impair this prominent form of synaptic plasticity. Cocaine self-administration alters glutamatergic synaptic transmission in the rat NAc shell, via an endocannabinoid mechanism (Ortinski et al. 2012). A single exposure to cocaine is sufficient to ablate eCB-LTD in NAc (Fourgeaud et al. 2004). A single in vivo exposure to the widely prescribed and abused opiate drug oxycodone prevents subsequent induction of eCB-LTD (Atwood et al. 2014a). This treatment also prevents opiate receptor-mediated LTD, in particular LTD mediated by the Mu opiate receptor. This loss of LTD persists for up to 4 days after drug treatment. Interestingly, eCB- and MOR-induced LTD are mutually occlusive and appear to share common mechanisms. These common mechanisms are a likely target of relatively long-lasting oxycodone impairment of normal synaptic plasticity.

5 Conclusion

It is clear that endocannabinoid signaling is critical to the function of the striatum. Components of the endocannabinoid system are found at numerous striatal synapses and participate in dictating synaptic transmission. The net effect of this signaling helps determine the strength of striatal output and thus basal ganglia-cortical circuit function and by extension behavioral output. As evidenced by the impact of cannabinoids (phytocannabinoids, synthetic cannabinoids, and modulators of endocannabinoid signaling) on striatum-dependent behaviors, it is clear that this signaling system indeed plays a crucial role in behavior. Of particular interest here, we have discussed how cannabinoids play a role in drug use and abuse. There is a bidirectional relationship between drugs of abuse and striatal endocannabinoid signaling. Drugs of abuse alter the function of the endogenous cannabinoid signaling players in the striatum, and endocannabinoid signaling may determine behavioral responding for drugs of abuse. It is conceivable that manipulating this relationship may reveal new and better therapeutics for drug abuse and addiction. Perhaps restoring endocannabinoid signaling altered by drugs of abuse will prove efficacious. As discussed here CB1 receptor antagonists can clearly impact behavioral responding for drugs of abuse. Perhaps targeting other members of the endocannabinoid system could better counter the behavioral impact of drugs of abuse. As we gain a better understanding of the role of the endocannabinoid system in striatal function, we will approach new avenues for treating drug abuse and addiction.

References

Ade KK, Lovinger DM (2007) Anandamide regulates postnatal development of long-term synaptic plasticity in the rat dorsolateral striatum. J Neurosci 27:2403–2409

Adermark L (2011) Modulation of endocannabinoid-mediated long-lasting disinhibition of striatal output by cholinergic interneurons. Neuropharmacology 61:1314–1320

Adermark L, Lovinger DM (2007a) Combined activation of L-type Ca2+ channels and synaptic transmission is sufficient to induce striatal long-term depression. J Neurosci 27:6781–6787

Adermark L, Lovinger DM (2007b) Retrograde endocannabinoid signaling at striatal synapses requires a regulated postsynaptic release step. Proc Natl Acad Sci U S A 104:20564–20569

Adermark L, Lovinger DM (2009) Frequency-dependent inversion of net striatal output by endocannabinoid-dependent plasticity at different synaptic inputs. J Neurosci 29:1375–1380

Adermark L, Talani G, Lovinger DM (2009) Endocannabinoid-dependent plasticity at GABAergic and glutamatergic synapses in the striatum is regulated by synaptic activity. Eur J Neurosci 29:32–41

Adermark L, Jonsson S, Ericson M, Soderpalm B (2011) Intermittent ethanol consumption depresses endocannabinoid-signaling in the dorsolateral striatum of rat. Neuropharmacology 61:1160–1165

Alger BE (2002) Retrograde signaling in the regulation of synaptic transmission: focus on endocannabinoids. Prog Neurobiol 68:247–286

Alvarez-Jaimes L, Polis I, Parsons LH (2008) Attenuation of cue-induced heroin-seeking behavior by cannabinoid CB1 antagonist infusions into the nucleus accumbens core and prefrontal cortex, but not basolateral amygdala. Neuropsychopharmacology 33(10):2483–2493

Ambrose LM, Unterwald EM, Van Bockstaele EJ (2004) Ultrastructural evidence for co-localization of dopamine D2 and micro-opioid receptors in the rat dorsolateral striatum. Anat Rec A Discov Mol Cell Evol Biol 279:583–591

Atwood BK, Mackie K (2010) CB2: a cannabinoid receptor with an identity crisis. Br J Pharmacol 160:467–479

Atwood BK, Kupferschmidt DA, Lovinger DM (2014a) Opioids induce dissociable forms of long-term depression of excitatory inputs to the dorsal striatum. Nat Neurosci 17:540–548

Atwood BK, Lovinger DM, Mathur BN (2014b) Presynaptic long-term depression mediated by G-coupled receptors. Trends Neurosci 37(11):663–673

Azad SC, Marsicano G, Eberlein I, Putzke J, Zieglgansberger W, Spanagel R, Lutz B (2001) Differential role of the nitric oxide pathway on delta(9)-THC-induced central nervous system effects in the mouse. Eur J Neurosci 13:561–568

Bagetta V, Picconi B, Marinucci S, Sgobio C, Pendolino V, Ghiglieri V, Fusco FR, Giampa C, Calabresi P (2011) Dopamine-dependent long-term depression is expressed in striatal spiny neurons of both direct and indirect pathways: implications for Parkinson's disease. J Neurosci 31:12513–12522

Bamford NS, Robinson S, Palmiter RD, Joyce JA, Moore C, Meshul CK (2004a) Dopamine modulates release from corticostriatal terminals. J Neurosci 24:9541–9552

Bamford NS, Zhang H, Schmitz Y, Wu NP, Cepeda C, Levine MS, Schmauss C, Zakharenko SS, Zablow L, Sulzer D (2004b) Heterosynaptic dopamine neurotransmission selects sets of corticostriatal terminals. Neuron 42:653–663

Beltramo M, Stella N, Calignano A, Lin SY, Makriyannis A, Piomelli D (1997) Functional role of high-affinity anandamide transport, as revealed by selective inhibition. Science 277:1094–1097

Bennett BD, Wilson CJ (1999) Spontaneous activity of neostriatal cholinergic interneurons in vitro. J Neurosci 19:5586–5596

Berretta S, Parthasarathy HB, Graybiel AM (1997) Local release of GABAergic inhibition in the motor cortex induces immediate-early gene expression in indirect pathway neurons of the striatum. J Neurosci 17:4752–4763

Bisogno T, Howell F, Williams G, Minassi A, Cascio MG, Ligresti A, Matias I, Schiano-Moriello-A, Paul P, Williams EJ, Gangadharan U, Hobbs C, Di Marzo V, Doherty P (2003) Cloning of the first sn1-DAG lipases points to the spatial and temporal regulation of endocannabinoid signaling in the brain. J Cell Biol 163:463–468

Bolam JP, Izzo PN, Graybiel AM (1988) Cellular substrate of the histochemically defined striosome/matrix system of the caudate nucleus: a combined Golgi and immunocytochemical study in cat and ferret. Neuroscience 24:853–875

Brown SJ, James S, Reddington M, Richardson PJ (1990) Both A1 and A2a purine receptors regulate striatal acetylcholine release. J Neurochem 55:31–38

Burattini C, Battistini G, Tamagnini F, Aicardi G (2014) Low-frequency stimulation evokes serotonin release in the nucleus accumbens and induces long-term depression via production of endocannabinoid. J Neurophysiol 111:1046–1055

Cadogan AK, Alexander SP, Boyd EA, Kendall DA (1997) Influence of cannabinoids on electrically evoked dopamine release and cyclic AMP generation in the rat striatum. J Neurochem 69:1131–1137

Caillé S, Parsons LH (2006) Cannabinoid modulation of opiate reinforcement through the ventral striatopallidal pathway. Neuropsychopharmacology 31(4):804–813

Calabresi P, Maj R, Mercuri NB, Bernardi G (1992a) Coactivation of D1 and D2 dopamine receptors is required for long-term synaptic depression in the striatum. Neurosci Lett 142:95–99

Calabresi P, Maj R, Pisani A, Mercuri NB, Bernardi G (1992b) Long-term synaptic depression in the striatum: physiological and pharmacological characterization. J Neurosci 12:4224–4233

Calabresi P, Mercuri NB, Sancesario G, Bernardi G (1993) Electrophysiology of dopamine-denervated striatal neurons. Implications for Parkinson's disease. Brain 116(Pt 2):433–452

Calabresi P, Pisani A, Mercuri NB, Bernardi G (1994) Post-receptor mechanisms underlying striatal long-term depression. J Neurosci 14:4871–4881

Calabresi P, Saiardi A, Pisani A, Baik JH, Centonze D, Mercuri NB, Bernardi G, Borrelli E (1997) Abnormal synaptic plasticity in the striatum of mice lacking dopamine D2 receptors. J Neurosci 17:4536–4544

Calabresi P, Centonze D, Gubellini P, Marfia GA, Bernardi G (1999a) Glutamate-triggered events inducing corticostriatal long-term depression. J Neurosci 19:6102–6110

Calabresi P, Gubellini P, Centonze D, Sancesario G, Morello M, Giorgi M, Pisani A, Bernardi G (1999b) A critical role of the nitric oxide/cGMP pathway in corticostriatal long-term depression. J Neurosci 19:2489–2499

Canales JJ (2005) Stimulant-induced adaptations in neostriatal matrix and striosome systems: transiting from instrumental responding to habitual behavior in drug addiction. Neurobiol Learn Mem 83:93–103

Castelli MP, Paola Piras A, D'Agostino A, Pibiri F, Perra S, Gessa GL, Maccarrone M, Pistis M (2007) Dysregulation of the endogenous cannabinoid system in adult rats prenatally treated with the cannabinoid agonist WIN 55,212-2. Eur J Pharmacol 573:11–19

Caulfield MP, Brown DA (1992) Cannabinoid receptor agonists inhibit Ca current in NG108-15 neuroblastoma cells via a pertussis toxin-sensitive mechanism. Br J Pharmacol 106:231–232

Ceccarini J, Kuepper R, Kemels D, van Os J, Henquet C, Van Laere K (2015) [18F]MK-9470 PET measurement of cannabinoid CB1 receptor availability in chronic cannabis users. Addict Biol 20(2):357–367

Centonze D, Grande C, Saulle E, Martin AB, Gubellini P, Pavon N, Pisani A, Bernardi G, Moratalla R, Calabresi P (2003a) Distinct roles of D1 and D5 dopamine receptors in motor activity and striatal synaptic plasticity. J Neurosci 23:8506–8512

Centonze D, Gubellini P, Pisani A, Bernardi G, Calabresi P (2003b) Dopamine, acetylcholine and nitric oxide systems interact to induce corticostriatal synaptic plasticity. Rev Neurosci 14:207–216

Centonze D, Battista N, Rossi S, Mercuri NB, Finazzi-Agro A, Bernardi G, Calabresi P, Maccarrone M (2004a) A critical interaction between dopamine D2 receptors and endocannabinoids mediates the effects of cocaine on striatal gabaergic Transmission. Neuropsychopharmacology 29:1488–1497

Centonze D, Gubellini P, Usiello A, Rossi S, Tscherter A, Bracci E, Erbs E, Tognazzi N, Bernardi G, Pisani A, Calabresi P, Borrelli E (2004b) Differential contribution of dopamine D2S and D2L receptors in the modulation of glutamate and GABA transmission in the striatum. Neuroscience 129:157–166

Cepeda C, Buchwald NA, Levine MS (1993) Neuromodulatory actions of dopamine in the neostriatum are dependent upon the excitatory amino acid receptor subtypes activated. Proc Natl Acad Sci U S A 90:9576–9580

Chen G, Cuzon Carlson VC, Wang J, Beck A, Heinz A, Ron D, Lovinger DM, Buck KJ (2011a) Striatal involvement in human alcoholism and alcohol consumption, and withdrawal in animal models. Alcohol Clin Exp Res 35(10):1739–1748

Chen M, Wan Y, Ade K, Ting J, Feng G, Calakos N (2011b) Sapap3 deletion anomalously activates short-term endocannabinoid-mediated synaptic plasticity. J Neurosci 31:9563–9573

Chepkova AN, Fleischer W, Kazmierczak T, Doreulee N, Haas HL, Sergeeva OA (2009) Developmental alterations of DHPG-induced long-term depression of corticostriatal synaptic transmission: switch from NMDA receptor-dependent towards CB1 receptor-dependent plasticity. Pflugers Arch 459:131–141

Choi S, Lovinger DM (1997a) Decreased frequency but not amplitude of quantal synaptic responses associated with expression of corticostriatal long-term depression. J Neurosci 17:8613–8620

Choi S, Lovinger DM (1997b) Decreased probability of neurotransmitter release underlies striatal long-term depression and postnatal development of corticostriatal synapses. Proc Natl Acad Sci U S A 94:2665–2670

Clarke RB, Adermark L (2010) Acute ethanol treatment prevents endocannabinoid-mediated long-lasting disinhibition of striatal output. Neuropharmacology 58:799–805

Cohen C, Perrault G, Voltz C, Steinberg R, Soubrié P (2002) SR141716, a central cannabinoid (CB (1)) receptor antagonist, blocks the motivational and dopamine-releasing effects of nicotine in rats. Behav Pharmacol 13(5–6):451–463

Concannon RM, Okine BN, Finn DP, Dowd E (2015) Differential upregulation of the cannabinoid CB(2) receptor in neurotoxic and inflammation-driven rat models of Parkinson's disease. Exp Neurol 269:133–141

Corbille AG, Valjent E, Marsicano G, Ledent C, Lutz B, Herve D, Girault JA (2007) Role of cannabinoid type 1 receptors in locomotor activity and striatal signaling in response to psychostimulants. J Neurosci 27:6937–6947

Corbit LH, Nie H, Janak PH (2012) Habitual alcohol seeking: time course and the contribution of subregions of the dorsal striatum. Biol Psychiatry 72(5):389–395

Cravatt BF, Giang DK, Mayfield SP, Boger DL, Lerner RA, Gilula NB (1996) Molecular characterization of an enzyme that degrades neuromodulatory fatty-acid amides. Nature 384:83–87

Cui Y, Paille V, Xu H, Genet S, Delord B, Fino E, Berry H, Venance L (2015) Endocannabinoids mediate bidirectional striatal spike-timing-dependent plasticity. J Physiol 593:2833–2849

Cuzon Carlson VC, Seabold GK, Helms CM, Garg N, Odagiri M, Rau AR, Daunais J, Alvarez VA, Lovinger DM, Grant KA (2011) Synaptic and morphological neuroadaptations in the putamen associated with long-term, relapsing alcohol drinking in primates. Neuropsychopharmacology 36(12):2513–2528

Davis MI, Puhl HL 3rd (2011) Nr4a1-eGFP is a marker of striosome-matrix architecture, development and activity in the extended striatum. PLoS One 6:e16619

Day TA, Rakhshan F, Deutsch DG, Barker EL (2001) Role of fatty acid amide hydrolase in the transport of the endogenous cannabinoid anandamide. Mol Pharmacol 59:1369–1375

De Luca MA, Valentini V, Bimpisidis Z, Cacciapaglia F, Caboni P, Di Chiara G (2014) Endocannabinoid 2-Arachidonoylglycerol self-administration by Sprague-Dawley rats and stimulation of in vivo Dopamine transmission in the nucleus accumbens shell. Front Psych 5:140

DePoy L, Daut R, Brigman JL, MacPherson K, Crowley N, Gunduz-Cinar O, Pickens CL, Cinar R, Saksida LM, Kunos G, Lovinger DM, Bussey TJ, Camp MC, Holmes A (2013) Chronic alcohol produces neuroadaptations to prime dorsal striatal learning. Proc Natl Acad Sci U S A 110:14783–14788

Desban M, Kemel ML, Glowinski J, Gauchy C (1993) Spatial organization of patch and matrix compartments in the rat striatum. Neuroscience 57:661–671

Deutsch DG, Glaser ST, Howell JM, Kunz JS, Puffenbarger RA, Hillard CJ, Abumrad N (2001) The cellular uptake of anandamide is coupled to its breakdown by fatty-acid amide hydrolase. J Biol Chem 276:6967–6973

Devane WA, Hanus L, Breuer A, Pertwee RG, Stevenson LA, Griffin G, Gibson D, Mandelbaum A, Etinger A, Mechoulam R (1992) Isolation and structure of a brain constituent that binds to the cannabinoid receptor. Science 258:1946–1949

Dhawan J, Deng H, Gatley SJ, Makriyannis A, Akinfeleye T, Bruneus M, Dimaio AA, Gifford AN (2006) Evaluation of the in vivo receptor occupancy for the behavioral effects of cannabinoids using a radiolabeled cannabinoid receptor agonist, R-[125/131I]AM2233. Synapse 60:93–101

Di Marzo V, Gianfrani C, De Petrocellis L, Milone A, Cimino G (1994) Polyunsaturated-fatty-acid oxidation in Hydra: regioselectivity, substrate-dependent enantioselectivity and possible biological role. Biochem J 300(Pt 2):501–507

Dinh TP, Carpenter D, Leslie FM, Freund TF, Katona I, Sensi SL, Kathuria S, Piomelli D (2002a) Brain monoglyceride lipase participating in endocannabinoid inactivation. Proc Natl Acad Sci U S A 99:10819–10824

Dinh TP, Freund TF, Piomelli D (2002b) A role for monoglyceride lipase in 2-arachidonoylglycerol inactivation. Chem Phys Lipids 121:149–158

Donoghue JP, Herkenham M (1986) Neostriatal projections from individual cortical fields conform to histochemically distinct striatal compartments in the rat. Brain Res 365:397–403

Doreulee N, Sergeeva OA, Yanovsky Y, Chepkova AN, Selbach O, Godecke A, Schrader J, Haas HL (2003) Cortico-striatal synaptic plasticity in endothelial nitric oxide synthase deficient mice. Brain Res 964:159–163

Dowie MJ, Grimsey NL, Hoffman T, Faull RL, Glass M (2014) Cannabinoid receptor CB2 is expressed on vascular cells, but not astroglial cells in the post-mortem human Huntington's disease brain. J Chem Neuroanat 59-60:62–71

Eblen F, Graybiel AM (1995) Highly restricted origin of prefrontal cortical inputs to striosomes in the macaque monkey. J Neurosci 15:5999–6013

Ebrahimi A, Pochet R, Roger M (1992) Topographical organization of the projections from physiologically identified areas of the motor cortex to the striatum in the rat. Neurosci Res 14:39–60

Egertova M, Elphick MR (2000) Localisation of cannabinoid receptors in the rat brain using antibodies to the intracellular C-terminal tail of CB. J Comp Neurol 422:159–171

Emson PC, Waldvogel HJ, Faull RLM (2010) Neurotransmitter receptors in the basal ganglia. In: Steiner H, Tseng KY (eds) Handbook of behavioral neuroscience. Handbook of basal ganglia structure and function, vol 20. pp 75–96

Everitt BJ, Robbins TW (2016) Drug addiction: Updating actions to habits to compulsions ten years on. Annu Rev Psychol 67:23–50

Fegley D, Kathuria S, Mercier R, Li C, Goutopoulos A, Makriyannis A, Piomelli D (2004a) Anandamide transport is independent of fatty-acid amide hydrolase activity and is blocked by the hydrolysis-resistant inhibitor AM1172. Proc Natl Acad Sci U S A 101:8756–8761

Ferre S, Fredholm BB, Morelli M, Popoli P, Fuxe K (1997) Adenosine-dopamine receptor-receptor interactions as an integrative mechanism in the basal ganglia. Trends Neurosci 20:482–487

Ferrer B, Bermudez-Silva FJ, Bilbao A, Alvarez-Jaimes L, Sanchez-Vera I, Giuffrida A, Serrano A, Baixeras E, Khaturia S, Navarro M, Parsons LH, Piomelli D, Rodriguez de Fonseca F (2007) Regulation of brain anandamide by acute administration of ethanol. Biochem J 404:97–104

Fino E, Glowinski J, Venance L (2005) Bidirectional activity-dependent plasticity at corticostriatal synapses. J Neurosci 25:11279–11287

Fino E, Paille V, Cui Y, Morera-Herreras T, Deniau JM, Venance L (2010) Distinct coincidence detectors govern the corticostriatal spike timing-dependent plasticity. J Physiol 588:3045–3062

Flores-Hernandez J, Galarraga E, Bargas J (1997) Dopamine selects glutamatergic inputs to neostriatal neurons. Synapse 25:185–195

Fourgeaud L, Mato S, Bouchet D, Hemar A, Worley PF, Manzoni OJ (2004) A single in vivo exposure to cocaine abolishes endocannabinoid-mediated long-term depression in the nucleus accumbens. J Neurosci 24:6939–6945

Freiman I, Anton A, Monyer H, Urbanski MJ, Szabo B (2006) Analysis of the effects of cannabinoids on identified synaptic connections in the caudate-putamen by paired recordings in transgenic mice. J Physiol 575:789–806

French ED, Dillon K, Wu X (1997) Cannabinoids excite dopamine neurons in the ventral tegmentum and substantia nigra. Neuroreport 8:649–652

Fusco FR, Martorana A, Giampa C, De March Z, Farini D, D'Angelo V, Sancesario G, Bernardi G (2004) Immunolocalization of CB1 receptor in rat striatal neurons: a confocal microscopy study. Synapse 53:159–167

Fuxe K, Marcellino D, Genedani S, Agnati L (2007) Adenosine A(2A) receptors, dopamine D (2) receptors and their interactions in Parkinson's disease. Mov Disord 22:1990–2017

Gao Y, Vasilyev DV, Goncalves MB, Howell FV, Hobbs C, Reisenberg M, Shen R, Zhang MY, Strassle BW, Lu P, Mark L, Piesla MJ, Deng K, Kouranova EV, Ring RH, Whiteside GT, Bates B, Walsh FS, Williams G, Pangalos MN, Samad TA, Doherty P (2010) Loss of retrograde endocannabinoid signaling and reduced adult neurogenesis in diacylglycerol lipase knock-out mice. J Neurosci 30:2017–2024

Gardner EL (2002) Addictive potential of cannabinoids: the underlying neurobiology. Chem Phys Lipids 121:267–290

Gardner EL (2005) Endocannabinoid signaling system and brain reward: emphasis on dopamine. Pharmacol Biochem Behav 81:263–284

Gerdeman G, Lovinger DM (2001) CB1 cannabinoid receptor inhibits synaptic release of glutamate in rat dorsolateral striatum. J Neurophysiol 85:468–471

Gerdeman GL, Ronesi J, Lovinger DM (2002) Postsynaptic endocannabinoid release is critical to long-term depression in the striatum. Nat Neurosci 5:446–451

Gerfen CR (1992) The neostriatal mosaic: multiple levels of compartmental organization. Trends Neurosci 15:133–139

Gerfen CR, Bolam JP (2010) The neuroanatomical organization of the basal ganglia. In: Steiner H, Tseng K.Y (eds) Handbook of behavioral neuroscience. Handbook of basal ganglia structure and function, vol 20. pp 3–28

Giuffrida A, Parsons LH, Kerr TM, Rodriguez de Fonseca F, Navarro M, Piomelli D (1999) Dopamine activation of endogenous cannabinoid signaling in dorsal striatum. Nat Neurosci 2:358–363

Glass M, Brotchie JM, Maneuf YP (1997) Modulation of neurotransmission by cannabinoids in the basal ganglia. Eur J Neurosci 9:199–203

Gonsiorek W, Lunn C, Fan X, Narula S, Lundell D, Hipkin RW (2000) Endocannabinoid 2-arachidonyl glycerol is a full agonist through human type 2 cannabinoid receptor: antagonism by anandamide. Mol Pharmacol 57:1045–1050

Goodman J, Packard MG (2014) Peripheral and intra-dorsolateral striatum injections of the cannabinoid receptor agonist WIN 55,212-2 impair consolidation of stimulus-response memory. Neuroscience 274:128–137

Goodman J, Packard MG (2015) The influence of cannabinoids on learning and memory processes of the dorsal striatum. Neurobiol Learn Mem 125:1–14

Gough AL, Olley JE (1978) Catalepsy induced by intrastriatal injections of delta9-THC and 11-OH-delta9-THC in the rat. Neuropharmacology 17(2):137–144

Gremel CM, Costa RM (2013) Orbitofrontal and striatal circuits dynamically encode the shift between goal-directed and habitual actions. Nat Commun 4:2264

Gremel C, Chancey J, Atwood B, Luo G, Neve R, Ramakrishnan C, Deisseroth K, Lovinger D, Costa R (2016) Endocannabinoid modulation of orbitostriatal circuits gates habit formation. Neuron 90(6):1312–1324

Grueter BA, Brasnjo G, Malenka RC (2010) Postsynaptic TRPV1 triggers cell type-specific long-term depression in the nucleus accumbens. Nat Neurosci 13:1519–1525

Gubellini P, Saulle E, Centonze D, Bonsi P, Pisani A, Bernardi G, Conquet F, Calabresi P (2001) Selective involvement of mGlu1 receptors in corticostriatal LTD. Neuropharmacology 40:839–846

Gubellini P, Picconi B, Bari M, Battista N, Calabresi P, Centonze D, Bernardi G, Finazzi-Agro A, Maccarrone M (2002) Experimental parkinsonism alters endocannabinoid degradation: implications for striatal glutamatergic transmission. J Neurosci 22:6900–6907

Guegan T, Cutando L, Gangarossa G, Santini E, Fisone G, Martinez A, Valjent E, Maldonado R, Martin M (2013) Operant behavior to obtain palatable food modifies ERK activity in the brain reward circuit. Eur Neuropsychopharmacol 23(3):240–252

Gulyas AI, Cravatt BF, Bracey MH, Dinh TP, Piomelli D, Boscia F, Freund TF (2004) Segregation of two endocannabinoid-hydrolyzing enzymes into pre- and postsynaptic compartments in the rat hippocampus, cerebellum and amygdala. Eur J Neurosci 20:441–458

Hart CL (2005) Increasing treatment options for cannabis dependence: a review of potential pharmacotherapies. Drug Alcohol Depend 80:147–159

Hashimotodani Y, Ohno-Shosaku T, Tanimura A, Kita Y, Sano Y, Shimizu T, Di Marzo V, Kano M (2013) Acute inhibition of diacylglycerol lipase blocks endocannabinoid-mediated retrograde signalling: evidence for on-demand biosynthesis of 2-arachidonoylglycerol. J Physiol 591:4765–4776

Haughey HM, Marshall E, Schacht JP, Louis A, Hutchison KE (2008) Marijuana withdrawal and craving: influence of the cannabinoid receptor 1 (CNR1) and fatty acid amide hydrolase (FAAH) genes. Addiction 103(10):1678–1686

Heifets BD, Chevaleyre V, Castillo PE (2008) Interneuron activity controls endocannabinoid-mediated presynaptic plasticity through calcineurin. Proc Natl Acad Sci U S A 105:10250–10255

Henriksson BG, Järbe TUC (1972) Δ9-Tetrahydrocannabinol used as discriminative stimulus for rats in position learning in a T-shaped water maze. Psychon Sci 27:25–26

Herkenham M, Lynn AB, de Costa BR, Richfield EK (1991a) Neuronal localization of cannabinoid receptors in the basal ganglia of the rat. Brain Res 547:267–274

Herkenham M, Lynn AB, Johnson MR, Melvin LS, de Costa BR, Rice KC (1991b) Characterization and localization of cannabinoid receptors in rat brain: a quantitative in vitro autoradiographic study. J Neurosci 11:563–583

Hernandez G, Cheer JF (2012) Effect of CB1 receptor blockade on food-reinforced responding and associated nucleus accumbens neuronal activity in rats. J Neurosci 32(33):11467–11477

Hernandez M, Berrendero F, Suarez I, Garcia-Gil L, Cebeira M, Mackie K, Ramos JA, Fernandez-Ruiz J (2000) Cannabinoid CB(1) receptors colocalize with tyrosine hydroxylase in cultured fetal mesencephalic neurons and their activation increases the levels of this enzyme. Brain Res 857:56–65

Hernandez G, Oleson EB, Gentry RN, Abbas Z, Bernstein DL, Arvanitogiannis A, Cheer JF (2014) Endocannabinoids promote cocaine-induced impulsivity and its rapid dopaminergic correlates. Biol Psychiatry 75(6):487–498

Hernandez-Lopez S, Tkatch T, Perez-Garci E, Galarraga E, Bargas J, Hamm H, Surmeier DJ (2000) D2 dopamine receptors in striatal medium spiny neurons reduce L-type Ca2+ currents and excitability via a novel PLC[beta]1-IP3-calcineurin-signaling cascade. J Neurosci 20:8987–8995

Hilario MR, Clouse E, Yin HH, Costa RM (2007) Endocannabinoid signaling is critical for habit formation. Front Integr Neurosci 1:6

Hillard CJ, Jarrahian A (2000) The movement of N-arachidonoylethanolamine (anandamide) across cellular membranes. Chem Phys Lipids 108:123–134

Hillard CJ, Edgemond WS, Jarrahian A, Campbell WB (1997) Accumulation of N-arachidonoylethanolamine (anandamide) into cerebellar granule cells occurs via facilitated diffusion. J Neurochem 69:631–638

Hoffman AF, Oz M, Caulder T, Lupica CR (2003) Functional tolerance and blockade of long-term depression at synapses in the nucleus accumbens after chronic cannabinoid exposure. J Neurosci 23:4815–4820

Hohmann AG, Herkenham M (2000) Localization of cannabinoid CB(1) receptor mRNA in neuronal subpopulations of rat striatum: a double-label in situ hybridization study. Synapse 37:71–80

Howe AR, Surmeier DJ (1995) Muscarinic receptors modulate N-, P-, and L-type Ca2+ currents in rat striatal neurons through parallel pathways. J Neurosci 15:458–469

Howlett AC, Barth F, Bonner TI, Cabral G, Casellas P, Devane WA, Felder CC, Herkenham M, Mackie K, Martin BR, Mechoulam R, Pertwee RG (2002) International Union of Pharmacology. XXVII. Classification of cannabinoid receptors. Pharmacol Rev 54:161–202

Hsu KS, Huang CC, Yang CH, Gean PW (1995) Presynaptic D2 dopaminergic receptors mediate inhibition of excitatory synaptic transmission in rat neostriatum. Brain Res 690:264–268

Hu SS, Mackie K (2015) Distribution of the Endocannabinoid system in the central nervous system. Handb Exp Pharmacol 231:59–93

Huang CC, Lo SW, Hsu KS (2001) Presynaptic mechanisms underlying cannabinoid inhibition of excitatory synaptic transmission in rat striatal neurons. J Physiol 532:731–748

Huang CC, Chen YL, Lo SW, Hsu KS (2002) Activation of cAMP-dependent protein kinase suppresses the presynaptic cannabinoid inhibition of glutamatergic transmission at corticostriatal synapses. Mol Pharmacol 61:578–585

Huang YC, Wang SJ, Chiou LC, Gean PW (2003) Mediation of amphetamine-induced long-term depression of synaptic transmission by CB1 cannabinoid receptors in the rat amygdala. J Neurosci 23:10311–10320

James S, Richardson PJ (1993) The subcellular distribution of [3H]-CGS 21680 binding sites in the rat striatum: copurification with cholinergic nerve terminals. Neurochem Int 23:115–122

Jarbe TU, Henriksson BG (1973) Effects of delta8-THC, and delta9-THC on the acquisition of a discriminative positional habit in rats. The transitions between normal and tetrahydrocannabinol-induced states on reversal learning. Psychopharmacologia 31:321–332

Joel D, Weiner I (2000) The connections of the dopaminergic system with the striatum in rats and primates: an analysis with respect to the functional and compartmental organization of the striatum. Neuroscience 96:451–474

Julian MD, Martin AB, Cuellar B, Rodriguez De Fonseca F, Navarro M, Moratalla R, Garcia-Segura LM (2003) Neuroanatomical relationship between type 1 cannabinoid receptors and dopaminergic systems in the rat basal ganglia. Neuroscience 119:309–318

Jung KM, Mangieri R, Stapleton C, Kim J, Fegley D, Wallace M, Mackie K, Piomelli D (2005) Stimulation of endocannabinoid formation in brain slice cultures through activation of group I metabotropic glutamate receptors. Mol Pharmacol 68:1196–1202

Kano M, Ohno-Shosaku T, Hashimotodani Y, Uchigashima M, Watanabe M (2009) Endocannabinoid-mediated control of synaptic transmission. Physiol Rev 89:309–380

Kathmann M, Bauer U, Schlicker E, Gothert M (1999) Cannabinoid CB1 receptor-mediated inhibition of NMDA- and kainate-stimulated noradrenaline and dopamine release in the brain. Naunyn Schmiedeberg's Arch Pharmacol 359:466–470

Katona I (2009) Endocannabinoid receptors: CNS localization of the CB(1) cannabinoid receptor. Curr Top Behav Neurosci 1:65–86

Kauer JA, Gibson HE (2009) Hot flash: TRPV channels in the brain. Trends Neurosci 32:215–224

Kawaguchi Y (1992) Large aspiny cells in the matrix of the rat neostriatum in vitro: physiological identification, relation to the compartments and excitatory postsynaptic currents. J Neurophysiol 67:1669–1682

Kawaguchi Y (1993) Physiological, morphological, and histochemical characterization of three classes of interneurons in rat neostriatum. J Neurosci 13:4908–4923

Kheirbek MA, Britt JP, Beeler JA, Ishikawa Y, McGehee DS, Zhuang X (2009) Adenylyl cyclase type 5 contributes to corticostriatal plasticity and striatum-dependent learning. J Neurosci 29:12115–12124

Kim DS, Palmiter RD (2008) Interaction of dopamine and adenosine receptor function in behavior: studies with dopamine-deficient mice. Front Biosci 13:2311–2318

Kincaid AE, Wilson CJ (1996) Corticostriatal innervation of the patch and matrix in the rat neostriatum. J Comp Neurol 374:578–592

Kita H (1993) GABAergic circuits of the striatum. Prog Brain Res 99:51–72

Kobayashi K, Manabe T, Takahashi T (1999) Calcium-dependent mechanisms involved in presynaptic long-term depression at the hippocampal mossy fibre-CA3 synapse. Eur J Neurosci 11:1633–1638

Kofalvi A, Rodrigues RJ, Ledent C, Mackie K, Vizi ES, Cunha RA, Sperlagh B (2005) Involvement of cannabinoid receptors in the regulation of neurotransmitter release in the rodent striatum: a combined immunochemical and pharmacological analysis. J Neurosci 25:2874–2884

Koles L, Garcao P, Zadori ZS, Ferreira SG, Pinheiro BS, da Silva-Santos CS, Ledent C, Kofalvi A (2013) Presynaptic TRPV1 vanilloid receptor function is age- but not CB1 cannabinoid receptor-dependent in the rodent forebrain. Brain Res Bull 97:126–135

Koos T, Tepper JM (1999) Inhibitory control of neostriatal projection neurons by GABAergic interneurons. Nat Neurosci 2:467–472

Kramer PF, Christensen CH, Hazelwood LA, Dobi A, Bock R, Sibley DR, Mateo Y, Alvarez VA (2011) Dopamine D2 receptor overexpression alters behavior and physiology in Drd2-EGFP mice. J Neurosci 31:126–132

Kreitzer AC, Malenka RC (2005) Dopamine modulation of state-dependent endocannabinoid release and long-term depression in the striatum. J Neurosci 25:10537–10545

Kreitzer AC, Malenka RC (2007) Endocannabinoid-mediated rescue of striatal LTD and motor deficits in Parkinson's disease models. Nature 445:643–647

Kreitzer AC, Malenka RC (2008) Striatal plasticity and basal ganglia circuit function. Neuron 60:543–554

Kreitzer AC, Regehr WG (2001) Retrograde inhibition of presynaptic calcium influx by endogenous cannabinoids at excitatory synapses onto Purkinje cells. Neuron 29:717–727

Kubota Y, Mikawa S, Kawaguchi Y (1993) Neostriatal GABAergic interneurones contain NOS, calretinin or parvalbumin. Neuroreport 5:205–208

Kushmerick C, Price GD, Taschenberger H, Puente N, Renden R, Wadiche JI, Duvoisin RM, Grandes P, von Gersdorff H (2004) Retroinhibition of presynaptic Ca2+ currents by endocannabinoids released via postsynaptic mGluR activation at a calyx synapse. J Neurosci 24:5955–5965

Langer LF, Graybiel AM (1989) Distinct nigrostriatal projection systems innervate striosomes and matrix in the primate striatum. Brain Res 498:344–350

Lerner TN, Kreitzer AC (2012) RGS4 is required for dopaminergic control of striatal LTD and susceptibility to parkinsonian motor deficits. Neuron 73:347–359

Lerner TN, Horne EA, Stella N, Kreitzer AC (2010) Endocannabinoid signaling mediates psychomotor activation by adenosine A2A antagonists. J Neurosci 30:2160–2164
Li ST, Kato K, Tomizawa K, Matsushita M, Moriwaki A, Matsui H, Mikoshiba K (2002) Calcineurin plays different roles in group II metabotropic glutamate receptor- and NMDA receptor-dependent long-term depression. J Neurosci 22:5034–5041
Ligresti A, Cascio MG, Di Marzo V (2005) Endocannabinoid metabolic pathways and enzymes. Curr Drug Targets CNS Neurol Disord 4:615–623
Liu J, Wang L, Harvey-White J, Osei-Hyiaman D, Razdan R, Gong Q, Chan AC, Zhou Z, Huang BX, Kim HY, Kunos G (2006) A biosynthetic pathway for anandamide. Proc Natl Acad Sci U S A 103:13345–13350
Lodge D, Tidball P, Mercier MS, Lucas SJ, Hanna L, Ceolin L, Kritikos M, Fitzjohn SM, Sherwood JL, Bannister N, Volianskis A, Jane DE, Bortolotto ZA, Collingridge GL (2013) Antagonists reversibly reverse chemical LTD induced by group I, group II and group III metabotropic glutamate receptors. Neuropharmacology 74:135–146
Lovinger DM, Tyler EC, Merritt A (1993) Short- and long-term synaptic depression in rat neostriatum. J Neurophysiol 70:1937–1949
Lovinger DM, Merritt A, Reyes D (1994) Involvement of N- and non-N-type calcium channels in synaptic transmission at corticostriatal synapses. Neuroscience 62:31–40
Lovinger DM, Davis MI, Costa RM (2010) Endocannabinoid signaling in the striatum. In: Steiner H, Tseng KY (eds) Handbook of behavioral neuroscience. Handbook of basal ganglia structure and function, vol 20. pp 167–186
Lupica CR, Riegel AC, Hoffman AF (2004) Marijuana and cannabinoid regulation of brain reward circuits. Br J Pharmacol 143:227–234
Maccarrone M, Attina M, Bari M, Cartoni A, Ledent C, Finazzi-Agro A (2001) Anandamide degradation and N-acylethanolamines level in wild-type and CB1 cannabinoid receptor knockout mice of different ages. J Neurochem 78:339–348
Maccarrone M, Gubellini P, Bari M, Picconi B, Battista N, Centonze D, Bernardi G, Finazzi-Agro A, Calabresi P (2003) Levodopa treatment reverses endocannabinoid system abnormalities in experimental parkinsonism. J Neurochem 85:1018–1025
Mackie K, Hille B (1992) Cannabinoids inhibit N-type calcium channels in neuroblastoma-glioma cells. Proc Natl Acad Sci U S A 89:3825–3829
Mackie K, Devane WA, Hille B (1993) Anandamide, an endogenous cannabinoid, inhibits calcium currents as a partial agonist in N18 neuroblastoma cells. Mol Pharmacol 44:498–503
Mackie K, Lai Y, Westenbroek R, Mitchell R (1995) Cannabinoids activate an inwardly rectifying potassium conductance and inhibit Q-type calcium currents in AtT20 cells transfected with rat brain cannabinoid receptor. J Neurosci 15:6552–6561
Mailleux P, Vanderhaeghen JJ (1992a) Age-related loss of cannabinoid receptor binding sites and mRNA in the rat striatum. Neurosci Lett 147:179–181
Mailleux P, Vanderhaeghen JJ (1992b) Distribution of neuronal cannabinoid receptor in the adult rat brain: a comparative receptor binding radioautography and in situ hybridization histochemistry. Neuroscience 48:655–668
Malach R, Graybiel AM (1986) Mosaic architecture of the somatic sensory-recipient sector of the cat's striatum. J Neurosci 6:3436–3458
Malinen H, Hyytiä P (2008) Ethanol self-administration is regulated by CB1 receptors in the nucleus accumbens and ventral tegmental area in alcohol-preferring AA rats. Alcohol Clin Exp Res 32(11):1976–1983
Malone DT, Kearn CS, Chongue L, Mackie K, Taylor DA (2008) Effect of social isolation on CB1 and D2 receptor and fatty acid amide hydrolase expression in rats. Neuroscience 152:265–272
Manzoni OJ, Bockaert J (2001) Cannabinoids inhibit GABAergic synaptic transmission in mice nucleus accumbens. Eur J Pharmacol 412:R3–R5
Marco EM, Echeverry-Alzate V, Lopez-Moreno JA, Gine E, Penasco S, Viveros MP (2014) Consequences of early life stress on the expression of endocannabinoid-related genes in the rat brain. Behav Pharmacol 25:547–556

Marshall K, Gowing L, Ali R, Le Foll B (2014) Pharmacotherapies for cannabis dependence. Cochrane Database Syst Rev 2:CD008940. doi:10.1002/14651858

Marsicano G, Lutz B (1999) Expression of the cannabinoid receptor CB1 in distinct neuronal subpopulations in the adult mouse forebrain. Eur J Neurosci 11:4213–4225

Martin BR, Compton DR, Thomas BF, Prescott WR, Little PJ, Razdan RK, Johnson MR, Melvin LS, Mechoulam R, Ward SJ (1991) Behavioral, biochemical, and molecular modeling evaluations of cannabinoid analogs. Pharmacol Biochem Behav 40:471–478

Martin HGS, Bernabeu A, Lassalle O, Bouille C, Beurrier C, Pelissier-Alicot A-L, Manzoni OJ (2015) Endocannabinoids mediate muscarinic acetylcholine receptor-dependent long-term depression in the adult medial prefrontal cortex. Front Cell Neurosci 9:457. doi:10.3389/fncel.2015.00457

Martire A, Tebano MT, Chiodi V, Ferreira SG, Cunha RA, Kofalvi A, Popoli P (2011) Pre-synaptic adenosine A2A receptors control cannabinoid CB1 receptor-mediated inhibition of striatal glutamatergic neurotransmission. J Neurochem 116:273–280

Mathur BN, Tanahira C, Tamamaki N, Lovinger DM (2013) Voltage drives diverse endocannabinoid signals to mediate striatal microcircuit-specific plasticity. Nat Neurosci 16:1275–1283

Mato S, Chevaleyre V, Robbe D, Pazos A, Castillo PE, Manzoni OJ (2004) A single in-vivo exposure to delta 9THC blocks endocannabinoid-mediated synaptic plasticity. Nat Neurosci 7 (6):585–586

Mato S, Robbe D, Puente N, Grandes P, Manzoni OJ (2005) Presynaptic homeostatic plasticity rescues long-term depression after chronic Delta 9-tetrahydrocannabinol exposure. J Neurosci 25:11619–11627

Mato S, Lafourcade M, Robbe D, Bakiri Y, Manzoni OJ (2008) Role of the cyclic-AMP/PKA cascade and of P/Q-type Ca++ channels in endocannabinoid-mediated long-term depression in the nucleus accumbens. Neuropharmacology 54:87–94

Matsuda LA, Lolait SJ, Brownstein MJ, Young AC, Bonner TI (1990) Structure of a cannabinoid receptor and functional expression of the cloned cDNA. Nature 346:561–564

Matsuda LA, Bonner TI, Lolait SJ (1993) Localization of cannabinoid receptor mRNA in rat brain. J Comp Neurol 327:535–550

Matyas F, Watanabe M, Mackie K, Katona I, Freund TF (2007) Molecular architecture of the cannabinoid signaling system in the core of the nucleus accumbens. Ideggyogy Sz 60:187–191

McCutcheon JE, Loweth JA, Ford KA, Marinelli M, Wolf ME, Tseng KY (2011) Group I mGluR activation reverses cocaine-induced accumulation of calcium-permeable AMPA receptors in nucleus accumbens synapses via a protein kinase C-dependent mechanism. J Neurosci 31 (41):14536–14541

Mechoulam R, Ben-Shabat S, Hanus L, Ligumsky M, Kaminski NE, Schatz AR, Gopher A, Almog S, Martin BR, Compton DR et al (1995) Identification of an endogenous 2-monoglyceride, present in canine gut, that binds to cannabinoid receptors. Biochem Pharmacol 50:83–90

Micale V, Cristino L, Tamburella A, Petrosino S, Leggio GM, Drago F, Di Marzo V (2009) Anxiolytic effects in mice of a dual blocker of fatty acid amide hydrolase and transient receptor potential vanilloid type-1 channels. Neuropsychopharmacology 34:593–606

Monory K, Blaudzun H, Massa F, Kaiser N, Lemberger T, Schutz G, Wotjak CT, Lutz B, Marsicano G (2007) Genetic dissection of behavioural and autonomic effects of Delta(9)-tetrahydrocannabinol in mice. PLoS Biol 5:e269

Narushima M, Hashimoto K, Kano M (2006a) Endocannabinoid-mediated short-term suppression of excitatory synaptic transmission to medium spiny neurons in the striatum. Neurosci Res 54:159–164

Narushima M, Uchigashima M, Hashimoto K, Watanabe M, Kano M (2006b) Depolarization-induced suppression of inhibition mediated by endocannabinoids at synapses from fast-spiking interneurons to medium spiny neurons in the striatum. Eur J Neurosci 24:2246–2252

Narushima M, Uchigashima M, Fukaya M, Matsui M, Manabe T, Hashimoto K, Watanabe M, Kano M (2007) Tonic enhancement of endocannabinoid-mediated retrograde suppression of inhibition by cholinergic interneuron activity in the striatum. J Neurosci 27:496–506

Naydenov AV, Sepers MD, Swinney K, Raymond LA, Palmiter RD, Stella N (2014) Genetic rescue of CB1 receptors on medium spiny neurons prevents loss of excitatory striatal synapses but not motor impairment in HD mice. Neurobiol Dis 71:140–150

Nazzaro C, Greco B, Cerovic M, Baxter P, Rubino T, Trusel M, Parolaro D, Tkatch T, Benfenati F, Pedarzani P, Tonini R (2012) SK channel modulation rescues striatal plasticity and control over habit in cannabinoid tolerance. Nat Neurosci 15:284–293

Nyilas R, Dudok B, Urban GM, Mackie K, Watanabe M, Cravatt BF, Freund TF, Katona I (2008) Enzymatic machinery for endocannabinoid biosynthesis associated with calcium stores in glutamatergic axon terminals. J Neurosci 28:1058–1063

Nyiri G, Cserep C, Szabadits E, Mackie K, Freund TF (2005) CB1 cannabinoid receptors are enriched in the perisynaptic annulus and on preterminal segments of hippocampal GABAergic axons. Neuroscience 136:811–822

Okamoto Y, Morishita J, Tsuboi K, Tonai T, Ueda N (2004) Molecular characterization of a phospholipase D generating anandamide and its congeners. J Biol Chem 279:5298–5305

Okamoto Y, Wang J, Morishita J, Ueda N (2007) Biosynthetic pathways of the endocannabinoid anandamide. Chem Biodivers 4:1842–1857

Olson PA, Tkatch T, Hernandez-Lopez S, Ulrich S, Ilijic E, Mugnaini E, Zhang H, Bezprozvanny I, Surmeier DJ (2005) G-protein-coupled receptor modulation of striatal CaV1.3 L-type Ca2+ channels is dependent on a Shank-binding domain. J Neurosci 25:1050–1062

Orio L, Edwards S, George O, Parsons LH, Koob GF (2009) A role for the endocannabinoid system in the increased motivation for cocaine in extended-access conditions. J Neurosci 29 (15):4846–4857

Ortinski PI, Vassoler FM, Carlson GC, Pierce RC (2012) Temporally dependent changes in cocaine-induced synaptic plasticity in the nucleus accumbens shell are reversed by D1-like dopamine receptor stimulation. Neuropsychopharmacology 37(7):1671–1682

Oude Ophuis RJ, Boender AJ, van Rozen AJ, Adan RA (2014) Cannabinoid, melanocortin and opioid receptor expression on DRD1 and DRD2 subpopulations in rat striatum. Front Neuroanat 8:14

Packard MG, McGaugh JL (1996) Inactivation of hippocampus or caudate nucleus with lidocaine differentially affects expression of place and response learning. Neurobiol Learn Mem 65:65–72

Palazuelos J, Aguado T, Pazos MR, Julien B, Carrasco C, Resel E, Sagredo O, Benito C, Romero J, Azcoitia I, Fernandez-Ruiz J, Guzman M, Galve-Roperh I (2009) Microglial CB2 cannabinoid receptors are neuroprotective in Huntington's disease excitotoxicity. Brain 132:3152–3164

Parsons LH, Hurd YL (2015) Endocannabinoid signalling in reward and addiction. Nat Rev Neurosci 16:579–594

Partridge JG, Janssen MJ, Chou DY, Abe K, Zukowska Z, Vicini S (2009) Excitatory and inhibitory synapses in neuropeptide Y-expressing striatal interneurons. J Neurophysiol 102:3038–3045

Perez-Burgos A, Perez-Rosello T, Salgado H, Flores-Barrera E, Prieto GA, Figueroa A, Galarraga E, Bargas J (2008) Muscarinic M(1) modulation of N and L types of calcium channels is mediated by protein kinase C in neostriatal neurons. Neuroscience 155:1079–1097

Perra S, Pillolla G, Melis M, Muntoni AL, Gessa GL, Pistis M (2005) Involvement of the endogenous cannabinoid system in the effects of alcohol in the mesolimbic reward circuit: electrophysiological evidence in vivo. Psychopharmacology 183(3):368–377

Pertwee RG (2015) Endocannabinoids and their pharmacological actions. Handb Exp Pharmacol 231:1–37

Pertwee RG, Wickens AP (1991) Enhancement by chlordiazepoxide of catalepsy induced in rats by intravenous or intrapallidal injections of enantiomeric cannabinoids. Neuropharmacology 30:237–244

Petryszyn S, Beaulieu JM, Parent A, Parent M (2014) Distribution and morphological characteristics of striatal interneurons expressing calretinin in mice: a comparison with human and nonhuman primates. J Chem Neuroanat 59-60:51–61

Picconi B, Bagetta V, Ghiglieri V, Paille V, Di Filippo M, Pendolino V, Tozzi A, Giampa C, Fusco FR, Sgobio C, Calabresi P (2011) Inhibition of phosphodiesterases rescues striatal long-term depression and reduces levodopa-induced dyskinesia. Brain 134:375–387

Pickel VM, Chan J, Kash TL, Rodriguez JJ, MacKie K (2004) Compartment-specific localization of cannabinoid 1 (CB1) and mu-opioid receptors in rat nucleus accumbens. Neuroscience 127:101–112

Pickel VM, Chan J, Kearn CS, Mackie K (2006) Targeting dopamine D2 and cannabinoid-1 (CB1) receptors in rat nucleus accumbens. J Comp Neurol 495:299–313

Pickel VM, Shobin ET, Lane DA, Mackie K (2012) Cannabinoid-1 receptors in the mouse ventral pallidum are targeted to axonal profiles expressing functionally opposed opioid peptides and contacting N-acylphosphatidylethanolamine-hydrolyzing phospholipase D terminals. Neuroscience 227:10–21

Piomelli D, Giuffrida A, Calignano A, Rodriguez de Fonseca F (2000) The endocannabinoid system as a target for therapeutic drugs. Trends Pharmacol Sci 21:218–224

Quiroz C, Lujan R, Uchigashima M, Simoes AP, Lerner TN, Borycz J, Kachroo A, Canas PM, Orru M, Schwarzschild MA, Rosin DL, Kreitzer AC, Cunha RA, Watanabe M, Ferre S (2009) Key modulatory role of presynaptic adenosine A2A receptors in cortical neurotransmission to the striatal direct pathway. ScientificWorldJournal 9:1321–1344

Robbe D, Alonso G, Duchamp F, Bockaert J, Manzoni OJ (2001) Localization and mechanisms of action of cannabinoid receptors at the glutamatergic synapses of the mouse nucleus accumbens. J Neurosci 21:109–116

Robbe D, Alonso G, Chaumont S, Bockaert J, Manzoni OJ (2002a) Role of p/q-Ca2+ channels in metabotropic glutamate receptor 2/3-dependent presynaptic long-term depression at nucleus accumbens synapses. J Neurosci 22:4346–4356

Robbe D, Kopf M, Remaury A, Bockaert J, Manzoni OJ (2002b) Endogenous cannabinoids mediate long-term synaptic depression in the nucleus accumbens. Proc Natl Acad Sci U S A 99:8384–8388

Roberts JC, Davis JB, Benham CD (2004) [3H]Resiniferatoxin autoradiography in the CNS of wild-type and TRPV1 null mice defines TRPV1 (VR-1) protein distribution. Brain Res 995:176–183

Rodriguez JJ, Mackie K, Pickel VM (2001) Ultrastructural localization of the CB1 cannabinoid receptor in mu-opioid receptor patches of the rat Caudate putamen nucleus. J Neurosci 21:823–833

Rodriguez-Moreno A, Banerjee A, Paulsen O (2010) Presynaptic NMDA receptors and spike timing-dependent depression at cortical synapses. Front Synaptic Neurosci 2:18

Romero J, Garcia-Palomero E, Castro JG, Garcia-Gil L, Ramos JA, Fernandez-Ruiz JJ (1997) Effects of chronic exposure to delta9-tetrahydrocannabinol on cannabinoid receptor binding and mRNA levels in several rat brain regions. Brain Res Mol Brain Res 46:100–108

Ronesi J, Lovinger DM (2005) Induction of striatal long-term synaptic depression by moderate frequency activation of cortical afferents in rat. J Physiol 562:245–256

Ronesi J, Gerdeman GL, Lovinger DM (2004) Disruption of endocannabinoid release and striatal long-term depression by postsynaptic blockade of endocannabinoid membrane transport. J Neurosci 24:1673–1679

Sagredo O, Gonzalez S, Aroyo I, Pazos MR, Benito C, Lastres-Becker I, Romero JP, Tolon RM, Mechoulam R, Brouillet E, Romero J, Fernandez-Ruiz J (2009) Cannabinoid CB2 receptor agonists protect the striatum against malonate toxicity: relevance for Huntington's disease. Glia 57:1154–1167

Seif T, Makriyannis A, Kunos G, Bonci A, Hopf FW (2011) The endocannabinoid 2-arachidonoylglycerol mediates D1 and D2 receptor cooperative enhancement of rat nucleus accumbens core neuron firing. Neuroscience 193:21–33

Sergeeva OA, Doreulee N, Chepkova AN, Kazmierczak T, Haas HL (2007) Long-term depression of cortico-striatal synaptic transmission by DHPG depends on endocannabinoid release and nitric oxide synthesis. Eur J Neurosci 26:1889–1894

Shen M, Thayer SA (1998) The cannabinoid agonist Win55,212-2 inhibits calcium channels by receptor-mediated and direct pathways in cultured rat hippocampal neurons. Brain Res 783:77–84

Shen M, Piser TM, Seybold VS, Thayer SA (1996) Cannabinoid receptor agonists inhibit glutamatergic synaptic transmission in rat hippocampal cultures. J Neurosci 16:4322–4334

Shen W, Flajolet M, Greengard P, Surmeier DJ (2008) Dichotomous dopaminergic control of striatal synaptic plasticity. Science 321:848–851

Shonesy BC, Wang X, Rose KL, Ramikie TS, Cavener VS, Rentz T, Baucum AJ 2nd, Jalan-Sakrikar N, Mackie K, Winder DG, Patel S, Colbran RJ (2013) CaMKII regulates diacylglycerol lipase-alpha and striatal endocannabinoid signaling. Nat Neurosci 16:456–463

Shonesy BC, Winder DG, Patel S, Colbran RJ (2015) The initiation of synaptic 2-AG mobilization requires both an increased supply of diacylglycerol precursor and increased postsynaptic calcium. Neuropharmacology 91:57–62

Sidlo Z, Reggio PH, Rice ME (2008) Inhibition of striatal dopamine release by CB1 receptor activation requires nonsynaptic communication involving GABA, H2O2, and KATP channels. Neurochem Int 52:80–88

Simon GM, Cravatt BF (2006) Endocannabinoid biosynthesis proceeding through glycerophospho-N-acyl ethanolamine and a role for alpha/beta-hydrolase 4 in this pathway. J Biol Chem 281:26465–26472

Simon GM, Cravatt BF (2008) Anandamide biosynthesis catalyzed by the phosphodiesterase GDE1 and detection of glycerophospho-N-acyl ethanolamine precursors in mouse brain. J Biol Chem 283:9341–9349

Simon GM, Cravatt BF (2010) Characterization of mice lacking candidate N-acyl ethanolamine biosynthetic enzymes provides evidence for multiple pathways that contribute to endocannabinoid production in vivo. Mol BioSyst 6:1411–1418

Singla S, Kreitzer AC, Malenka RC (2007) Mechanisms for synapse specificity during striatal long-term depression. J Neurosci 27:5260–5264

Smart D, Jerman JC (2000) Anandamide: an endogenous activator of the vanilloid receptor. Trends Pharmacol Sci 21:134

Song WJ, Tkatch T, Surmeier DJ (2000) Adenosine receptor expression and modulation of Ca(2+) channels in rat striatal cholinergic interneurons. J Neurophysiol 83:322–332

Soria-Gomez E, Matias I, Rueda-Orozco PE, Cisneros M, Petrosino S, Navarro L, Di Marzo V, Prospero-Garcia O (2007) Pharmacological enhancement of the endocannabinoid system in the nucleus accumbens shell stimulates food intake and increases c-Fos expression in the hypothalamus. Br J Pharmacol 151:1109–1116

Stanton PK, Gage AT (1996) Distinct synaptic loci of Ca2+/calmodulin-dependent protein kinase II necessary for long-term potentiation and depression. J Neurophysiol 76:2097–2101

Starowicz K, Cristino L, Di Marzo V (2008) TRPV1 receptors in the central nervous system: potential for previously unforeseen therapeutic applications. Curr Pharm Des 14:42–54

Stella N, Schweitzer P, Piomelli D (1997) A second endogenous cannabinoid that modulates long-term potentiation. Nature 388:773–778

Stromberg I, Popoli P, Muller CE, Ferre S, Fuxe K (2000) Electrophysiological and behavioural evidence for an antagonistic modulatory role of adenosine A2A receptors in dopamine D2 receptor regulation in the rat dopamine-denervated striatum. Eur J Neurosci 12:4033–4037

Sugiura T, Kondo S, Sukagawa A, Nakane S, Shinoda A, Itoh K, Yamashita A, Waku K (1995) 2-Arachidonoylglycerol: a possible endogenous cannabinoid receptor ligand in brain. Biochem Biophys Res Commun 215:89–97

Sung KW, Choi S, Lovinger DM (2001) Activation of group I mGluRs is necessary for induction of long-term depression at striatal synapses. J Neurophysiol 86:2405–2412

Svenningsson P, Le Moine C, Kull B, Sunahara R, Bloch B, Fredholm BB (1997) Cellular expression of adenosine A2A receptor messenger RNA in the rat central nervous system with special reference to dopamine innervated areas. Neuroscience 80:1171–1185

Szabo B, Dorner L, Pfreundtner C, Norenberg W, Starke K (1998) Inhibition of GABAergic inhibitory postsynaptic currents by cannabinoids in rat corpus striatum. Neuroscience 85:395–403

Szabo B, Muller T, Koch H (1999) Effects of cannabinoids on dopamine release in the corpus striatum and the nucleus accumbens in vitro. J Neurochem 73:1084–1089

Tanda G, Munzar P, Goldberg SR (2000) Self-administration behavior is maintained by the psychoactive ingredient of marijuana in squirrel monkeys. Nat Neurosci 3:1073–1107

Tang K, Low MJ, Grandy DK, Lovinger DM (2001) Dopamine-dependent synaptic plasticity in striatum during in vivo development. Proc Natl Acad Sci U S A 98:1255–1260

Tanimura A, Yamazaki M, Hashimotodani Y, Uchigashima M, Kawata S, Abe M, Kita Y, Hashimoto K, Shimizu T, Watanabe M, Sakimura K, Kano M (2010) The endocannabinoid 2-arachidonoylglycerol produced by diacylglycerol lipase alpha mediates retrograde suppression of synaptic transmission. Neuron 65:320–327

Tepper JM, Bolam JP (2004) Functional diversity and specificity of neostriatal interneurons. Curr Opin Neurobiol 14:685–692

Tepper JM, Abercrombie ED, Bolam JP (2007) Basal ganglia macrocircuits. Prog Brain Res 160:3–7

Tepper JM, Tecuapetla F, Koos T, Ibanez-Sandoval O (2010) Heterogeneity and diversity of striatal GABAergic interneurons. Front Neuroanat 4:150

Thiemann G, van der Stelt M, Petrosino S, Molleman A, Di Marzo V, Hasenohrl RU (2008) The role of the CB1 cannabinoid receptor and its endogenous ligands, anandamide and 2-arachidonoylglycerol, in amphetamine-induced behavioural sensitization. Behav Brain Res 187:289–296

Tozzi A, Tscherter A, Belcastro V, Tantucci M, Costa C, Picconi B, Centonze D, Calabresi P, Borsini F (2007) Interaction of A2A adenosine and D2 dopamine receptors modulates corticostriatal glutamatergic transmission. Neuropharmacology 53:783–789

Tozzi A, de Iure A, Di Filippo M, Tantucci M, Costa C, Borsini F, Ghiglieri V, Giampa C, Fusco FR, Picconi B, Calabresi P (2011) The distinct role of medium spiny neurons and cholinergic interneurons in the D(2)/A(2)A receptor interaction in the striatum: implications for Parkinson's disease. J Neurosci 31:1850–1862

Tozzi A, de Iure A, Marsili V, Romano R, Tantucci M, Di Filippo M, Costa C, Napolitano F, Mercuri NB, Borsini F, Giampa C, Fusco FR, Picconi B, Usiello A, Calabresi P (2012) A2A adenosine receptor antagonism enhances synaptic and motor effects of cocaine via CB1 cannabinoid receptor activation. PLoS One 7:e38312

Tsou K, Brown S, Sanudo-Pena MC, Mackie K, Walker JM (1998) Immunohistochemical distribution of cannabinoid CB1 receptors in the rat central nervous system. Neuroscience 83:393–411

Twitchell W, Brown S, Mackie K (1997) Cannabinoids inhibit N- and P/Q-type calcium channels in cultured rat hippocampal neurons. J Neurophysiol 78:43–50

Uchigashima M, Narushima M, Fukaya M, Katona I, Kano M, Watanabe M (2007) Subcellular arrangement of molecules for 2-arachidonoyl-glycerol-mediated retrograde signaling and its physiological contribution to synaptic modulation in the striatum. J Neurosci 27:3663–3676

Ueda N, Kurahashi Y, Yamamoto S, Tokunaga T (1995) Partial purification and characterization of the porcine brain enzyme hydrolyzing and synthesizing anandamide. J Biol Chem 270:23823–23827

Valjent E, Maldonado R (2000) A behavioural model to reveal place preference to delta 9-tetrahydrocannabinol in mice. Psychopharmacology (Berl) 147:436–438

Van Waes V, Beverley JA, Siman H, Tseng KY, Steiner H (2012) CB1 cannabinoid receptor expression in the striatum: association with corticostriatal circuits and developmental regulation. Front Pharmacol 3:21

Volkow ND, Wang GJ, Telang F, Fowler JS, Alexoff D, Logan J, Jayne M, Wong C, Tomasi D (2014) Decreased dopamine brain reactivity in marijuana abusers is associated with negative emotionality and addiction severity. Proc Natl Acad Sci U S A 111(30):E3149–E3156

Walsh JP (1993) Depression of excitatory synaptic input in rat striatal neurons. Brain Res 608:123–128

Wang H, Pickel VM (2002) Dopamine D2 receptors are present in prefrontal cortical afferents and their targets in patches of the rat caudate-putamen nucleus. J Comp Neurol 442:392–404

Wang Z, Kai L, Day M, Ronesi J, Yin HH, Ding J, Tkatch T, Lovinger DM, Surmeier DJ (2006) Dopaminergic control of corticostriatal long-term synaptic depression in medium spiny neurons is mediated by cholinergic interneurons. Neuron 50:443–452

Wang W, Dever D, Lowe J, Storey GP, Bhansali A, Eck EK, Nitulescu I, Weimer J, Bamford NS (2012) Regulation of prefrontal excitatory neurotransmission by dopamine in the nucleus accumbens core. J Physiol 590:3743–3769

Wilson RI, Nicoll RA (2001) Endogenous cannabinoids mediate retrograde signalling at hippocampal synapses. Nature 410:588–592

Wilson RI, Nicoll RA (2002) Endocannabinoid signaling in the brain. Science 296:678–682

Winters BD, Kruger JM, Huang X, Gallaher ZR, Ishikawa M, Czaja K, Krueger JM, Huang YH, Schluter OM, Dong Y (2012) Cannabinoid receptor 1-expressing neurons in the nucleus accumbens. Proc Natl Acad Sci U S A 109:E2717–E2725

Wong MY, Borgkvist A, Choi SJ, Mosharov EV, Bamford NS, Sulzer D (2015) Dopamine-dependent corticostriatal synaptic filtering regulates sensorimotor behavior. Neuroscience 290:594–607

Wu YW, Kim JI, Tawfik VL, Lalchandani RR, Scherrer G, Ding JB (2015) Input- and cell-type-specific endocannabinoid-dependent LTD in the striatum. Cell Rep 10:75–87

Xi ZX, Gilbert JG, Peng XQ, Pak AC, Li X, Gardner EL (2006) Cannabinoid CB1 receptor antagonist AM251 inhibits cocaine-primed relapse in rats: role of glutamate in the nucleus accumbens. J Neurosci 26(33):8531–8536

Xi ZX, Peng XQ, Li X, Song R, Zhang HY, Liu QR, Yang HJ, Bi GH, Li J, Gardner EL (2011) Brain cannabinoid CB_2 receptors modulate cocaine's actions in mice. Nat Neurosci 14 (9):1160–1166

Xia JX, Li J, Zhou R, Zhang XH, Ge YB, Ru Yuan X (2006) Alterations of rat corticostriatal synaptic plasticity after chronic ethanol exposure and withdrawal. Alcohol Clin Exp Res 30:819–824

Yin HH, Knowlton BJ (2006) The role of the basal ganglia in habit formation. Nat Rev Neurosci 7 (6):464–476

Yin HH, Lovinger DM (2006) Frequency-specific and D2 receptor-mediated inhibition of glutamate release by retrograde endocannabinoid signaling. Proc Natl Acad Sci U S A 103:8251–8256

Yin HH, Davis MI, Ronesi JA, Lovinger DM (2006) The role of protein synthesis in striatal long-term depression. J Neurosci 26:11811–11820

Yin HH, Adermark L, Lovinger DM (2008) Neurotensin reduces glutamatergic transmission in the dorsolateral striatum via retrograde endocannabinoid signaling. Neuropharmacology 54:79–86

Zangen A, Solinas M, Ikemoto S, Goldberg SR, Wise RA (2006) Two brain sites for cannabinoid reward. J Neurosci 26(18):4901–4907

Zhang X, Feng ZJ, Chergui K (2015) Induction of cannabinoid- and N-methyl-D-aspartate receptor-mediated long-term depression in the nucleus accumbens and dorsolateral striatum is region and age dependent. Int J Neuropsychopharmacol 18(4). doi:10.1093/ijnp/pyu052

Zhou FM, Wilson CJ, Dani JA (2002) Cholinergic interneuron characteristics and nicotinic properties in the striatum. J Neurobiol 53:590–605

Zygmunt PM, Petersson J, Andersson DA, Chuang H, Sorgard M, Di Marzo V, Julius D, Hogestatt ED (1999) Vanilloid receptors on sensory nerves mediate the vasodilator action of anandamide. Nature 400:452–457

Lipid Mediators in the Regulation of Emotions, Memory, and Cognitive Functions

Beat Lutz

Abstract This chapter aims at highlighting the diverse roles of endocannabinoids in the coordination of balanced neuronal activities, which finally set the basis for the organism's characteristics to store and remember important and useful things, to forget non-useful things, and to cope with new challenges. Altogether, the fine-tuned regulation of these processes is crucial for optimal life and survival. The endocannabinoid system appears to be a central intrinsic homeostatic factor in the organism, modulating these processes. Receptors for (endo)cannabinoids are also targets for exogenous cannabinoids, putting also relevance of external substances in the interference with these processes. The genetic dissection of the endocannabinoid system together with the many pharmacological, biochemical, behavioral, and electrophysiological approaches has led to relevant insights into this lipid signaling system in the regulation of emotions, memory, and cognitive functions. Recent progresses are discussed in this chapter, particularly the question on cell type- and region-specific involvements of endocannabinoids and the CB1 receptor-dependent regulation of distinct neuronal pathways in the context of fear behaviors and stress-induced effects on memory formation.

1 Introduction

The very widespread and diversely acting lipid signaling class of endocannabinoids has gained an intensive interest both in basic and in applied medical sciences but also in societal aspects of the steadily discussed issue of *Cannabis* legalization (Bloomfield et al. 2016; Volkow et al. 2016). Numerous recent reviews have elaborated on various aspects of the endocannabinoids anandamide (AEA, arachidonoyl ethanolamide) and 2-arachidonoyl glycerol (2-AG), their corresponding receptors, cannabinoid type 1 (CB1) receptor and type 2 (CB2) receptor, and the synthesizing and degrading enzymes of endocannabinoids, comprising the endocannabinoid system (Soltesz et al. 2015; Maccarrone et al. 2014;

B. Lutz (✉)
Institute of Physiological Chemistry, University Medical Center of the Johannes Gutenberg University, Duesbergweg 6, 55128 Mainz, Germany
e-mail: blutz@uni-mainz.de

M. Melis (ed.), *Endocannabinoids and Lipid Mediators in Brain Functions*,
DOI 10.1007/978-3-319-57371-7_6

Lutz et al. 2015). The intrinsic complexity of the endocannabinoid system and also the continuous technical advancements in analytical approaches, together with posing novel questions that require an open-minded and an unbiased view, have enabled to reveal surprising and highly interesting new insights into the functioning of this lipid signaling system. This chapter is not meant to be complete nor to be well balanced, but wishes to discuss some topics that may be relevant for the further progress in understanding this signaling system.

A recurrent theme in understanding a signaling system is how the signal is generated at a particular site; where exactly the signal is received, i.e., where the receptors are located; and, finally, how the signal is further propagated and then how it is terminated. The fact that endocannabinoids are lipids and cannot be detected by high-resolution methods and that both CB1 and CB2 receptor proteins are often expressed at extremely low levels (e.g., CB1 receptor in astrocytes, oligodendrocytes, mitochondria; CB2 receptor in neurons) and in different cellular compartments (e.g., pre- and post-synapse, soma, mitochondria, lysosomes), but nevertheless revealing detectable and physiologically relevant functions, gives the promise that we will continue to gain new and exciting insights, thus making (endo) cannabinoid research a rewarding endeavor. Studies on the endogenous functions of the endocannabinoid system will fuel the research on the mechanistic understandings on how the different pharmacologically active plant-derived compounds (e.g., from *Cannabis sativa*) mediate their effects.

2 The Regulation of Fear Memory and Extinction

Due to high relevance in psychiatric disorders, such as posttraumatic stress disorders, phobias, and anxiety disorders, the scientific community has shown great interest on the impact of psychological and physical stress on emotional memory and, specifically, on how memory of traumatic events is "burnt" into neural systems and how to relieve patients from these intruding memories (Singewald et al. 2015). Numerous works have investigated the role of the endocannabinoid system in fear memory formation and, in particular, in fear extinction (Lutz et al. 2015; Morena and Campolongo 2014).

The preferred behavioral paradigm used has been the auditory and contextual fear conditioning where both memory formation and fear extinction have been investigated. Broadly speaking, CB1 receptor is necessary for extinction of aversive memories and for reduction of fear responses and is involved in fear-coping strategies (Lutz et al. 2015). Specifically, complete loss both of CB1 receptor (Marsicano et al. 2002; Lutz et al. 2015) and of the 2-AG-synthesizing enzyme DAGLα (Jenniches et al. 2016), thereby interrupting 2-AG/CB1 receptor signaling, leads to impaired fear extinction and increased anxiety. On the other hand, increased AEA signaling by genetic alterations or pharmacological interventions of the AEA-degrading enzyme fatty acid amide hydrolase (FAAH) resulted in the facilitation of fear extinction (Dincheva et al. 2015; Gunduz-Cinar et al. 2013),

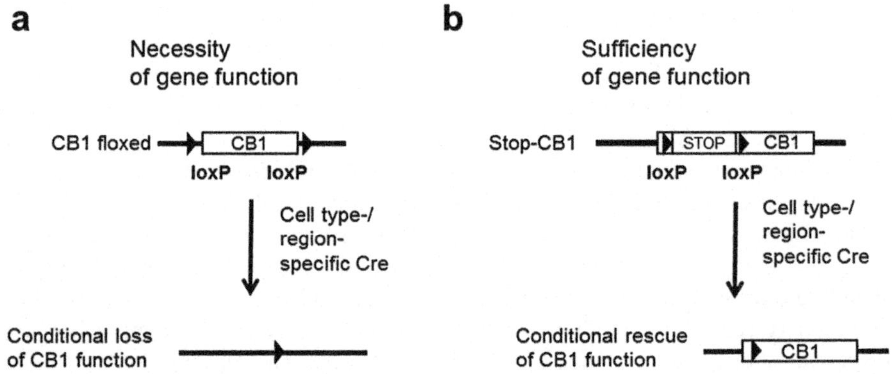

Fig. 1 Schematic representation of genetic analysis of CB1 receptor in mouse using the Cre/loxP system. (**a**) To assess necessary gene function, CB1-floxed mouse line, containing two loxP sites flanking an essential gene region (in this example, the coding region of the CB1 receptor gene) but retaining wild-type gene function, is crossed with a particular Cre recombinase-expressing transgenic mouse line or is stereotaxically injected with virus expressing Cre recombinase in a region-/cell type-specific manner, leading to a cell type- and/or region-specific loss of gene function. (**b**) To assess sufficient gene function, Stop-CB1 mouse line contains a transcriptional stop cassette in the untranslated 5′ region of the CB1 gene, flanked by two loxP sites, leading to the inactivation of CB1 gene function. Crossing of Stop-CB1 line with a particular Cre recombinase-expressing transgenic mouse line or stereotaxically injected with virus expressing Cre recombinase in a region-/cell type-specific manner leads to a cell type- and/or region-specific reactivation of gene function, thus a rescue of gene function in a cell type- and/or region-specific manner

while genetic inactivation of the 2-AG-degrading enzyme monoacylglycerol lipase (MAGL) tended to show slower contextual fear extinction (Kishimoto et al. 2015).

To date, however, it is still unclear where exactly in the fear circuitry the CB1 receptor is centrally involved in order to regulate fear response, fear memory formation, and fear extinction. Conditional mutagenesis in mice using the Cre/loxP system has enabled the generation of CB1 receptor deficiency in particular cellular subpopulations or brain regions and thereby assessed the necessity of this receptor (Fig. 1a).

Using this strategy, the question is put forward under which circumstances the system "breaks" down and CB1 receptor-dependent behaviors are dysregulated. To this end, CB1 receptor expressed on GABAergic and glutamatergic neurons has been investigated in details (Dubreucq et al. 2012; Jacob et al. 2012; Lutz et al. 2015). Here, CB1 receptor expressed in dorsal telencephalic glutamatergic neurons is necessary for appropriate fear memory extinction, while CB1 receptor on forebrain GABAergic neurons plays a minor role in the regulation of this behavior. CB1 receptor expressed in adrenergic/noradrenergic cells does not seem to play a role in fear expression and fear extinction (Busquets-Garcia et al. 2016). Most advanced insights into the role of CB1 receptor on a distinct projection with regard to a particular behavior were gained recently. Presynaptic CB1 receptor controls the expression of aversive memories by selectively modulating cholinergic transmission at medial habenular synapses projecting onto target postsynaptic interpeduncular neurons (Soria-Gomez et al. 2015). Further elucidation of CB1 receptor function in these behaviors will require advanced genetic methods, whereby, e.g., the CB1 receptor gene is specifically inactivated only in a distinct neural cell type in a particular brain region. This requires the development of very specific Cre recombinase-expressing transgenic driver mouse lines (which are mostly not yet available), of newly developed viral constructs directing Cre recombinase expressing into distinct cell types, and possibly also the consideration of novel genetic methods, such as the CRISPR/Cas technique (Platt et al. 2014).

In the meantime, another genetic approach has been developed, aiming at shedding light onto the cell type-specific function of CB1 receptor in fear memory and anxiety (Ruehle et al. 2013). Based on the stringent genetic approaches and power in other model systems, e.g., yeast, *C. elegans*, and *Drosophila*, a complementary approach has been undertaken. Here, the aim is to reconstruct the "normal" behavior by reintroducing CB1 receptor only in distinct cell types in an otherwise CB1 receptor-deficient background, thereby investigating the sufficiency of CB1 receptor function for a distinct behavior. This approach is schematized in Fig. 1b.

By applying this approach, dorsal telencephalic glutamatergic CB1 receptor was sufficient for several CB1 receptor-dependent behaviors, including normal anxiety-like behavior, matching with the results obtained from experiments using conditional CB1 receptor deletion in this neuronal population (Ruehle et al. 2013). However, the major unexpected deviation was the freezing during cued fear extinction. Here, fear extinction behavior was worsened in the rescue mice as compared to mice with complete loss of CB1 receptor function. In continuation of this approach, mice were generated with a specific rescue of CB1 receptor in forebrain GABAergic interneurons and striatal GABAergic neurons. Again, while anxiety-like behavior was rescued, no substantial sufficiency of GABAergic CB1 receptor was observed for fear extinction (Remmers et al. 2017). While for the rescue of anxiety-like behaviors, both glutamatergic and GABAergic CB1 receptors, respectively, are sufficient, the pathways involved in fear extinction appear to be diverse and more complex. The currently known pathways (Tovote et al. 2015) will form a basis for further studies unraveling these CB1 receptor-containing pathways.

Figure 2 illustrates the two approaches of necessity and sufficiency in gene function, by using a simple representation and arguing here that a distinct sensory input (here, 8) will need a complex processing in neural networks, leading to the perception of the number 8 with all its meaning and implications. To address necessity, the loss of distinct component(s) in the information processing might lead to the perception of a wrong and misleading number (here, 5) or to information that might not be meaningful and interpretable. To address sufficiency, on a (genetic) null background, distinct components are reintroduced into the complex system, aiming at finding the minimal requirement for an information processing leading to the correct and discernable number 8. In this process, also other information might be generated (the inverted number 3).

However, considering the high complexity of brain neural networks and the high degree of self-organizing capacity and compensatory mechanisms of the brain after

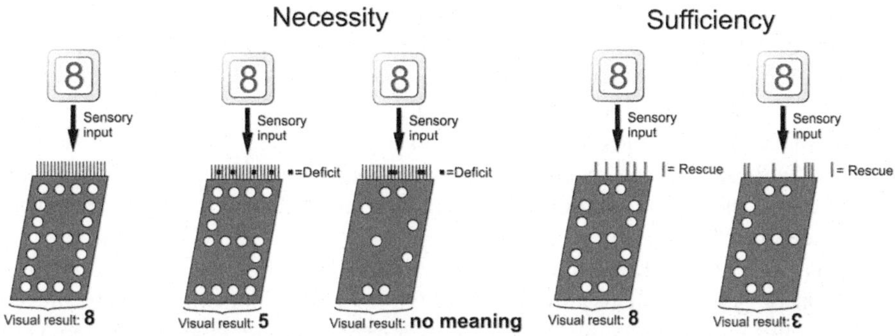

Fig. 2 Schematic representation to visualize the analyses of necessity (second and third panel) and sufficiency of gene function (fourth and fifth panel), using the example of information processing leading to the LED display of numbers in an electronic device. In order to test necessity, distinct inputs and processing are inactivated, leading to altered information (second and third panels, number 5 or complete loss of information). While the initial sensory information is always the same, sufficient quantity and quality of inputs and processing must be present, in order to retain the clear recognition of the number 8 (fourth panel). This approach can also lead to wrong and/or nonsense information (fifth panel, inverted 3)

distinct functional failures aiming at ensuring survival, if possible, the genetic approaches addressing necessity and sufficiency of gene function in distinct cell types and/or neuronal projections are very challenging and not without pitfalls. It might be argued that despite a correct rescue of the number 8, not the normally active and important networks were engaged but compensatory and self-organizing, and thereby rescuing processes have been activated. Therefore, the genetic approaches require the assistance and complementation by other methods, such as pharmacology, optogenetics, and electrophysiology. In this combination of methods, recent investigations have enabled novel insights into CB1 receptor functions, as detailed below in two examples.

3 Impact of Stress on Emotional Memory

It is well established that the emotional state effects memory consolidation processes (de Quervain et al. 2017; Luksys and Sandi 2011). Acute stressful encounters activate the sympathetic-adrenal system and the hypothalamic-pituitary-adrenal (HPA) axis, thereby initiating the peripheral production and release of hormones such as adrenaline, noradrenaline, and cortisol, to trigger both peripheral physiological responses, such as increase of blood pressure, and also responses in the central nervous system. Noradrenaline produced by the sympathetic nerve terminals, and adrenalin and cortisol in the adrenal glands, reaches the bloodstream. These hormones evoke pronounced effects on the cognitive capacity, by exerting their effects on different brain regions involved in the formation of memories (Gazarini et al. 2013; Singewald et al. 2015; Mizrachi Zer-Aviv et al. 2016). On one hand, emotions can help us to remember important events of our life, but on the other hand, they can also be responsible for failures or exaggeration in "recording" of the memories.

The endocannabinoid system is a signaling system that helps to reduce the release of stress hormones when a stressful event triggers this response. In this process, the organism can recover to the conditions prior to the stress through the activation of the endocannabinoid system and tuning down the stress response (Morena et al. 2016).

It has been reported that the enhancement of AEA levels by pharmacological inhibition or genetic inactivation of the AEA-degrading enzyme FAAH relieves anxiety states induced, e.g., by chronic stressor exposure (Hill et al. 2013; Lomazzo et al. 2015) and alleviates impaired fear extinction after chronic stress (Laricchiuta et al. 2013). Two possible cellular processes are at play and may explain the beneficial effects of experimentally enhanced AEA levels in the alleviation of anxiety after chronic stress (Lutz et al. 2015). (i) It has been reported in rodents that CB1 receptor signaling is decreased after stress in several brain regions involved in emotional processing, including the hippocampus, nucleus accumbens, prefrontal cortex, and basolateral amygdala (BLA). (ii) Chronic stress increases FAAH activity and, thus, decreases AEA levels. The underlying mechanisms leading to decreased CB1 receptor signaling or increased FAAH activity after stress have not yet been elucidated. Recently, it was proposed that, in the BLA, postsynaptic activation of the corticotropin-releasing hormone receptor type 1 (CRHR1) initiates a pathway leading to the increased activity of FAAH, resulting in dysregulated synaptic function after chronic stress (Gray et al. 2015). CRHR1 in the amygdala seems also to be involved in the effect of stress regarding 2-AG signaling, which is dysfunctional after stress (Qin et al. 2015), and can be rescued by tyrosine phosphatase PTP1B inhibition or by glucocorticoid receptor antagonism.

4 Impact of Stress on Nonemotional Memory

The mechanisms underlying the influence of stress on the formation of nonemotional memories, such as episodic and autobiographic events and spatial information, show still a large gap in knowledge. Based on the involvement of the endocannabinoid system in the negative feedback of stress response, the question rises on the role of the CB1 receptor and the endocannabinoids in a behavioral paradigm of stress-induced impairment of consolidation of nonemotional memory. To this end, the novel object recognition test was applied, a behavioral paradigm where the hippocampus is centrally involved in (Busquets-Garcia et al. 2016). Mice were habituated to the V-maze, and after 24 h, mice were exposed for 9 min with two identical objects, each placed at the end of the arms. 20 min after this memory formation process, mice were acutely stressed by a foot shock. After 24 h, to assess memory consolidation, mice were placed into the same V-maze, where one of the two objects was changed with a new, non-familiar object. Exploration for the new object in relation to the old, familiar object was expressed as a discrimination index, whereby high discrimination index means strong previously acquired memory consolidation. Using this behavioral paradigm, several interesting and unexpected observations were obtained. (i) While the acute stress after memory formation (day 1) did not impair short-term memory, when the discrimination index was determined 1 hour after memory formation, stress did impair memory consolidation in a CB1 receptor-dependent manner as measured on day 2. Remarkably, complete loss of CB1 receptor function led to alleviation of the stress-induced impairment of memory consolidation. The use of mutants with conditional CB1 receptor loss in forebrain GABAergic, in dorsal telencephalic glutamatergic, in forebrain GABAergic and dorsal telencephalic glutamatergic, and in central serotonergic neurons did not reveal CB1 receptor function in these cell types for this behavior. However, the loss in adrenergic/noradrenergic cells (both in the central and peripheral nervous system and in the adrenal gland) alleviated the stress-induced impairment of memory consolidation, similarly as the complete CB1 receptor-deficiency. (ii) Systemic pharmacological treatment with a peripherally restricted CB1 receptor antagonist (AM6545) prior to stress or local injection of CB1 receptor antagonist rimonabant (SR1) into the hippocampus led to the recovery of this memory impairment. (iii) Conditional rescue mutants expressing CB1 receptor only on adrenergic/noradrenergic cells regained the sensitivity to shock-induced impairment of memory consolidation, indicating that CB1 receptor expressed only in these cell types is sufficient for mediating stress-induced memory impairment. This is remarkable in the light of the complexity of the stress response and its effect on memory consolidation. (iv) Plasma levels of adrenalin and noradrenalin were increased at 90 min after memory formation in the AM6545 group as compared to vehicle group, suggesting that these increased levels in the phase of memory consolidation are important for prohibiting the stress-induced impairment of memory consolidation. (v) Sotalol, a peripherally restricted β-adrenergic receptor antagonist, was able to prevent the rescuing effect of rimonabant regarding stress-induced impairment of memory consolidation, pointing to a key role of peripheral β-adrenergic receptor signaling downstream of CB1 receptor blockade. Altogether, these data allow the proposition that only peripheral blockade of CB1 receptor can be beneficial for alleviation of stress-induced impairment of nonemotional memory and, furthermore, indicate an intriguing cross talk between peripheral and central mechanisms underlying the effect of stress on nonemotional memory consolidation and, in consequence, on potential new approaches in the treatment of cognitive interference on stress-related disorders. This is very attractive in the light that CB1 receptor blockade in the central nervous system may lead to serious side effects (Fig. 3).

Interestingly, another study showed that the anorexic and anxiogenic effects of the CB1 receptor antagonist rimonabant required the peripheral activation of the sympathetic system, as shown by the application of sotalol (Bellocchio et al. 2013). Both investigations strongly suggest the need of increasing the focus on the role of the endocannabinoid system in the interactions between peripheral and central

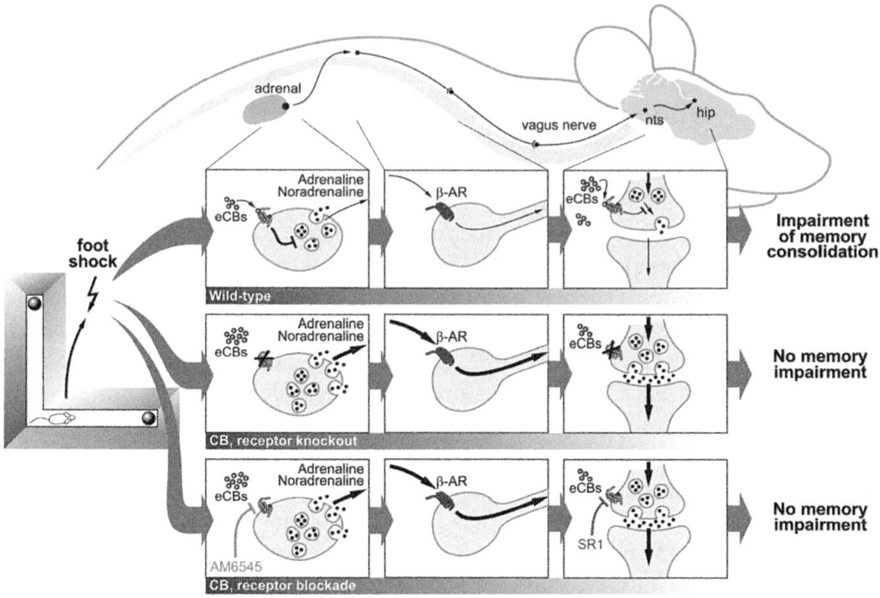

Fig. 3 Summary of the role of CB1 receptor and the endocannabinoid system in the mediation of stress-induced impairment of nonemotional memory consolidation. CB1 receptor both in hippocampal noradrenergic terminals and in the adrenal gland and sympathetic neuron is important for this effect, as both pharmacological blockade of peripheral CB1 receptor (using AM6545) and intrahippocampal (hip) blockade of CB1 receptor (using SR1) led to the alleviation of stress-induced memory impairment. In this memory consolidation-promoting effect, the peripheral β-adrenergic receptor (β-AR) must be functional. It is suggested that sustained adrenergic/noradrenergic signaling via peripheral nerves (e.g., vagus) and the nucleus of the tractus solitarius (nts) are involved

processes. It is obvious that the behavioral output (e.g., stress response, memory impairment, etc.) originates in the actions of central nervous system-guided processes, but they are strongly influenced by peripheral signaling systems. This has been recognized, of course, in the regulation and action of the HPA axis, where the central effects of cortisol and the interaction of the endocannabinoid system have been described in numerous studies. However, it remains to be investigated how the peripheral adrenergic/noradrenergic signals reach the brain and influence memory consolidation. Furthermore, acute stress enhances endocannabinoid levels in the hippocampus during the consolidation window in the object recognition task. Accordingly, intrahippocampal administration of rimonabant blocked the stress-induced impairment nonemotional memory both in wild-type mice and in mice re-expressing (rescuing) CB1 receptor only in the adrenergic/noradrenergic cell. Further experiments are needed to detail the function of CB1 receptor in these cell types and in the modulation of neurotransmitter release, e.g., in the hippocampus.

5 Fear Response to Unpredictable Threat

The fear reaction of a mammal to threat can be quick and rapidly dissipating or be more long-lasting and may turn into a pathological state in its extreme. While fear circuits activated by discrete stimuli (e.g., contextual and cued fear conditioning) have been understood in good details, the mechanisms underlying the shift from fear to an anxiety state have been barely understood (Tovote et al. 2015). Recent investigations proposed these two states to be operationalized as phasic and sustained fear in response to predictable and unpredictable threat, respectively (Davis et al. 2010; Seidenbecher et al. 2016). This gave an important entry point for translational research. Animals exposed to unpredictable, but not to predictable threats show behaviors resembling anxiety symptoms in humans, underlying the importance of unpredictability in anxiety disorders. Recent investigations suggest that the extended amygdala, including the central amygdala (CeA) and the bed nucleus of stria terminalis (BNST), is key in the control of these behaviors (Davis et al. 2010). The presence of a functional endocannabinoid system in CeA (Kamprath et al. 2011; Ramikie et al. 2014) and BNST (Puente et al. 2011) led to the motivation to investigate the role of endocannabinoids and CB1 receptor in a mouse model of sustained fear (Seidenbecher et al. 2016). Unpredictability of the threat and functional CB1 receptor in the BNST are required to develop sustained fear (Lange et al. 2016). Conditional genetic loss of function experiments, in combination with conditional genetic rescue experiments, and BNST-specific applications of CB1 receptor blocker, together with in vitro optogenetic and electrophysiological experiments, were able to propose a mechanism on how CB1 receptor regulates the emergence of sustained fear (Fig. 4).

Using this behavioral paradigm and advanced genetic interference methods should allow the further description of the CB1 receptor-containing projections from the amygdala to the alBNST and how the receiving neurons in the alBNST converge the two signals from BA and CeA, thus leading to output signals, which determine behavioral response of sustained fear. It appears that the integration of two signals in one neuron in the BNST requires time-dependent mechanisms, whereby the production and action of endocannabinoids have key roles.

6 CB2 Receptor and Emotions

In the very recent years, a possible involvement of CB2 receptor in neural functions has been proposed. CB2 receptor appears to be expressed in the hippocampus (Li and Kim 2015; Stempel et al. 2016), and genetic loss of CB1 receptor leads to reduced synaptic transmission and long-term potentiation (Li and Kim 2016b). Detailed studies on CB2 receptor function in the hippocampus led to the proposal that CB2 receptor is involved in a self-regulatory manner and regulates input/output functions and modulates gamma oscillations (Stempel et al. 2016). In fact, loss of

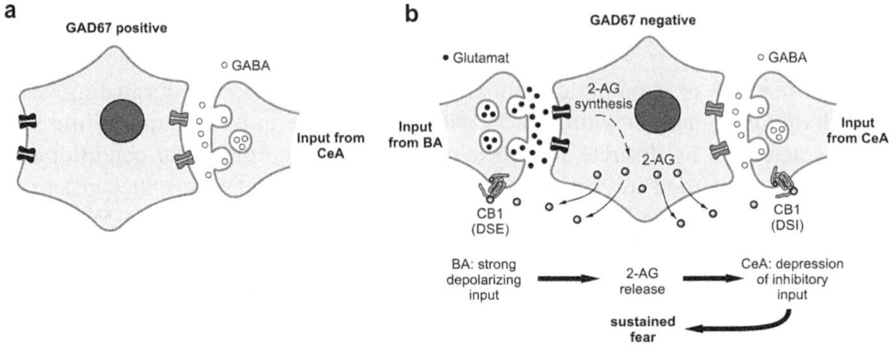

Fig. 4 Schematic drawing of CB1 receptor-regulated pathways from the basal amygdala (BA) and from the central amygdala (CeA) to the anterolateral bed nucleus of the stria terminalis (alBNST). (a) While in the alBNST, GABAergic neurons (identified as glutamic acid decarboxylase 67, GAD67-positive cells) do obtain GABAergic synaptic input from CeA, but in a CB1 receptor-independent manner, (b) GAD67-negative neurons (presumably glutamatergic neurons) receive glutamatergic inputs from BA and GABAergic inputs from CeA, both of them contain presynaptic CB1 receptor. Therefore, depolarization-induced suppression of excitation (DSE) and inhibition (DSI) can occur at both synapses. In its temporal sequence, BA input leads to activation of the alBNST GAD67-negative neuron; thereupon 2-AG is generated, which in turn suppresses the glutamatergic (from BA) and GABAergic inputs (from CeA), leading to the depression of the inhibitory input (a tonic disinhibition) and the slow activation in the alBNST neurons, mediating finally sustainment of fear response

CB2 impairs contextual fear memory, while cued fear conditioning was not altered (Li and Kim 2016a) and also reduces the consolidation of aversive memory as tested in the step-down inhibitory avoidance task (Garcia-Gutierrez et al. 2013). Further investigations using conditional CB2 receptor deletions will enable to uncover the underlying mechanisms involved in CB2 receptor-mediated functions in learning and memory.

Recently, the natural CB2 receptor agonist β-caryophyllene has been tested in a mouse model of Alzheimer disease in order to alleviate cognitive impairments. In fact, β-caryophyllene, given orally, prevented cognitive impairment in the Alzheimer mouse model APP/PS1, and this positive cognitive effect was associated with reduced β-amyloid burden in the hippocampus and cerebral cortex. Additionally, β-caryophyllene reduced astrogliosis and microglial activation as well as the levels of COX-2 protein and the mRNA levels of the proinflammatory cytokines tumor necrosis factor-α and interleukin-1β in the cerebral cortex. The use of the CB2 antagonist AM630 or the PPARγ antagonist GW9662 significantly reversed the protective effects of β-caryophyllene on APP/PS1 mice (Cheng et al. 2014). These are interesting results with high therapeutic potential, as inflammatory processes are central to many facets of brain dysregulation.

7 Therapeutic Interventions

The pharmacological intervention in order to alleviate fear expression or to promote fear extinction has attracted a long-standing interest. It was reported that at least in the rodent model system, the antidepressant drug fluoxetine promotes fear extinction (Karpova et al. 2011). A recent investigation showed that chronic fluoxetine treatment produced a significant and selective increase in levels of anandamide in the BLA and an associated decrease in activity of the AEA-catabolizing enzyme FAAH. Slice electrophysiological recordings showed that fluoxetine-induced increases in AEA were associated with the amplification of endocannabinoid-mediated tonic constraint of inhibitory, but not excitatory, transmission in the BLA. Behaviorally, chronic fluoxetine facilitated extinction retrieval in a manner that was prevented by systemic or BLA-specific blockade of CB1 receptor (Gunduz-Cinar et al. 2016). This fits to the proposition that enhanced AEA signaling in the amygdala beneficially affects fear extinction (Gunduz-Cinar et al. 2013). In contrast, pharmacological enhancement of 2-AG signaling with JZL184 leads to impairment of short-term fear extinction (Hartley et al. 2016) and even enhanced fear expression, presumably through CB1 receptor on GABAergic neurons (Llorente-Berzal et al. 2015).

Another drug to modulate fear memories in a beneficial manner is the natural component of *Cannabis sativa*, cannabidiol (CBD). Repeated microinjections of CBD into the infralimbic cortex (IL) facilitated fear extinction, as indicated by reduced levels of freezing during extinction test. Systemic administration of the CB1 receptor antagonist rimonabant blocked the effects of intra-IL CBD, suggesting that CBD acts through CB1 receptor to facilitate fear extinction (Do Monte et al. 2013). Another study showed that under conditions of strong fear conditioning, CBD reduced contextual fear memory expression both acutely during the extinction session and later at a fear retention test (Song et al. 2016). In contrast, when initial conditioning was weaker, CBD had an effect to increase freezing at the fear retention test relative to vehicle controls. This bidirectional effect of CBD may be related to stress levels induced by conditioning and evoked at retrieval during extinction, rather than the strength of the memory *per se*, implicating that the therapeutic application must be very thoughtfully considered. Mechanistic studies have recently demonstrated the ability of intra-nucleus accumbens injections of CBD to block the formation of conditioned freezing behaviors, which was dependent on intra-ventral tegmental GABAergic transmission (Norris et al. 2016). In summary, CBD seems to be an interesting drug for the treatment of dysregulation in fear and anxiety, though the underlying mechanisms have to be elucidated (Jurkus et al. 2016).

8 Conclusions

Due to the versatile roles of the endocannabinoid system in fear, anxiety, stress, and cognition and based on the data obtained in the recent years, further research on this lipid signaling system will enable us to obtain many novel insights into numerous central questions in biology, neuroscience, and even computational sciences. It will be a challenging task to uncover the precise roles of endocannabinoids and their receptors at distinct synapses and in the control of neural network regulation, finally regulating behavioral outputs. This knowledge can then be rationalized in translational neurosciences, aiming at, e.g., defining novel pharmacological interventions for fear and anxiety disorders and cognitive disabilities, where both drugs are specifically designed to target components of the endocannabinoid system, but also plant-derived compounds, which may have decreased specificity but overall may show high beneficial effects with low unwanted side effects.

References

Bellocchio L, Soria-Gomez E, Quarta C, Metna-Laurent M, Cardinal P, Binder E, Cannich A, Delamarre A, Haring M, Martin-Fontecha M, Vega D, Leste-Lasserre T, Bartsch D, Monory K, Lutz B, Chaouloff F, Pagotto U, Guzman M, Cota D, Marsicano G (2013) Activation of the sympathetic nervous system mediates hypophagic and anxiety-like effects of CB(1) receptor blockade. Proc Natl Acad Sci U S A 110:4786–4791

Bloomfield MA, Ashok AH, Volkow ND, Howes OD (2016) The effects of Delta9-tetrahydrocannabinol on the dopamine system. Nature 539:369–377

Busquets-Garcia A, Gomis-Gonzalez M, Srivastava RK, Cutando L, Ortega-Alvaro A, Ruehle S, Remmers F, Bindila L, Bellocchio L, Marsicano G, Lutz B, Maldonado R, Ozaita A (2016) Peripheral and central CB1 cannabinoid receptors control stress-induced impairment of memory consolidation. Proc Natl Acad Sci U S A 113:9904–9909

Cheng Y, Dong Z, Liu S (2014) beta-Caryophyllene ameliorates the Alzheimer-like phenotype in APP/PS1 Mice through CB2 receptor activation and the PPARgamma pathway. Pharmacology 94:1–12

Davis M, Walker DL, Miles L, Grillon C (2010) Phasic vs sustained fear in rats and humans: role of the extended amygdala in fear vs anxiety. Neuropsychopharmacology 35:105–135

de Quervain D, Schwabe L, Roozendaal B (2017) Stress, glucocorticoids and memory: implications for treating fear-related disorders. Nat Rev Neurosci 18:7–19

Dincheva I, Drysdale AT, Hartley CA, Johnson DC, Jing D, King EC, Ra S, Gray JM, Yang R, DeGruccio AM, Huang C, Cravatt BF, Glatt CE, Hill MN, Casey BJ, Lee FS (2015) FAAH genetic variation enhances fronto-amygdala function in mouse and human. Nat Commun 6:6395

Do Monte FH, Souza RR, Bitencourt RM, Kroon JA, Takahashi RN (2013) Infusion of cannabidiol into infralimbic cortex facilitates fear extinction via CB1 receptors. Behav Brain Res 250:23–27

Dubreucq S, Matias I, Cardinal P, Haring M, Lutz B, Marsicano G, Chaouloff F (2012) Genetic dissection of the role of cannabinoid type-1 receptors in the emotional consequences of repeated social stress in mice. Neuropsychopharmacology 37:1885–1900

Garcia-Gutierrez MS, Ortega-Alvaro A, Busquets-Garcia A, Perez-Ortiz JM, Caltana L, Ricatti MJ, Brusco A, Maldonado R, Manzanares J (2013) Synaptic plasticity alterations associated with memory impairment induced by deletion of CB2 cannabinoid receptors. Neuropharmacology 73:388–396

Gazarini L, Stern CA, Carobrez AP, Bertoglio LJ (2013) Enhanced noradrenergic activity potentiates fear memory consolidation and reconsolidation by differentially recruiting alpha1- and beta-adrenergic receptors. Learn Mem 20:210–219

Gray JM, Vecchiarelli HA, Morena M, Lee TT, Hermanson DJ, Kim AB, McLaughlin RJ, Hassan KI, Kuhne C, Wotjak CT, Deussing JM, Patel S, Hill MN (2015) Corticotropin-releasing hormone drives anandamide hydrolysis in the amygdala to promote anxiety. J Neurosci 35:3879–3892

Gunduz-Cinar O, Hill MN, McEwen BS, Holmes A (2013) Amygdala FAAH and anandamide: mediating protection and recovery from stress. Trends Pharmacol Sci 34:637–644

Gunduz-Cinar O, Flynn S, Brockway E, Kaugars K, Baldi R, Ramikie TS, Cinar R, Kunos G, Patel S, Holmes A (2016) Fluoxetine facilitates fear extinction through amygdala endocannabinoids. Neuropsychopharmacology 41:1598–1609

Hartley ND, Gunduz-Cinar O, Halladay L, Bukalo O, Holmes A, Patel S (2016) 2-arachidonoylglycerol signaling impairs short-term fear extinction. Transl Psychiatry 6:e749

Hill MN, Kumar SA, Filipski SB, Iverson M, Stuhr KL, Keith JM, Cravatt BF, Hillard CJ, Chattarji S, McEwen BS (2013) Disruption of fatty acid amide hydrolase activity prevents the effects of chronic stress on anxiety and amygdalar microstructure. Mol Psychiatry 18:1125–1135

Jacob W, Marsch R, Marsicano G, Lutz B, Wotjak CT (2012) Cannabinoid CB1 receptor deficiency increases contextual fear memory under highly aversive conditions and long-term potentiation in vivo. Neurobiol Learn Mem 98:47–55

Jenniches I, Ternes S, Albayram O, Otte DM, Bach K, Bindila L, Michel K, Lutz B, Bilkei-Gorzo A, Zimmer A (2016) Anxiety, stress, and fear response in mice with reduced endocannabinoid levels. Biol Psychiatry 79:858–868

Jurkus R, Day HL, Guimaraes FS, Lee JL, Bertoglio LJ, Stevenson CW (2016) Cannabidiol regulation of learned fear: implications for treating anxiety-related disorders. Front Pharmacol 7:454

Kamprath K, Romo-Parra H, Haring M, Gaburro S, Doengi M, Lutz B, Pape HC (2011) Short-term adaptation of conditioned fear responses through endocannabinoid signaling in the central amygdala. Neuropsychopharmacology 36:652–663

Karpova NN, Pickenhagen A, Lindholm J, Tiraboschi E, Kulesskaya N, Agustsdottir A, Antila H, Popova D, Akamine Y, Bahi A, Sullivan R, Hen R, Drew LJ, Castren E (2011) Fear erasure in mice requires synergy between antidepressant drugs and extinction training. Science 334:1731–1734

Kishimoto Y, Cagniard B, Yamazaki M, Nakayama J, Sakimura K, Kirino Y, Kano M (2015) Task-specific enhancement of hippocampus-dependent learning in mice deficient in monoacylglycerol lipase, the major hydrolyzing enzyme of the endocannabinoid 2-arachidonoylglycerol. Front Behav Neurosci 9:134

Lange MD, Daldrup T, Remmers F, Szkudlarek HJ, Lesting J, Guggenhuber S, Ruehle S, Jungling K, Seidenbecher T, Lutz B, Pape HC (2016) Cannabinoid CB1 receptors in distinct circuits of the extended amygdala determine fear responsiveness to unpredictable threat. Mol Psychiatry. doi:10.1038/mp.2016.156

Laricchiuta D, Centonze D, Petrosini L (2013) Effects of endocannabinoid and endovanilloid systems on aversive memory extinction. Behav Brain Res 256:101–107

Li Y, Kim J (2015) Neuronal expression of CB2 cannabinoid receptor mRNAs in the mouse hippocampus. Neuroscience 311:253–267

Li Y, Kim J (2016a) CB2 Cannabinoid receptor knockout in mice impairs contextual long-term memory and enhances spatial working memory. Neural Plast 2016:9817089

Li Y, Kim J (2016b) Deletion of CB2 cannabinoid receptors reduces synaptic transmission and long-term potentiation in the mouse hippocampus. Hippocampus 26:275–281

Llorente-Berzal A, Terzian AL, Di Marzo V, Micale V, Viveros MP, Wotjak CT (2015) 2-AG promotes the expression of conditioned fear via cannabinoid receptor type 1 on GABAergic neurons. Psychopharmacology (Berl) 232:2811–2825

Lomazzo E, Bindila L, Remmers F, Lerner R, Schwitter C, Hoheisel U, Lutz B (2015) Therapeutic potential of inhibitors of endocannabinoid degradation for the treatment of stress-related hyperalgesia in an animal model of chronic pain. Neuropsychopharmacology 40:488–501

Luksys G, Sandi C (2011) Neural mechanisms and computations underlying stress effects on learning and memory. Curr Opin Neurobiol 21:502–508

Lutz B, Marsicano G, Maldonado R, Hillard CJ (2015) The endocannabinoid system in guarding against fear, anxiety and stress. Nat Rev Neurosci 16:705–718

Maccarrone M, Guzman M, Mackie K, Doherty P, Harkany T (2014) Programming of neural cells by (endo)cannabinoids: from physiological rules to emerging therapies. Nat Rev Neurosci 15:786–801

Marsicano G, Wotjak CT, Azad SC, Bisogno T, Rammes G, Cascio MG, Hermann H, Tang J, Hofmann C, Zieglgansberger W, Di Marzo V, Lutz B (2002) The endogenous cannabinoid system controls extinction of aversive memories. Nature 418:530–534

Mizrachi Zer-Aviv T, Segev A, Akirav I (2016) Cannabinoids and post-traumatic stress disorder: clinical and preclinical evidence for treatment and prevention. Behav Pharmacol 27:561–569

Morena M, Campolongo P (2014) The endocannabinoid system: an emotional buffer in the modulation of memory function. Neurobiol Learn Mem 112:30–43

Morena M, Patel S, Bains JS, Hill MN (2016) Neurobiological interactions between stress and the endocannabinoid system. Neuropsychopharmacology 41:80–102

Norris C, Loureiro M, Kramar C, Zunder J, Renard J, Rushlow W, Laviolette SR (2016) Cannabidiol modulates fear memory formation through interactions with serotonergic transmission in the mesolimbic system. Neuropsychopharmacology 41:2839–2850

Platt RJ et al (2014) CRISPR-Cas9 knockin mice for genome editing and cancer modeling. Cell 159:440–455

Puente N, Cui Y, Lassalle O, Lafourcade M, Georges F, Venance L, Grandes P, Manzoni OJ (2011) Polymodal activation of the endocannabinoid system in the extended amygdala. Nat Neurosci 14:1542–1547

Qin Z, Zhou X, Pandey NR, Vecchiarelli HA, Stewart CA, Zhang X, Lagace DC, Brunel JM, Beique JC, Stewart AF, Hill MN, Chen HH (2015) Chronic stress induces anxiety via an amygdalar intracellular cascade that impairs endocannabinoid signaling. Neuron 85:1319–1331

Ramikie TS, Nyilas R, Bluett RJ, Gamble-George JC, Hartley ND, Mackie K, Watanabe M, Katona I, Patel S (2014) Multiple mechanistically distinct modes of endocannabinoid mobilization at central amygdala glutamatergic synapses. Neuron 81:1111–1125

Remmers F, Lange MD, Hamann M, Ruehle S, Pape HC, Lutz B (2017) Addressing sufficiency of the CB1 receptor for endocannabinoid-mediated functions through conditional genetic rescue in forebrain GABAergic neurons. Brain Struct Funct. doi:10.1007/s00429-017-1411-5

Ruehle S, Remmers F, Romo-Parra H, Massa F, Wickert M, Wortge S, Haring M, Kaiser N, Marsicano G, Pape HC, Lutz B (2013) Cannabinoid CB1 receptor in dorsal telencephalic glutamatergic neurons: distinctive sufficiency for hippocampus-dependent and amygdala-dependent synaptic and behavioral functions. J Neurosci 33:10264–10277

Seidenbecher T, Remmes J, Daldrup T, Lesting J, Pape HC (2016) Distinct state anxiety after predictable and unpredictable fear training in mice. Behav Brain Res 304:20–23

Singewald N, Schmuckermair C, Whittle N, Holmes A, Ressler KJ (2015) Pharmacology of cognitive enhancers for exposure-based therapy of fear, anxiety and trauma-related disorders. Pharmacol Ther 149:150–190

Soltesz I, Alger BE, Kano M, Lee SH, Lovinger DM, Ohno-Shosaku T, Watanabe M (2015) Weeding out bad waves: towards selective cannabinoid circuit control in epilepsy. Nat Rev Neurosci 16:264–277

Song C, Stevenson CW, Guimaraes FS, Lee JL (2016) Bidirectional effects of cannabidiol on contextual fear memory extinction. Front Pharmacol 7:493

Soria-Gomez E, Busquets-Garcia A, Hu F, Mehidi A, Cannich A, Roux L, Louit I, Alonso L, Wiesner T, Georges F, Verrier D, Vincent P, Ferreira G, Luo M, Marsicano G (2015) Habenular CB1 receptors control the expression of aversive memories. Neuron 88:306–313

Stempel AV, Stumpf A, Zhang HY, Ozdogan T, Pannasch U, Theis AK, Otte DM, Wojtalla A, Racz I, Ponomarenko A, Xi ZX, Zimmer A, Schmitz D (2016) Cannabinoid type 2 receptors mediate a cell type-specific plasticity in the hippocampus. Neuron 90:795–809

Tovote P, Fadok JP, Luthi A (2015) Neuronal circuits for fear and anxiety. Nat Rev Neurosci 16:317–331

Volkow ND, Swanson JM, Evins AE, DeLisi LE, Meier MH, Gonzalez R, Bloomfield MA, Curran HV, Baler R (2016) Effects of cannabis use on human behavior, including cognition, motivation, and psychosis: a review. JAMA Psychiat 73:292–297

The Endocannabinoid System in Prefrontal Synaptopathies

Andrew F. Scheyer, Henry G.S. Martin, and Olivier J. Manzoni

Abstract Endocannabinoid (eCB) function in the prefrontal cortex (PFC) presents as a highly connected, widespread hub for the modulation of a vast array of synaptic functions. Given the known role of the PFC in a range of disorders from mental retardation, neurodegenerative diseases, and schizophrenia to drug addiction, stress, and anxiety, all of which are characterized by aberrant function at the synaptic compartment (here termed "synaptopathies"), it follows that eCB dysfunction is implicated as a significant causal factor in a plethora of disease states. Genetic and environmental manipulation of the eCB system in the PFC has thus been extensively studied with regard to many of these diseases and highlighted significant causal factors in their development and expression. Here, we examine the current body of literature as well as extensions of theoretical implications regarding the role of eCBs in the generation and presenting characteristics of a range of disorders. Elucidating the functions of eCBs in disease states will participate to the understanding of their etiology and the development of innovative pharmacotherapies.

1 Introduction

The eCB Hub System In mammals, the prefrontal cortex (PFC) is the most highly evolved brain region. The PFC is classically described as functioning in a manner essential to working memory, reasoning, action planning, cognitive flexibility, and emotionally guided behaviors (Goldman-Rakic 1990; Seamans et al. 1995). The power of the PFC with regard to regulation of our thoughts, actions, and emotions is guided by its complex internal circuit organization and extensive connectivity with other brain regions. Thus, a current hypothesis is that the PFC is a major functional network hub in the brain.

According to the hub theory of information processing, there is however a dark side to being a hub in a highly ordered network. Due to their high connectivity, compromised hubs are extremely deleterious to the integrity and information

A.F. Scheyer (✉) • H.G.S. Martin • O.J. Manzoni
INSERM U901, INMED and UMR S901, Aix-Marseille University, Marseille 13009, France
e-mail: andrew.scheyer@inserm.fr

M. Melis (ed.), *Endocannabinoids and Lipid Mediators in Brain Functions*,
DOI 10.1007/978-3-319-57371-7_7

processing in the whole network. Thus it is unsurprising that PFC malfunctions are a common denominator in several neuropsychiatric diseases including (but not limited to) mental retardation, autism, schizophrenia, depression, and addiction (Goto et al. 2010).

At the synaptic level, the neural code is processed by postsynaptic molecular machines made of macromolecular complexes of interacting proteins organized into a scale-free network (Grant 2003). The "small-world" nature of multiprotein complexes implies highly organized interconnected structures where changes in one protein readily alter the functions of many others. Therefore, proteins with a higher number of connections, the so-called hub proteins, are of particular significance to network integrity, synaptic functions, and plasticity (Song and Singh 2013).

One of the most important signaling machines at central excitatory synapses is the endocannabinoid (eCB system). The "eCB hub system" is made of some of the most highly connected synaptic proteins (e.g., mGluR1/mGluR5, NMDAR, etc.; Melis and Pistis 2012). Logically, the "eCB hub system" participates in a wide array of physiological functions, many of which also implicate the PFC (e.g., memory, addiction, planning, etc.). This chapter will focus on the role of the "eCB hub system" in the etiology and expression of neuropsychiatric synaptic diseases (the so-called synaptopathies) implicating the eCB system within the PFC brain hub.

Prefrontal eCB System The eCB system is highly promiscuous; functional at inhibitory, excitatory, and neuromodulatory synapses throughout the central nervous system (Kano et al. 2009; Katona and Freund 2012). Similar to other brain regions, modulation of neurotransmission by eCBs in the PFC is common, but also restricted. CB1 receptors (CB1R) and other eCB molecular components show neuron-specific expression and selective subcellular localization, permitting the expression of homosynaptic and heterosynaptic modulation of neurotransmission (Katona and Freund 2012).

In the PFC, the highest concentrations of CB1R mRNA are found in interneurons, particularly those positive for cholecystokinin and calbindin in both deep and superficial cortical layers of the rat PFC (Marsicano and Lutz 1999). However the correlation between the degree of eCB modulation of synaptic transmission and the relative abundance of CB1R mRNA and protein is not necessarily direct. In both the hippocampus and somatosensory cortex where CB1R expression is also predominantly restricted to interneurons, it is equally excitatory synaptic transmission that is sensitive to CB1R knockout (Domenici et al. 2006; Monory et al. 2006). Furthermore, CB1R immunoreactivity is detected in fibers across the PFC surrounding pyramidal neurons in deep (layers 5/6) and superficial cortical (layer 2) layers suggesting that distant afferent connections may be equally modulated by eCBs (Egertová and Elphick 2000; Lafourcade et al. 2007). A similar pattern of CB1R immunohistochemistry is found in primate and human PFC tissue, where a strong CB1R immunoreactivity is also found in layer 4 (Eggan and Lewis 2007; Eggan et al. 2010; Long et al. 2012). However, direct laminar comparison between the primate and rat PFC is difficult due to distinct underlying cytoarchitecture

(Uylings et al. 2003). Electron microscopic studies of layer 5 synapses have identified CB1Rs at both symmetrical and asymmetrical synapses, with the receptor generally localized in the perisynaptic region of the presynapse (Eggan and Lewis 2007; Lafourcade et al. 2007). It is also worth noting that CB1Rs may be present in neuromodulatory afferents which are widely found in the PFC. Thus eCBs have the potential not only to modulate excitatory and inhibitory synaptic transmission in the PFC but also neuromodulatory tone (Katona and Freund 2012; Lee et al. 2015).

Other partners of the synaptic eCB system are broadly expressed in the PFC. mGluR5 shows dominant expression over mGluR1 and is found throughout the PFC layers (Hayashi et al. 1993; Lafourcade et al. 2007), where its activation triggers multiple forms of synaptic plasticity via the induction of eCB signaling (Lafourcade et al. 2007; Sheinin et al. 2008). Ultrastructural analysis of layer 5/6 asymmetrical synapses has identified co-localization of the 2-AG/mGluR5 signaling complex system (i.e., postsynaptic perisynaptic mGluR5 and DAGLα) with associated presynaptic CB1R (Jung et al. 2012). In contrast to some other brain regions, components of both the anandamide (AEA) and 2-arachidonylglycerol (2-AG) synthesis/degradation molecular machinery are found, including NAPE-PLD, MAGLα, and FAAH along with ABHD6 (Dinh et al. 2002; Egertová et al. 2003; Long et al. 2012; Marrs et al. 2010).

Notwithstanding the high expression of CB1R in PFC interneurons, eCB modulation of excitatory neurotransmission appears more prevalent at least in deep layers. Short-term plasticity of glutamatergic synapses (DSE) mediated by eCBs is inducible in layer 5 pyramidal rat neurons (Fortin and Levine 2007; Heng et al. 2011) (Fig. 1). The DSE is mGluR-independent and calcium sensitive; however, in

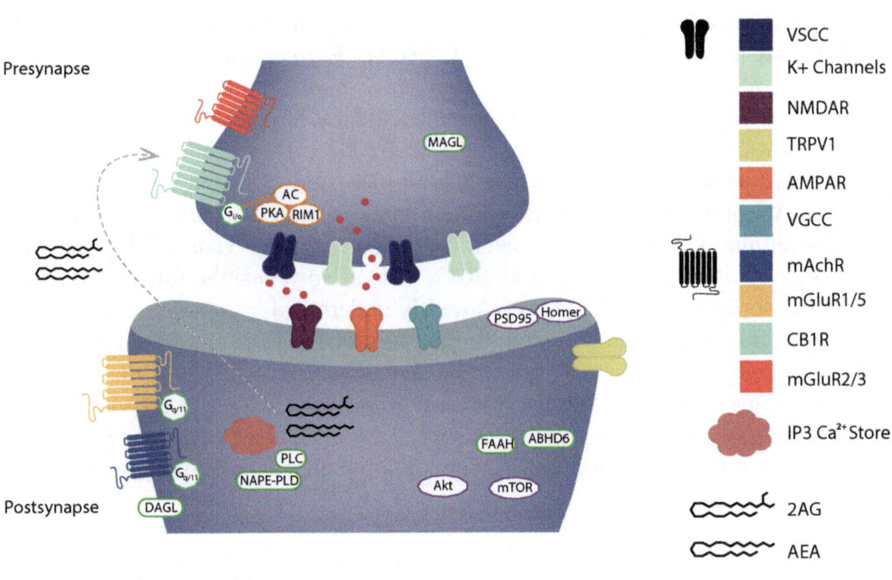

Fig. 1 Schematic representation of the principal components of the endocannabinoid-signaling complex at glutamatergic synapse of the CNS

contrast to the hippocampus, DSE does not appear to be universally expressed, and the success of induction may depend on pyramidal neuron subtype, species, developmental age, and layer specificity (Fortin and Levine 2007; Heng et al. 2011; Marrs et al. 2010). Notably, responses evoked in superficial layers are insensitive to CB1R agonists, in contrast to responses from basal dendrites, arguing for an input specificity (Fortin and Levine 2007). Activity-driven long-term depression, the phenomenon of lasting reduction in the efficacy of synaptic signaling following a specific pattern of activity, (eCB-LTD) is inducible in layer 5 pyramidal neurons in response to moderate frequency (10 Hz) stimulation, dependent on 2-AG signaling and mGluR5 activation (Lafourcade et al. 2007; Lovelace et al. 2014; Marrs et al. 2010). Alternatively, optogenetic-induced synaptic release of endogenous acetylcholine (ACh) and activation of M_1 muscarinic ACh receptors can also engage the eCB system at these synapses with concurrent synaptic transmission (Martin et al. 2015). In contrast eCB modulation of inhibitory transmission alone has not been reported in these neurons, and standard short-term plasticity (DSI; in which depolarization of the postsynaptic neuron induces a transient suppression of inhibitory transmission) protocols fail to produce transient depression (Fortin and Levine 2007; Marrs et al. 2010). However, a significant portion of inhibitory synapses (>30 %) co-express CB1R and D2R and appear sensitive to eCB modulation upon D2R co-activation (Chiu et al. 2010). Mechanistically, this reduction in inhibitory neurotransmission relies on postsynaptic mGluR5 activation. Thus, under conditions of increased DA drive, this may permit disinhibition of a subset of neurons receiving salient excitatory input.

eCB-mediated long-term plasticity has not been reported in superficial layers of the PFC; however, DSI is present. The DSI again appears limited to a subpopulation of GABA synapses, perhaps reflecting interneuron subtype, and is dependent on 2-AG/CB1R signaling (Yoshino et al. 2011). Furthermore, action potentials induced at natural firing frequencies may also induce short-term depression of inhibitory transmission. This may be mechanistically important in PFC-dependent functional working memory (Carter and Wang 2007). Although not strictly synaptic, an unusual form of eCB-mediated inhibition may also play a role in these neurons. Action potential firing is reported to activate a chloride current mediated by intracellular CB2R in these neurons, which in turn leads to a short-term depression of firing (den Boon et al. 2012, 2014). However, the role of CB2Rs in synaptic disorders of the PFC is not currently understood.

How eCB signaling modulates information processing in the PFC is currently unknown. Agonists of the eCB system enhance cortical up-states in the PFC, and antagonists increase down-states (Pava and Woodward 2014), likely via modulation of inhibitory neurotransmission in superficial layers. However, the role of the eCB system in the integration of PFC signaling remains an area of active research.

2 Drug Abuse and Addiction

The PFC and its synaptic connectivity with a variety of target brain structures has been identified as a crucial target for the development of rewarding behaviors including those which underlie drug abuse and addiction (Goldstein and Volkow 2011; Kalivas and Volkow 2005). Indeed, dysfunctional or abnormal activity in the PFC has been linked to control over drug-seeking behavior (for reviews, see George and Koob 2010; Goldstein and Volkow 2011; Lüscher and Malenka 2011).

Alcohol Density and reactivity of CB1R in the cortex, in addition to several other brain regions, are decreased following chronic ethanol exposure in animal models (Vinod et al. 2006). Whole-brain analyses suggest that this may be due to decreased CB1R mRNA induced by ethanol exposure (Basavarajappa and Hungund 1999). Similarly, human PET imaging following chronic ethanol abuse indicates long-term downregulation of CB1R surface expression (Hirvonen et al. 2013; Ceccarini et al. 2014). This phenomenon is accompanied by reduced activity of the AEA hydrolase, FAAH (Vinod and Hungund 2006).

Pharmacological manipulation of the eCB system has also been shown to alter rates and likelihood of ethanol consumption, suggesting a critical role for this system in the development or maintenance of addiction. Specifically, CB1R agonists consistently increase ethanol intake and motivation for consumption (Colombo et al. 2007; Gallate et al. 1999) as well as propensity toward relapse (Alen et al. 2008). Increasing eCB levels through FAAH inhibition mimicked this effect (Blednov et al. 2007; Hansson et al. 2007; Vinod et al. 2008). Conversely, CB1R antagonists or inverse agonists reduce intake (Arnone et al. 1997; Colombo et al. 2007; Freedland et al. 2001), a phenomenon that has been localized to the PFC (Hansson et al. 2007). Similarly, CB1R antagonist or inverse agonist administration to ethanol-preferring mice effectively attenuates sensitization (Marinho et al. 2015) in mice and presents a novel route toward the treatment of alcoholism in humans (Colombo et al. 2007; Maccioni et al. 2010; Vasiljevik et al. 2013).

Genetic evidence further suggests that eCB signaling in the PFC contributes to ethanol consumption/preference. Namely, ethanol-preferring rats exhibit decreased CB1R function, as well as altered levels of eCBs in the PFC (Hansson et al. 2007). Similarly, genetic changes leading to increased CB1R density in the PFC are correlated with enhanced ethanol-elicited subjective reward and brain activation in human subjects (Hutchison et al. 2008). Neuroimaging of alcoholic subjects confirms this as a functional deficit by revealing decreased CB1R receptor binding (Hirvonen et al. 2013), which may trigger the compensatory increase in CB1R density. Indeed, postmortem study of alcoholics reveals increased CB1R levels in the PFC (Erdozain et al. 2015) in addition to decreased AEA levels (Lehtonen et al. 2010). Additionally, downstream signaling assays reveal that CB1R activation is impaired in the PFC of alcoholics, suggesting a twofold reduction in function. This reduction is further confounded by reduced MAGL activity (Erdozain et al. 2015), which has also been observed in ethanol-preferring rats (Hansson et al. 2007). These results may further indicate compensatory changes in the eCB system, as decreased CB1R signaling function is accompanied by decreased enzymatic degradation (and therefore likely increased circulating concentrations) of 2-AG.

Finally, converging evidence suggests a significant role for mGluR5 in the development or occurrence of alcohol dependence. Specifically, mGluR5-lacking mice exhibit an increased conditioned place preference at low doses of ethanol (Bird et al. 2008), in addition to increased rates of consumption during the dark cycle (Eisenhardt et al. 2015). Given the role of mGluR5 signaling in the production of on-demand eCBs, these data suggest that attenuated eCB signaling may contribute to the development of ethanol dependence. Clearly, the eCB system is an important and valuable target for the continued understanding and treatment of alcohol abuse and addiction, and continued research will be necessary to further elucidate the underlying mechanisms and progress toward potential targeted pharmacotherapeutic interventions for this disorder.

Cocaine The effects of cocaine on eCB signaling in the PFC are substantial and varied and depend largely upon the type or duration of administration as well as timing in relation to administration or consumption. During or immediately following cocaine self-administration, eCB-LTD in the PFC is suppressed or abolished (Kasanetz et al. 2013), an effect that is also seen in the nucleus accumbens following a single in vivo exposure (Fourgeaud et al. 2004). Intriguingly, while certain alterations in eCB function are induced directly as a result of cocaine exposure or self-administration, others may appear only during the transition to addiction, such as the suppression of mGluR2/3-LTD in the PFC (Kasanetz et al. 2013). Indeed, manipulation of altered eCB transmission may prove to be a valuable route toward the treatment of cocaine addiction, as evidenced by abolished reinstatement of cocaine seeking in rats following pretreatment with a CB1R antagonist (McReynolds et al. 2015).

Levels of 2-AG are increased in the PFC following withdrawal from self-administered or noncontingent cocaine exposure, while levels of AEA in the PFC are reduced following noncontingent but not self-administered cocaine (Bystrowska et al. 2014), further suggesting a role for alterations in the eCB system in the development of addiction-related behaviors. Additionally, both experimenter- and self-administered cocaine result in increases in *N*-acylethanolamines (Bystrowska et al. 2014) including the AEA function-enhancing fatty acid *N*-palmitoylethanolamine (Ho et al. 2008) and the putative eCB *N*-oleoylethanolamine (Overton et al. 2006).

Chronic exposure in both humans and mice has been shown to result in decreased surface expression of CB1R with increased cytosolic levels, suggesting a downregulation of function in the system. Furthermore, chronic cocaine exposure or intake decreases downstream signaling from the activated CB1R (Alváro-Bartolomé and Garcià-Sevilla 2013). In this vein, the cocaine sensitization paradigm has further elucidated discrepancies between repeated and acute cocaine administration. While cocaine-sensitized mice exhibit increased levels of CB1R as well as both FAAH and MAGL in the PFC, acute cocaine only alters the ratios of eCB precursors and their respective enzymatic partners, suggesting that chronic cocaine treatment induces either entirely different or more robust changes in the eCB system than acute exposure (Blanco et al. 2015). It is apparent, then, that cocaine self-administration in human addicts as well as animal models substantially impacts eCB function through manipulation of eCBs, their enzymatic pathways, and receptors.

Cannabis Cannabis abuse, especially during adolescence, has profound consequences on the functions of the eCB system in the PFC. Adults exhibit reduced or absent vulnerability to many of the metaplastic effects of cannabis exposure (Realini et al. 2011; Schneider and Koch 2003); however, the impact of adolescent exposure may prove relevant in the development of such aberrant behaviors as addiction and psychosis (for reviews, see Caballero and Tseng 2012; Hurd et al. 2014; Rubino and Parolaro 2008). PFC-dependent cognitive functions such as working memory are impaired in adulthood following cannabis exposure during development in animal models (Lovelace et al. 2015; Raver et al. 2013). Indeed, cannabis exposure during adolescence is positively correlated with the development of psychosis (see section: Schizophrenia) and increased risk of illicit drug use in humans (Ellgren et al. 2008; Fergusson and Horwood 2000; Yamaguchi and Kandel 1984; Lynskey et al. 2003; Argawal et al. 2004).

Interestingly, the impact of exposure to exogenous cannabinoids may be gender-specific (Lovelace et al. 2015; Realini et al. 2011; Rubino et al. 2008c, 2009). Reductions in CB1R expression in adulthood following adolescent exposure emerge differentially between genders, being more profound and widespread in female over male rats. Similarly, low PFC CREB activity is a consequence of adolescent THC exposure in adult female rats but not male (Rubino et al. 2008c). These differences are paralleled by the behavioral manifestations of such altered activity, including markers of depression (Rubino and Parolaro 2008) and addiction vulnerability (Higuera-Matas et al. 2008). Together, these results identify significant differences in the consequences of cannabinoid-induced deregulation of PFC function between genders.

In light of the cannabinoid-induced perturbations in PFC function, it is not surprising that repeated exposure to cannabinoid agonists during adolescence has profound and lasting effects on the function of the local eCB system. Animals exposed to a CB1R agonist subchronically during adolescence exhibit numerous deficits in eCB function as measured during adulthood, including desensitized CB1Rs, attenuation of the eCB-independent mGluR2/3-LTD, and a deficit in L2/3-L5 eCB-LTD (Lovelace et al. 2015). This adolescent exposure also results in aberrant development of other PFC proteins such as the μ-type opioid receptor that may also contribute to prefrontal synaptic dysfunction (Ellgren et al. 2008).

In humans, chronic cannabis use beginning during adolescence and early adulthood results in decreased CB1R expression as measured via PET imaging (Ceccarini et al. 2015), suggesting that cannabis use and sustained dysfunction in the eCB system is a common phenomenon. Thus, consumption of cannabis during adolescence results in persistent, substantial perturbations in eCB function that may underlie many of the aforementioned behavioral abnormalities resultant of this exposure.

Other Drugs Various alterations in eCB function have been further noted with other drugs of abuse. Briefly, withdrawal following morphine sensitization is correlated with increased AEA levels as well as potentially impaired FAAH function in the PFC. Trends toward increased levels of AEA and 2-AG following acute exposure were also noted, suggesting a role for the eCB system in multiple stages of morphine abuse (Viganò et al. 2004). Significantly, there is overlap in CB1R and opioid receptor expression patterns (Rodriguez et al. 2001; Salio et al. 2001), and CB1R and mu-opioid receptors may form heterodimers (Rios et al. 2006; Hojo et al. 2008). Indeed, CB1R and opioid agonists exhibit significant cross-tolerance (Maldonado 2002; Thorat and Bhargava 1994; Viganò et al. 2005), suggesting further potential links between the development of opiate addiction and alterations in the eCB system.

Other drugs, such as methylphenidate, have been shown to induce lasting effects on synaptic plasticity in the PFC following cessation of use, including those previously exhibited to be eCB-dependent such as HFS-LTD (Burgos et al. 2015). Similarly, withdrawal from chronically administered nicotine induces increased levels of AEA in the PFC (Cippitelli et al. 2011). Interestingly, certain behavioral effects of nicotine withdrawal may be attenuated via inhibition of FAAH, further indicating a role for eCBs in nicotine abuse or withdrawal (Cippitelli et al. 2011). The involvement of the eCB system in the effects of such varied drugs of abuse is a clear indicator that further research must be conducted to elucidate the full role played by this molecular machinery and crucial synaptic modulator in drug abuse and addiction. Future research into the etiology and synaptic consequences of drug use and addiction must therefore consider the contributions and alterations of the eCB system in order to fully understand the cellular and molecular underpinnings of substance abuse disorders.

3 Alzheimer and Huntington Diseases

Cortical areas involved in cognitive flexibility, learning, memory formation, and storage are prime targets of the neurodegenerative and functional deficits accompanying Alzheimer disease (AD) and Huntington disease (HD). The complex multifactorial etiology of AD and HD and how the therapeutic opportunities provided by the fast-moving eCB pharmacology could help treat these neurodegenerative disorders are beyond the scope of this chapter. Excellent current updates can be found in Bedse et al. (Bedse et al. 2015) and Aso and Ferrer who noted:

> Cannabinoids may target in parallel several processes that play key roles in AD, including Aβ and tau aberrant processing, chronic inflammatory responses, excitotoxicity, mitochondrial dysfunction and oxidative stress among others. Clinical data also reveal an

improvement in behavioral in patients with AD after treatment with cannabinoids. (Aso and Ferrer 2014; Fagan and Campbell 2014; Kluger et al. 2015)

Alzheimer Disease It is clear that the PFC eCB signaling system is not spared in AD, where the disease presents with such symptoms as a decline in executive function and memory and behaviors with a known dependence upon eCB function. In human patients, there is an increased [35 S]GTPS binding stimulated by the CB1R agonist WIN55,212-2 in PFC layers 5 and 6 of AD I–IV stages, while CB1R density is enhanced in layer 6 only at the later stages of the disease (Manuel et al. 2014). Accordingly, the binding of the CB1R radioligand [125I]SD7015 was found to be enhanced in the PFC of human patients (Farkas et al. 2012). Reminiscent of the human data, CB1R mRNA levels, but not protein levels, are augmented in the PFC of 3 x Tg-AD mice (Bedse et al. 2014). Furthermore, in the Aβ PPswe/PS1Δ E9 mouse model of AD, there is higher CB1R-effector coupling compared to wild-type mice in the frontal cortex, but no difference in PFC-linked associative memory nor in lipidomic profiles (Maroof et al. 2014). In the same model, social recognition (a behavior where PFC contribution is likely) deficits were prevented by long-term cannabidiol (CBD) treatment (Cheng et al. 2014). Emphasizing the therapeutic potential of CBD in AD, substantial cognitive improvements were reported by a combination of THC and CBD (Aso et al. 2015).

In addition to CB1R, evidence points to a role for the CB2R in AD and associated neuroinflammation (Schmöle et al. 2015) (for a review, see Ahmed et al. 2015). CB2R mRNA and protein levels are increased in the cortex of human AD patients or Aβ PPswe/PS1Δ E9 mice (Bedse et al. 2015), while increased CB2R expression in the PFC of later-stage AD patients correlates with amyloid Aβ levels (Solas et al. 2013).

These data indicate that targeting the eCB system is a viable therapeutic strategy in AD-related PFC deficits. Nonetheless, how AD impacts the synaptic functions of the eCB system in the PFC remains unknown.

Huntington Disease HD is a neurodegenerative disease caused by a CAG repeat expansion in the huntingtin gene. Executive dysfunction may precede clinical manifestation of the movement disorder. Reduced CB1R binding in the basal ganglia and substantia nigra is an early signature of HD (Glass et al. 1993, 2000; Richfield and Herkenham 1994), and in animal models of HD, both CB1R expression and eCB levels are reduced (Dowie et al. 2009; Glass 2001; Lastres-Becker et al. 2001, 2002). While evidence for a correlation between HD and the basal ganglia's eCB system is prevalent, the data linking HD progression to the eCB in the PFC are scarce (Maccarrone et al. 2007). In the cortex of presymptomatic R6/R2 mice, 2-AG levels are decreased by 25%, while in the cortex of symptomatic R6/2mice, a 28% decrease of 2-AG levels is combined with a 50% increase in AEA levels (Bari et al. 2013; Bisogno et al. 2008). In R6/2 mice, Centonze et al. (2005) showed a reduction in CB1R expression and CB1R-mediated synaptic responses at striatal GABAergic synapses. Based on the aforementioned biochemical evidences of PFC damages in HD and recent human PET data inversely correlating the disease burden with CB1R availability in the PFC (Van Laere et al. 2010), it is tempting to hypothesize that executive deficits in HD implicate alterations of the synaptic functions of the eCB system in the PFC.

4 Schizophrenia

Schizophrenia is a severe synaptopathy characterized by disordered thought processes and both positive and negative cognitive maladies which often appear first during early adulthood and impact a wide array of cognitive and behavioral functions (Lewis and Lieberman 2000). Dysfunction in the PFC has been identified as a major contributing factor in the development and manifestation of this disease (Tsai and Coyle 2002), and dysregulation of the eCB system in particular has been identified as a possible cause (Tan et al. 2014; Ujike and Morita 2004; Volk and Lewis 2015). Here, we will examine evidence suggesting that dysfunction of eCB signaling in the PFC may be both a contributing factor to the development of schizophrenia as well as a point of vulnerability in those predisposed to the disease.

PFC eCB Dysfunction in Schizophrenia Schizophrenic humans show increased CB1R binding density in the PFC (Dalton et al. 2011; Dean et al. 2001; Jenko et al. 2012; Newell et al. 2006; Zavitsanou et al. 2004) as well as increased AEA levels in circulating CSF and blood (De Marchi et al. 2003; Giuffrida et al. 2004; Koethe et al. 2009; Leweke et al. 1999). PET studies utilizing a CB1R-selective ligand have confirmed that CB1R binding is elevated in schizophrenic patients and inversely correlated with negative symptoms (Wong et al. 2010). Furthermore, a link with CNR1, the gene encoding for the CB1R, has been noted as altered in schizophrenic patients (Leroy et al. 2001; Ujike et al. 2002). However, a more recent study identified a notable decrease in CB1R mRNA in the PFC of schizophrenic subjects (Eggan et al. 2008). This contradiction was attributed to possible allosteric binding of previously used radioligands and highlights a need for further, and more technically consistent, studies in determining this abnormality. Interestingly, a mutation in the CNR2 gene encoding for CB2R that reduces its function has also been found to be elevated in a postmortem study of schizophrenic brains (Ishiguro et al. 2010). In vitro modeling of this mutant CB2R showed decreased responding to 2-AG, and indeed in a model of schizophrenic symptoms induced by the administration of the NMDAR antagonist MK-801 (for a review, see Thornberg and Saklad 1996), the use of a CB2R inverse agonist exacerbated symptomology, confirming a potential role for CB2R dysfunction in schizophrenia.

While mRNA levels for some enzymes responsible for the synthesis and degradation of eCBs are unaltered in schizophrenics (Volk et al. 2010), recent evidence suggests that levels of the serine hydrolase α-β-hydrolase domain 6 (ABHD6) which regulates 2-AG signaling and metabolism in the PFC are increased in young schizophrenic patients compared to healthy controls (Volk et al. 2013). Interestingly, postmortem brain tissue analysis has revealed increased levels of 2-AG in the PFC of schizophrenic patients (Koethe et al. 2009), suggesting that this elevated 2-AG metabolism may be a compensatory mechanism for the otherwise enhanced levels of circulating eCBs in schizophrenic individuals.

Animal models of schizophrenia similarly show significant alterations in eCB system function. Following chronic-intermittent phencyclidine (PCP) administration to model schizophrenic symptoms, CB1R-activated GTPyS binding is reduced, accompanied by increased levels of 2-AG in the PFC (Vigano et al. 2009). Not surprisingly, the symptoms of schizophrenia in this model are enhanced by THC administration, which is similarly observed in human schizophrenic subjects (Chakraborty et al. 2014; Negrete 1989; Turner and Tsuang 1990; Ujike and Morita 2004; Treffert 1978). Interestingly, certain aspects of the disease in a similar model of PCP-induced schizophrenic symptomology can be rescued using a systemically administered FAAH inhibitor or a direct CB1R agonist (Seillier et al. 2013). The contradictory findings between animal models and human studies may be due to compensatory changes in eCB function in response to the pharmacological generation of the disease model. Together, these results depict a multifaceted relationship between eCB function and schizophrenia, which may highlight avenues toward the identification of biomarkers of the disease. The sometimes contradictory and often complex interactions demonstrate a necessity for continued investigation.

Cannabis Use and Schizophrenia Enhanced deficits from cannabis use in schizophrenic subjects provide further evidence supporting a role for dysfunction in the eCB system. Compared to control individuals, schizophrenic humans show significantly increased responses to THC exposure with regard to multiple symptoms of the disease (D'Souza et al. 2005). Accordingly, the frequency and intensity of psychotic episodes in schizophrenic individuals is increased following cannabis use (Linzen et al. 1994; Negrete and Knapp 1986). This interaction is likely attributable to preexisting deficits in PFC function (Hambrecht and Häfner 2000; Knable and Weinberger 1997). Patients who continue to use cannabis show increased frequency and intensity of psychotic episodes (Foti et al. 2010), providing further evidence for an ongoing interaction between aberrant function in the eCB system and the course of the disease. Furthermore, gamma- and theta-band oscillations in the PFC are attenuated in rats exposed to a CB1R agonist (Hajós et al. 2008), where similar deficits have been identified in human schizophrenic subjects (Caballero and Tseng 2012; Spencer et al. 2003; Uhlhaas and Singer 2010). Together, these findings draw a clear correlation between altered eCB functions and the prevalence, timing, and severity of schizophrenia.

Cannabis consumption during adolescence, a critical period of development for the eCB system (particularly in the prefrontal cortex; Long et al. 2012), is positively correlated with an increased likelihood and severity of the emergence of schizophrenia (for reviews, see Arseneault et al. 2004; Casadio et al. 2011). While previous cannabis consumption is a significant predictor for the development of schizophrenia in predisposed populations (Compton et al. 2009), cannabis consumption prior to the age of 18 resulted in a more than doubled risk of schizophrenia precipitation in a large cohort of non-predisposed individuals, which is elevated to nearly sixfold with "heavy" use (Andréasson et al. 1987). The increased risk of schizophrenia due to cannabis use is similar to so-called high-risk individuals (i.e., those with a strong family history of schizophrenia) which may cause as much as a sixfold increase in the likelihood of developing psychosis (Miller et al. 2001). Nevertheless, PCP-induced psychotic symptoms are enhanced in CB1R-KO mice (Haller et al. 2005). The available evidence clearly highlights a profound relationship between eCB function and schizophrenia, which warrants continued investigation both for elucidating the underlying mechanisms of disease development and potential pharmacotherapeutic interventions (Saito et al. 2013).

eCB-Based Pharmacotherapies Given the clear link between eCB dysfunction and schizophrenic symptomology, manipulation of the eCB system provides a clear and potentially valuable route toward the development of targeted pharmacotherapies for the treatment of schizophrenia. One existing candidate with such potential is CBD, a naturally occurring component of cannabis that acts as an indirect antagonist at both CBR1 and CB2R (for a review of pharmacological properties, see Mechoulam et al. 2007).

CBD administration attenuates several symptoms in animal models of schizophrenia including MK-801-induced hyperactivity as well as deficits in prepulse inhibition and social withdrawal (Gururajan et al. 2011; Long et al. 2006). Similarly, CBD has been shown to be equally efficacious as the atypical antipsychotic clozapine in attenuating chronic-MK801-induced deficits in social interaction and novel object recognition, two hallmarks of schizophrenic symptomology in animal models (Gomes et al. 2015b), possibly via reduced inflammation and inhibition of microglia in the PFC. These results are further supported by CBD's ability to attenuate MK-801-induced deficits in prepulse inhibition via reversal of elevated Delta-FosB and parvalbumin expression in the PFC (Gomes et al. 2015a).

In human subjects, CBD has been shown to attenuate the anxiety-inducing effects of THC (Crippa et al. 2009) as well as its psychotic symptoms as measured by the Positive and Negative Syndrome scale. These properties have been suggested to arise from the ability of CBD to disrupt connectivity between the amygdala and anterior cingulate (Fusar-Poli et al. 2009), in addition to reducing activation of the striatum, medial temporal cortex, and PFC (Bhattacharyya et al. 2012) in a manner opposite to THC.

Several clinical studies support the potential for CBD in the targeted pharmacotherapeutic treatment of schizophrenia (for a full review, see Iseger and Bossong 2015). Case studies of schizophrenic patients have shown promise in treatment via CBD (Zuardi et al. 1995), including treatment-resistant symptoms (Zuardi et al. 2006). Indeed, comparing the effects of CBD to the atypical antipsychotic amisulpride, Leweke and colleagues found no difference in efficacy in reducing psychotic symptomology. However, the therapeutic effects were accompanied by significantly reduced side effects (Leweke et al. 2012). Clearly, the evidence for preexisting or cannabis-induced eCB dysfunction in schizophrenia, combined with promising evidence for the treatment of its symptomology with CBD, indicates great promise in the development of eCB-targeting therapy.

However, further research is required to elucidate the mechanisms underlying these effects and develop more targeting approaches to pharmacological intervention.

5 Autism Spectrum Disorders

Autism spectrum disorders (ASDs) represent a group of both genetically and environmentally induced defects, which share behavioral phenotypes, notably deficits in social reciprocity and stereotyped behaviors. A common neurological nexus among these spectrum disorders is the PFC, which is linked genetically, anatomically, and functionally to ASD (Courchesne et al. 2011; Willsey et al. 2013). Yet, save Fragile X syndrome, there is surprisingly little clinical data on the eCB system in ASD and only indirect peripheral measurements of eCBs (Chakrabarti et al. 2015). Notwithstanding, there has been significant popular interest in the use of medical marijuana in the treatment of ASDs. Beyond the anecdotal however, there is limited current evidence supporting such an intervention (Hadland et al. 2015).

Fragile X Syndrome FXS is an X-chromosome-linked hereditary disorder characterized by intellectual disability and variable changes in social behavior often linked to autism (Garber et al. 2008). In most affected individuals, there is transcriptional silencing of the fragile X mental retardation gene (*FMR1*) due to expansion of a trinucleotide CGG repeat in the 5′-untranslated region and a corresponding loss or reduction in the FMR1 protein (FMRP; Penagarikano et al. 2007). Since FMRP is an RNA-binding protein, this in turn leads to the loss of regulated translation of a large subset of associated mRNAs notably rich in encoded synaptic proteins (Darnell et al. 2011). The large and diverse set of FMRP-associated mRNAs means that FXS probably has a highly complex developmentally linked etiology; however, an important and much explored hub in the expression and treatment of FXS has been the pathological enhancement of mGluR5 function (Darnell and Klann 2013). Treatment with mGluR5 antagonists in animal models of FXS recovers many of the associated physiological and behavioral deficits (Michalon et al. 2012). However, subsequent clinical trials have proven less promising, perhaps due to the difficulties in selectively targeting a FXS associated deficits without harming general mGluR5 function in the brain (Mullard 2015).

Many of the behavioral signatures of FXS are linked to PFC-directed actions: loss of executive control, deficits in working memory, and abnormal social interaction (Cornish et al. 2008). Likewise the *Fmr1* knockout mouse model shows alterations in both PFC-dependent learning and social behavior, although baseline and mGluR5 antagonist effects are rather inconsistent (Gantois et al. 2013; Kramvis et al. 2013; Krueger and Bear 2011; Liu and Smith 2009; Mao et al. 2013; Sidorov et al. 2014). Since eCB mobilization is a major downstream effector of mGluR5 signaling, it is unsurprising that there have been reports of changes in the eCB system in FXS. In the hippocampus where enhanced mGluR5 function is most robustly reported, there is a parallel upregulation of the eCB signaling at both excitatory and inhibitory synapse (Tang and Alger 2015; Zhang and Alger 2010), and consequently drugs targeting the CB1R have proven successful in restoring hippocampal-based behavior in the *Fmr1* KO mouse model (Busquets-Garcia et al. 2013). In contrast, current evidence points to a deficit in the eCB signaling in the PFC. Similar to results in the nucleus accumbens, synaptic induction of eCB-LTD is impaired in layer 5 pyramidal neurons in adult *Fmr1* KO mice (Jung et al. 2012). This loss in eCB-LTD does not appear to be linked to a direct loss of CB1R function, but rather a deficit in coupling between mGluR5 and DAGL-α and thus the on-demand biosynthesis of 2-AG. Notably DAGL-α mRNA is the most prominent component of the eCB system thus far shown to be a FMRP target (Darnell et al. 2011). This opposite regulation in eCB signaling between the PFC and hippocampus appears to reflect a common phenomenon in FXS, where neuronal properties and signaling are changed in a region-specific manner (Contractor et al. 2015). Thus, pharmacological treatment of *Fmr1* KO animals with a drug that enhances eCB signaling, JZL184, has been shown to restore selective behavioral deficits (Jung et al. 2012). Therefore, although targeting the eCB system in the treatment of FXS may solve some of the global risks of directly targeting mGluR5, the heterogeneity in eCB deficits in FXS significantly complicates this approach.

Sodium Valproate Environmental insults leading to ASD have extensively been examined in the sodium valproate (VPA) rat model. Here, a single in utero exposure to the teratogen valproate results in offspring which exhibit several of the hallmarks of ASD (Roullet et al. 2013). VPA pups show abnormal PFC connectivity and changes in pre- and postsynaptic function under basal conditions (Rinaldi et al. 2008). Furthermore, measurement of correlated activity in layer 5 neurons suggests a loss of muscarinic acetylcholine receptor (mAChR)-mediated decorrelation in these animals, indicating a deficit in attention microcircuitry (Luongo et al. 2015). Clinical studies have reported abnormalities in mAChR function in the autistic forebrain (Perry et al. 2001), and recent work has linked M_1 mAChR modulation of synaptic function to the eCB system in the PFC (Martin et al. 2015). Direct measurement of eCB synaptic plasticity, however, suggests that eCB-LTD is intact in the VPA rat at later developmental stages (Martin et al. 2015). These studies however focused on the function of the CB1R in the PFC. In contrast, direct measurement of eCBs and their associated receptors in the VPA pups have instead identified changes in other eCB receptors (GPR55, PPARα) which may play an important role in ASD (Kerr et al. 2013). Notably, AEA, a putative agonist of GPR55 and PPAR, has recently shown to be mobilized by oxytocin receptors in cortical and subcortical regions linking pro-social signaling and the eCB system (Ninan 2011; Wei et al. 2015).

Neuroligins The investigation of eCBs in genetic models of ASD, other than in the Fmr1 knockout mouse, is similarly in its infancy. The established neuroligin 3 knockout mouse model and the neuroligin 3 mutant (arginine 451 → cysteine substitution) ASD gain of function mutation have been studied in the hippocampus.

In both models, deficits in interneuron GABA release linked to the eCB system are reported. Notably, at the cholecystokinin basket cell to CA1 pyramidal synapse, a selective loss in CB1R tonic signaling, but not phasic signaling, is found (Foldy et al. 2013). A similar enhancement in inhibitory drive is found in the somatosensory cortex which may also be linked to changes in tonic CB1R signaling in these same mice (Speed et al. 2015). Interestingly, a contrasting knockout of neuroligin partner β-neurexins led to an increase in CB1R tonic signaling at excitatory synapses and depressed glutamate release (Anderson et al. 2015). These data suggest that the eCB system may be integral to the changes in excitatory/inhibitory balance proposed to occur in ASD. However, currently neither the neuroligin nor nerexin model has been investigated in the PFC; thus, caution should be taken in the interpretation of these findings in the context of PFC physiology.

6 Intellectual Disability

The use of cannabis results in acute and chronic deficits in cognition. Perhaps unsurprisingly there is also evidence that mPFC deficits in eCB signaling are also linked to genetic intellectual disability (ID) disorders. As discussed above, one of the most consistent phenotypes in FXS is moderate ID. It appears that interventions which modulate mGluR5 signaling may normalize cognitive performance and learning in PFC-dependent tasks (Chen et al. 2014; Xu et al. 2012). However, whether this is linked to eCB signaling is unclear.

Down syndrome (DS), caused by a trisomy of chromosome 21, is the most common genetic cause of ID. Typical of other forms of ID, both dendritic and synaptic alterations in cortical neurons are found. In concert with these morphological changes, there are also DS-linked deficits in synaptic regulation (Garner and Wetmore 2012). An important DS-linked gene is DYRK1A. Increase of the genetic expression of Dyrk1A in mouse models recapitulates many of the synaptic deficits found in DS. In the mPFC of Dyrk1A mice, there is a loss of both LTP and eCB-LTD in deep layer 5 principal neurons (Thomazeau et al. 2014). This loss of eCB-LTD is not associated with reduced CB1R function, but instead involves reduced retrograde 2-AG signaling. Similar to findings in the Fmr1 KO mouse, enhancement of 2-AG with the MAGL inhibitor JZL184 effectively restores eCB-LTD. Promisingly, treatment of mice with JZL184 also restores some of the cognitive deficits found in the Ts65Dn mouse model of DS (Lysenko et al. 2014), suggesting the eCB system may in general be a therapeutic target in DS.

Currently it is unknown if eCB synaptic function is altered in other genetic disorders showing phenotypic ID (Rett syndrome, neurofibromatosis, tuberous sclerosis). However, they along with FXS and DS often share a commonality of deficits in postsynaptic protein expression and regulation (Castrén et al. 2012). Given that postsynaptic coupling to the eCB system in the PFC is defective in both FXS and DS, it will be interesting to see if the eCB system is likewise altered in other models of ID.

7 Stress and Anxiety

The involvement of the eCB system in the emotional states of stress and anxiety is readily apparent. While not identical, a great deal of overlap exists between these mental conditions and the current research addressing them. Therefore, stress and anxiety are presented here in tandem. Indeed, a great wealth of evidence points to a bidirectional relationship between eCB dysfunction and the intensity, duration, and frequency of such events as well as the impact of such states on eCB function itself. This is not surprising, given the dense distribution of CB1Rs in numerous regions of the brain involved in emotional states such as the PFC, amygdala, hippocampus, thalamus, and hypothalamus in both humans (Mato and Pazos 2004) and rodents (Herkenham 1992; Herkenham et al. 1990, 1991). Here, we will examine the evidence supporting a role for eCB dysfunction in anxious states as well as the impact of such states on the eCB system. There exists a significant overlap in research paradigms for animal models of chronic stress and those for depression, the former of which may precipitate symptomology resembling the latter (Willner 2005). As such, a further section on depression will focus primarily on animal models of depression and the results of chronic stress, while this section's primary focus is acute stress and anxiety. For a thorough review of the eCB system's role in stress, see chapter by Hill and McLaughlin, as well as a relevant review by McEwan et al. (2015).

eCB Dysfunction and Stress Phenotypes Dysfunctional eCB signaling has been linked to a variety of disordered emotional states in humans including elevated levels of anxiety. Recapitulation of this signaling dysfunction in animal models has yielded similar behavioral phenotypes, indicating a foundational relationship between the system-level aberration and the expression of elevated stress or anxiety. Indeed, genetic predisposition toward anxiety has been exhibited in animal models using a CB1R knockout mouse (Aso et al. 2008; Martin et al. 2002; Steiner et al. 2008). These CB1R knockout mice also show increased anxiety under basal conditions as compared to control mice, in addition to exhibiting exaggerated responses to subsequent stress (Hill et al. 2011a). Furthermore, genetic deletion of MAGL, which induces increased 2-AG levels in the mPFC, results in an anxiogenic phenotype through desensitized CB1R signaling (Imperatore et al. 2015), further supporting a correlation between attenuated eCB signaling and anxiety-like states. In humans, FAAH gene variation linked with decreased expression has been associated with reduced stress reactivity in line with that seen in animal models (Gunduz-Cinar et al. 2013). Finally, aberrant eCB signaling in animals fed a diet deficient in $n - 3$ polyunsaturated fats, which elevates levels of 2-AG (Watanabe et al. 2003) and induces a desensitized or uncoupled state in presynaptic CB1Rs (Lafourcade et al. 2011; Larrieu et al. 2012) in the PFC, has been shown to increase anxiety-like behaviors. Clearly, dysregulation of eCB function in the PFC significantly impacts the expression of anxiety as determined by a variety of animal models.

Pharmacological manipulation of eCB function can readily recapitulate the outcomes of the aforementioned dysfunctions and has led to a vastly increased understanding of the relationship between the eCB system and the emotional states of stress and anxiety. Indeed, while direct activation of CB1R has been repeatedly shown to produce anxiolytic effects in animals exposed to stressful conditions (Adamczyk et al. 2008; Bambico et al. 2007; Hill and Gorzalka 2005; Rutkowska and Jachimczuk 2004), CB1R antagonists induce an anxiogenic state which mirrors that of chronic stress in animals (Beyer et al. 2010) as well as humans (Hill and Gorzalka 2009; Nissen et al. 2008). Indeed, pretreatment with a CB1R antagonist significantly enhances stress-induced activation of neurons in the PFC (Patel et al. 2005), further supporting the observation that CB1R activation is inversely correlated with anxiogenic states in models of acute stress.

Conversely, direct activation of CB1Rs in the PFC induces a state that is functionally protective against stressful conditions (Rubino et al. 2008a, b). Additionally, enhancing CB1R function in the PFC via direct infusion of a CB1R agonist, FAAH inhibitor or reuptake inhibitor enhances the extinction of both learned and innate fear responses (Lin et al. 2009). Furthermore, enhancement of AEA levels through local infusion of a FAAH inhibitor reduces anxiety-like behaviors, while overexpression of FAAH conversely reduces AEA levels and therefore produces anxiogenic symptoms (Rubino et al. 2008b). Similar effects have been noted following direct infusion of an AEA reuptake inhibitor in the prelimbic PFC which reduces anxious responses in the elevated plus maze and Vogel conflict test, further supporting the notion that CB1 signaling in the PFC modulates both innate and learned fear behavior (Lisboa et al. 2015). These effects are mirrored following inhibition of AEA hydrolysis or reuptake prior to other stressful conditions such as the forced swim test (Adamczyk et al. 2008; Gobbi et al. 2005; Hill and Gorzalka 2005) and the tail suspension test (Naidu et al. 2007). More directly, local infusion of a CB1R agonist attenuates the stress response to the forced swim test. This effect was determined to act via a 5-HT-dependent mechanism (Bambico et al. 2007), lending support to the idea that eCB function in states of stress and anxiety is largely dependent upon modulation of 5-HT, GABA, or glucocorticoid signaling (Hill et al. 2011b). Aside from activation of CB1Rs, TRPV1 channel activation may contribute to the anxiety-modulating effects of local AEA signal enhancement, as TRPV1 blockade in the PFC has been shown to elicit an anxiolytic response (Aguiar et al. 2009). Given these consistent findings, it is clear that the eCB signaling and modulation of multiple neurotransmitter systems in the PFC is fundamentally tied to the emotional states of stress and anxiety, though further research is required to elucidate the exact mechanisms underlying this relationship.

Stress-Induced Modulation of eCB Function Stressful experience has been repeatedly shown to alter eCB signaling in the PFC. Chronic or repeated stress in rodents decreases AEA signaling, possibly through increased clearance of the molecule by its degrading enzyme FAAH (Bortolato et al. 2007; Rademacher et al. 2008). Interestingly, the same procedures result in increased 2-AG signaling, though without any effect on CB1R receptor density (Rademacher et al. 2008). A similar increase in 2-AG signaling in the PFC is found following multiple, but not single, exposures to restraint stress (Patel et al. 2005). Furthermore, CB1R mRNA is downregulated in the PFC following predator threat stress exposure (Campos et al. 2013). An elegant study attempting to piece together the multiple changes in the eCB system during stress showed that exposure to stress increases 2-AG levels in the mPFC via a glucocorticoid-dependent mechanism as a means of regulating activation of the stress-mediating HPA axis (Hill et al. 2011b).

Taken together, the currently available data in the field indicate that the impact of stress on eCB functioning is largely negative (i.e., anxiogenic experiences decrease eCB signaling). However, the final outcome measures and resultant changes induced by stress provide an unclear picture of the multiple stages of regulation and compensatory changes that occur. In part, it is apparent that these discrepancies owe to the variance in protocol for induction, including the degree of predictability. Nonetheless, it is clear that stress-induced regulation of the eCB system is a complex and time-dependent phenomenon that requires increased and more consistent research before it is fully elucidated.

8 Depression

Humans have used cannabis for its multiple effects on mood for thousands of years. However, exposure to THC during critical development periods such as adolescence induces synaptic and behavioral abnormalities that suggest a depression-like state (Rubino et al. 2008c, 2009, 2015). The participation of CB1R in the physiological processing of emotions is well documented (Draycott et al. 2014; Martin et al. 2002; Rubino et al. 2015; Witkin et al. 2005), and a role for eCB dysfunction in major depression has long been postulated (Gorzalka and Hill 2011; Hillard and Liu 2014; Serra and Fratta 2007).

Individuals suffering from major depression exhibit a variety of abnormal functions in the PFC as determined by neuroimaging studies (for a review, see Rive et al. 2013). Notably, depressed suicide victims exhibit increased CB1R density, elevated levels of eCBs, and enhanced agonist-stimulated CB1R activity (Vinod and Hungund 2006). Similar postmortem studies have confirmed abnormal eCB function in the PFC of depressed subjects (Koethe et al. 2007), including elevated CB1R levels and enhanced CB1R-mediated signaling in the PFC of alcoholic suicide victims as compared to non-suicidal alcoholics (Vinod et al. 2005). This finding is further supported by elevated CB1R function in the brains of those diagnosed with depression (Choi et al. 2012) as well as increased levels of AEA and 2-AG found in the PFCs of depressed suicide victims (Hungund et al. 2004) and in the serum of depressed patients (Hill and Gorzalka 2009).

In addition to a wealth of measures in human depression subjects, early animal studies indicated a role for eCB signaling in mood disorders, notably the CB1R knockout mouse, which shows anxiety-like behaviors in such depression-linked assays as the elevated plus-maze (EPM), open-field, and light-dark box (Haller et al. 2002; Maccarrone et al. 2002; Martin et al. 2002; Urigüen et al. 2002). These conditions are accompanied by aberrant neurochemical states known to be associated with mood disorders such as increased corticosterone responses and HPA axis activity (Urigüen et al. 2002). Similar effects are seen in nongenetically altered mice administered CB1R antagonists (Marsicano et al. 2002), while conversely, administration of Δ-9-THC in a model of depression in rats decreases depression-like symptomatology (Elbatsh et al. 2012). As detailed in depth below, these findings highlight a substantial role for the eCB system in the generation, expression, and treatment of mood disorders such as depression.

Animal Models Induction of a depression-like state in rodents achieved via chronic unpredictable stress leads to decreased levels of AEA in the PFC (Hill et al. 2008a). A similar effect is seen following the forced swim test, wherein the stressor induces a reduction in AEA levels in the PFC, which can be rescued (along with resultant depression-like symptomology) via direct application of an FAAH inhibitor (McLaughlin and Gobbi 2012). Additionally, repeated depression-inducing stressors elevate CB1R expression and mRNA in the PFC (Bortolato et al. 2007; Lee and Hill 2013; McLaughlin et al. 2013; Zoppi et al. 2011). This elevation in CB1R expression is likely a compensatory mechanism in response to reduced AEA levels following the induction of stressful states (Hill and Gorzalka 2009; McLaughlin et al. 2014), indicating a dynamic response in PFC eCB function. These data may offer some explanation for competing or contradictory results in measurements of the eCB system in depression models. Interestingly, local CB1R blockade in the PFC increases immobility responses to the forced swim test following chronic unpredictable stress (McLaughlin et al. 2013), indicating that negative modulation of CB1R signaling in the PFC contributes strongly to the development of anxiogenic responding.

Pharmacological manipulation of eCB signaling has provided further insights into the role of eCBs in emotional behavior. Increasing levels of AEA via FAAH inhibition reduces immobility in the forced swim test, an indication of antidepressant-like activity (Adamczyk et al. 2008; Hill and Gorzalka 2005; Hill et al. 2006; Umathe et al. 2011). Similar dose-dependent antidepressant effects in this test are also seen following intra-PFC infusion of a CB1R agonist (Bambico et al. 2007) or FAAH inhibitor (McLaughlin et al. 2012). Conversely, genetic deletion of MAGL, leading to increased levels of 2-AG, increases anxiogenic responding (Imperatore et al. 2015). These findings further highlight the complex regulation of emotion by the eCB system, wherein low levels of cannabinoids inhibit glutamatergic transmission and thus induce an anxiolytic effect, while higher levels inhibit GABAergic transmission and generate anxiogenic behaviors (Rey et al. 2012).

Perhaps not surprisingly given the known alterations in eCB signaling during development, chronic stress differentially alters eCB signaling in the PFC in adolescent rats as compared to adult rats. Repeated stress in adult or adolescent rats causes increased CB1R binding in the PFC; however, while eCB signaling in adult rats normalizes over a subsequent 40-day recovery period, adolescents exhibit sustained CB1R downregulation following stress exposure (Lee and Hill 2013). Indeed, adolescent rats exposed to chronic stress conditions exhibit lasting deficits in PFC function and exaggerated anxiogenic responses 30 days after the cessation of treatment, which could be ameliorated when the animals were treated with a CB1R agonist prior to stress sessions (Abush and Akirav 2013). The time course of impact for lasting consequences resultant of pharmacological enhancement or attenuation of the eCB system is clearly a topic requiring a substantial degree of additional research, especially considering the potential for eCB-targeting pharmacotherapies for targeting emotional states which appear during multiple stages of development.

CB1R-deficient mice are commonly employed as an animal model of depression (for a review, see Valverde and Torrens 2012). Such hallmarks of depressive behavior include passive coping (Aso et al. 2008; Choi et al. 2012; Mato et al. 2007; Steiner et al. 2008) and anhedonia (Martin et al. 2002; Sanchis-Segura et al. 2004). These mice also show such PFC-localized neurochemical signs of depression as increased extracellular levels of 5-HT (Aso et al. 2008), an effect which is also seen with the administration of a CB1R antagonist (Tzavara et al. 2003). Interestingly, activation of CB1Rs in the mPFC increases 5-HT neuron activity (Bambico et al. 2007; McLaughlin et al. 2012) and results in a hyperactive HPA axis (Hill et al. 2011b). This contradiction may be due to compensatory changes in signaling in CB1R-deficient models. The alteration of 5-HT levels in CB1R knockout mice is further accompanied by dysfunction in 5-HT receptor signaling in the PFC (Mato et al. 2007). While further research is required to elucidate the exact mechanisms underlying these changes, the evidence clearly indicates a role for eCB signaling in the regulation of serotonergic neuronal communication in the PFC.

eCB-Targeting Pharmacotherapies While traditional pharmacotherapeutic approaches to the treatment of depression have focused on regulation of monoaminergic synaptic transmission, the role of eCBs in modulating serotonergic, dopaminergic, GABAergic, and noradrenergic neurotransmission has poised the eCB system as a logical and valuable target for treating the disorder (Wyrofsky et al. 2015). Indeed, CB1R antagonists exert antidepressant-like effects in animal models (Griebel et al. 2005; Shearman et al. 2003; Witkin et al. 2005) in a manner similar to traditional antidepressants, in that application of CB1R antagonists increases PFC levels of 5-HT, DA, and NA (Need et al. 2006; Tzavara et al. 2003). However, given the known complications with human administration of CB1R antagonists such as increased risk of anxiety, depressive mood disorders, and risk of self-harm (Christensen et al. 2007), this is unlikely to prove an applicable route for the treatment of depression.

Monoamine system-targeted antidepressants have been shown to similarly alter eCBs in the PFC (Hill et al. 2008a; Manna and Umathe 2012; Smaga et al. 2014). These effects are replicated in multiple classes of antidepressant, lending additional evidence to an interplay between eCB signaling and monoamine regulation. Indeed, altered CB1R transmission in the PFC in an animal model of depression is effectively reversed following treatment with the 5-HT reuptake inhibitor (SSRI) fluoxetine (Rodriguez-Gaztelumendi et al. 2009), which has also been shown to increase CB1R signaling in the PFC (Mato et al. 2010). This finding is not surprising, given the previously demonstrated increase in CB1R density in other brain regions following treatment with an SSRI (Hill et al. 2006). Additionally, treatment with the MAOI tranylcypromine leads to decreased AEA levels and increased CB1R binding density in the PFC (Hill et al. 2008b). Preventative treatment with the tricyclic antidepressant imipramine during stress exposure prevents the normally observed increase in PFC CB1R density (Hill et al. 2008a). Finally, electroconvulsive shock therapy, which has been validated as a beneficial treatment for major depression, restores normal eCB function in the PFC (Hill et al. 2007). Together, these findings suggest significant, bidirectional interactivity between eCB signaling and monoamine regulation.

In humans, a number of studies utilizing direct CB1R modulators have been conducted for such mood disorders as seasonal affective disorder (Bergamaschi et al. 2011) and post-traumatic stress disorder (PTSD; Cameron et al. 2014; Fraser 2009; Passie et al. 2012; Roitman et al. 2014). Indeed, Δ-9-THC administration reduces negative responding and enhances positive association in a facial visualization task, suggesting an effective antidepressant-like activity in emotional processing (Bossong et al. 2013). The potential antidepressant effects of cannabinoids are not surprising given the preponderance of self-medication with cannabis in subjects diagnosed with generalized anxiety disorder, depression, and PTSD (Bowers and Ressler 2015), the latter of which is also associated with genetic risk factors related to abnormal eCB function (Dincheva et al. 2015; Gunduz-Cinar et al. 2013; Lu et al. 2008; Pardini et al. 2012). Less direct approaches to the treatment of mood disorders via eCB-targeting therapies such as inhibition of FAAH have been proposed and have shown significant promise in preclinical models of anxiety (for a full review, see Table 1 in Fowler 2015) and are currently under testing in a number of human clinical trials (Trials: NCT01665573, NCT01964651, NCT01650597, NCT02065739, NCT02169973).

In addition to CB1R activity, recent evidence has also illustrated a potential role for CB2Rs in depression. A CB2R gene polymorphism has been linked to depression in human subjects (Onaivi et al. 2008). Additionally, administration of a selective CB2 agonist has been shown to reduce depression-like immobility behavior in a rodent model (Hu et al. 2009). AEA signaling through TRPV1 receptors in the PFC has also been indicated in the generation of anxiolytic responses (Aguiar et al. 2009). Indeed, increasing concentrations of AEA in the PFC induces a shift from primarily CB1R-mediated anxiolytic response to TRPV1-mediated anxiogenic response (Rubino et al. 2008b), illuminating a delicate balance in targeting the eCB system with regard to the treatment of mood disorders. While TRPV1 knockout mice exhibit decreased immobility responses (You et al. 2012), direct TRPV1 activation has been shown to elicit antidepressant-like effects (Hayase 2011; Kasckow et al. 2004) which has been suggested as a possible mechanism of action contributing to the actions of such classical antidepressant drugs as fluoxetine (Manna and Umathe 2012). Clearly, regulation of eCB function presents a valuable, though complicated, route toward the pharmacotherapeutic treatment of such mood disorders as depression.

9 Conclusion

In the PFC, the eCB system functions as a hub composed by highly organized synaptic proteins. The "eCB hub system" physiologically connects most, if not all, parts of the PFC circuitry to regulate synaptic transmission and ultimately participates to the expression of PFC-dependent behaviors. The central importance of the eCB hub system in PFC functions is highlighted by its covariance with synaptic and behavioral deficits in a wide array of etiologically diverse synaptopathies, from neurodegenerative to genetic and environmental. While immensely complicated, due in part to its widespread expression, the eCB hub system presents a relatively unexploited target of therapeutic opportunity for the recovery of synaptic function and plasticity in a wide range of disorders wherein restoring network integrity may ameliorate both the symptomology and, in many cases, underlying etiology of such maladies.

Acknowledgments This work was supported by the Institut National de la Santé et de la Recherche Médicale (INSERM), a NIDA/INSERM fellowship to AFS and grants from Agence Nationale de la Recherche (CYFIP-Aut to HSM and OM), the FRAXA Research Foundation; and the Fondation Jêrome Lejeune (to HSM and OM). The authors thank Laura F. Scheyer for the design and execution of the chapter's schematic diagram.

References

Abush H, Akirav I (2013) Cannabinoids ameliorate impairments induced by chronic stress to synaptic plasticity and short-term memory. Neuropsychopharmacology 38(8):1521–1534. doi:10.1038/npp.2013.51

Adamczyk P, Golda A, McCreary AC, Filip M, Przegaliński E (2008) Activation of endocannabinoid transmission induces antidepressant-like effects in rats. J Physiol Pharmacol 59(2):217–228

Aguiar DC, Terzian AL, Guimaraes FS, Moreira FA (2009) Anxiolytic-like effects induced by blockade of transient receptor potential vanilloid type 1 (TRPV1) channels in the medial prefrontal cortex of rats. Psychopharmacology 205(2):217–225. doi:10.1007/s00213-009-1532-5

Ahmed A, van der Marck MA, van den Elsen G, Olde Rikkert M (2015) Cannabinoids in late-onset Alzheimer's disease. Clin Pharmacol Ther 97(6):597–606. doi:10.1002/cpt.117

Alen F, Moreno-Sanz G, Isabel de Tena A, Brooks RD, Lopez-Jimenez A, Navarro M, Lopez-Moreno JA (2008) Pharmacological activation of CB1 and D2 receptors in rats: predominant role of CB1 in the increase of alcohol relapse. Eur J Neurosci 27(12):3292–3298. doi:10.1111/j.1460-9568.2008.06302.x

Alváro-Bartolomé M, Garcià-Sevilla JA (2013) Dysregulation of cannabinoid CB1 receptor and associated signaling networks in brains of cocaine addicts and cocaine-treated rodents. Neuroscience 247:294–308. doi:10.1016/j.neuroscience.2013.05.035

Anderson GR, Aoto J, Tabuchi K, Foldy C, Covy J, Yee AX, Wu D, Lee SJ, Chen L, Malenka RC, Sudhof TC (2015) beta-Neurexins control neural circuits by regulating synaptic endocannabinoid signaling. Cell 162(3):593–606. doi:10.1016/j.cell.2015.06.056

Andréasson S, Allebeck P, Engström A, Rydber U (1987) Cannabis and schizophrenia. A longitudinal study of Swedish conscripts. Lancet 2(8574):1483–1486

Argawal A, Neale MC, Prescott CA, Kendler KS (2004) A twin study of early cannabis use and subsequent use and abuse/dependence of other illicit drugs. Psychol Med 34(7):1227–1237

Arnone M, Maruani J, Chaperon F, Thiébot MH, Poncelot M, Soubrié P, Le Fur G (1997) Selective inhibition of sucrose and ethanol intake by SR 141716, an antagonist of central cannabinoid (CB1) receptors. Psychopharmacology 132(1):104–106

Arseneault L, Cannon M, Witton J, Murray RM (2004) Causal association between cannabis and psychosis: examination of the evidence. Br J Psychiatry 184:110–117

Aso E, Ferrer I (2014) Cannabinoids for treatment of Alzheimer's disease: moving toward the clinic. Front Pharmacol 5:37. doi:10.3389/fphar.2014.00037

Aso E, Ozaita A, Valdizan EM, Ledent C, Pazos A, Maldonado R, Valverde O (2008) BDNF impairment in the hippocampus is related to enhanced despair behavior in CB1 knockout mice. J Neurochem 105(2):565–572. doi:10.1111/j.1471-4159.2007.05149.x

Aso E, Sanchez-Pla A, Vegas-Lozano E, Maldonado R, Ferrer I (2015) Cannabis-based medicine reduces multiple pathological processes in AbetaPP/PS1 mice. J Alzheimers Dis 43 (3):977–991. doi:10.3233/JAD-141014

Bambico FR, Katz N, Debonnel G, Gobbi G (2007) Cannabinoids elicit antidepressant-like behavior and activate serotonergic neurons through the medial prefrontal cortex. J Neurosci 27(43):11700–11711. doi:10.1523/JNEUROSCI.1636-07.2007

Bari M, Battista N, Valenza M, Mastrangelo N, Malaponti M, Catanzaro G, Centonze D, Finazzi-Agro A, Cattaneo E, Maccarrone M (2013) In vitro and in vivo models of Huntington's disease show alterations in the endocannabinoid system. FEBS J 280(14):3376–3388. doi:10.1111/febs.12329

Basavarajappa BS, Hungund BL (1999) Down-regulation of cannabinoid receptor agonist-stimulated [35S]GTP gamma S binding in synaptic plasma membrane from chronic ethanol exposed mouse. Brain Res 815(1):89–97

Bedse G, Romano A, Cianci S, Lavecchia AM, Lorenzo P, Elphick MR, Laferla FM, Vendemiale G, Grillo C, Altieri F, Cassano T, Gaetani S (2014) Altered expression of the CB1 cannabinoid receptor in the triple transgenic mouse model of Alzheimer's disease. J Alzheimers Dis 40(3):701–712. doi:10.3233/JAD-131910

Bedse G, Romano A, Lavecchia AM, Cassano T, Gaetani S (2015) The role of endocannabinoid signaling in the molecular mechanisms of neurodegeneration in Alzheimer's disease. J Alzheimers Dis 43(4):1115–1136. doi:10.3233/JAD-141635

Bergamaschi MM, Queiroz RH, Chagas MH, de Oliveira DC, De Martinis BS, Kapczinski F, Quevedo J, Roesler R, Schroder N, Nardi AE, Martin-Santos R, Hallak JE, Zuardi AW, Crippa JA (2011) Cannabidiol reduces the anxiety induced by simulated public speaking in treatment-naive social phobia patients. Neuropsychopharmacology 36(6):1219–1226. doi:10.1038/npp.2011.6

Beyer CE, Dwyer JM, Piesla MJ, Platt BJ, Shen R, Rahman Z, Chan K, Manners MT, Samad TA, Kennedy JD, Bingham B, Whiteside GT (2010) Depression-like phenotype following chronic CB1 receptor antagonism. Neurobiol Dis 39(2):148–155. doi:10.1016/j.nbd.2010.03.020

Bhattacharyya S, Crippa JA, Allen P, Martin-Santos R, Borgwardt S, Fusar-Poli P, Rubia K, Kambeitz J, O'Carroll C, Seal ML, Giampietro V, Brammer M, Zuardi AW, Atakan Z, McGuire PK (2012) Induction of psychosis by Δ9-tetrahydrocannabinol reflects modulation of prefrontal and striatal function during attentional salience processing. Arch Gen Psychiatry 69(1):27–36. doi:10.1001/archgenpsychiatry.2011.161

Bird MK, Kirchhoff J, Djouma E, Lawrence AJ (2008) Metabotropic glutamate 5 receptors regulate sensitivity to ethanol in mice. Int J Neuropsychopharmacol 11(6):765–774. doi:10.1017/S1461145708008572

Bisogno T, Martire A, Petrosino S, Popoli P, Di Marzo V (2008) Symptom-related changes of endocannabinoid and palmitoylethanolamide levels in brain areas of R6/2 mice, a transgenic model of Huntington's disease. Neurochem Int 52(1-2):307–313. doi:10.1016/j.neuint.2007.06.031

Blanco E, Pavon FJ, Palomino A, Luque-Rojas MJ, Serrano A, Rivera P, Bilbao A, Alen F, Vida M, Suarez J, Rodriguez de Fonseca F (2015) Cocaine-induced behavioral sensitization is associated with changes in the expression of endocannabinoid and glutamatergic signaling systems in the mouse prefrontal cortex. Int J Neuropsychopharmacol 18(1). doi:10.1093/ijnp/pyu024

Blednov YA, Cravatt BF, Boehm SL 2nd, Walker D, Harris RA (2007) Role of endocannabinoids in alcohol consumption and intoxication: studies of mice lacking fatty acid amide hydrolase. Neuropsychopharmacology 32(7):1570–1582. doi:10.1038/sj.npp.1301274

Bortolato M, Mangieri RA, Fu J, Kim JH, Arguello O, Duranti A, Tontini A, Mor M, Tarzia G, Piomelli D (2007) Antidepressant-like activity of the fatty acid amide hydrolase inhibitor URB597 in a rat model of chronic mild stress. Biol Psychiatry 62(10):1103–1110. doi:10.1016/j.biopsych.2006.12.001

Bossong MG, van Hell HH, Jager G, Kahn RS, Ramsey NF, Jansma JM (2013) The endocannabinoid system and emotional processing: a pharmacological fMRI study with 9-tetrahydrocannabinol. Eur Neuropsychopharmacol 23(12):1687–1697. doi:10.1016/j.euroneuro.2013.06.009

Bowers ME, Ressler KJ (2015) An overview of translationally informed treatments for posttraumatic stress disorder: animal models of pavlovian fear conditioning to human clinical trials. Biol Psychiatry 78(5):E15–E27. doi:10.1016/j.biopsych.2015.06.008

Burgos H, Cofre C, Hernandez A, Saez-Briones P, Agurto R, Castillo A, Morales B, Zeise ML (2015) Methylphenidate has long-lasting metaplastic effects in the prefrontal cortex of adolescent rats. Behav Brain Res 291:112–117. doi:10.1016/j.bbr.2015.05.009

Busquets-Garcia A, Gomis-Gonzalez M, Guegan T, Agustin-Pavon C, Pastor A, Mato S, Perez-Samartin A, Matute C, de la Torre R, Dierssen M, Maldonado R, Ozaita A (2013) Targeting the endocannabinoid system in the treatment of fragile X syndrome. Nat Med 19(5):603–607. doi:10.1038/nm.3127

Bystrowska B, Smaga I, Frankowska M, Filip M (2014) Changes in endocannabinoid and N-acylethanolamine levels in rat brain structures following cocaine self-administration and extinction training. Prog Neuro-Psychopharmacol Biol Psychiatry 50:1–10. doi:10.1016/j.pnpbp.2013.12.002

Caballero A, Tseng KY (2012) Association of cannabis use during adolescence, prefrontal CB1 receptor signaling, and schizophrenia. Front Pharmacol 3:101. doi:10.3389/fphar.2012.00101

Cameron C, Watson D, Robinson J (2014) Use of a synthetic cannabinoid in a correctional population for posttraumatic stress disorder-related insomnia and nightmares, chronic pain, harm reduction, and other indications: a retrospective evaluation. J Clin Psychopharmacol 34(5):559–564. doi:10.1097/JCP.0000000000000180

Campos AC, Ferreira FR, da Silva WA Jr, Guimaraes FS (2013) Predator threat stress promotes long lasting anxiety-like behaviors and modulates synaptophysin and CB1 receptors expression in brain areas associated with PTSD symptoms. Neurosci Lett 533:34–38. doi:10.1016/j.neulet.2012.11.016

Carter E, Wang XJ (2007) Cannabinoid-mediated disinhibition and working memory: dynamical interplay of multiple feedback mechanisms in a continuous attractor model of prefrontal cortex. Cereb Cortex 17(Suppl 1):i16–i26. doi:10.1093/cercor/bhm103

Casadio P, Fernandes C, Murray RM, Di Forti M (2011) Cannabis use in young people: the risk for schizophrenia. Neurosci Biobehav Rev 35(8):1779–1787. doi:10.1016/j.neubiorev.2011.04.007

Castrén E, Elgersma Y, Maffei L, Hagerman R (2012) Treatment of neurodevelopmental disorders in adulthood. J Neurosci 32(41):14074–14079. doi:10.1523/JNEUROSCI.3287-12.2012

Ceccarini J, Hompes T, Verhaeghen A, Casteels C, Peuskens H, Bormans G, Claes S, Van Laere K (2014) Changes in cerebral CB1 receptor availability after acute and chronic alcohol abuse and monitored abstinence. J Neurosci 34(8):2822–2831. doi:10.1523/JNEUROSCI.0849-13.2014

Ceccarini J, Kuepper R, Kemels D, van Os J, Henquet C, Van Laere K (2015) [18F]MK-9470 PET measurement of cannabinoid CB1 receptor availability in chronic cannabis users. Addict Biol 20(2):357–367. doi:10.1111/adb.12116

Centonze D, Rossi S, Prosperetti C, Tscherter A, Bernardi G, Maccarrone M, Calabresi P (2005) Abnormal sensitivity to cannabinoid receptor stimulation might contribute to altered gamma-aminobutyric acid transmission in the striatum of R6/2 Huntington's disease mice. Biol Psychiatry 57(12):1583–1589. doi:10.1016/j.biopsych.2005.03.008

Chakrabarti B, Persico A, Battista N, Maccarrone M (2015) Endocannabinoid signaling in autism. Neurotherapeutics 12(4):837–847. doi:10.1007/s13311-015-0371-9

Chakraborty R, Chatterjee A, Chaudhury S (2014) Impact of substance use disorder on presentation and short-term course of schizophrenia. Psychiatry J 2014:280243. doi:10.1155/2014/280243

Chen T, Lu JS, Song Q, Liu MG, Koga K, Descalzi G, Li YQ, Zhuo M (2014) Pharmacological rescue of cortical synaptic and network potentiation in a mouse model for fragile X syndrome. Neuropsychopharmacology 39(8):1955–1967. doi:10.1038/npp.2014.44

Cheng D, Spiro AS, Jenner AM, Garner B, Karl T (2014) Long-term cannabidiol treatment prevents the development of social recognition memory deficits in Alzheimer's disease transgenic mice. J Alzheimers Dis 42(4):1383–1396. doi:10.3233/JAD-140921

Chiu CQ, Puente N, Grandes P, Castillo PE (2010) Dopaminergic modulation of endocannabinoid-mediated plasticity at GABAergic synapses in the prefrontal cortex. J Neurosci 30 (21):7236–7248. doi:10.1523/JNEUROSCI.0736-10.2010

Choi K, Le T, McGuire J, Xing G, Zhang L, Li H, Parker CC, Johnson LR, Ursano RJ (2012) Expression pattern of the cannabinoid receptor genes in the frontal cortex of mood disorder patients and mice selectively bred for high and low fear. J Psychiatr Res 46(7):882–889. doi:10.1016/j.jpsychires.2012.03.021

Christensen R, Kristensen PK, Bartels EM, Bliddal H, Astrup A (2007) Efficacy and safety of the weight-loss drug rimonabant: a meta-analysis of randomised trials. Lancet 370 (9600):1706–1713. doi:10.1016/S0140-6736(07)61721-8

Cippitelli A, Astarita G, Duranti A, Caprioli G, Ubaldi M, Stopponi S, Kallupi M, Sagratini G, Rodriguez de Fonseca F, Piomelli D, Ciccocioppo R (2011) Endocannabinoid regulation of acute and protracted nicotine withdrawal: effect of FAAH inhibition. PLoS One 6(11):e28142. doi:10.1371/journal.pone.0028142

Colombo G, Orru A, Lai P, Cabras C, Maccioni P, Rubio M, Gessa GL, Carai MA (2007) The cannabinoid CB1 receptor antagonist, rimonabant, as a promising pharmacotherapy for alcohol dependence: preclinical evidence. Mol Neurobiol 36(1):102–112. doi:10.1007/s12035-007-0017-y

Compton MT, Chien VH, Bollini AM (2009) Associations between past alcohol, cannabis, and cocaine use and current schizotypy among first-degree relatives of patients with schizophrenia and non-psychiatric controls. Psychiatry Q 80(3):143–154. doi:10.1007/s11126-009-9102-x

Contractor A, Klyachko VA, Portera-Cailliau C (2015) Altered neuronal and circuit excitability in fragile X syndrome. Neuron 87(4):699–715. doi:10.1016/j.neuron.2015.06.017

Cornish KM, Li L, Kogan CS, Jacquemont S, Turk J, Dalton A, Hagerman RJ, Hagerman PJ (2008) Age-dependent cognitive changes in carriers of the fragile X syndrome. Cortex 44 (6):628–636. doi:10.1016/j.cortex.2006.11.002

Courchesne E, Mouton PR, Calhoun ME, Semendeferi K, Ahrens-Barbeau C, Hallet MJ, Barnes CC, Pierce K (2011) Neuron number and size in prefrontal cortex of children with autism. JAMA 306(18):2001–2010

Crippa JA, Zuardi AW, Martin-Santos R, Bhattacharyya S, Atakan Z, McGuire P, Fusar-Poli P (2009) Cannabis and anxiety: a critical review of the evidence. Hum Psychopharmacol 24 (7):515–523. doi:10.1002/hup.1048

Dalton VS, Long LE, Weickert CS, Zavitsanou K (2011) Paranoid schizophrenia is characterized by increased CB1 receptor binding in the dorsolateral prefrontal cortex. Neuropsychopharmacology 36(8):1620–1630. doi:10.1038/npp.2011.43

Darnell JC, Klann E (2013) The translation of translational control by FMRP: therapeutic targets for FXS. Nat Neurosci 16(11):1530–1536. doi:10.1038/nn.3379

Darnell JC, Van Driesche SJ, Zhang C, Hung KY, Mele A, Fraser CE, Stone EF, Chen C, Fak JJ, Chi SW, Licatalosi DD, Richter JD, Darnell RB (2011) FMRP stalls ribosomal translocation on mRNAs linked to synaptic function and autism. Cell 146(2):247–261

De Marchi N, De Petrocellis L, Orlando P, Daniele F, Fezza F, Di Marzo V (2003) Endocannabinoid signalling in the blood of patients with schizophrenia. Lipids Health Dis 19(2):5

Dean B, Sundram S, Bradbury R, Scarr E, Copolov D (2001) Studies on [3H]CP-55940 binding in the human central nervous system: regional specific changes in density of cannabinoid-1 receptors associated with schizophrenia and cannabis use. Neuroscience 103(1):9–15

den Boon FS, Chameau P, Schaafsma-Zhao Q, van Aken W, Bari M, Oddi S, Kruse CG, Maccarrone M, Wadman WJ, Werkman TR (2012) Excitability of prefrontal cortical pyramidal neurons is modulated by activation of intracellular type-2 cannabinoid receptors. Proc Natl Acad Sci USA 109(9):3534–3549

den Boon FS, Chameau P, Houthuijs K, Bolijn S, Mastrangelo N, Kruse CG, Maccarrone M, Wadman WJ, Werkman TR (2014) Endocannabinoids produced upon action potential firing evoke a Cl(-) current via type-2 cannabinoid receptors in the medial prefrontal cortex. Pflugers Arch 466(12):2257–2268. doi:10.1007/s00424-014-1502-6

Dincheva I, Drysdale AT, Hartley CA, Johnson DC, Jing D, King EC, Ra S, Gray JM, Yang R, DeGruccio AM, Huang C, Cravatt BF, Glatt CE, Hill MN, Casey BJ, Lee FS (2015) FAAH genetic variation enhances fronto-amygdala function in mouse and human. Nat Commun 6:6395. doi:10.1038/ncomms7395

Dinh TP, Freund TF, Piomelli D (2002) A role for monoglyceride lipase in 2-arachidonoylglycerol inactivation. Chem Phys Lipids 121(1–2):149–158

Domenici MR, Azad SC, Marsicano G, Shierloh A, Wotjak CT, Dodt HU, Zieglgänsberger W, Lutz B, Rammes G (2006) Cannabinoid receptor type 1 located on presynaptic terminals of principal neurons in the forebrain controls glutamatergic synaptic transmission. j. Neuroscience 26(21):5794–5799

Dowie MJ, Bradshaw HB, Howard ML, Nicholson LF, Faull RL, Hannan AJ, Glass M (2009) Altered CB1 receptor and endocannabinoid levels precede motor symptom onset in a transgenic mouse model of Huntington's disease. Neuroscience 163(1):456–465. doi:10.1016/j.neuroscience.2009.06.014

Draycott B, Loureiro M, Ahmad T, Tan H, Zunder J, Laviolette SR (2014) Cannabinoid transmission in the prefrontal cortex bi-phasically controls emotional memory formation via functional interactions with the ventral tegmental area. J Neurosci 34(39):13096–13109. doi:10.1523/JNEUROSCI.1297-14.2014

D'Souza DC, Abi-Saab WM, Madonick S, Forselius-Bielen K, Doersch A, Braley G, Gueorguieva R, Cooper TB, Krystal JH (2005) Delta-9-tetrahydrocannabinol effects in schizophrenia: implications for cognition, psychosis, and addiction. Biol Psychiatry 57(6):594–608. doi:10.1016/j.biopsych.2004.12.006

Egertová M, Elphick MR (2000) Localisation of cannabinoid receptors in the rat brain using antibodies to the intracellular C-terminal tail of CB. J Comp Neurol 422(2):159–171

Egertová M, Cravatt BF, Elphick MR (2003) Comparative analysis of fatty acid amide hydrolase and cb1 cannabinoid receptor expression in the mouse brain: evidence of a widespread role for fatty acid amide hydrolase in regulation of endocannabinoid signaling. Neuroscience 119 (2):481–496. doi:10.1016/s0306-4522(03)00145-3

Eggan SM, Lewis DA (2007) Immunocytochemical distribution of the cannabinoid CB1 receptor in the primate neocortex: a regional and laminar analysis. Cereb Cortex 17(1):175–191. doi:10.1093/cercor/bhj136

Eggan SM, Hashimoto T, Lewis DA (2008) Reduced cortical cannabinoid 1 receptor messenger RNA and protein expression in schizophrenia. Arch Gen Psychiatry 65(7):772–784. doi:10.1001/archpsyc.65.7.772

Eggan SM, Mizoguchi Y, Stoyak SR, Lewis DA (2010) Development of cannabinoid 1 receptor protein and messenger RNA in monkey dorsolateral prefrontal cortex. Cereb Cortex 20 (5):1164–1174. doi:10.1093/cercor/bhp179

Eisenhardt M, Leixner S, Spanagel R, Bilbao A (2015) Quantification of alcohol drinking patterns in mice. Addict Biol. doi:10.1111/adb.12325

Elbatsh MM, Moklas MA, Marsden CA, Kendall DA (2012) Antidepressant-like effects of Delta (9)-tetrahydrocannabinol and rimonabant in the olfactory bulbectomised rat model of depression. Pharmacol Biochem Behav 102(2):357–365. doi:10.1016/j.pbb.2012.05.009

Ellgren M, Artmann A, Tkalych O, Gupta A, Hansen HS, Hansen SH, Devi LA, Hurd YL (2008) Dynamic changes of the endogenous cannabinoid and opioid mesocorticolimbic systems during adolescence: THC effects. Eur Neuropsychopharmacol 18(11):826–834. doi:10.1016/j.euroneuro.2008.06.009

Erdozain AM, Rubio M, Valdizan EM, Pazos A, Meana JJ, Fernandez-Ruiz J, Alexander SP, Callado LF (2015) The endocannabinoid system is altered in the post-mortem prefrontal cortex of alcoholic subjects. Addict Biol 20(4):773–783. doi:10.1111/adb.12160

Fagan SG, Campbell VA (2014) The influence of cannabinoids on generic traits of neurodegeneration. Br J Pharmacol 171(6). doi:10.1111/bph.2014.171.issue-6

Farkas S, Nagy K, Palkovits M, Kovacs GG, Jia Z, Donohue S, Pike V, Halldin C, Mathe D, Harkany T, Gulyas B, Csiba L (2012) [(1)(2)(5)I]SD-7015 reveals fine modalities of CB (1) cannabinoid receptor density in the prefrontal cortex during progression of Alzheimer's disease. Neurochem Int 60(3):286–291. doi:10.1016/j.neuint.2011.11.004

Fergusson DM, Horwood LJ (2000) Does cannabis use encourage other forms of illicit drug use? Addiction 95(4):502–520

Foldy C, Malenka RC, Sudhof TC (2013) Autism-associated neuroligin-3 mutations commonly disrupt tonic endocannabinoid signaling. Neuron 78(3):498–509. doi:10.1016/j.neuron.2013.02.036

Fortin DA, Levine ES (2007) Differential effects of endocannabinoids on glutamatergic and GABAergic inputs to layer 5 pyramidal neurons. Cereb Cortex 17(1):163–174. doi:10.1093/cercor/bhj133

Foti DJ, Kotov R, Guey LT, Bromet EJ (2010) Cannabis use and the course of schizophrenia: 10-year follow-up after first hospitalization. Am J Psychiatry 167(8):987–993. doi:10.1176/appi.ajp.2010.09020189

Fourgeaud L, Mato S, Bouchet D, Hemar A, Worley PF, Manzoni OJ (2004) A single in vivo exposure to cocaine abolishes endocannabinoid-mediated long-term depression in the nucleus accumbens. J Neurosci 24(31):6939–6945. doi:10.1523/JNEUROSCI.0671-04.2004

Fowler CJ (2015) The potential of inhibitors of endocannabinoid metabolism as anxiolytic and antidepressive drugs – a practical view. Eur Neuropsychopharmacol 25(6):749–762. doi:10.1016/j.euroneuro.2015.02.005

Fraser GA (2009) The use of a synthetic cannabinoid in the management of treatment-resistant nightmares in posttraumatic stress disorder (PTSD). CNS Neurosci Ther 15(1):84–88. doi:10.1111/j.1755-5949.2008.00071.x

Freedland CS, Sharpe AL, Samson HH, Porrino LJ (2001) Effects of SR141716A on ethanol and sucrose self-administration. Alcohol Clin Exp Res 25(2):277–282

Fusar-Poli P, Crippa JA, Bhattacharyya S, Borgwardt SJ, Allen P, Martin-Santos R, Seal M, Surguladze SA, O'Carroll C, Atakan Z, Zuardi AW, McGuire PK (2009) Distinct effects of {delta}9-tetrahydrocannabinol and cannabidiol on neural activation during emotional processing. Arch Gen Psychiatry 66(1):95–105. doi:10.1001/archgenpsychiatry.2008.519

Gallate JE, Saharov T, Mallet PE, McGregor IS (1999) Increased motivation for beer in rats following administration of a cannabinoid CB1 receptor agonist. Eur J Pharmacol 370 (3):233–240

Gantois I, Pop AS, de Esch CE, Buijsen RA, Pooters T, Gomez-Mancilla B, Gasparini F, Oostra BA, D'Hooge R, Willemsen R (2013) Chronic administration of AFQ056/Mavoglurant restores social behaviour in Fmr1 knockout mice. Behav Brain Res 239:72–79. doi:10.1016/j.bbr.2012.10.059

Garber KB, Visootsak J, Warren ST (2008) Fragile X syndrome. Eur J Hum Genet 16(6):666–672. doi:10.1038/ejhg.2008.61

Garner GC, Wetmore DZ (2012) Synaptic pathology of down syndrome. Adv Exp Med Biol 970:451–468

George O, Koob GF (2010) Individual differences in prefrontal cortex function and the transition from drug use to drug dependence. Neurosci Biobehav Rev 35(2):232–247. doi:10.1016/j.neubiorev.2010.05.002

Giuffrida A, Leweke FM, Gerth CW, Schreiber D, Koethe D, Faulhaber J, Klosterkotter J, Piomelli D (2004) Cerebrospinal anandamide levels are elevated in acute schizophrenia and are inversely correlated with psychotic symptoms. Neuropsychopharmacology 29 (11):2108–2114. doi:10.1038/sj.npp.1300558

Glass M (2001) The role of cannabinoids in neurodegenerative diseases. Prog Neuro-Psychopharmacol Biol Psychiatry 25(4):743–765

Glass M, Faull RL, Dragunow M (1993) Loss of cannabinoid receptors in the substantia nigra in Huntington's disease. Neuroscience 56(3):523–527

Glass M, Dragunow M, Faull RL (2000) The pattern of neurodegeneration in Huntington's disease: a comparative study of cannabinoid, dopamine, adenosine and GABA(A) receptor alterations in the human basal ganglia in Huntington's disease. Neuroscience 97(3):505–519

Gobbi G, Bambico FR, Mangieri R, Bortolato M, Campolongo P, Solinas M, Cassano T, Morgese MG, Debonnel G, Duranti A, Tontini A, Tarzia G, Mor M, Trezza V, Goldberg SR, Cuomo V, Piomelli D (2005) Antidepressant-like activity and modulation of brain monoaminergic transmission by blockade of anandamide hydrolysis. Proc Natl Acad Sci USA 102 (51):18620–18625. doi:10.1073/pnas.0509591102

Goldman-Rakic PS (1990) Cellular and circuit basis of working memory in prefrontal cortex of nonhuman primates. Prog Brain Res 85:325–335

Goldstein RZ, Volkow ND (2011) Dysfunction of the prefrontal cortex in addiction: neuroimaging findings and clinical implications. Nat Rev Neurosci 12(11):652–669. doi:10.1038/nrn3119

Gomes FV, Issy AC, Ferreira FR, Viveros MP, Del Bel EA, Guimaraes FS (2015a) Cannabidiol attenuates sensorimotor gating disruption and molecular changes induced by chronic antagonism of NMDA receptors in mice. Int J Neuropsychopharmacol 18(5). doi:10.1093/ijnp/pyu041

Gomes FV, Llorente R, Del Bel EA, Viveros MP, Lopez-Gallardo M, Guimaraes FS (2015b) Decreased glial reactivity could be involved in the antipsychotic-like effect of cannabidiol. Schizophr Res 164(1-3):155–163. doi:10.1016/j.schres.2015.01.015

Gorzalka BB, Hill MN (2011) Putative role of endocannabinoid signaling in the etiology of depression and actions of antidepressants. Prog Neuro-Psychopharmacol Biol Psychiatry 35 (7):1575–1585. doi:10.1016/j.pnpbp.2010.11.021

Goto Y, Yang CR, Otani S (2010) Functional and dysfunctional synaptic plasticity in prefrontal cortex: roles in psychiatric disorders. Biol Psychiatry 67(3):199–207. doi:10.1016/j.biopsych.2009.08.026

Grant SG (2003) Synapse signalling complexes and networks: machines underlying cognition. Bioessays 25(12):1229–1235. doi:10.1002/bies.10381

Griebel G, Stemmelin J, Scatton B (2005) Effects of the cannabinoid CB1 receptor antagonist rimonabant in models of emotional reactivity in rodents. Biol Psychiatry 57(3):261–267. doi:10.1016/j.biopsych.2004.10.032

Gunduz-Cinar O, MacPherson KP, Cinar R, Gamble-George J, Sugden K, Williams B, Godlewski G, Ramikie TS, Gorka AX, Alapafuja SO, Nikas SP, Makriyannis A, Poulton R, Patel S, Hariri AR, Caspi A, Moffitt TE, Kunos G, Holmes A (2013) Convergent translational evidence of a role for anandamide in amygdala-mediated fear extinction, threat processing and stress-reactivity. Mol Psychiatry 18(7):813–823. doi:10.1038/mp.2012.72

Gururajan A, Taylor DA, Malone DT (2011) Effect of cannabidiol in a MK-801-rodent model of aspects of schizophrenia. Behav Brain Res 222(2):299–308. doi:10.1016/j.bbr.2011.03.053

Hadland SE, Knight JR, Harris SK (2015) The knowledge gaps for medical marijuana in pediatric conditions. J Dev Behav Pediatr 36(9):767–768

Hajós M, Hoffmann WE, Kocsis B (2008) Activation of cannabinoid-1 receptors disrupts sensory gating and neuronal oscillation: relevance to schizophrenia. Biol Psychiatry 63 (11):1075–1083. doi:10.1016/j.biopsych.2007.12.005

Haller J, Bakos N, Szirmay M, Ledent C, Freund TF (2002) The effects of genetic and pharmacological blockade of the CB1 cannabinoid receptor on anxiety. Eur J Neurosci 16 (7):1395–1398. doi:10.1046/j.1460-9568.2002.02192.x

Haller J, Szirmai M, Varga B, Ledent C, Freund TF (2005) Cannabinoid CB1 receptor dependent effects of the NMDA antagonist phencyclidine in the social withdrawal model of schizophrenia. Behav Pharmacol 16(5-6):415–422

Hambrecht M, Häfner H (2000) Cannabis, vulnerability, and the onset of schizophrenia: an epidemiological perspective. Aust N Z J Psychiatry 34(3):468–475

Hansson AC, Bermudez-Silva FJ, Malinen H, Hyytia P, Sanchez-Vera I, Rimondini R, Rodriguez de Fonseca F, Kunos G, Sommer WH, Heilig M (2007) Genetic impairment of frontocortical endocannabinoid degradation and high alcohol preference. Neuropsychopharmacology 32 (1):117–126. doi:10.1038/sj.npp.1301034

Hayase T (2011) Differential effects of TRPV1 receptor ligands against nicotine-induced depression-like behaviors. BMC Pharmacol 11:6. doi:10.1186/1471-2210-11-6

Hayashi Y, Momiyama A, Takahashi H, Ohishi J, Ogawa-Meguro R, Shigemoto R, Mizuno N, Nakanishi S (1993) Role of a metabotropic glutamate receptor in synaptic modulation in the accessory olfactory bulb. Nature 366(6456):687–690

Heng L, Beverley JA, Steiner H, Tseng KY (2011) Differential developmental trajectories for CB1 cannabinoid receptor expression in limbic/associative and sensorimotor cortical areas. Synapse 65(4):278–286. doi:10.1002/syn.20844

Herkenham M (1992) Cannabinoid receptor localization in brain: relationship to motor and reward systems. Ann N Y Acad Sci 654:19–32

Herkenham M, Lynn AB, Little MD, Johnson MR, Melvin LS, de Costa BR, Rice KC (1990) Cannabinoid receptor localization in brain. Proc Natl Acad Sci USA 87(5):1932–1936

Herkenham M, Lynn AB, Johnson MR, Melvin LS, de Costa BR, Rice KC (1991) Characterization and localization of cannabinoid receptors in rat brain: a quantitative in vitro autoradiographic study. J Neurosci 11(2):563–583

Higuera-Matas A, Soto-Montenegro ML, del Olmo N, Miguens M, Torres I, Vaquero JJ, Sanchez J, Garcia-Lecumberri C, Desco M, Ambrosio E (2008) Augmented acquisition of cocaine self-administration and altered brain glucose metabolism in adult female but not male rats exposed to a cannabinoid agonist during adolescence. Neuropsychopharmacology 33 (4):806–813. doi:10.1038/sj.npp.1301467

Hill MN, Gorzalka BB (2005) Pharmacological enhancement of cannabinoid CB1 receptor activity elicits an antidepressant-like response in the rat forced swim test. Eur Neuropsychopharmacol 15(6):593–599. doi:10.1016/j.euroneuro.2005.03.003

Hill MN, Gorzalka BB (2009) Impairments in endocannabinoid signaling and depressive illness. JAMA 301(11):1165–1166. doi:10.1001/jama.2009.369

Hill MN, Ho WS, Sinopoli KJ, Viau V, Hillard CJ, Gorzalka BB (2006) Involvement of the endocannabinoid system in the ability of long-term tricyclic antidepressant treatment to suppress stress-induced activation of the hypothalamic-pituitary-adrenal axis. Neuropsychopharmacology 31(12):2591–2599. doi:10.1038/sj.npp.1301092

Hill MN, Barr AM, Ho WS, Carrier EJ, Gorzalka BB, Hillard CJ (2007) Electroconvulsive shock treatment differentially modulates cortical and subcortical endocannabinoid activity. J Neurochem 103(1):47–56. doi:10.1111/j.1471-4159.2007.04688.x

Hill MN, Carrier EJ, McLaughlin RJ, Morrish AC, Meier SE, Hillard CJ, Gorzalka BB (2008a) Regional alterations in the endocannabinoid system in an animal model of depression: effects of concurrent antidepressant treatment. J Neurochem 106(6):2322–2336. doi:10.1111/j.1471-4159.2008.05567.x

Hill MN, Ho WS, Hillard CJ, Gorzalka BB (2008b) Differential effects of the antidepressants tranylcypromine and fluoxetine on limbic cannabinoid receptor binding and endocannabinoid contents. J Neural Transm (Vienna) 115(12):1673–1679. doi:10.1007/s00702-008-0131-7

Hill MN, Hillard CJ, McEwen BS (2011a) Alterations in corticolimbic dendritic morphology and emotional behavior in cannabinoid CB1 receptor-deficient mice parallel the effects of chronic stress. Cereb Cortex 21(9):2056–2064. doi:10.1093/cercor/bhq280

Hill MN, McLaughlin RJ, Pan B, Fitzgerald ML, Roberts CJ, Lee TT, Karatsoreos IN, Mackie K, Viau V, Pickel VM, McEwen BS, Liu QS, Gorzalka BB, Hillard CJ (2011b) Recruitment of prefrontal cortical endocannabinoid signaling by glucocorticoids contributes to termination of the stress response. J Neurosci 31(29):10506–10515. doi:10.1523/JNEUROSCI.0496-11.2011

Hillard CJ, Liu QS (2014) Endocannabinoid signaling in the etiology and treatment of major depressive illness. Curr Pharm Des 20(23):39

Hirvonen J, Zanotti-Fregonara P, Umhau JC, George DT, Rallis-Frutos D, Lyoo CH, Li CT, Hines CS, Sun H, Terry GE, Morse C, Zoghbi SS, Pike VW, Innis RB, Heilig M (2013) Reduced cannabinoid CB1 receptor binding in alcohol dependence measured with positron emission tomography. Mol Psychiatry 18(8):916–921. doi:10.1038/mp.2012.100

Ho WS, Barrett DA, Randall MD (2008) 'Entourage' effects of N-palmitoylethanolamide and N-oleoylethanolamide on vasorelaxation to anandamide occur through TRPV1 receptors. Br J Pharmacol 155(6):837–846. doi:10.1038/bjp.2008.324

Hojo S, Sudo Y, Ando Y, Minami K, Takada M, Matsubara T, Kanaide M, Taniyama K, Sumikaway K, Uezono Y (2008) mu-Opioid receptor forms a functional heterodimer with cannabinoid CB1 receptor: electrophysiological and FRET assay analysis. J Pharmacol Sci 108 (3):308–319

Hu B, Doods H, Treede RD, Ceci A (2009) Depression-like behaviour in rats with mononeuropathy is reduced by the CB2-selective agonist GW405833. Pain 143(3):206–212. doi:10.1016/j.pain.2009.02.018

Hungund BL, Vinod KY, Kassir SA, Basavarajappa BS, Yalamanchili R, Cooper TB, Mann JJ, Arango V (2004) Upregulation of CB1 receptors and agonist-stimulated [35S]GTPgammaS binding in the prefrontal cortex of depressed suicide victims. Mol Psychiatry 9(2):184–190. doi:10.1038/sj.mp.4001376

Hurd YL, Michaelides M, Miller ML, Jutras-Aswad D (2014) Trajectory of adolescent cannabis use on addiction vulnerability. Neuropharmacology 76(Pt B):416–424. doi:10.1016/j.neuropharm.2013.07.028

Hutchison KE, Haughey H, Niculescu M, Schacht J, Kaiser A, Stitzel J, Horton WJ, Filbey F (2008) The incentive salience of alcohol: translating the effects of genetic variant in CNR1. Arch Gen Psychiatry 65(7):841–850. doi:10.1001/archpsyc.65.7.841

Imperatore R, Morello G, Luongo L, Taschler U, Romano R, De Gregorio D, Belardo C, Maione S, Di Marzo V, Cristino L (2015) Genetic deletion of monoacylglycerol lipase leads to impaired cannabinoid receptor CB1 R signaling and anxiety-like behavior. J Neurochem 135 (4):799–813. doi:10.1111/jnc.13267

Iseger TA, Bossong MG (2015) A systematic review of the antipsychotic properties of cannabidiol in humans. Schizophr Res 162(1-3):153–161. doi:10.1016/j.schres.2015.01.033

Ishiguro H, Horiuchi Y, Ishikawa M, Koga M, Imai K, Suzuki Y, Morikawa M, Inada T, Watanabe Y, Takahashi M, Someya T, Ujike H, Iwata N, Ozaki N, Onaivi ES, Kunugi H, Sasaki T, Itokawa M, Arai M, Niizato K, Iritani S, Naka I, Ohashi J, Kakita A, Takahashi H,

Nawa H, Arinami T (2010) Brain cannabinoid CB2 receptor in schizophrenia. Biol Psychiatry 67(10):974–982. doi:10.1016/j.biopsych.2009.09.024

Jenko KJ, Hirvonen J, Henter ID, Anderson KB, Zoghbi SS, Hyde TM, Deep-Soboslay A, Innis RB, Kleinman JE (2012) Binding of a tritiated inverse agonist to cannabinoid CB1 receptors is increased in patients with schizophrenia. Schizophr Res 141(2–3):185–188. doi:10.1016/j.schres.2012.07.021

Jung KM, Sepers M, Henstridge CM, Lassalle O, Neuhofer D, Martin H, Ginger M, Frick A, DiPatrizio NV, Mackie K, Katona I, Piomelli D, Manzoni OJ (2012) Uncoupling of the endocannabinoid signalling complex in a mouse model of fragile X syndrome. Nat Commun 3:1080. doi:10.1038/ncomms2045

Kalivas PW, Volkow ND (2005) The neural basis of addiction: a pathology of motivation and choice. Am J Psychiatry 162(8):1403–1413

Kano M, Ohno-Shosaku T, Hashimotodani Y, Uchigashima M, Watanabe M (2009) Endocannabinoid-mediated control of synaptic transmission. Physiol Rev 89(1):309–380. doi:10.1152/physrev.00019.2008

Kasanetz F, Lafourcade M, Deroche-Gamonet V, Revest JM, Berson N, Balado E, Fiancette JF, Renault P, Piazza PV, Manzoni OJ (2013) Prefrontal synaptic markers of cocaine addiction-like behavior in rats. Mol Psychiatry 18(6):729–737. doi:10.1038/mp.2012.59

Kasckow JW, Mulchahey JJ, Geracioti TD Jr (2004) Effects of the vanilloid agonist olvanil and antagonist capsazepine on rat behaviors. Prog Neuro-Psychopharmacol Biol Psychiatry 28 (2):291–295. doi:10.1016/j.pnpbp.2003.10.007

Katona I, Freund TF (2012) Multiple functions of endocannabinoid signaling in the brain. Annu Rev Neurosci 35:529–558. doi:10.1146/annurev-neuro-062111-150420

Kerr DM, Downey L, Conboy M, Finn DP, Roche M (2013) Alterations in the endocannabinoid system in the rat valproic acid model of autism. Behav Brain Res 249:124–132. doi:10.1016/j.bbr.2013.04.043

Kluger B, Triolo P, Jones W, Jankovic J (2015) The therapeutic potential of cannabinoids for movement disorders. Mov Disord 30(3):313–327. doi:10.1002/mds.26142

Knable MB, Weinberger DR (1997) Dopamine, the prefrontal cortex and schizophrenia. Synapse 25(3):306–308

Koethe D, Llenos IC, Dulay JR, Hoyer C, Torrey EF, Leweke FM, Weis S (2007) Expression of CB1 cannabinoid receptor in the anterior cingulate cortex in schizophrenia, bipolar disorder, and major depression. J Neural Transm (Vienna) 114(8):1055–1063. doi:10.1007/s00702-007-0660-5

Koethe D, Giuffrida A, Schreiber D, Hellmich M, Schultze-Lutter F, Ruhrmann S, Klosterkotter J, Piomelli D, Leweke FM (2009) Anandamide elevation in cerebrospinal fluid in initial prodromal states of psychosis. Br J Psychiatry 194(4):371–372. doi:10.1192/bjp.bp.108.053843

Kramvis I, Mansvelder HD, Loos M, Meredith R (2013) Hyperactivity, perseveration and increased responding during attentional rule acquisition in the Fragile X mouse model. Front Behav Neurosci 7:172. doi:10.3389/fnbeh.2013.00172

Krueger DD, Bear MF (2011) Toward fulfilling the promise of molecular medicine in fragile X syndrome. Annu Rev Med 62:411–429. doi:10.1146/annurev-med-061109-134644

Lafourcade M, Elezgarai I, Mato S, Bakiri Y, Grandes P, Manzoni OJ (2007) Molecular components and functions of the endocannabinoid system in mouse prefrontal cortex. PLoS One 2(8): e709. doi:10.1371/journal.pone.0000709

Lafourcade M, Larrieu T, Mato S, Duffaud A, Sepers M, Matias I, De Smedt-Peyrusse V, Labrousse VF, Bretillon L, Matute C, Rodriguez-Puertas R, Laye S, Manzoni OJ (2011) Nutritional omega-3 deficiency abolishes endocannabinoid-mediated neuronal functions. Nat Neurosci 14(3):345–350. doi:10.1038/nn.2736

Larrieu T, Madore C, Joffre C, Layé S (2012) Nutritional n-3 polyunsaturated fatty acids deficiency alters cannabinoid receptor signaling pathway in the brain and associated anxiety-like behavior in mice. J Physiol Biochem 68(4):671–681

Lastres-Becker I, Fezza F, Cabeira M, Bisogno T, Ramos JA, Milone A, Fernàndez-Ruiz J, Di Marzo V (2001) Changes in endocannabinoid transmission in the basal ganglia in a rat model of Huntington's disease. Neuroreport 12(10):2125–2129

Lastres-Becker I, Gomez M, De Miguel R, Ramos JA, Fernandez-Ruiz J (2002) Loss of cannabinoid CB(1) receptors in the basal ganglia in the late akinetic phase of rats with experimental Huntington's disease. Neurotox Res 4(7-8):601–608. doi:10.1080/10298420290030514

Lee TT, Hill MN (2013) Age of stress exposure modulates the immediate and sustained effects of repeated stress on corticolimbic cannabinoid CB(1) receptor binding in male rats. Neuroscience 249:106–114. doi:10.1016/j.neuroscience.2012.11.017

Lee SH, Ledri M, Toth B, Marchionni I, Henstridge CM, Dudok B, Kenesei K, Barna L, Szabo SI, Renkecz T, Oberoi M, Watanabe M, Limoli CL, Horvai G, Soltesz I, Katona I (2015) Multiple forms of endocannabinoid and endovanilloid signaling regulate the tonic control of GABA release. J Neurosci 35(27):10039–10057. doi:10.1523/JNEUROSCI.4112-14.2015

Lehtonen M, Storvik M, Tupala E, Hyytia P, Tiihonen J, Callaway JC (2010) Endogenous cannabinoids in post-mortem brains of Cloninger type 1 and 2 alcoholics. Eur Neuropsychopharmacol 20(4):245–252. doi:10.1016/j.euroneuro.2009.12.008

Leroy S, Griffon N, Bourdel MC, Olie JP, Poirier MF, Krebs MO (2001) Schizophrenia and the cannabinoid receptor type 1 (CB1): association study using a single-base polymorphism in coding exon 1. Am J Med Genet 105(8):749–752

Leweke FM, Giuffrida A, Wurster U, Emrich HM, Piomelli D (1999) Elevated endogenous cannabinoids in schizophrenia. Neuroreport 10(8):1665–1669

Leweke FM, Piomelli D, Pahlisch F, Muhl D, Gerth CW, Hoyer C, Klosterkotter J, Hellmich M, Koethe D (2012) Cannabidiol enhances anandamide signaling and alleviates psychotic symptoms of schizophrenia. Transl Psychiatry 2:e94. doi:10.1038/tp.2012.15

Lewis DA, Lieberman JA (2000) Catching up on schizophrenia: natural history and neurobiology. Neuron 28(2):325–334

Lin HC, Mao SC, Su CL, Gean PW (2009) The role of prefrontal cortex CB1 receptors in the modulation of fear memory. Cereb Cortex 19(1):165–175. doi:10.1093/cercor/bhn075

Linzen DH, Dingemans PM, Lenior ME (1994) Cannabis abuse and the course of recent-onset schizophrenic disorders. Arch Gen Psychiatry 51(4):273–279

Lisboa SF, Borges AA, Nejo P, Fassini A, Guimaraes FS, Resstel LB (2015) Cannabinoid CB1 receptors in the dorsal hippocampus and prelimbic medial prefrontal cortex modulate anxiety-like behavior in rats: additional evidence. Prog Neuro-Psychopharmacol Biol Psychiatry 59:76–83. doi:10.1016/j.pnpbp.2015.01.005

Liu ZH, Smith CB (2009) Dissociation of social and nonsocial anxiety in a mouse model of fragile X syndrome. Neurosci Lett 454(1):62–66. doi:10.1016/j.neulet.2009.02.066

Long LE, Malone DT, Taylor DA (2006) Cannabidiol reverses MK-801-induced disruption of prepulse inhibition in mice. Neuropsychopharmacology 31(4):795–803. doi:10.1038/sj.npp.1300838

Long LE, Lind J, Webster M, Weickert CS (2012) Developmental trajectory of the endocannabinoid system in human dorsolateral prefrontal cortex. BMC Neurosci 13(87):1–14

Lovelace JW, Vieira PA, Corches A, Mackie K, Korzus E (2014) Impaired fear memory specificity associated with deficient endocannabinoid-dependent long-term plasticity. Neuropsychopharmacology 39(7):1685–1693. doi:10.1038/npp.2014.15

Lovelace JW, Corches A, Vieira PA, Hiroto AS, Mackie K, Korzus E (2015) An animal model of female adolescent cannabinoid exposure elicits a long-lasting deficit in presynaptic long-term plasticity. Neuropharmacology 99:242–255. doi:10.1016/j.neuropharm.2015.04.034

Lu AT, Ogdie MN, Järvelin MR, Moilanen IK, Loo SK, McCracken JT, McGough JJ, Yang MH, Peltonen L, Nelson SF, Cantor RM, Smalley SL (2008) Association of the cannabinoid receptor gene (CNR1) with ADHD and post-traumatic stress disorder. Am J Med Genet B Neuropsychiatr Genet 147B(8):1488–1494. doi:10.1002/ajmg.b

Luongo FJ, Horn ME, Sohal VS (2015) Putative microcircuit-level substrates for attention are disrupted in mouse models of autism. Biol Psychiatry. doi:10.1016/j.biopsych.2015.04.014

Lüscher C, Malenka RC (2011) Drug-evoked synaptic plasticity in addiction: from molecular changes to circuit remodeling. Neuron 69(4):650–663. doi:10.1016/j.neuron.2011.01.017

Lynskey MT, Heath AC, Bucholz KK, Slutske WS, Madden PA, Nelson EC, Statham DJ, Martin NG (2003) Escalation of drug use in early-onset cannabis users vs co-twin controls. JAMA 289 (4):427–433

Lysenko LV, Kim J, Henry C, Tyrtyshnaia A, Kohnz RA, Madamba F, Simon GM, Kleschevnikova NE, Nomura DK, Ezekowitz RA, Kleschevnikov AM (2014) Monoacylglycerol lipase inhibitor JZL184 improves behavior and neural properties in Ts65Dn mice, a model of down syndrome. PLoS One 9(12):e114521. doi:10.1371/journal.pone.0114521

Maccarrone M, Valverde O, Barbaccia ML, Castañé A, Maldonado R, Ledent C, Parmentier M, Finazzi-Agrò A (2002) Age-related changes of anandamide metabolism in CB1 receptor knockout mice: correlation with behaviour. Eur J Neurosci 15(7):1178–1186

Maccarrone M, Battista N, Centonze D (2007) The endocannabinoid pathway in Huntington's disease: a comparison with other neurodegenerative diseases. Prog Neurobiol 81 (5-6):349–379. doi:10.1016/j.pneurobio.2006.11.006

Maccioni P, Colombo G, Carai MA (2010) Blockade of the cannabinoid CB1 receptor and alcohol dependence: preclinical evidence and preliminary clinical data. CNS Neurol Disord Drug Targets 9(1):55–59

Maldonado R (2002) Study of cannabinoid dependence in animals. Pharmacol Ther 95 (2):153–164

Manna SS, Umathe SN (2012) A possible participation of transient receptor potential vanilloid type 1 channels in the antidepressant effect of fluoxetine. Eur J Pharmacol 685(1-3):81–90. doi:10.1016/j.ejphar.2012.04.023

Manuel I, Gonzalez de San Roman E, Giralt MT, Ferrer I, Rodriguez-Puertas R (2014) Type-1 cannabinoid receptor activity during Alzheimer's disease progression. J Alzheimers Dis 42 (3):761–766. doi:10.3233/JAD-140492

Mao SC, Chang CH, Wu CC, Orejarena MJ, Manzoni OJ, Gean PW (2013) Inhibition of spontaneous recovery of fear by mGluR5 after prolonged extinction training. PLoS One 8 (3):e59580. doi:10.1371/journal.pone.0059580

Marinho EA, Oliveira-Lima AJ, Santos R, Hollais AW, Baldaia MA, Wuo-Silva R, Yokoyama TS, Takatsu-Coleman AL, Patti CL, Longo BM, Berro LF, Frussa-Filho R (2015) Effects of rimonabant on the development of single dose-induced behavioral sensitization to ethanol, morphine and cocaine in mice. Prog Neuro-Psychopharmacol Biol Psychiatry 58:22–31. doi:10.1016/j.pnpbp.2014.11.010

Maroof N, Ravipati S, Pardon MC, Barrett DA, Kendall DA (2014) Reductions in endocannabinoid levels and enhanced coupling of cannabinoid receptors in the striatum are accompanied by cognitive impairments in the AbetaPPswe/PS1DeltaE9 mouse model of Alzheimer's disease. J Alzheimers Dis 42(1):227–245. doi:10.3233/JAD-131961

Marrs WR, Blankman JL, Horne EA, Thomazeau A, Lin YH, Coy J, Bodor AL, Muccioli GG, Hu SS, Woodruff G, Fung S, Lafourcade M, Alexander JP, Long JZ, Li W, Xu C, Moller T, Mackie K, Manzoni OJ, Cravatt BF, Stella N (2010) The serine hydrolase ABHD6 controls the accumulation and efficacy of 2-AG at cannabinoid receptors. Nat Neurosci 13(8):951–957. doi:10.1038/nn.2601

Marsicano G, Lutz B (1999) Expression of the cannabinoid receptor CB1 in distinct neuronal subpopulations in the adult mouse forebrain. Eur J Neurosci 11(12):4213–4225

Marsicano G, Wotjak CT, Azad SC, Bisogno T, Rammes G, Cascio MG, Hermann H, Tang J, Hofmann C, Zieglgänsberger W, Di Marzo V, Lutz B (2002) The endogenous cannabinoid system controls excitation of aversive memories. Nature 418(6897):530–534

Martin M, Ledent C, Parmentier M, Maldonado R, Valverde O (2002) Involvement of CB1 cannabinoid receptors in emotional behaviour. Psychopharmacology 159(4):379–387. doi:10.1007/s00213-001-0946-5

Martin HG, Bernabeu A, Lassalle O, Bouille C, Beurrier C, Pelissier-Alicot AL, Manzoni OJ (2015) Endocannabinoids mediate muscarinic acetylcholine receptor-dependent long-term depression in the adult medial prefrontal cortex. Front Cell Neurosci 9(457):1–11. doi:10.3389/fncel.2015.00457

Mato S, Pazos A (2004) Influence of age, postmortem delay and freezing storage period on cannabinoid receptor density and functionality in human brain. Neuropharmacology 46 (5):716–726. doi:10.1016/j.neuropharm.2003.11.004

Mato S, Aso E, Castro E, Martin M, Valverde O, Maldonado R, Pazos A (2007) CB1 knockout mice display impaired functionality of 5-HT1A and 5-HT2A/C receptors. J Neurochem 103 (5):2111–2120. doi:10.1111/j.1471-4159.2007.04961.x

Mato S, Vidal R, Castro E, Diaz A, Pazos A, Valdizan EM (2010) Long-term fluoxetine treatment modulates cannabinoid type 1 receptor-mediated inhibition of adenylyl cyclase in the rat prefrontal cortex through 5-hydroxytryptamine 1A receptor-dependent mechanisms. Mol Pharmacol 77(3):424–434. doi:10.1124/mol.109.060079

McEwan BS, Bowles NP, Gray JD, Hill MN, Hunter RG, Karatsoreos IN, Nasca C (2015) Mechanisms of stress in the brain. Nat Neurosci 18(10):135301363. doi:10.1038/nn.4086

McLaughlin RJ, Gobbi G (2012) Cannabinoids and emotionality: a neuroanatomical perspective. Neuroscience 204:134–144. doi:10.1016/j.neuroscience.2011.07.052

McLaughlin RJ, Hill MN, Bambico FR, Stuhr KL, Gobbi G, Hillard CJ, Gorzalka BB (2012) Prefrontal cortical anandamide signaling coordinates coping responses to stress through a serotonergic pathway. Eur Neuropsychopharmacol 22(9):664–671. doi:10.1016/j.euroneuro.2012.01.004

McLaughlin RJ, Hill MN, Dang SS, Wainwright SR, Galea LA, Hillard CJ, Gorzalka BB (2013) Upregulation of CB(1) receptor binding in the ventromedial prefrontal cortex promotes proactive stress-coping strategies following chronic stress exposure. Behav Brain Res 237:333–337. doi:10.1016/j.bbr.2012.09.053

McLaughlin RJ, Hill MN, Gorzalka BB (2014) A critical role for prefrontocortical endocannabinoid signaling in the regulation of stress and emotional behavior. Neurosci Biobehav Rev 42:116–131. doi:10.1016/j.neubiorev.2014.02.006

McReynolds JR, Doncheck EM, Vranjkovic O, Ganzman GS, Baker DA, Hillard CJ, Mantsch JR (2015) CB1 receptor antagonism blocks stress-potentiated reinstatement of cocaine seeking in rats. Psychopharmacology. doi:10.1007/s00213-015-4092-x

Mechoulam R, Peters M, Murillo-Rodriguez E, Hanus LO (2007) Cannabidiol – recent advances. Chem Biodivers 4(8):1678–1692

Melis M, Pistis M (2012) Hub and switches: endocannabinoid signalling in midbrain dopamine neurons. Philos Trans R Soc Lond Ser B Biol Sci 367(1607):3276–3285. doi:10.1098/rstb.2011.0383

Michalon A, Sidorov M, Ballard TM, Ozmen L, Spooren W, Wettstein JG, Jaeschke G, Bear MF, Lindemann L (2012) Chronic pharmacological mGlu5 inhibition corrects fragile X in adult mice. Neuron 74(1):49–56. doi:10.1016/j.neuron.2012.03.009

Miller P, Lawrie SM, Hodges A, Clafferty RA, Cunningham Owens DG, Johnstone EC (2001) Genetic liability, illicit drug use, life stress and psychotic symptoms: preliminary findings from the Edinburgh study of people at high risk for schizophrenia. Br J Psychiatry 178 (524–530):524

Monory K, Massa F, Egertová M, Eder M, Blaudzun H, Westenbroek R, Kelsch W, Jacob W, Marsch R, Ekker M, Long JZ, Rubenstein JL, Goebbels S, Nave KA, During M, Klugmann M, Wölfel B, Dodt HU, Zieglgänsberger W, Wotjak CT, Mackie K, Elphick MR, Marsicano G, Lutz B (2006) The endocannabinoid system controls key epileptogenic circuits in the hippocampus. Neuron 51(4):455–466

Mullard A (2015) Fragile X disappointments upset autism ambitions. Nat Rev Drug Discov 14 (3):151–153. doi:10.1038/nrd4555

Naidu PS, Varvel SA, Ahn K, Cravatt BF, Martin BR, Lichtman AH (2007) Evaluation of fatty acid amide hydrolase inhibition in murine models of emotionality. Psychopharmacology 192 (1):61–70. doi:10.1007/s00213-006-0689-4

Need AB, Davis RJ, Alexander-Chacko JT, Eastwood B, Chernet E, Phebus LA, Sindelar DK, Nomikos GG (2006) The relationship of in vivo central CB1 receptor occupancy to changes in cortical monoamine release and feeding elicited by CB1 receptor antagonists in rats. Psychopharmacology 184(1):26–35. doi:10.1007/s00213-005-0234-x

Negrete JC (1989) Cannabis and schizophrenia. Br J Addict 84(4):349–351

Negrete JC, Knapp WP (1986) The effects of cannabis use on the clinical condition of schizophrenics. NIDA Res Monogr 67:321–327

Newell KA, Deng C, Huang XF (2006) Increased cannabinoid receptor density in the posterior cingulate cortex in schizophrenia. Exp Brain Res 172(4):556–560. doi:10.1007/s00221-006-0503-x

Ninan I (2011) Oxytocin suppresses basal glutamatergic transmission but facilitates activity-dependent synaptic potentiation in the medial prefrontal cortex. J Neurochem 119 (2):324–331. doi:10.1111/j.1471-4159.2011.07430.x

Nissen SE, Nicholls SJ, Wolski K, Rodés-Cabau J, Cannon CP, Deanfield JE, Deprés JP, Kastelein JJ, Steinhubl SR, Kapadia S, Yasin M, Ruzyllo W, Gaudin C, Job B, Hu B, Bhatt DL, Lincoff AM, Tuzcu EM, investigators S (2008) Effect of rimonabant on progression of atherosclerosis in patients with abdominal obesity and coronary artery disease: the STRADIVARIUS randomized controlled trial. JAMA 299 (13):1547-1560. doi:10.1001/jama.299.13.1547

Onaivi ES, Ishiguro H, Gong JP, Patel S, Meozzi PA, Myers L, Perchuk A, Mora Z, Tagliaferro PA, Gardner E, Brusco A, Akinshola BE, Hope B, Lujilde J, Inada T, Iwasaki S, Macharia D, Teasenfitz L, Arinami T, Uhl GR (2008) Brain neuronal CB2 cannabinoid receptors in drug abuse and depression: from mice to human subjects. PLoS One 3(2):e1640. doi:10.1371/journal.pone.0001640

Overton HA, Babbs AJ, Doel SM, Fyfe MC, Gardner LS, Griffin G, Jackson HC, Procter MJ, Rasamison CM, Tang-Christensen M, Widdowson PS, Williams GM, Reynet C (2006) Deorphanization of a G protein-coupled receptor for oleoylethanolamide and its use in the discovery of small-molecule hypophagic agents. Cell Metab 3(3):167–175. doi:10.1016/j.cmet.2006.02.004

Pardini M, Krueger F, Koenigs M, Raymont V, Hodgkinson C, Zoubak S, Goldman D, Grafman J (2012) Fatty-acid amide hydrolase polymorphisms and post-traumatic stress disorder after penetrating brain injury. Transl Psychiatry 2:e75. doi:10.1038/tp.2012.1

Passie T, Emrich HM, Karst M, Brandt SD, Halpern JH (2012) Mitigation of post-traumatic stress symptoms by Cannabis resin: a review of the clinical and neurobiological evidence. Drug Test Anal 4(7-8):649–659. doi:10.1002/dta.1377

Patel S, Roelke CT, Rademacher DJ, Hillard CJ (2005) Inhibition of restraint stress-induced neural and behavioural activation by endogenous cannabinoid signalling. Eur J Neurosci 21 (4):1057–1069. doi:10.1111/j.1460-9568.2005.03916.x

Pava MJ, Woodward JJ (2014) Chronic ethanol alters network activity and endocannabinoid signaling in the prefrontal cortex. Front Integr Neurosci 8(58):1–12. doi:10.3389/fnint.2014.00058

Penagarikano O, Mulle JG, Warren ST (2007) The pathophysiology of fragile x syndrome. Annu Rev Genomics Hum Genet 8:109–129. doi:10.1146/annurev.genom.8.080706.092249

Perry EK, Lee ML, Martin-Ruiz CM, Court JA, Voisen SG, Merrit J, Folly E, Iversen PE, Bauman ML, Perry RH, Wenk GL (2001) Cholinergic activity in autism: abnormalities in the cerebral cortex and basal forebrain. Am J Psychiatry 158(7):1058–1066

Rademacher DJ, Meier SE, Shi L, Ho WS, Jarrahian A, Hillard CJ (2008) Effects of acute and repeated restraint stress on endocannabinoid content in the amygdala, ventral striatum, and medial prefrontal cortex in mice. Neuropharmacology 54(1):108–116. doi:10.1016/j.neuropharm.2007.06.012

Raver SM, Haughwout SP, Keller A (2013) Adolescent cannabinoid exposure permanently suppresses cortical oscillations in adult mice. Neuropsychopharmacology 38(12):2338–2347. doi:10.1038/npp.2013.164

Realini N, Viganò D, Guidali C, Zamberletti E, Rubino T, Parolaro D (2011) Chronic URB597 treatment at adulthood reverted most depressive-like symptoms induced by adolescent exposure to THC in female rats. Neuropharmacology 60(2-3):235–243. doi:10.1016/j.neuropharm.2010.09.003

Rey AA, Purrio M, Viveros MP, Lutz B (2012) Biphasic effects of cannabinoids in anxiety responses: CB1 and GABA(B) receptors in the balance of GABAergic and glutamatergic neurotransmission. Neuropsychopharmacology 37(12):2624–2634. doi:10.1038/npp.2012.123

Richfield EK, Herkenham M (1994) Selective vulnerability in Huntington's disease: preferential loss of cannabinoid receptors in lateral globus pallidus. Ann Neurol 36(4):577–584

Rinaldi T, Perrodin C, Markram H (2008) Hyper-connectivity and hyper-plasticity in the medial prefrontal cortex in the valproic Acid animal model of autism. Front Neural Circuits 2:4. doi:10.3389/neuro.04.004.2008

Rios C, Gomes I, Devi LA (2006) mu opioid and CB1 cannabinoid receptor interactions: reciprocal inhibition of receptor signaling and neuritogenesis. Br J Pharmacol 148 (4):387–395. doi:10.1038/sj.bjp.0706757

Rive MM, van Rooijen G, Veltman DJ, Phillips ML, Schene AH, Ruhe HG (2013) Neural correlates of dysfunctional emotion regulation in major depressive disorder. A systematic review of neuroimaging studies. Neurosci Biobehav Rev 37(10 Pt 2):2529–2553. doi:10.1016/j.neubiorev.2013.07.018

Rodriguez JJ, Mackie K, Pickel VM (2001) Ultrastructural localization of the CB1 cannabinoid receptor in mu-opioid receptor patches of the rat Caudate putamen nucleus. J Neurosci 21 (3):823–833

Rodriguez-Gaztelumendi A, Rojo ML, Pazos A, Diaz A (2009) Altered CB receptor-signaling in prefrontal cortex from an animal model of depression is reversed by chronic fluoxetine. J Neurochem 108(6):1423–1433. doi:10.1111/j.1471-4159.2009.05898.x

Roitman P, Mechoulam R, Cooper-Kazaz R, Shalev A (2014) Preliminary, open-label, pilot study of add-on oral Delta9-tetrahydrocannabinol in chronic post-traumatic stress disorder. Clin Drug Investig 34(8):587–591. doi:10.1007/s40261-014-0212-3

Roullet FI, Lai JK, Foster JA (2013) In utero exposure to valproic acid and autism – a current review of clinical and animal studies. Neurotoxicol Teratol 36:47–56. doi:10.1016/j.ntt.2013.01.004

Rubino T, Parolaro D (2008) Long lasting consequences of cannabis exposure in adolescence. Mol Cell Endocrinol 286(1–2 Suppl 1):S108–S113. doi:10.1016/j.mce.2008.02.003

Rubino T, Guidali C, Vigano D, Realini N, Valenti M, Massi P, Parolaro D (2008a) CB1 receptor stimulation in specific brain areas differently modulate anxiety-related behaviour. Neuropharmacology 54(1):151–160. doi:10.1016/j.neuropharm.2007.06.024

Rubino T, Realini N, Castiglioni C, Guidali C, Vigano D, Marras E, Petrosino S, Perletti G, Maccarrone M, Di Marzo V, Parolaro D (2008b) Role in anxiety behavior of the endocannabinoid system in the prefrontal cortex. Cereb Cortex 18(6):1292–1301. doi:10.1093/cercor/bhm161

Rubino T, Viganò D, Realini N, Guidali C, Braida D, Capurro V, Castiglioni C, Cherubino F, Romualdi P, Candeletti S, Sala M, Parolaro D (2008c) Chronic delta 9-tetrahydrocannabinol during adolescence provokes sex-dependent changes in the emotional profile in adult rats: behavioral and biochemical correlates. Neuropsychopharmacology 33(11):2760–2771. doi:10.1038/sj.npp.1301664

Rubino T, Realini N, Braida D, Alberio T, Capurro V, Viganò D, Guidali C, Sala M, Fasano M, Parolaro D (2009) The depressive phenotype induced in adult female rats by adolescent exposure to THC is associated with cognitive impairment and altered neuroplasticity in the prefrontal cortex. Neurotox Res 15(4):291–302. doi:10.1007/s12640-009-9031-3

Rubino T, Zamberletti E, Parolaro D (2015) Endocannabinoids and Mental Disorders. Handb Exp Pharmacol 231:261–283. doi:10.1007/978-3-319-20825-1_9

Rutkowska M, Jachimczuk O (2004) Antidepressant--like properties of ACEA (arachidonyl-2-chloroethylamide), the selective agonist of CB1 receptors. Acta Pol Pharm 61(2):165–167

Saito A, Ballinger MD, Pletnikov MV, Wong DF, Kamiya A (2013) Endocannabinoid system: potential novel targets for treatment of schizophrenia. Neurobiol Dis 53:10–17. doi:10.1016/j.nbd.2012.11.020

Salio C, Fischer J, Franzoi MF, Mackie K, Kaneko T, Conrath M (2001) CB1-cannabinoid and mu-opioid receptor co-localization on postsynaptic target in the rat dorsal horn. Neuroreport 12 (17):3689–3692

Sanchis-Segura C, Cline BH, Marsicano G, Lutz B, Spanagel R (2004) Reduced sensitivity to reward in CB1 knockout mice. Psychopharmacology 176(2):223–232. doi:10.1007/s00213-004-1877-8

Schmöle AC, Lundt R, Ternes S, Albayram O, Ulas T, Schultze JL, Bano D, Nicotera P, Alferink J, Zimmer A (2015) Cannabinoid receptor 2 deficiency results in reduced neuroinflammation in an Alzheimer's disease mouse model. Neurobiol Aging 36(2):710–719. doi:10.1016/j.neurobiolaging.2014.09.019

Schneider M, Koch M (2003) Chronic pubertal, but not adult chronic cannabinoid treatment impairs sensorimotor gating, recognition memory, and the performance in a progressive ratio task in adult rats. Neuropsychopharmacology 28(10):1760–1769. doi:10.1038/sj.npp.1300225

Seamans JK, Floresco SB, Phillips AG (1995) Functional differences between the prelimbic and anterior cingulate regions of the rat prefrontal cortex. Behav Neurosci 109(6):1063–1073

Seillier A, Martinez AA, Giuffrida A (2013) Phencyclidine-induced social withdrawal results from deficient stimulation of cannabinoid CB(1) receptors: implications for schizophrenia. Neuropsychopharmacology 38(9):1816–1824. doi:10.1038/npp.2013.81

Serra G, Fratta W (2007) A possible for for the endocannabinoid system in the neurobiology of depression. Clin Pract Epidemiol Ment Health 3(25):11. doi:10.1186/1745-0179-3-

Shearman LP, Rosko KM, Fleischer R, Wang J, Xu S, Tong XS, Rocha BA (2003) Antidepressant-like and anorectic effects of the cannabinoid CB1 receptor inverse agonist AM251 in mice. Behav Pharmacol 14(8):573–582. doi:10.1097/01.fbp.0000104880.69384.38

Sheinin A, Talani G, Davis MI, Lovinger DM (2008) Endocannabinoid- and mGluR5-dependent short-term synaptic depression in an isolated neuron/bouton preparation from the hippocampal CA1 region. J Neurophysiol 100(2):1041–1052. doi:10.1152/jn.90226.2008

Sidorov MS, Krueger DD, Taylor M, Gisin E, Osterweil EK, Bear MF (2014) Extinction of an instrumental response: a cognitive behavioral assay in Fmr1 knockout mice. Genes Brain Behav 13(5):451–458. doi:10.1111/gbb.12137

Smaga I, Bystrowska B, Gawlinski D, Pomierny B, Stankowicz P, Filip M (2014) Antidepressants and changes in concentration of endocannabinoids and N-acylethanolamines in rat brain structures. Neurotox Res 26(2):190–206. doi:10.1007/s12640-014-9465-0

Solas M, Francis PT, Franco R, Ramirez MJ (2013) CB2 receptor and amyloid pathology in frontal cortex of Alzheimer's disease patients. Neurobiol Aging 34(3):805–808. doi:10.1016/j.neurobiolaging.2012.06.005

Song J, Singh M (2013) From hub proteins to hub modules: the relationship between essentiality and centrality in the yeast interactome at different scales of organization. PLoS Comput Biol 9 (2):e1002910. doi:10.1371/journal.pcbi.1002910

Speed HE, Masiulis I, Gibson JR, Powell CM (2015) Increased cortical inhibition in autism-linked neuroligin-3R451C mice is due in part to loss of endocannabinoid signaling. PLoS One 10 (10):1–16

Spencer KM, Nestor PG, Niznikiewicz MA, Salisbury DF, Shenton ME, McCarley RW (2003) Abnormal neural synchrony in schizophrenia. J Neurosci 23(19):7407–7411

Steiner MA, Wanisch K, Monory K, Marsicano G, Borroni E, Bachli H, Holsboer F, Lutz B, Wotjak CT (2008) Impaired cannabinoid receptor type 1 signaling interferes with stress-coping behavior in mice. Pharmacogenomics J 8(3):196–208. doi:10.1038/sj.tpj.6500466

Tan H, Ahmad T, Loureiro M, Zunder J, Laviolette SR (2014) The role of cannabinoid transmission in emotional memory formation: implications for addiction and schizophrenia. Front Psychiatry 5:73. doi:10.3389/fpsyt.2014.00073

Tang AH, Alger BE (2015) Homer protein-metabotropic glutamate receptor binding regulates endocannabinoid signaling and affects hyperexcitability in a mouse model of fragile X syndrome. J Neurosci 35(9):3938–3945. doi:10.1523/JNEUROSCI.4499-14.2015

Thomazeau A, Lassalle O, Iafrati J, Souchet B, Guedj F, Janel N, Chavis P, Delabar J, Manzoni OJ (2014) Prefrontal deficits in a murine model overexpressing the down syndrome candidate gene dyrk1a. J Neurosci 34(4):1138–1147. doi:10.1523/JNEUROSCI.2852-13.2014

Thorat SN, Bhargava HN (1994) Evidence for a bidirectional cross-tolerance between morphine and delta 9-tetrahydrocannabinol in mice. Eur J Pharmacol 260(1):5–13

Thornberg SA, Saklad SR (1996) A review of NMDA receptors and the phencyclidine model of schizophrenia. Pharmacotherapy 16(1):82–93

Treffert DA (1978) Marijuana use in schizophrenia: a clear hazard. Am J Psychiatry 135 (10):1213–1215

Tsai G, Coyle JT (2002) Glutamatergic mechanisms in schizophrenia. Annu Rev Pharmacol Toxicol 42:165–179

Turner WM, Tsuang MT (1990) Impact of substance abuse on the course and outcome of schizophrenia. Schizophr Bull 16(1):87–95

Tzavara ET, Davis RJ, Perry KW, Li X, Salhoff C, Bymaster FP, Witkin JM, Nomikos GG (2003) The CB1 receptor antagonist SR141716A selectively increases monoaminergic neurotransmission in the medial prefrontal cortex: implications for therapeutic actions. Br J Pharmacol 138(4):544–553. doi:10.1038/sj.bjp.0705100

Uhlhaas PJ, Singer W (2010) Abnormal neural oscillations and synchrony in schizophrenia. Nat Rev Neurosci 11(2):100–113. doi:10.1038/nrn2774

Ujike H, Morita Y (2004) New perspectives in the studies on endocannabinoid and cannabis: cannabinoid receptors and schizophrenia. J Pharmacol Sci 96(4):376–381

Ujike H, Takaki M, Nakata K, Tanaka Y, Takeda T, Kodama M, Fujiwara Y, Sakai A, Kuroda S (2002) CNR1, central cannabinoid receptor gene, associated with susceptibility to hebephrenic schizophrenia. Mol Psychiatry 7(5). doi:10.1038/sj.mp.4001029

Umathe SN, Manna SS, Jain NS (2011) Involvement of endocannabinoids in antidepressant and anti-compulsive effect of fluoxetine in mice. Behav Brain Res 223(1):125–134. doi:10.1016/j.bbr.2011.04.031

Urigüen L, Fernández B, Romero EM, De Pedro N, Delgado MJ, Guaza C, Schmidhammer H, Viveros MP (2002) Effects of 14-methoxymetopon, a potent opioid agonist, on the responses to the tail electric stimulation test and plus-maze activity in male rats: neuroendocrine correlates. Brain Res Bull 57(5):661–666

Uylings HBM, Groenewegen HJ, Kolb B (2003) Do rats have a prefrontal cortex? Behav Brain Res 146(1-2):3–17. doi:10.1016/j.bbr.2003.09.028

Valverde O, Torrens M (2012) CB1 receptor-deficient mice as a model for depression. Neuroscience 204:193–206. doi:10.1016/j.neuroscience.2011.09.031

Van Laere K, Casteels C, Dhollander I, Goffin K, Grachev I, Bormans G, Vandenberghe W (2010) Widespread decrease of type 1 cannabinoid receptor availability in Huntington disease in vivo. J Nucl Med 51(9):1413–1417. doi:10.2967/jnumed.110.077156

Vasiljevik T, Franks LN, Ford BM, Douglas JT, Prather PL, Fantegrossi WE, Prisinzano TE (2013) Design, synthesis, and biological evaluation of aminoalkylindole derivatives as cannabinoid receptor ligands with potential for treatment of alcohol abuse. J Med Chem 56 (11):4537–4550. doi:10.1021/jm400268b

Viganò D, Valenti M, Cascio MG, Di Marzo V, Parolaro D, Rubino T (2004) Changes in endocannabinoid levels in a rat model of behavioural sensitization to morphine. Eur J Neurosci 20(7):1849–1857. doi:10.1111/j.1460-9568.2004.3645.x

Viganò D, Rubino T, Vaccani A, Bianchessi S, Marmorato P, Castiglioni C, Parolaro D (2005) Molecular mechanisms involved in the asymmetric interaction between cannabinoid and opioid systems. Psychopharmacology 182(4):527–536. doi:10.1007/s00213-005-0114-4

Vigano D, Guidali C, Petrosino S, Realini N, Rubino T, Di Marzo V, Parolaro D (2009) Involvement of the endocannabinoid system in phencyclidine-induced cognitive deficits modelling schizophrenia. Int J Neuropsychopharmacol 12(5):599–614. doi:10.1017/S1461145708009371

Vinod KY, Hungund BL (2006) Role of the endocannabinoid system in depression and suicide. Trends Pharmacol Sci 27(10):539–545. doi:10.1016/j.tips.2006.08.006

Vinod KY, Arango V, Xie S, Kassir SA, Mann JJ, Cooper TB, Hungund BL (2005) Elevated levels of endocannabinoids and CB1 receptor-mediated G-protein signaling in the prefrontal cortex of alcoholic suicide victims. Biol Psychiatry 57(5):480–486. doi:10.1016/j.biopsych.2004.11.033

Vinod KY, Yalamanchili R, Xie S, Cooper TB, Hungund BL (2006) Effect of chronic ethanol exposure and its withdrawal on the endocannabinoid system. Neurochem Int 49(6):619–625. doi:10.1016/j.neuint.2006.05.002

Vinod KY, Sanguino E, Yalamanchili R, Manzanares J, Hungund BL (2008) Manipulation of fatty acid amide hydrolase functional activity alters sensitivity and dependence to ethanol. J Neurochem 104(1):233–243. doi:10.1111/j.1471-4159.2007.04956.x

Volk DW, Lewis DA (2015) The role of endocannabinoid signaling in cortical inhibitory neuron dysfunction in schizophrenia. Biol Psychiatry. doi:10.1016/j.biopsych.2015.06.015

Volk DW, Eggan SM, Lewis DA (2010) Alterations in metabotropic glutamate receptor 1alpha and regulator of G protein signaling 4 in the prefrontal cortex in schizophrenia. Am J Psychiatry 167(12):1489–1498. doi:10.1176/appi.ajp.2010.10030318

Volk DW, Siegel BI, Verrico CD, Lewis DA (2013) Endocannabinoid metabolism in the prefrontal cortex in schizophrenia. Schizophr Res 147(1):53–57. doi:10.1016/j.schres.2013.02.038

Watanabe S, Doshi M, Hamazaki T (2003) n-3 Polyunsaturated fatty acid (PUFA) deficiency elevates and n-3 PUFA enrichment reduces brain 2-arachidonoylglycerol level in mice. Prostaglandins Leukot Essent Fatty Acids 69(1):51–59

Wei D, Lee D, Cox CD, Karsten CA, Peñagarikano O, Geschwind DH, Gall CM, Piomelli D (2015) Endocannabinoid signaling mediates oxytocin-driven social reward. Proc Natl Acad Sci USA 112(45):14084–14089

Willner P (2005) Chronic mild stress (CMS) revisited: consistency and behavioural-neurobiological concordance in the effects of CMS. Neuropsychobiology 52(2):90–110. doi:10.1159/000087097

Willsey AJ, Sanders SJ, Li M, Dong S, Tebbenkamp AT, Muhle RA, Reilly SK, Lin L, Fertuzinhos S, Miller JA, Murtha MT, Bichsel C, Niu W, Cotney J, Ercan-Sencicek AG, Gockley J, Gupta AR, Han W, He X, Hoffman EJ, Klei L, Lei J, Liu W, Liu L, Lu C, Xu X, Zhu Y, Mane SM, Lein ES, Wei L, Noonan JP, Roeder K, Devlin B, Sestan N, State MW (2013) Coexpression networks implicate human midfetal deep cortical projection neurons in the pathogenesis of autism. Cell 155(5):997–1007. doi:10.1016/j.cell.2013.10.020

Witkin JM, Tzavara ET, Davis RJ, Li X, Nomikos GG (2005) A therapeutic role for cannabinoid CB1 receptor antagonists in major depressive disorders. Trends Pharmacol Sci 26 (12):609–617. doi:10.1016/j.tips.2005.10.006

Wong DF, Kuwabara H, Horti AG, Raymont V, Brasic J, Guevara M, Ye W, Dannals RF, Ravert HT, Nandi A, Rahmim A, Ming JE, Grachev I, Roy C, Cascella N (2010) Quantification of cerebral cannabinoid receptors subtype 1 (CB1) in healthy subjects and schizophrenia by the novel PET radioligand [11C]OMAR. NeuroImage 52(4):1505–1513. doi:10.1016/j.neuroimage.2010.04.034

Wyrofsky R, McGonigle P, Van Bockstaele EJ (2015) Drug discovery strategies that focus on the endocannabinoid signaling system in psychiatric disease. Expert Opin Drug Discovery 10 (1):17–36. doi:10.1517/17460441.2014.966680

Xu ZH, Yang Q, Feng B, Liu SB, Zhang N, Xing JH, Li XQ, Wu YM, Gao GD, Zhao MG (2012) Group I mGluR antagonist rescues the deficit of D1-induced LTP in a mouse model of fragile X syndrome. Mol Neurodegener 7(24):1–14

Yamaguchi K, Kandel DB (1984) Patterns of drug use from adolescence to young adulthood: III. Predictors of progression. Am J Public Health 74(7):673–681

Yoshino H, Miyamae T, Hansen G, Zambrowicz B, Flynn M, Pedicord D, Blat Y, Westphal RS, Zaczek R, Lewis DA, Gonzalez-Burgos G (2011) Postsynaptic diacylglycerol lipase mediates retrograde endocannabinoid suppression of inhibition in mouse prefrontal cortex. J Physiol 589 (Pt 20):4857–4884. doi:10.1113/jphysiol.2011.212225

You IJ, Jung YH, Kim MJ, Kwon SH, Hong SI, Lee SY, Jang CG (2012) Alterations in the emotional and memory behavioral phenotypes of transient receptor potential vanilloid type 1-deficient mice are mediated by changes in expression of 5-HT(1)A, GABA(A), and NMDA receptors. Neuropharmacology 62(2):1034–1043. doi:10.1016/j.neuropharm.2011.10.013

Zavitsanou K, Garrick T, Huang XF (2004) Selective antagonist [3H]SR141716A binding to cannabinoid CB1 receptors is increased in the anterior cingulate cortex in schizophrenia. Prog Neuro-Psychopharmacol Biol Psychiatry 28(2):355–360. doi:10.1016/j.pnpbp.2003.11.005

Zhang L, Alger BE (2010) Enhanced endocannabinoid signaling elevates neuronal excitability in fragile X syndrome. J Neurosci 30(16):5724–5729. doi:10.1523/JNEUROSCI.0795-10.2010

Zoppi S, Perez Nievas BG, Madrigal JL, Manzanares J, Leza JC, Garcia-Bueno B (2011) Regulatory role of cannabinoid receptor 1 in stress-induced excitotoxicity and neuroinflammation. Neuropsychopharmacology 36(4):805–818. doi:10.1038/npp.2010.214

Zuardi AW, Morais SL, Guimarães FS, Mechoulam R (1995) Antipsychotic effect of cannabidiol. J Clin Psychiatry 56(10):485–486

Zuardi AW, Hallak JE, Dursun SM, Morais SL, Sanches RF, Musty RE, Crippa JA (2006) Cannabidiol monotherapy for treatment-resistant schizophrenia. J Psychopharmacol 20 (5):683–686

Cannabinoids and Mitochondria

Etienne Hebert-Chatelain, Giovanni Marsicano, and Tifany Desprez

Abstract Mitochondria are key organelles providing energy supply and many other vital functions to cells. Shortly after the discovery of plant-derived cannabinoid compounds, some studies indicated their impact onto mitochondrial functions. The later identification of cannabinoid receptors as classical seven-transmembrane G protein-coupled receptors suggested that these mitochondrial effects might be due to unspecific membrane-altering properties of cannabinoids. However, the recent discovery that brain mitochondria contain significant amounts of functional type-1 cannabinoid receptors (CB_1) shed new light on cannabinoid physiology and pharmacology. In this chapter, we will summarize historical and recent evidence of the cannabinoid impact on mitochondrial functions in peripheral and central organs of the body.

1 Introduction

Type-1 cannabinoid receptor CB_1 is a G protein-coupled receptor (GPCR), widely expressed in the brain. In the central nervous system, CB_1 receptor activation at both presynaptic and postsynaptic sites results in the regulation of neuronal excitability, neurotransmitter release, and synaptic plasticity. In turn, CB_1 receptor regulates numerous physiological and behavioral processes, including memory, pain perception, food intake, motor control, and others. However, the cellular

E. Hebert-Chatelain (✉)
Département de Biologie, Université de Moncton, Moncton, NB, Canada

INSERM U1215, NeuroCentre Magendie, 146, rue Léo Saignat, Bordeaux, France

Université Bordeaux, Bordeaux, France

NeuroCentre Magendie, Bordeaux, France
e-mail: etienne.hebert.chatelain@umoncton.ca

G. Marsicano • T. Desprez
INSERM U1215, NeuroCentre Magendie, 146, rue Léo Saignat, Bordeaux, France

Université Bordeaux, Bordeaux, France

NeuroCentre Magendie, Bordeaux, France

M. Melis (ed.), *Endocannabinoids and Lipid Mediators in Brain Functions*,
DOI 10.1007/978-3-319-57371-7_8

mechanisms of CB_1-mediated control of these functions are only partially understood. Although CB_1 receptors primarily signal at the plasma membrane (pmCB_1), a portion of these receptors is also functionally present at mitochondrial membranes (mtCB_1). This chapter describes how cannabinoids and mtCB_1 receptor regulate brain mitochondrial functions and how these mechanisms might participate in the processing of neuronal information.

2 The Endocannabinoid System

The endocannabinoid system (ECS) has been identified as the target of the main psychoactive component of the plant *Cannabis sativa,* the Δ^9-tetrahydrocannabinol, abbreviated as THC. Its structure was partially identified by Roger Adams and colleagues in the 1940s (Adams 1942) and then fully elucidated by Raphael Mechoulam and colleagues in the 1960s (Gaoni and Mechoulam 1964). In 1990, the first cannabinoid receptor was cloned, named CB_1 receptor and found to be highly expressed in the brain (Matsuda et al. 1990). The identification and the cloning of a cannabinoid receptor that responds to THC (Matsuda et al. 1990; Devane et al. 1992) gave the birth to an explosion of interest in the cannabinoid research that continues nowadays. The ECS is schematically composed by at least two receptors, the cannabinoid receptors 1 and 2 (CB_1 and CB_2, respectively), their endogenous ligands called endocannabinoids (eCBs), such as anandamide (AEA) and 2-arachidonoylglycerol (2-AG), and the enzymes involved in their synthesis and inactivation (Piomelli 2003; Di Marzo and De Petrocellis 2012). CB_1 receptors were initially identified mainly in the central nervous system (CNS) (Herkenham et al. 1990), whereas it is now clear that also many peripheral organs contain the receptor (Pagotto et al. 2006). CB_2 receptors are expressed primarily by immune and hematopoietic cells (Munro et al. 1993) and in microglial brain cells (Nunez et al. 2004), whereas their presence in neurons is still under scrutiny, due to the lack of suitable tools (Onaivi et al. 2006; Onaivi et al. 2012). However, it is now well established that most of central effects of cannabinoids in the brain are due to CB_1 receptor activation (Ledent et al. 1999; Zimmer et al. 1999).

CB_1 is a seven-transmembrane G protein-coupled receptor, and it is presently considered as the most abundant metabotropic receptor in the brain (Herkenham et al. 1991). The comprehension of the complex roles of the CB_1 protein in the physiological-behavioral effects of cannabinoids requires a precise mapping of their sites of action at the regional, cellular, and subcellular level. These receptors are particularly rich in certain areas of the CNS such as cerebellum and basal ganglia (Katona et al. 1999; Pettit et al. 1998; Tsou et al. 1998). CB_1 receptors are also dense in the hippocampus, cerebral cortex, hypothalamus, olfactory system, spinal cord, and other regions (Moldrich and Wenger 2000).

Despite the wide expression of CB_1 receptors in the CNS, their cellular distribution appears to be restricted to certain cell types. Within cortical regions, CB_1 is mainly found in GABAergic interneurons co-expressing glutamic acid decarboxylase and cholecystokinin (Marsicano and Lutz 1999). Cortical glutamatergic neurons also contain CB_1 receptors but in lower amount as compared to GABAergic interneurons (Marsicano and Lutz 1999; Mailleux et al. 1992; Marsicano et al. 2003; Marsicano and Kuner 2008). Despite the minority of CB_1 receptors on glutamatergic cells, these receptors play significant roles in functions modulated by (endo)cannabinoids such as resistance to excitotoxic insults in the hippocampus (Monory et al. 2006), fear coping, stress and anxiety (Dubreucq et al. 2012; Metna-Laurent et al. 2012), and food intake (Bellocchio et al. 2010; Soria-Gomez et al. 2014). In addition, serotonin-releasing neurons projecting from the raphe nuclei to the basolateral amygdala and the hippocampal CA3 region also express CB_1 receptors (Haring et al. 2007). Despite its well-known neuronal localization, the low but functional presence of CB_1 receptors on astrocytes was recently revealed using pharmacological tools and conditional mutants (Bosier et al. 2013; Han et al. 2012; Navarrete and Araque 2008). Indeed, astroglial CB_1 receptors were shown to regulate astrocyte leptin signaling (Bosier et al. 2013) and to be responsible for the impairment in working memory performance induced by cannabinoids (Han et al. 2012). Thus, the anatomical and functional data shortly summarized above show that CB_1 receptors expressed in specific cell types play important physiological and pharmacological roles independently of the relative levels of expression. In this way, cellular localization is likely a better predictor of the functions of CB_1 receptors than their levels of expression (Marsicano and Kuner 2008).

CB_1 receptors are preferentially targeted to plasma membranes of presynaptic terminals and axons, where they likely mediate the well-known inhibitory actions on neurotransmitter release (Katona et al. 1999; Chevaleyre and Castillo 2003; Wilson and Nicoll 2001). Because of their presence within different neuronal types, CB_1 receptors modulate several major neurotransmitters such as glutamate, γ-aminobutyric acid (GABA), acetylcholine, noradrenaline, serotonin, and cholecystokinin, among others (Howlett 2002; Pertwee et al. 2002; Szabo and Schlicker 2005). In astroglial cells, CB_1 receptors were also shown to contribute to the regulation of synaptic transmission, possibly through the regulation of gliotransmitters (Navarrete and Araque 2008, 2010).

Besides their presynaptic distribution, CB_1 receptors were also observed in somatodendritic neuronal compartments in several brain regions, where they appear to be rather intracellular than located at plasma membranes (Ong and Mackie 1999; Pickel et al. 2004; Sierra et al. 2014; Wilson-Poe et al. 2012). The classical paradigm of GPCR functioning postulates that GPCRs are confined at the cell surface where, once activated by their agonists, they activate G proteins signaling and initiate various intracellular pathways. Their presence in intracellular compartments, such as the endoplasmic reticulum (ER), endosomes, or lysosomes, is therefore assimilated to a nonfunctional state corresponding either to their synthesis and trafficking toward cell surface or to their internalization (Ferguson et al. 2001). However, this idea was recently challenged by emerging evidence showing the functionality of different GPCRs at intracellular membranes, including ER, endosomes, nuclei, and mitochondria (Belous et al. 2004; Irannejad and von Zastrow 2014). CB_1 receptors were recently shown to share such unconventional intracellular localization. Rozenfeld and colleagues showed that CB_1 receptors present in endosomal/lysosomal compartments are able to activate signaling (Rozenfeld and Devi 2008). Early studies in vitro showed the accumulation of cannabinoid-binding probes in subcellular fractions, including mitochondria (Colburn et al. 1974). More recent studies reported also anatomical evidence pointing to the presence of CB_1 receptor immunolabeling in intracellular compartments and in particular on mitochondria (Ong and Mackie 1999; Rodriguez et al. 2001). The mitochondrion was indeed described as one of the most commonly labeled organelles by CB_1-positive immunoparticles (Rodriguez et al. 2001).

3 Mitochondria

Although the brain represents only 2% of the body mass, it consumes 20% of the total body energy at rest (Erecinska and Silver 2001; Kety 1957; Rolfe and Brown 1997; Sokoloff 1960). Given that mitochondria are the main cellular converter of the energy contained in nutrients into a useable form, the adenosine triphosphate (ATP), mitochondria are essential for brain functions. Moreover, these organelles play several other crucial physiological processes important for brain physiology, including buffering of intracellular calcium (Ca^{2+}), modulation of reactive oxygen species (ROS), production of several neurotransmitters, and metabolism of steroid. Accordingly, mitochondrial dysfunctions can contribute to the development of neurodegenerative diseases (Mattson et al. 2008).

3.1 Mitochondrial Structure

Mitochondria are double-membrane enclosed organelles consisting in an outer (OMM) and an inner mitochondrial membranes (IMM) (Frey and Mannella 2000). These membranes create two different mitochondrial compartments: the internal lumen called the matrix and a much narrower intermembrane space (IMS). The OMM appears similar to other cellular membranes, whereas the IMM is enriched in cardiolipin and is subdivided into two compartments: the inner-boundary membrane close to the OMM and the invaginations, called *cristae* (Herrmann and Riemer 2010). The inner-boundary membrane can be physically linked to the OMM. Several proteins, including the outer membrane voltage-dependent anion channel (VDAC), the inner membrane adenine nucleotide translocator (ANT), and the mitochondrial benzodiazepine receptor, form contact sites between the outer and the inner membranes, which are involved in mitochondria permeability transition and apoptosis (Brdiczka et al. 2006). Cristae are heavily enriched in complexes of the electron transport system (ETS; see below). Their structure, which can differ greatly according to cell types and physiological state, has important impact on mitochondrial functions such as apoptosis and ATP production (Hackenbrock 1966; Scorrano et al. 2002). For instance, alterations of cristae structure is impaired in several human diseases including several neurodegenerative diseases (Trimmer et al. 2000; Moreira et al. 2001). Mitochondria are the only organelles to possess their own DNA (the mitochondrial DNA, or mtDNA), which is located in the matrix. mtDNA is the basis of the endosymbiotic hypothesis of mitochondria origin (Sagan 1967; Behnke 1977), and it encodes for 13 proteins that are subunit of the ETS (Tuppen et al. 2010; Taylor and Turnbull 2005). Not surprisingly, deleterious mtDNA mutations lead to several neuropathies and myopathies (Tuppen et al. 2010; Taylor and Turnbull 2005).

3.2 ATP Production

Mitochondria are considered as the bioenergetic powerhouses of cells because they are producing most of the ATP consumed by cells via the oxidative phosphorylation processes (OXPHOS). Briefly, proteins and carbohydrates are first oxidized in the cytosol, forming metabolites that translocate into mitochondria, whereas fatty acids are metabolized through β-oxidation directly in mitochondria. Here, these metabolites undergo sequential enzymatic reactions trough the tricarboxylic acid (TCA) cycle allowing the production of electron donors. These reducing equivalents (NADH and $FADH_2$) feed in turn the ETS. This process consists of a series of four enzymatic complexes acting as electron carriers. Electrons enter in this respiratory chain either via the complex I (NADH-ubiquinone oxidoreductase) or complex II (succinate-ubiquinone oxidoreductase) and are subsequently transferred to complex III (ubiquinol-cytochrome *c* reductase) and finally to complex IV (cytochrome *c* oxidase) where they permit the reduction of oxygen (O_2) to form water (H_2O). The course of electrons through the complex I, III, and IV is coupled to the translocation of protons from the mitochondrial matrix to the IMS, inducing an electrochemical gradient ($\Delta\Psi$) across the mitochondrial IM. The complex V (F_0F_1 ATP synthase) exploits this mitochondrial membrane potential to produce ATP (Fig. 1; for a detailed description of OXPHOS process, please see (Navarro and Boveris 2007; Chance and Williams 1955; Mitchell 1961)). The ATP is then translocated out of mitochondria by the adenine nucleotide translocator (ANT, or ADP/ATP translocator) to ensure cellular processes. Given that the brain is a highly oxidative organ, most of the energy is derived from oxidative reactions (Cai et al. 2011). OXPHOS is therefore crucial for maintaining energy-dependent physiological processes such as axonal growth and branching, cytoskeletal remodeling, reversal of the ion influxes for the generation of action potentials and synaptic transmission (Harris et al. 2012), as well as other mitochondrial functions including ROS production and Ca^{2+} buffering.

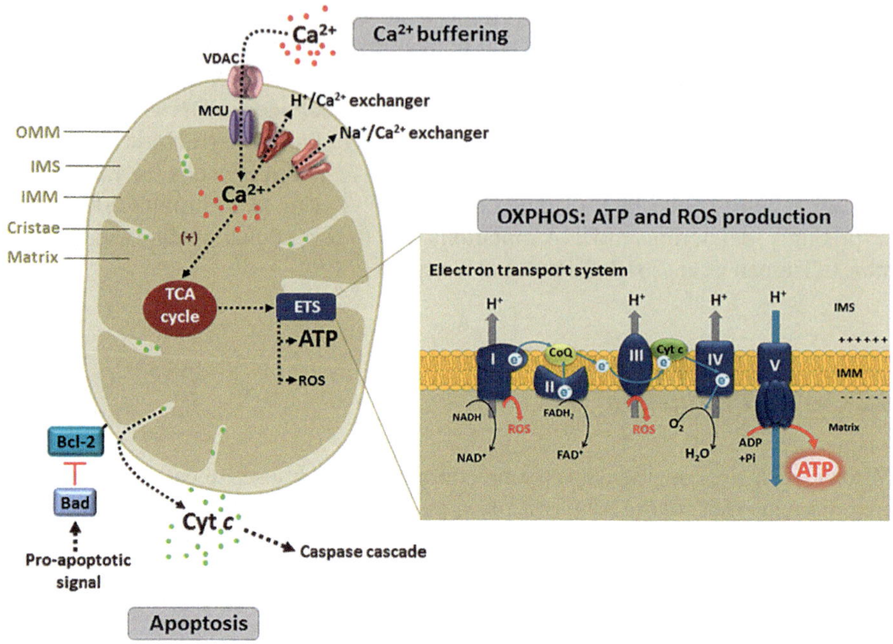

Fig. 1 Mitochondrial functions. Nutrients (carbohydrates, fatty acids, and amino acids) are oxidized in the cytosol and the TCA cycle which generates electron donors (NADH and $FADH_2$). These compounds then feed the electron transport system (ETS) which consists of a series of four enzymatic complexes which are enriched in mitochondria cristae. During the course of electrons across the complexes I, III, and IV, protons (H^+) are translocated from the matrix to the intermembrane space (IMS). This electrochemical gradient across the IMM allows the ATP synthase (complex V) producing ATP and represents a strong driving force for the import of Ca^{2+} into mitochondria. This process named oxidative phosphorylation (OXPHOS) generates most of the ATP consumed by cells and generates reactive oxygen species (ROS). Mitochondria regulate Ca^{2+} homeostasis. Ca^{2+} is imported across the outer mitochondrial membrane (OMM) through the large voltage-dependent anion channel (VDAC). Then, the mitochondrial Ca^{2+} uniporter (MCU) allows the import of Ca^{2+} across the inner mitochondrial membrane (IMM), whereas H^+/Ca^{2+} and Na^{2+}/Ca^{2+} exchangers allow the extrusion of Ca^{2+} from mitochondrial matrix. Ca^{2+} increases the activity of three different dehydrogenases of the tricarboxylic acid (TCA) cycle, thereby regulating mitochondrial metabolism. Mitochondria are also key players in programmed cell death or apoptosis. Several pro-apoptotic proteins, including cytochrome *c* (cyt *c*), are located inside mitochondria. Upon apoptotic stimuli, the pro-apoptotic protein Bad translocates to mitochondria and inhibits the anti-apoptotic protein Bcl-2. This process leads to the release of cyt *c* into the cytosol, the activation of the nonspecific caspases and ultimately cell death

3.3 ROS Production

During mitochondrial respiration, significant amounts of ROS are formed, resulting from the incomplete reduction of the O_2 molecule into water (Fig. 1). Between 0.1 and 4% of O_2 is partly reduced by mitochondria to produce the anion superoxide (O_2^-). Mitochondria are considered as the major sites of ROS production in the cell

(Cadenas and Davies 2000). Much of the superoxide seems produced by complexes I and III (Fig. 1) and is then converted to hydrogen peroxide (H_2O_2) by the mitochondrial antioxidant manganese superoxide dismutase (MnSOD). H_2O_2 can also generate the highly reactive hydroxyl radical OH˙. Excessive levels of ROS can damage all cellular components, including DNA, lipids, and proteins, leading to DNA mutagenesis, apoptosis, alteration of redox signaling, and accelerated aging (Marchi et al. 2012). Notably, high ROS production is linked to several neurodegenerative diseases (Hroudova and Fisar 2013). Besides their damaging impact, ROS produced by mitochondria are now considered as important signal transducers in physiological processes including synaptic plasticity (Lee et al. 2010; Knapp and Klann 2002; Ma et al. 2011), learning and memory (Olsen et al. 2013; Kishida and Klann 2007), and regulation of hypothalamic functions (Horvath et al. 2009).

3.4 Ca^{2+} Homeostasis

Beyond their role in generating ATP and ROS, mitochondria contribute to Ca^{2+} homeostasis (Fig. 1). Considering their capacity to remove Ca^{2+} from the cytoplasm and store it in their matrix, mitochondria are crucial regulators of intracellular Ca^{2+} dynamics and of several Ca^{2+}-dependent signaling processes such as synaptic transmission (Giorgi et al. 2012; Billups and Forsythe 2002). Mitochondrial Ca^{2+} uptake depends on the strong driving force created by the mitochondrial membrane potential across the IM that is built by the ETS (Bianchi et al. 2004). Ca^{2+} transport across the OMM occurs through VDAC. In contrast, the IMM contains different proteins able to modulate mitochondrial Ca^{2+} uptake and/or efflux, including the mitochondrial Ca^{2+} uniporter (MCU) which transfer Ca^{2+} into the mitochondrial matrix and Na^+/Ca^{2+} and H^+/Ca^{2+} antiporters that move Ca^{2+} out of the mitochondria (Giorgi et al. 2012) (Fig. 1). While Ca^{2+} uptake can stimulate the mitochondrial metabolism, excessive amounts of mitochondrial Ca^{2+} can be detrimental for the cell and lead to excitotoxicity and apoptosis (Benedict et al. 2012; Hajnoczky et al. 2006). A "normal" increase of Ca^{2+} in the mitochondrial matrix increases the production of ATP due to direct activation of three mitochondrial dehydrogenases (pyruvate dehydrogenase, isocitrate dehydrogenase, and α-ketoglutarate dehydrogenase) thereby linking increased activity of neurons to higher mitochondrial ATP production (Denton 2009; Gunter et al. 2004) (Fig. 1). More recently, it was shown that Ca^{2+} entry into the mitochondrion induces the upregulation of mitochondrial cAMP/PKA signaling through sAC, which results in higher ATP synthesis (Di Benedetto et al. 2013). It was suggested that Ca^{2+} could be one of the positive regulators of mitochondrial energy metabolism at nerve terminals by potentially acting as a feed-forward mechanism to boost ATP synthesis in order to prevent energy drop during synaptic activity (Rangaraju et al. 2014). However, mitochondrial Ca^{2+} overload lead to the release of Ca^{2+} out of mitochondria, swelling of the matrix, and rupture of the outer membrane (Nicholls 2005). The process, named permeability transition, was proposed as one mechanism to initiate the release of cytochrome *c* and other pro-apoptotic protein and seems responsible for the neuronal death following pathological activation of NMDA-selective glutamate receptors (Nicholls 2005).

3.5 Apoptosis

Mitochondria are the main actors of programmed cell death or apoptosis (Fig. 1). Different pro-apoptotic proteins are indeed contained within mitochondria, including cytochrome *c*, the serine protease OMI/HtrA2, apoptosis-inducing factor, endonuclease G, and Smac/Diablo (Suen et al. 2008; Gorman et al. 2000). Upon apoptotic stimuli, pro-apoptotic proteins Bad, Bax, or Bid translocate to mitochondria and inhibit the action of anti-apoptotic Bcl-2 proteins, inducing the release of pro-apoptotic proteins from mitochondria, probably through the permeability transition pore (Czabotar et al. 2014) (Fig. 1). Once in the cytosol, cytochrome *c* interact with Apaf-1 to form the apoptosome and induce the activation of the nonspecific caspases and ultimately cell death (Fig. 1). In the nervous system, apoptotic mechanisms have been associated with various cellular processes, such as synaptic plasticity and neurodegenerative disorders (Mattson et al. 2008; Li and Sheng 2012). For instance, the mitochondrial apoptotic pathway is involved in long-term depression of synaptic transmission (LTD) (Li et al. 2010). The impaired synaptic plasticity observed in Alzheimer's disease was shown to involve the mitochondrial pathway of apoptosis (Olsen and Sheng 2012). Apoptosis deficiency was also proposed to cause brain overgrowth by controlling neural cell numbers in the developing brain (Kuan et al. 2000; Nonomura et al. 2013).

4 The Interplay Between Cannabinoids and Mitochondrial Functions

The identification of the chemical structures of cannabinoids (Adams 1942; Gaoni and Mechoulam 1964) rose a great interest to characterize the biological mechanisms underlying the effects of this class of plant-derived molecules. For instance, it was early proposed that THC could potentially alter the metabolism of monoamines (e.g., dopamine) in the brain by acting on the enzyme MAO, an oxidoreductase located in the mitochondrial OM and responsible for degradation of monoamine neurotransmitters (Shih et al. 1999). Once released into the synaptic cleft, monoamines' action is terminated by their reuptake into the presynaptic terminal, where they can be recycled into synaptic vesicles or degraded by MAO (Mukherjee and Yang 1999). Brain monoamines such as dopamine, serotonin, and norepinephrine play a key role in the regulation of brain functions, including motor control, mood, or cognitive functions (Koob and Le Moal 2001). As both MAO inhibitors and CB_1

receptor agonists exert antidepressant-like effect (Fiedorowicz and Swartz 2004; Hill and Gorzalka 2005), the interplay between MAO and cannabinoid system was tested. Early studies found that cannabinoid compounds inhibit the activity of MAO in isolated brain mitochondria (Schurr and Livne 1975, 1976; Schurr et al. 1978). The inhibitory effect of CB_1 receptor agonists (THC, WIN55,212-2 [WIN], and AEA) on MAO activity with serotonin as a substrate was confirmed in a crude mitochondrial fraction isolated from mammalian brain cortex (Fisar 2010).

Other cannabinoid-induced mitochondrial effects were early described. Already in the 1970s of the last century, different studies reported effects of cannabinoids on mitochondria, including decrease of complex I or V activities and changes of mitochondrial ultrastructure (Schurr and Livne 1975; Bartova and Birmingham 1976; Mahoney and Harris 1972; Chari-Bitron and Bino 1971; Bino et al. 1972). The impact of cannabinoids on OXPHOS was further described in the last 15 years, and nowadays it seems well accepted that most cannabinoids inhibit mitochondrial O_2 consumption, decrease ATP production, and disrupt mitochondrial membrane potential in different cell types and tissues (Badawy et al. 2009; Whyte et al. 2010; Zaccagnino et al. 2011; Athanasiou et al. 2007; Rossato et al. 2005; Sarafian et al. 2003). Moreover, blockade or activation of CB_1 receptors alters mitochondrial functions in peripheral adipocytes (Tedesco et al. 2008, 2010).

Different cannabinoids, including THC, cannabidiol, 2-AG, and AEA, are also known as apoptotic inducers in different types of cells, with mechanisms often involving ROS production, release of cytochrome *c*, and caspase activation (Shrivastava et al. 2011; Gallily et al. 2003; Pellerito et al. 2014; Massi et al. 2006; Lee et al. 2008; Szoke et al. 2002; Ligresti et al. 2006; Campbell 2001; Mato et al. 2010; Rimmerman et al. 2013; Ryan et al. 2009; Catanzaro et al. 2009; Siegmund et al. 2007).

Cannabinoids also participate in the regulation of mitochondrial transport along neuronal processes (Boesmans et al. 2009). The number of mitochondria transported in enteric neuronal fibers is decreased by CB_1 receptor agonists and conversely enhanced by CB_1 receptor inhibition, indicating a role of CB_1 receptors in slowing down mitochondrial trafficking (Boesmans et al. 2009) (Fig. 2).

CB_1 receptors activation can lead to robust feeding despite of the animal being sated (Bermudez-Silva et al. 2012). Interestingly, recent studies revealed that such cannabinoid-mediated feeding behavior involves changes in mitochondrial activity. Activation of CB_1 receptors by arachidonyl-2'-chloroethylamide (ACEA) alters hypothalamic mitochondrial respiration and increases ROS levels in hypothalamic pro-opiomelanocortin (POMC) neurons (Koch et al. 2015). The feeding effects induced by activation of CB_1 receptors were shown to depend on the mitochondrial uncoupling protein 2 (UCP2) (Koch et al. 2015).

Overall, these results emphasized the impact of cannabinoids on mitochondrial functions and regulation. But how do cannabinoids interact with mitochondrial functions? Several potential molecular mechanisms underlying this interaction were proposed during the last decades, all maintaining a certain degree of likelihood nowadays, depending on the experimental conditions (doses, modes of treatment, etc.), the types of cells, and/or the mitochondrial functions under scrutiny.

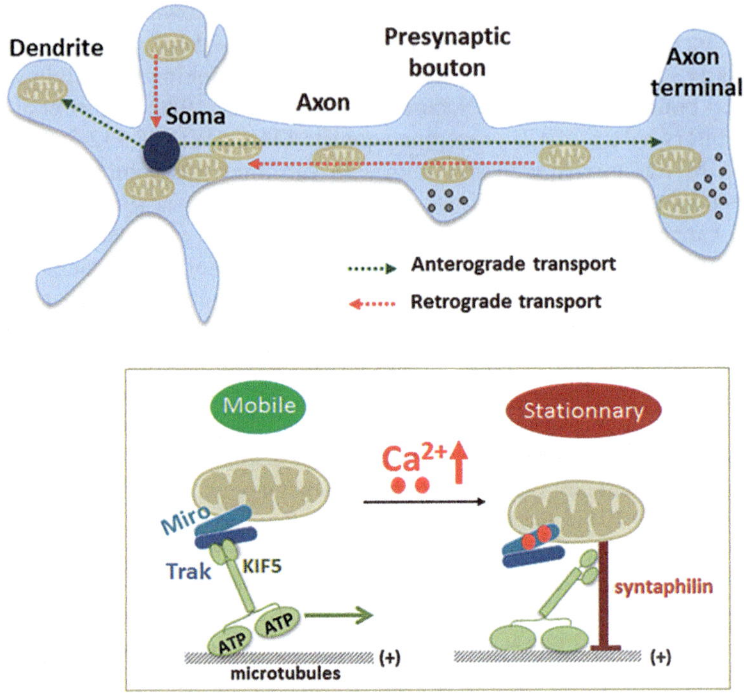

Fig. 2 Activity-dependent regulation of mitochondrial transport. The Miro–Trak adaptor complex mediates KIF5-driven mitochondrial transport. Ca^{2+} binding to Miro causes the release of KIF5 motors from mitochondria. Thus, Ca^{2+} influx after synaptic activity arrests mobile mitochondria at activated synapses. Additionally, a Miro-Ca^{2+}-sensing pathway triggers the binding switch of KIF5 motors from the Miro–Trak adaptor complex to anchoring protein syntaphilin, which immobilizes axonal mitochondria via inhibiting motor ATPase activity

Cannabinoids were initially proposed to act on mitochondria because of their lipophilic structure, which could lead to alteration of mitochondrial membrane properties (Bartova and Birmingham 1976). To illustrate a receptor-independent mechanism, it was found that high doses of 2-AG induces cell death (Siegmund et al. 2007). Another cannabinoid, cannabidiol (CBD), which has low agonistic activity for CB_1 and CB_2 receptors also modulate mitochondrial functions (Shrivastava et al. 2011; Gallily et al. 2003; Massi et al. 2006; Lee et al. 2008; Mato et al. 2010; Rimmerman et al. 2013). The mechanisms underlying CBD interaction with mitochondria are not clarified. CBD impacts on mitochondrial Ca^{2+}homeostasis through its physical interaction with VDAC, decreasing its conductance and regulating intracellular Ca^{2+} levels (Shoshan-Barmatz et al. 2010). Therefore, it is possible that the antiepileptic and neuroprotective properties of CBD (Rimmerman et al. 2013; Cunha et al. 1980) involve this specific interaction of CBD and VDAC to limit increase of cytoplasmic Ca^{2+} levels.

After the identification of CB_1 receptors as typical plasma membrane GPCRs, it was proposed that cannabinoids might alter mitochondrial functions via indirect

CB_1 receptor-dependent signaling (Campbell 2001). For instance, the THC-induced apoptosis observed in cultured cortical neurons is blocked by the CB_1 receptor antagonist AM251 and pertussis toxin (PTX), suggesting that CB_1-dependent apoptosis via mitochondria involves receptor-mediated activation of the G-protein subtypes Gi/o (Campbell 2001). The THC-induced apoptosis in Jurkat cells seems to be mediated by downregulation of the Raf-1/mitogen-activated protein kinase/ERK kinase pathway and translocation of the pro-apoptotic Bad protein to mitochondria, and it is blocked by CB_1 and CB_2 antagonists (Jia et al. 2006). Moreover, anandamide induces receptor-dependent apoptosis involving ceramide synthesis and p38 mitogen-activated protein kinase activity (Fonseca et al. 2013). Activation of CB_1 receptors by ACEA activates p38 MAPK and reduces AMPK phosphorylation, which in turn decreases mitochondrial biogenesis in mouse white adipose tissue, muscle, and liver (Tedesco et al. 2010). These findings support the idea that cannabinoids might affect mitochondrial functions through pmCB_1 receptor-dependent modulation of cytoplasmic signaling pathways.

Therefore, until few years ago, cannabinoid effects on mitochondrial functions were fully ascribed to direct unspecific membrane disturbance induced by these lipid molecules or to indirect signal transduction originating from cannabinoid receptors located on plasma membranes, or both. Moreover, although mitochondrial staining with CB_1 receptor antisera was a common observation, it was considered as unspecific background for many years. However, recent results challenged this idea, indicating that CB_1 receptors are present at mitochondrial membranes in the periphery, such as in spermatozoa (Aquila et al. 2010), in skeletal muscles (Mendizabal-Zubiaga et al. 2016), and in the brain, where they directly regulate mitochondrial OXPHOS activity (Benard et al. 2012; Hebert-Chatelain et al. 2014a, 2016; Vallee et al. 2014).

5 The Functional Presence of Mitochondrial CB_1 Receptors in the Brain

In 2012, a study from our laboratory demonstrated the functional presence of CB_1 receptors on brain mitochondrial membranes, identifying mitochondrial CB_1 receptors (mtCB_1) as a direct modulator of bioenergetics cellular processes in the brain (Benard et al. 2012) (Fig. 3). Pharmacological activation of CB_1 receptors accompanied by rigorous controls using mutant mice fully lacking CB_1 expression ($CB_1^{-/-}$ mice) showed that mtCB_1 receptors directly impact on endogenous respiration of brain mitochondria (Benard et al. 2012).

In more details, electron immunohistochemistry showed that approximately 10 to 15% of total CB_1 receptors in the CA1 hippocampal region are located in mitochondria of wild-type mice, largely above background levels quantified in tissues from $CB_1^{-/-}$ mutants (Benard et al. 2012). MtCB_1 receptors are more densely present in GABAergic interneurons than on glutamatergic neurons in the

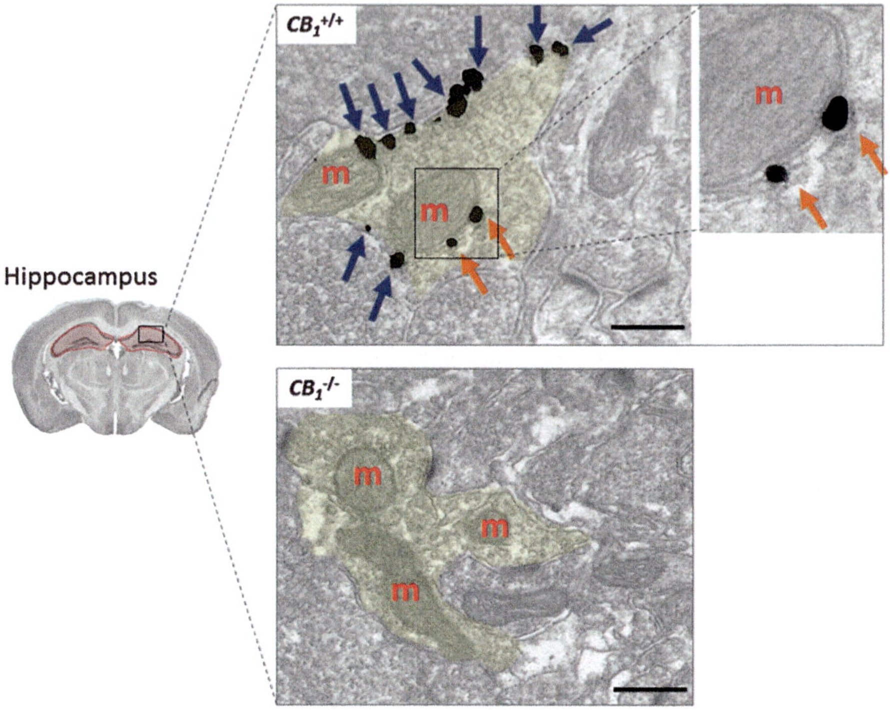

Fig. 3 CB_1 receptors in hippocampal mitochondria. Electron immunogold detection of CB_1 receptors within mitochondrial membranes of hippocampal cells in the CA1 region of wild-type $CB_1^{+/+}$ and knockout $CB_1^{-/-}$ mice. Inset, detail of single CB_1-positive mitochondria. *Blue arrows*, plasma membrane CB_1 receptors; *orange arrows*, mtCB_1 receptors. Scale bar, 0.25 μm. Adapted from Benard et al. (2012)

hippocampus, which goes in parallel with the well-known higher densities of "GABAergic" CB_1 receptors as compared to "glutamatergic" ones. At the subcellular level, mtCB_1 receptors are equally distributed to both dendrites and axon terminals, making their localization different from the pool localized at plasma membranes (pmCB_1), almost exclusively found at presynaptic terminals (Benard et al. 2012). Further experiments suggested that mtCB_1 receptors are present at the level of OMMs and they have a topological orientation with a cytoplasmic protein N-terminus and the C-terminus facing the interior of the organelle (Benard et al. 2012), enabling the mtCB_1 receptor-dependent signaling to reach proteins within the mitochondrial matrix, including proteins regulated by the intramitochondrial cAMP-PKA pathway (see below). However, the precise location of mtCB_1 receptors within mitochondrial membranes (i.e., IMM, OMM, and/or IMM-OMM contact sites (Frey and Mannella 2000)) is being investigated but remains unknown. Interestingly, several Gα_i proteins were observed in IMM-OMM contact sites in mitochondria of human placenta (Kuyznierewicz and Thomson 2002). Considering that mtCB_1 affects mitochondrial functions through these proteins

(Hebert-Chatelain et al. 2016), we can hypothesize that mtCB$_1$ receptors might be partially inserted within these contact sites. Since this first study, mtCB$_1$ receptors were also observed in different neuronal populations by other laboratories (Koch et al. 2015; Morozov et al. 2013; Ma et al. 2015). Nevertheless, it was shown that antibodies recognizing the CB$_1$ receptor could bind to nonspecific target such as the mitochondrial protein stomatin-like protein 2 (Morozov et al. 2013), indicating that identification of mtCB$_1$ receptors must be performed using $CB_1^{-/-}$ tissues as negative controls and strict quantification procedures (see references, Hebert-Chatelain et al. (2014a, b), Morozov et al. (2014), for methodological discussions). Indeed, despite the general problems arisen from the use of antibodies, the adoption of appropriate methodological approaches has recently led to the agreement that mitochondrial localization of CB$_1$ receptors is present in the brain, including also hypothalamic cells (Koch et al. 2015).

The direct impact of mtCB$_1$ receptors on respiratory functions of mitochondria was tested in purified brain mitochondria from wild-type and $CB_1^{-/-}$ mice, in CB$_1$-transfected primary mouse fibroblast (MFs) from $CB_1^{-/-}$ mice, and in transfected HEK293 cells (Benard et al. 2012; Hebert-Chatelain et al. 2014a; Vallee et al. 2014). Exogenous application of the CB$_1$ receptor agonists THC, WIN, or HU210 decreases endogenous mitochondrial respiration, whereas no changes were observed in control-transfected cells or brain mitochondria from $CB_1^{-/-}$ mice, indicating a direct regulation of respiration by cannabinoids through CB$_1$ receptors (Benard et al. 2012; Hebert-Chatelain et al. 2014a; Vallee et al. 2014). Similarly, a different laboratory observed that the CB$_1$ receptor antagonist AM251 is able to block the decrease of mitochondrial respiration induced by low doses of cannabinoids in brain mitochondria (Fisar et al. 2014). Our team also observed that treatment with synthetic cannabinoids HU210 and WIN decreases mobility of neuronal mitochondria (Hebert-Chatelain et al. 2016).

CB$_1$ receptor signaling is at the crossroad of the pharmacological effects of cannabinoids and the physiological roles of the ECS. Thus, an important question to address was whether the ECS physiologically modulates mitochondrial activity via mtCB$_1$ receptors. FAAH, the primary degradative enzyme for AEA, is densely present in the mitochondria (Benard et al. 2012), and we observed that purified mitochondria contain AEA- and 2-AG-degrading activity that is ascribed to FAAH and MAGL, representing approximately 18% and 12% from the total cellular enzyme activity, respectively (Benard et al. 2012). Pharmacological inhibition of MAGL in purified mitochondria increases 2-AG levels and decreases respiration (Benard et al. 2012). Interestingly, a strong inverse correlation was found between the levels of endogenous 2-AG in mitochondria and O_2 consumption (Benard et al. 2012), suggesting that endogenous mtCB$_1$ receptor signaling within mitochondria control the respiratory activity of the organelles.

To discriminate the respective influence of pmCB$_1$ versus intracellular CB$_1$ receptors on these effects, indirect pharmacological tools were developed. A biotinylated version of the lipophilic CB$_1$ receptor agonist HU210 (HU210-biotin, hereafter HU-biot), in which the presence of the hydrophilic biotin extension prevents the cell penetration, was synthesized and used in vitro (Fig 4a). Interestingly,

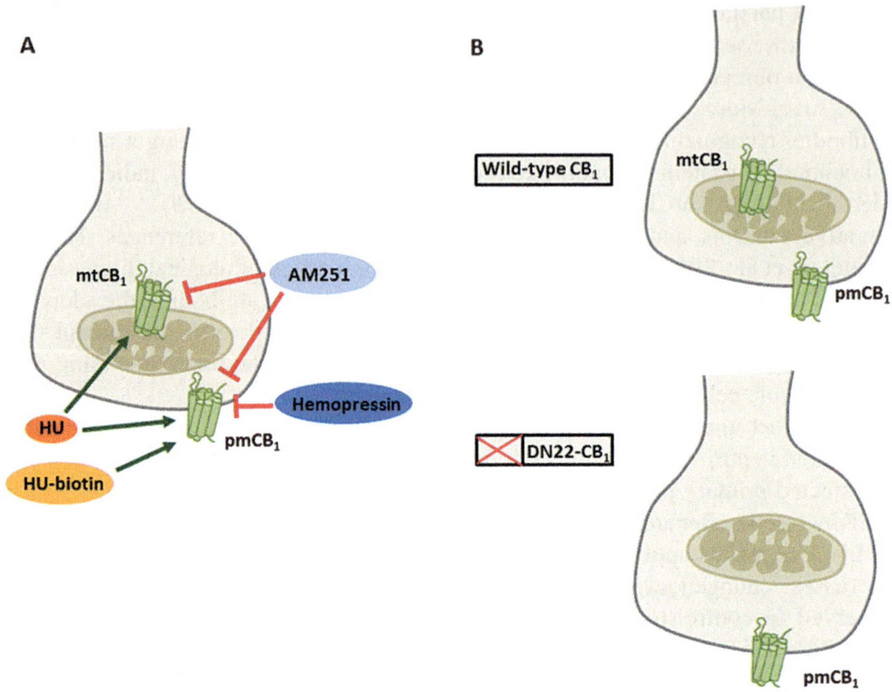

Fig. 4 Pharmacological and genetic tools to study the role of $mtCB_1$ receptors in brain physiology. (**a**) The CB_1 agonist HU210 (HU) is cell permeable and can activate both plasma membrane CB_1 (pmCB1) and intracellular CB_1 receptors such as $mtCB_1$. The addition of hydrophilic biotin to HU210 (HU-biot) blocks the capacity of this agonist to penetrate cell and activate $mtCB_1$. The CB_1 antagonist AM251 is cell permeant and can block the effect of both $pmCB_1$ and $mtCB_1$, whereas hemopressin (an antagonist of CB_1 that does not penetrate within the cell) only blocks the effect of $pmCB_1$. (**b**) The deletion of the 22 N-terminal amino acids of CB_1 (DN22-CB_1) excludes the receptor from mitochondria. These tools are then useful to describe the role of $mtCB_1$ on cannabinoid-dependent regulation of mitochondrial activity, brain physiology, and behavior

HU-biot treatment does not alter O_2 consumption in intact CB_1-expressing cells, whereas the effect appears when CB_1-MFs are permeabilized (Benard et al. 2012). Similarly, the cell-impermeant CB_1 antagonist hemopressin was not able to block the decrease of respiration induced by HU210 (Fig. 4a), unless cell was permeabilized (Benard et al. 2012). Altogether, the results demonstrate that reduction of the respiratory activity in living cells is limited to the action of intracellular CB_1 receptors and most likely through $mtCB_1$ receptors. Later, we developed genetic tools to target more specifically $mtCB_1$ receptors. We observed that deletion of its first 22 N-terminal amino acids (DN22-CB_1 mutant protein) excludes CB_1 receptors from mitochondria (Fig. 4b). Injection of an adeno-associated virus expressing wild-type CB_1 or DN22-CB_1 into the hippocampus of null $CB_1^{-/-}$ mice induced the same total amount of total functional CB_1 protein (Hebert-Chatelain et al. 2016). However, the proportion of $mtCB_1$ receptors in $CB_1^{-/-}$

mice injected with viral DN22-CB_1 was equivalent to the amount of $mtCB_1$ observed in $CB_1^{-/-}$ mice injected with GFP (Hebert-Chatelain et al. 2016), showing that the first 22 N-terminal amino acids of the CB_1 protein are essential to import the receptor to mitochondrial membranes. Cannabinoids are not able to impair mitochondrial respiration and mitochondrial trafficking in cells expressing the mutant DN22-CB_1 (Hebert-Chatelain et al. 2016), confirming that reduction of mitochondrial activity by cannabinoids occurs mainly through $mtCB_1$ receptors.

Our team recently characterized the molecular mechanisms linking activation of $mtCB_1$ receptors and decreased mitochondrial activity. Briefly, activated $mtCB_1$ receptors release Gα proteins, which then bind to the mitochondrial cAMP-generating enzyme soluble adenylyl cyclase (sAC). This interaction lowers mitochondrial cAMP levels and the activity of intramitochondrial PKA (Fig. 5a). Different inhibitors targeting the key steps of this biochemical cascade (i.e., the $G\alpha_{i/o}$ inhibitor pertussis toxin, the sAC inhibitor KH7, and the PKA inhibitor H89) block the effect of cannabinoids on complex I activity, mitochondrial respiration, and mitochondrial mobility (Hebert-Chatelain et al. 2016). Therefore, activation of $mtCB_1$ receptors decreases mitochondrial activity through an intramitochondrial Gα-sAC-cAMP-PKA signaling pathway (Fig 5a). Considering that PKA can phosphorylate several mitochondrial proteins (Di Benedetto et al. 2013; Acin-Perez et al. 2009; Palmisano et al. 2007; Papa et al. 1996; Sardanelli et al. 2006; Technikova-Dobrova et al. 2001), we characterized the PKA-dependent mitochondrial phosphoproteome and observed that the complex I subunit NDUFS2 is less phosphorylated by PKA upon treatment with cannabinoids in an $mtCB_1$-dependent manner (Hebert-Chatelain et al. 2016). Strikingly, cannabinoids were not able to decrease mitochondrial activity in cells expressing a phospho-mimetic mutant of NDUFS2 (Hebert-Chatelain et al. 2016), indicating that $mtCB_1$ regulates mitochondrial respiration by decreasing the PKA-dependent phosphorylation of the OXPHOS component NDUFS2. Studying the impact of cannabinoids on the mitochondrial phosphoproteome will certainly benefit to our understanding of the processes by which cannabinoids impact on mitochondrial physiology and brain functions.

The idea that $mtCB_1$ receptors might participate in eCB-dependent modulation of synaptic transmission was further tested (Fig. 5b). Agonists of CB_1 receptors reduce excitatory synaptic transmission in hippocampal slices from wild-type animals, but not from $CB_1^{-/-}$ mice (Kano et al. 2009). Activation of $mtCB_1$ receptors seems a key step in this process since cannabinoids are not able to reduce excitatory synaptic transmission in (1) hippocampal slices pretreated with KH7 that blocks sAC and $mtCB_1$-dependent signaling (see above) and (2) hippocampal slices from $CB_1^{-/-}$ mice expressing viral DN22-CB_1 (Hebert-Chatelain et al. 2016). CB_1 receptors likely also participate in depolarization-induced suppression of inhibition (DSI), a form of short-term plasticity of inhibitory neurotransmission. Briefly, the depolarization of a postsynaptic cell leads to eCB mobilization, which retrogradely activates presynaptic CB_1 receptors and transiently decreases GABAergic inhibitory neurotransmission (Kano et al. 2009; Alger 2002). DSI levels are proportional to the duration of the depolarization step, with longer depolarization inducing

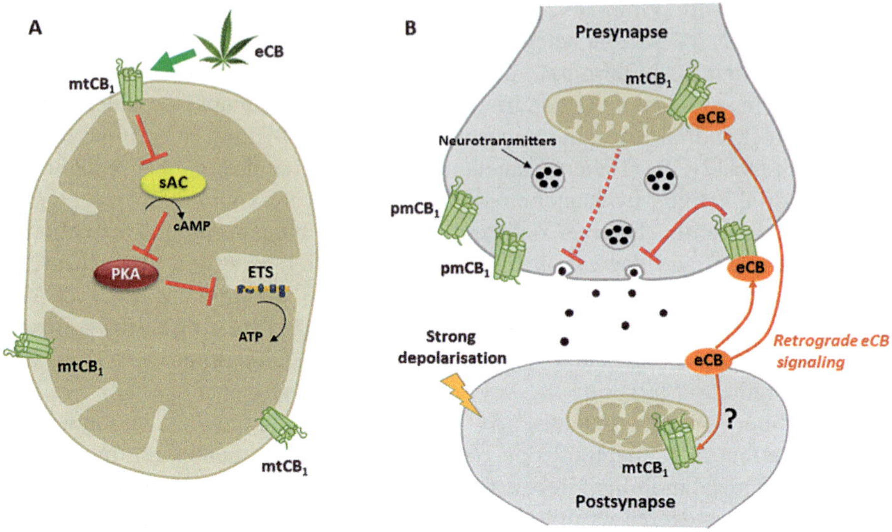

Fig. 5 Mitochondrial CB_1 receptor regulates oxidative phosphorylation and contributes to synaptic plasticity. (**a**) Possible molecular mechanism linking activation of $mtCB_1$ receptor and lower oxidative phosphorylation. Activation of $mtCB_1$ receptors by (endo)cannabinoids (eCB) decreases the activity of soluble adenylyl cyclase (sAC) and consequently reduces mitochondrial cAMP levels and PKA activity. This inhibition is associated with lower activity of the complex I of the electron transport system (ETS) and of oxidative phosphorylation process, suggesting that $mtCB_1$ receptors could control cellular bioenergetics through the mitochondrial cAMP-PKA pathway. (**b**) Intracellular CB_1 receptor, possibly $mtCB_1$, seems to participate in depolarization-induced suppression of inhibition (DSI), an endocannabinoid-dependent form of short-term plasticity of inhibitory transmission in the hippocampus. Briefly, a strong postsynaptic depolarization induces the release of eCBs which cross the synaptic cleft and activates presynaptic plasma membrane CB_1 receptors ($pmCB_1$). This process inhibits the release of the neurotransmitter GABA which results in reduced inhibitory neurotransmission. Results obtained in our laboratory suggest that activation of $mtCB_1$ receptor is involved in this process. For instance, the cell-impermeant CB_1 receptor agonist HU210-biotin only partially occluded DSI, whereas the inhibitor of the complex I of the ETS rotenone potentiated weak DSI (see text for a detailed description of these processes)

higher levels of presynaptic inhibition (Kano et al. 2009; Alger 2002). DSI is blocked by CB_1 receptor antagonists, but it is also prevented (occluded) by CB_1 receptor agonists, which occupy CB_1 receptors and impede further actions of endogenously mobilized eCBs (Kano et al. 2009; Alger 2002). In hippocampal slices, HU210 completely occluded "strong DSI" induced by 5 sec depolarization steps, whereas HU-biot at saturating doses had only a partial effect (approximately 50% occlusion). Similarly, the cell-impermeant CB_1 receptor antagonist, the peptide hemopressin (Heimann et al. 2007), partially reduced "strong DSI," whereas the membrane-permeant CB_1 receptor antagonist AM251 abolished it. Finally, rotenone, a mitochondrial complex I inhibitor, was ineffective on "strong DSI," but potentiated "weak DSI," suggesting that mitochondrial processes participate in the expression of "strong DSI" (Fig. 5b). Altogether, these results suggest that $mtCB_1$ receptors are important in eCB-dependent synaptic plasticity. A recent study showed that mtCB$_1$ receptors might be a potential target for the treatment of brain ischemic injury. This work showed that ACEA restored cell viability, inhibited generation of ROS, reduced apoptosis (both in vitro and in vivo), and improved OXPHOS enzyme activities after cerebral ischemia/reperfusion injury (Ma et al. 2015). Interestingly, these benefits were blocked by the cell-permeant AM251, but only partially reversed by the cell-impermeant hemopressin, suggesting that mtCB$_1$ receptors might be involved in the beneficial effects of ACEA during I/R (Ma et al. 2015).

Agonists of CB$_1$ receptors can impair high brain functions such as learning and memory. For instance, cannabinoids induce amnesia in the novel object recognition test in wild-type mice but not in CB$_1^{-/-}$ mice (Puighermanal et al. 2009). The modulation of mitochondrial activity and synaptic transmission by mtCB$_1$ receptors suggest that cannabinoids could impact on high brain functions at least partly through mtCB$_1$ receptors. Indeed, we observed that the cannabinoid-induced impairment of memory performance was rescued in $CB_1^{-/-}$ mice expressing hippocampal wild-type CB$_1$ but not DN22-CB$_1$ (Hebert-Chatelain et al. 2016). Modulation of mtCB$_1$-dependent signaling can also modify the impact of cannabinoids on memory performance. Indeed, intra-hippocampal injection of the sAC inhibitor KH7 or expression of a constitutively active mutant of PKA specifically targeted to mitochondria blocks the decrease of memory performance induced by systemic injection of WIN (Hebert-Chatelain et al. 2016). Similarly, viral expression of the phospho-mimetic mutant of NDUFS2 in mouse hippocampi also partly blocks the amnesic effect of WIN (Hebert-Chatelain et al. 2016). Altogether, these results indicate that cannabinoids modulate brain physiology and functions through mtCB$_1$-dependent regulation of mitochondrial activity. Future work will be needed to understand the role of mtCB$_1$ receptors in the numerous brain functions and behaviors that are modulated by (endo)cannabinoids.

Today, it appears that cannabinoids can have both receptor-dependent and receptor-independent effects on mitochondria. Considering that both pmCB$_1$ and mtCB$_1$ receptors impact on mitochondria, we can speculate that both receptors could cooperate to modulate mitochondrial functions. For instance, activation of pmCB$_1$ receptors could lead to long-term decrease of mitochondrial biogenesis and activity, as observed in white adipose tissue, muscle, and the liver (Tedesco et al. 2010), whereas activation of mtCB$_1$ receptors induces short-term and acute changes of mitochondrial activity (Benard et al. 2012; Hebert-Chatelain et al. 2014a). Further studies are needed to investigate the specific impact of pmCB$_1$ and mtCB$_1$ receptors on synaptic transmission, brain functions, and behavior.

6 Conclusion

It is well established that bioenergetics processes, including mitochondrial respiration, support neurotransmission and brain functions. Brain mtCB$_1$ receptor activation directly regulates mitochondrial respiration and ATP production. This function seems to be regulated through an intramitochondrial Gα-sAC-cAMP-PKA signaling pathway. Thus, $mtCB_1$ receptors might participate in brain functions by acutely altering mitochondrial respiration, thereby modulating energy supply, which is a key limiting factor of neuronal signaling. The correct functioning of mitochondria is fundamental for the generation of most cellular ATP, and its long-term impairment has been implicated in many neurological and psychiatric diseases. Further studies are needed to investigate the role of acute regulation of mitochondrial activity through $mtCB_1$ receptors in brain physiological and behavioral effects of cannabinoids. To this aim, the design and development of powerful tools that functionally discriminate between the functions played by $mtCB_1$ from the ones played by CB_1 receptors at other cellular locations (such as DN22-CB_1) are necessary. Finally, cannabinoid drugs are endowed with several therapeutic potentials via activation of CB_1 receptors, unfortunately limited by important side effects (Piomelli 2003). The selective targeting of specific subcellular populations of CB_1 receptors in the brain might become a very interesting way to obtain safer and more efficient therapeutic means against several brain disorders.

Acknowledgments The work in Marsicano's lab is supported by INSERM, EU–Fp7 (REPROBESITY, HEALTH–F2–2008–223713 and PAINCAGE, HEALTH-603191), European Research Council (Endofood, ERC–2010–StG–260515, and CannaPreg, ERC-2014-PoC-640923), Fondation pour la Recherche Medicale (DRM20101220445, DPP20151033974), Human Frontiers Science Program, Region Aquitaine, and Agence Nationale de la Recherche (ANR Blanc NeuroNutriSens ANR-13-BSV4-0006 and BRAIN ANR-10-LABX-0043). We thank all the members of Marsicano's lab for the useful discussions. The work of Etienne Hebert-Chatelain is supported by the Natural Sciences and Engineering Research Council of Canada (Discovery grant: RGPIN-2015-05880), Alzheimer Society of Canada (New Investigator Award: 17-09), the Canadian Breast Cancer Foundation (Research grant 16880-2015-17342), Beatrice Hunter Cancer Research Institute (Seed Fund), New Brunswick Health Research Foundation, New Brunswick Innovation Foundation, and Université de Moncton.

References

Acin-Perez R et al (2009) Cyclic AMP produced inside mitochondria regulates oxidative phosphorylation. Cell Metab 9:265–276. doi:10.1016/j.cmet.2009.01.012

Adams R (1942) Marihuana: Harvey Lecture, February 19, 1942. Bull NY Acad Med 18:705–730

Alger BE (2002) Retrograde signaling in the regulation of synaptic transmission: focus on endocannabinoids. Prog Neurobiol 68:247–286

Aquila S et al (2010) Human sperm anatomy: ultrastructural localization of the cannabinoid1 receptor and a potential role of anandamide in sperm survival and acrosome reaction. Anat Rec (Hoboken) 293:298–309. doi:10.1002/ar.21042

Athanasiou A et al (2007) Cannabinoid receptor agonists are mitochondrial inhibitors: a unified hypothesis of how cannabinoids modulate mitochondrial function and induce cell death. Biochem Biophys Res Commun 364:131–137. doi:10.1016/j.bbrc.2007.09.107

Badawy ZS et al (2009) Cannabinoids inhibit the respiration of human sperm. Fertil Steril 91:2471–2476. doi:10.1016/j.fertnstert.2008.03.075

Bartova A, Birmingham MK (1976) Effect of delta9-tetrahydrocannabinol on mitochondrial NADH-oxidase activity. J Biol Chem 251:5002–5006

Behnke HD (1977) The origin of plastids and mitochondria. The endosymbiotic hypothesis. MMW Munchener medizinische Wochenschrift 119:317–318
Bellocchio L et al (2010) Bimodal control of stimulated food intake by the endocannabinoid system. Nat Neurosci 13:281–283. doi:10.1038/nn.2494
Belous A et al (2004) Mitochondrial P2Y-Like receptors link cytosolic adenosine nucleotides to mitochondrial calcium uptake. J Cell Biochem 92:1062–1073. doi:10.1002/jcb.20144
Benard G et al (2012) Mitochondrial CB(1) receptors regulate neuronal energy metabolism. Nat Neurosci 15:558–564. doi:10.1038/nn.3053
Benedict AL et al (2012) Neuroprotective effects of sulforaphane after contusive spinal cord injury. J Neurotrauma 29:2576–2586. doi:10.1089/neu.2012.2474
Bermudez-Silva FJ, Cardinal P, Cota D (2012) The role of the endocannabinoid system in the neuroendocrine regulation of energy balance. J Psychopharmacol 26:114–124. doi:10.1177/0269881111408458
Bianchi C, Genova ML, Parenti Castelli G, Lenaz G (2004) The mitochondrial respiratory chain is partially organized in a supercomplex assembly: kinetic evidence using flux control analysis. J Biol Chem 279:36562–36569. doi:10.1074/jbc.M405135200
Billups B, Forsythe ID (2002) Presynaptic mitochondrial calcium sequestration influences transmission at mammalian central synapses. J Neurosci 22:5840–5847
Bino T, Chari-Bitron A, Shahar A (1972) Biochemical effects and morphological changes in rat liver mitochondria exposed to 1-tetrahydrocannabinol. Biochim Biophys Acta 288:195–202
Boesmans W, Ameloot K, van den Abbeel V, Tack J, Vanden Berghe P (2009) Cannabinoid receptor 1 signalling dampens activity and mitochondrial transport in networks of enteric neurones. Neurogastroenterol Motil 21:958–e977. doi:10.1111/j.1365-2982.2009.01300.x
Bosier B et al (2013) Astroglial CB1 cannabinoid receptors regulate leptin signaling in mouse brain astrocytes. Mol Metab 2(4):393–404. doi:10.1016/j.molmet.2013.08.001
Brdiczka DG, Zorov DB, Sheu SS (2006) Mitochondrial contact sites: their role in energy metabolism and apoptosis. Biochim Biophys Acta 1762:148–163. doi:10.1016/j.bbadis.2005.09.007
Cadenas E, Davies KJ (2000) Mitochondrial free radical generation, oxidative stress, and aging. Free Radic Biol Med 29:222–230
Cai Q, Davis ML, Sheng ZH (2011) Regulation of axonal mitochondrial transport and its impact on synaptic transmission. Neurosci Res 70:9–15. doi:10.1016/j.neures.2011.02.005
Campbell VA (2001) Tetrahydrocannabinol-induced apoptosis of cultured cortical neurones is associated with cytochrome c release and caspase-3 activation. Neuropharmacology 40:702–709
Catanzaro G, Rapino C, Oddi S, Maccarrone M (2009) Anandamide increases swelling and reduces calcium sensitivity of mitochondria. Biochem Biophys Res Commun 388:439–442. doi:10.1016/j.bbrc.2009.08.037
Chance B, Williams GR (1955) A method for the localization of sites for oxidative phosphorylation. Nature 176:250–254
Chari-Bitron A, Bino T (1971) Effect of 1-tetrahydrocannabinol on ATPase activity of rat liver mitochondria. Biochem Pharmacol 20:473–475
Chevaleyre V, Castillo PE (2003) Heterosynaptic LTD of hippocampal GABAergic synapses: a novel role of endocannabinoids in regulating excitability. Neuron 38:461–472
Colburn RW, Ng LK, Lemberger L, Kopin IJ (1974) Subcellular distribution of delta9-tetrahydrocannabinol in rat brain. Biochem Pharmacol 23:873–877
Cunha JM et al (1980) Chronic administration of cannabidiol to healthy volunteers and epileptic patients. Pharmacology 21:175–185
Czabotar PE, Lessene G, Strasser A, Adams JM (2014) Control of apoptosis by the BCL-2 protein family: implications for physiology and therapy. Nat Rev Mol Cell Biol 15:49–63. doi:10.1038/nrm3722
Denton RM (2009) Regulation of mitochondrial dehydrogenases by calcium ions. Biochim Biophys Acta 1787:1309–1316. doi:10.1016/j.bbabio.2009.01.005

Devane WA et al (1992) A novel probe for the cannabinoid receptor. J Med Chem 35:2065–2069
Di Benedetto G, Scalzotto E, Mongillo M, Pozzan T (2013) Mitochondrial Ca(2)(+) uptake induces cyclic AMP generation in the matrix and modulates organelle ATP levels. Cell Metab 17:965–975. doi:10.1016/j.cmet.2013.05.003
Di Marzo V, De Petrocellis L (2012) Why do cannabinoid receptors have more than one endogenous ligand? Philos Trans R Soc Lond B Biol Sci 367:3216–3228. doi:10.1098/rstb.2011.0382
Dubreucq S et al (2012) Genetic dissection of the role of cannabinoid type-1 receptors in the emotional consequences of repeated social stress in mice. Neuropsychopharmacology 37:1885–1900. doi:10.1038/npp.2012.36
Erecinska M, Silver IA (2001) Tissue oxygen tension and brain sensitivity to hypoxia. Respir Physiol 128:263–276
Ferguson CJ et al (2001) Cellular localization of divalent metal transporter DMT-1 in rat kidney. Am J Physiol Renal Physiol 280:F803–F814
Fiedorowicz JG, Swartz KL (2004) The role of monoamine oxidase inhibitors in current psychiatric practice. J Psychiatr Pract 10:239–248
Fisar Z (2010) Inhibition of monoamine oxidase activity by cannabinoids. Naunyn Schmiedebergs Arch Pharmacol 381:563–572. doi:10.1007/s00210-010-0517-6
Fisar Z, Singh N, Hroudova J (2014) Cannabinoid-induced changes in respiration of brain mitochondria. Toxicol Lett 231:62–71. doi:10.1016/j.toxlet.2014.09.002
Fonseca BM, Correia-da-Silva G, Teixeira NA (2013) The endocannabinoid anandamide induces apoptosis of rat decidual cells through a mechanism involving ceramide synthesis and p38 MAPK activation. Apoptosis 18:1526–1535. doi:10.1007/s10495-013-0892-9
Frey TG, Mannella CA (2000) The internal structure of mitochondria. Trends Biochem Sci 25:319–324
Gallily R et al (2003) Gamma-irradiation enhances apoptosis induced by cannabidiol, a non-psychotropic cannabinoid, in cultured HL-60 myeloblastic leukemia cells. Leuk lymphoma 44:1767–1773. doi:10.1080/1042819031000103917
Gaoni Y, Mechoulam R (1964) Isolation, structure, and partial synthesis of an active constituent of hashish. J Am Chem Soc 86:1646–1647. doi:10.1021/ja01062a046
Giorgi C et al (2012) Mitochondrial calcium homeostasis as potential target for mitochondrial medicine. Mitochondrion 12:77–85. doi:10.1016/j.mito.2011.07.004
Gorman AM, Ceccatelli S, Orrenius S (2000) Role of mitochondria in neuronal apoptosis. Dev Neurosci 22:348–358. doi:10.1159/000017460
Gunter TE, Yule DI, Gunter KK, Eliseev RA, Salter JD (2004) Calcium and mitochondria. FEBS Lett 567:96–102. doi:10.1016/j.febslet.2004.03.071
Hackenbrock CR (1966) Ultrastructural bases for metabolically linked mechanical activity in mitochondria. I. Reversible ultrastructural changes with change in metabolic steady state in isolated liver mitochondria. J Cell Biol 30:269–297
Hajnoczky G et al (2006) Mitochondrial calcium signalling and cell death: approaches for assessing the role of mitochondrial Ca2+ uptake in apoptosis. Cell Calcium 40:553–560. doi:10.1016/j.ceca.2006.08.016
Han J et al (2012) Acute cannabinoids impair working memory through astroglial CB1 receptor modulation of hippocampal LTD. Cell 148:1039–1050. doi:10.1016/j.cell.2012.01.037
Haring M, Marsicano G, Lutz B, Monory K (2007) Identification of the cannabinoid receptor type 1 in serotonergic cells of raphe nuclei in mice. Neuroscience 146:1212–1219. doi:10.1016/j.neuroscience.2007.02.021
Harris JJ, Jolivet R, Attwell D (2012) Synaptic energy use and supply. Neuron 75:762–777. doi:10.1016/j.neuron.2012.08.019
Hebert-Chatelain E et al (2014a) Cannabinoid control of brain bioenergetics: exploring the subcellular localization of the CB1 receptor. Mol Metab 3:495–504. doi:10.1016/j.molmet.2014.03.007

Hebert-Chatelain E et al (2014b) Studying mitochondrial CB1 receptors: yes we can. Mol Metab 3:339. doi:10.1016/j.molmet.2014.03.008
Hebert-Chatelain E et al (2016) A cannabinoid link between mitochondria and memory. Nature. doi:10.1038/nature20127
Heimann AS et al (2007) Hemopressin is an inverse agonist of CB1 cannabinoid receptors. Proc Natl Acad Sci USA 104:20588–20593. doi:10.1073/pnas.0706980105
Herkenham M et al (1990) Cannabinoid receptor localization in brain. Proc Natl Acad Sci USA 87:1932–1936
Herkenham M et al (1991) Characterization and localization of cannabinoid receptors in rat brain: a quantitative in vitro autoradiographic study. J Neurosci 11:563–583
Herrmann JM, Riemer J (2010) The intermembrane space of mitochondria. Antioxid Redox Signal 13:1341–1358. doi:10.1089/ars.2009.3063
Hill MN, Gorzalka BB (2005) Pharmacological enhancement of cannabinoid CB1 receptor activity elicits an antidepressant-like response in the rat forced swim test. Eur Neuropsychopharmacol 15:593–599. doi:10.1016/j.euroneuro.2005.03.003
Horvath TL, Andrews ZB, Diano S (2009) Fuel utilization by hypothalamic neurons: roles for ROS. Trends Endocrinol Metab 20:78–87. doi:10.1016/j.tem.2008.10.003
Howlett AC (2002) The cannabinoid receptors. Prostaglandins Other Lipid Mediat 68–69:619–631
Hroudova J, Fisar Z (2013) Control mechanisms in mitochondrial oxidative phosphorylation. Neural Regen Res 8:363–375. doi:10.3969/j.issn.1673-5374.2013.04.009
Irannejad R, von Zastrow M (2014) GPCR signaling along the endocytic pathway. Curr Opin Cell Biol 27:109–116. doi:10.1016/j.ceb.2013.10.003
Jia W et al (2006) Delta9-tetrahydrocannabinol-induced apoptosis in Jurkat leukemia T cells is regulated by translocation of Bad to mitochondria. Mol Cancer Res 4:549–562. doi:10.1158/1541-7786.MCR-05-0193
Kano M, Ohno-Shosaku T, Hashimotodani Y, Uchigashima M, Watanabe M (2009) Endocannabinoid-mediated control of synaptic transmission. Physiol Rev 89:309–380. doi:10.1152/physrev.00019.2008
Katona I et al (1999) Presynaptically located CB1 cannabinoid receptors regulate GABA release from axon terminals of specific hippocampal interneurons. J Neurosci 19:4544–4558
Kety SS (1957) Determinants of tissue oxygen tension. Fed Proc 16:666–671
Kishida KT, Klann E (2007) Sources and targets of reactive oxygen species in synaptic plasticity and memory. Antioxid Redox Signal 9:233–244. doi:10.1089/ars.2007.9.ft-8
Knapp LT, Klann E (2002) Potentiation of hippocampal synaptic transmission by superoxide requires the oxidative activation of protein kinase C. J Neurosci 22:674–683
Koch M et al (2015) Hypothalamic POMC neurons promote cannabinoid-induced feeding. Nature 519:45–50. doi:10.1038/nature14260
Koob GF, Le Moal M (2001) Drug addiction, dysregulation of reward, and allostasis. Neuropsychopharmacology 24:97–129. doi:10.1016/S0893-133X(00)00195-0
Kuan CY, Roth KA, Flavell RA, Rakic P (2000) Mechanisms of programmed cell death in the developing brain. Trends Neurosci 23:291–297
Kuyznierewicz I, Thomson M (2002) GTP-binding proteins G(salpha), G(ialpha), and Ran identified in mitochondria of human placenta. Cell Biol Int 26:99–108. doi:10.1006/cbir.2001.0823
Ledent C et al (1999) Unresponsiveness to cannabinoids and reduced addictive effects of opiates in CB1 receptor knockout mice. Science 283:401–404
Lee CY et al (2008) A comparative study on cannabidiol-induced apoptosis in murine thymocytes and EL-4 thymoma cells. Int immunopharmacol 8:732–740. doi:10.1016/j.intimp.2008.01.018
Lee KY, Chung K, Chung JM (2010) Involvement of reactive oxygen species in long-term potentiation in the spinal cord dorsal horn. J Neurophysiol 103:382–391. doi:10.1152/jn.90906.2008
Li Z, Sheng M (2012) Caspases in synaptic plasticity. Mol Brain 5:15. doi:10.1186/1756-6606-5-15

Li Z et al (2010) Caspase-3 activation via mitochondria is required for long-term depression and AMPA receptor internalization. Cell 141:859–871. doi:10.1016/j.cell.2010.03.053

Ligresti A et al (2006) Antitumor activity of plant cannabinoids with emphasis on the effect of cannabidiol on human breast carcinoma. J Pharmacol Exp Ther 318:1375–1387. doi:10.1124/jpet.106.105247

Ma T et al (2011) Amyloid beta-induced impairments in hippocampal synaptic plasticity are rescued by decreasing mitochondrial superoxide. J Neurosci 31:5589–5595. doi:10.1523/JNEUROSCI.6566-10.2011

Ma L et al (2015) Mitochondrial CB1 receptor is involved in ACEA-induced protective effects on neurons and mitochondrial functions. Sci Rep 5:12440. doi:10.1038/srep12440

Mahoney JM, Harris RA (1972) Effect of 9-tetrahydrocannabinol on mitochondrial processes. Biochem Pharmacol 21:1217–1226

Mailleux P, Parmentier M, Vanderhaeghen JJ (1992) Distribution of cannabinoid receptor messenger RNA in the human brain: an in situ hybridization histochemistry with oligonucleotides. Neurosci Lett 143:200–204

Marchi S et al (2012) Mitochondria-ros crosstalk in the control of cell death and aging. J Signal Transduct 2012:329635. doi:10.1155/2012/329635

Marsicano G, Kuner R (2008) Anatomical distribution of receptors, ligands and enzymes in the brain and in the spinal cord: circuitries and neurochemistry. Cannabinoids Brain. doi:10.1007/978-0-387-74349-3_10

Marsicano G, Lutz B (1999) Expression of the cannabinoid receptor CB1 in distinct neuronal subpopulations in the adult mouse forebrain. Eur J Neurosci 11:4213–4225

Marsicano G et al (2003) CB1 cannabinoid receptors and on-demand defense against excitotoxicity. Science 302:84–88. doi:10.1126/science.1088208

Massi P et al (2006) The non-psychoactive cannabidiol triggers caspase activation and oxidative stress in human glioma cells. Cell Mol Life Sci 63:2057–2066. doi:10.1007/s00018-006-6156-x

Mato S, Victoria Sanchez-Gomez M, Matute C (2010) Cannabidiol induces intracellular calcium elevation and cytotoxicity in oligodendrocytes. Glia 58:1739–1747. doi:10.1002/glia.21044

Matsuda LA, Lolait SJ, Brownstein MJ, Young AC, Bonner TI (1990) Structure of a cannabinoid receptor and functional expression of the cloned cDNA. Nature 346:561–564. doi:10.1038/346561a0

Mattson MP, Gleichmann M, Cheng A (2008) Mitochondria in neuroplasticity and neurological disorders. Neuron 60:748–766. doi:10.1016/j.neuron.2008.10.010

Mendizabal-Zubiaga J et al (2016) Cannabinoid CB1 receptors are localized in striated muscle mitochondria and regulate mitochondrial respiration. Front Physiol 7:476. doi:10.3389/fphys.2016.00476

Metna-Laurent M et al (2012) Bimodal control of fear-coping strategies by CB(1) cannabinoid receptors. J Neurosci 32:7109–7118. doi:10.1523/JNEUROSCI.1054-12.2012

Mitchell P (1961) Coupling of phosphorylation to electron and hydrogen transfer by a chemi-osmotic type of mechanism. Nature 191:144–148

Moldrich G, Wenger T (2000) Localization of the CB1 cannabinoid receptor in the rat brain. An immunohistochemical study. Peptides 21:1735–1742

Monory K et al (2006) The endocannabinoid system controls key epileptogenic circuits in the hippocampus. Neuron 51:455–466. doi:10.1016/j.neuron.2006.07.006

Moreira PI, Santos MS, Moreno A, Oliveira C (2001) Amyloid beta-peptide promotes permeability transition pore in brain mitochondria. Biosci Rep 21:789–800

Morozov YM et al (2013) Antibodies to cannabinoid type 1 receptor co-react with stomatin-like protein 2 in mouse brain mitochondria. Eur J Neurosci 38:2341–2348. doi:10.1111/ejn.12237

Morozov YM, Horvath TL, Rakic P (2014) A tale of two methods: identifying neuronal CB1 receptors. Mol Metab 3:338. doi:10.1016/j.molmet.2014.03.006

Mukherjee J, Yang ZY (1999) Monoamine oxidase A inhibition by fluoxetine: an in vitro and in vivo study. Synapse 31:285–289. doi:10.1002/(SICI)1098-2396(19990315)31:4<285::AID-SYN6>3.0.CO;2-5

Munro S, Thomas KL, Abu-Shaar M (1993) Molecular characterization of a peripheral receptor for cannabinoids. Nature 365(6441):61–65. doi:10.1038/365061a0

Navarrete M, Araque A (2008) Endocannabinoids mediate neuron-astrocyte communication. Neuron 57:883–893. doi:10.1016/j.neuron.2008.01.029

Navarrete M, Araque A (2010) Endocannabinoids potentiate synaptic transmission through stimulation of astrocytes. Neuron 68:113–126. doi:10.1016/j.neuron.2010.08.043

Navarro A, Boveris A (2007) The mitochondrial energy transduction system and the aging process. Am J Physiol Cell Physiol 292:C670–C686. doi:10.1152/ajpcell.00213.2006

Nicholls DG (2005) Mitochondria and calcium signaling. Cell Calcium 38:311–317. doi:10.1016/j.ceca.2005.06.011

Nonomura K et al (2013) Local apoptosis modulates early mammalian brain development through the elimination of morphogen-producing cells. Dev Cell 27:621–634. doi:10.1016/j.devcel.2013.11.015

Nunez E et al (2004) Cannabinoid CB2 receptors are expressed by perivascular microglial cells in the human brain: an immunohistochemical study. Synapse 53:208–213. doi:10.1002/syn.20050

Olsen KM, Sheng M (2012) NMDA receptors and BAX are essential for Abeta impairment of LTP. Sci Rep 2:225. doi:10.1038/srep00225

Olsen RH, Johnson LA, Zuloaga DG, Limoli CL, Raber J (2013) Enhanced hippocampus-dependent memory and reduced anxiety in mice over-expressing human catalase in mitochondria. J Neurochem 125:303–313. doi:10.1111/jnc.12187

Onaivi ES et al (2006) Discovery of the presence and functional expression of cannabinoid CB2 receptors in brain. Ann NY Acad Sci 1074:514–536. doi:10.1196/annals.1369.052

Onaivi ES, Ishiguro H, Gu S, Liu QR (2012) CNS effects of CB2 cannabinoid receptors: beyond neuro-immuno-cannabinoid activity. J Psychopharmacol 26:92–103. doi:10.1177/0269881111400652

Ong WY, Mackie K (1999) A light and electron microscopic study of the CB1 cannabinoid receptor in the primate spinal cord. J Neurocytol 28:39–45

Pagotto U, Marsicano G, Cota D, Lutz B, Pasquali R (2006) The emerging role of the endocannabinoid system in endocrine regulation and energy balance. Endocr Rev 27:73–100. doi:10.1210/er.2005-0009

Palmisano G, Sardanelli AM, Signorile A, Papa S, Larsen MR (2007) The phosphorylation pattern of bovine heart complex I subunits. Proteomics 7:1575–1583. doi:10.1002/pmic.200600801

Papa S et al (1996) The nuclear-encoded 18 kDa (IP) AQDQ subunit of bovine heart complex I is phosphorylated by the mitochondrial cAMP-dependent protein kinase. FEBS Lett 379:299–301. doi:10.1016/0014-5793(95)01532-9

Pellerito O et al (2014) WIN induces apoptotic cell death in human colon cancer cells through a block of autophagic flux dependent on PPARgamma down-regulation. Apoptosis 19:1029–1042. doi:10.1007/s10495-014-0985-0

Pertwee RG, Ross RA, Craib SJ, Thomas A (2002) (-)-Cannabidiol antagonizes cannabinoid receptor agonists and noradrenaline in the mouse vas deferens. Eur J Pharmacol 456:99–106

Pettit DA, Harrison MP, Olson JM, Spencer RF, Cabral GA (1998) Immunohistochemical localization of the neural cannabinoid receptor in rat brain. J Neurosci Res 51:391–402

Pickel VM, Chan J, Kash TL, Rodriguez JJ, MacKie K (2004) Compartment-specific localization of cannabinoid 1 (CB1) and mu-opioid receptors in rat nucleus accumbens. Neuroscience 127:101–112. doi:10.1016/j.neuroscience.2004.05.015

Piomelli D (2003) The molecular logic of endocannabinoid signalling. Nat Rev Neurosci 4:873–884. doi:10.1038/nrn1247

Puighermanal E et al (2009) Cannabinoid modulation of hippocampal long-term memory is mediated by mTOR signaling. Nature Neurosci 12:1152–1158. doi:10.1038/nn.2369

Rangaraju V, Calloway N, Ryan TA (2014) Activity-driven local ATP synthesis is required for synaptic function. Cell 156:825–835. doi:10.1016/j.cell.2013.12.042

Rimmerman N et al (2013) Direct modulation of the outer mitochondrial membrane channel, voltage-dependent anion channel 1 (VDAC1) by cannabidiol: a novel mechanism for cannabinoid-induced cell death. Cell Death Dis 4:e949. doi:10.1038/cddis.2013.471

Rodriguez JJ, Mackie K, Pickel VM (2001) Ultrastructural localization of the CB1 cannabinoid receptor in mu-opioid receptor patches of the rat Caudate putamen nucleus. J Neurosci 21:823–833

Rolfe DF, Brown GC (1997) Cellular energy utilization and molecular origin of standard metabolic rate in mammals. Physiol Rev 77:731–758

Rossato M, Ion Popa F, Ferigo M, Clari G, Foresta C (2005) Human sperm express cannabinoid receptor Cb1, the activation of which inhibits motility, acrosome reaction, and mitochondrial function. J Clin Endocrinol Metab 90:984–991. doi:10.1210/jc.2004-1287

Rozenfeld R, Devi LA (2008) Regulation of CB1 cannabinoid receptor trafficking by the adaptor protein AP-3. FASEB J 22:2311–2322. doi:10.1096/fj.07-102731

Ryan D, Drysdale AJ, Lafourcade C, Pertwee RG, Platt B (2009) Cannabidiol targets mitochondria to regulate intracellular Ca2+ levels. J Neurosci 29:2053–2063. doi:10.1523/JNEUROSCI.4212-08.2009

Sagan L (1967) On the origin of mitosing cells. J Theor Biol 14:255–274

Sarafian TA, Kouyoumjian S, Khoshaghideh F, Tashkin DP, Roth MD (2003) Delta 9-tetrahydrocannabinol disrupts mitochondrial function and cell energetics. Am J Physiol Lung Cell Mol Physiol 284:L298–L306. doi:10.1152/ajplung.00157.2002

Sardanelli AM et al (2006) Occurrence of A-kinase anchor protein and associated cAMP-dependent protein kinase in the inner compartment of mammalian mitochondria. FEBS Lett 580:5690–5696. doi:10.1016/j.febslet.2006.09.020

Schurr A, Livne A (1975) Proceedings: differential inhibition of mitochondrial monoamine oxidase from brain by hashish components. Isr J Med Sci 11:1188

Schurr A, Livne A (1976) Differential inhibition of mitochondrial monoamine oxidase from brain by hashish components. Biochem Pharmacol 25:1201–1203

Schurr A, Porath O, Krup M, Livne A (1978) The effects of hashish components and their mode of action on monoamine oxidase from the brain. Biochem Pharmacol 27:2513–2517

Scorrano L et al (2002) A distinct pathway remodels mitochondrial cristae and mobilizes cytochrome c during apoptosis. Dev Cell 2:55–67

Shih JC, Chen K, Ridd MJ (1999) Monoamine oxidase: from genes to behavior. Annu Rev Neurosci 22:197–217. doi:10.1146/annurev.neuro.22.1.197

Shoshan-Barmatz V et al (2010) VDAC, a multi-functional mitochondrial protein regulating cell life and death. Mol Asp Med 31:227–285. doi:10.1016/j.mam.2010.03.002

Shrivastava A, Kuzontkoski PM, Groopman JE, Prasad A (2011) Cannabidiol induces programmed cell death in breast cancer cells by coordinating the cross-talk between apoptosis and autophagy. Mol Cancer Ther 10:1161–1172. doi:10.1158/1535-7163.MCT-10-1100

Siegmund SV et al (2007) The endocannabinoid 2-arachidonoyl glycerol induces death of hepatic stellate cells via mitochondrial reactive oxygen species. FASEB J 21:2798–2806. doi:10.1096/fj.06-7717com

Sierra S et al (2014) Detection of cannabinoid receptors CB1 and CB2 within basal ganglia output neurons in macaques: changes following experimental parkinsonism. Brain Struct Funct. doi:10.1007/s00429-014-0823-8

Sokoloff L (1960) Quantitative measurements of cerebral blood flow in man. Methods Med Res 8:253–261

Soria-Gomez E et al (2014) The endocannabinoid system controls food intake via olfactory processes. Nat Neurosci 17:407–415. doi:10.1038/nn.3647

Suen DF, Norris KL, Youle RJ (2008) Mitochondrial dynamics and apoptosis. Genes Dev 22:1577–1590. doi:10.1101/gad.1658508

Szabo B, Schlicker E (2005) Effects of cannabinoids on neurotransmission. Handb Exp Pharmacol 168:327–365

Szoke E, Czeh G, Szolcsanyi J, Seress L (2002) Neonatal anandamide treatment results in prolonged mitochondrial damage in the vanilloid receptor type 1-immunoreactive B-type neurons of the rat trigeminal ganglion. Neuroscience 115:805–814

Taylor RW, Turnbull DM (2005) Mitochondrial DNA mutations in human disease. Nat Rev Genet 6:389–402. doi:10.1038/nrg1606

Technikova-Dobrova Z et al (2001) Cyclic adenosine monophosphate-dependent phosphorylation of mammalian mitochondrial proteins: enzyme and substrate characterization and functional role. Biochemistry 40:13941–13947. doi:10.1021/bi011066p

Tedesco L et al (2008) Cannabinoid type 1 receptor blockade promotes mitochondrial biogenesis through endothelial nitric oxide synthase expression in white adipocytes. Diabetes 57:2028–2036. doi:10.2337/db07-1623

Tedesco L et al (2010) Cannabinoid receptor stimulation impairs mitochondrial biogenesis in mouse white adipose tissue, muscle, and liver: the role of eNOS, p38 MAPK, and AMPK pathways. Diabetes 59:2826–2836. doi:10.2337/db09-1881

Trimmer PA et al (2000) Abnormal mitochondrial morphology in sporadic Parkinson's and Alzheimer's disease cybrid cell lines. Exp Neurol 162:37–50. doi:10.1006/exnr.2000.7333

Tsou K, Brown S, Sanudo-Pena MC, Mackie K, Walker JM (1998) Immunohistochemical distribution of cannabinoid CB1 receptors in the rat central nervous system. Neuroscience 83:393–411

Tuppen HA, Blakely EL, Turnbull DM, Taylor RW (2010) Mitochondrial DNA mutations and human disease. Biochim Biophys Acta 1797:113–128. doi:10.1016/j.bbabio.2009.09.005

Vallee M et al (2014) Pregnenolone can protect the brain from cannabis intoxication. Science 343:94–98. doi:10.1126/science.1243985

Whyte DA et al (2010) Cannabinoids inhibit cellular respiration of human oral cancer cells. Pharmacology 85:328–335. doi:10.1159/000312686

Wilson RI, Nicoll RA (2001) Endogenous cannabinoids mediate retrograde signalling at hippocampal synapses. Nature 410:588–592. doi:10.1038/35069076

Wilson-Poe AR, Morgan MM, Aicher SA, Hegarty DM (2012) Distribution of CB1 cannabinoid receptors and their relationship with mu-opioid receptors in the rat periaqueductal gray. Neuroscience 213:191–200. doi:10.1016/j.neuroscience.2012.03.038

Zaccagnino P, Corcelli A, Baronio M, Lorusso M (2011) Anandamide inhibits oxidative phosphorylation in isolated liver mitochondria. FEBS Lett 585:429–434. doi:10.1016/j.febslet.2010.12.032

Zimmer A, Zimmer AM, Hohmann AG, Herkenham M, Bonner TI (1999) Increased mortality, hypoactivity, and hypoalgesia in cannabinoid CB1 receptor knockout mice. Proc Natl Acad Sci USA 96:5780–5785

Susceptibility to Psychiatric Diseases After Cannabis Abuse in Adolescence: Animal Models

Tiziana Rubino and Daniela Parolaro

Abstract Epidemiological evidence suggests that adolescent exposure to delta-9-tetrahydrocanabinol, the psychoactive component of *Cannabis*, confers an increased risk for developing psychiatric disorders later in life. However, epidemiological studies have a correlative nature and are limited by ethical problems which can be circumvented by animal models.

After a preliminary section describing the dynamic changes that occur in the endocannabinoid system during adolescence, in the following sections, the more relevant animal data on the long-lasting consequences of adolescent exposure to cannabinoids is reported. Behavioral evidence in terms of impairment in cognition, vulnerability to mood disorders, schizophrenia, and subsequent drug abuse as well as the underlying neurobiological aspects are summarized.

The arising picture seems to support the hypothesis that adolescent exposure to cannabinoids might represent a risk factor for the development of psychiatric-like symptoms in adulthood since the external stimulation of the endocannabinoid system can profoundly alter its physiological role, thus triggering alterations in the maturational events occurring in the adolescent brain.

Quantifying the relative adverse and beneficial effects of cannabis and its constituent cannabinoids is becoming a priority particularly considering that several countries are legalizing its use. Although it is still a matter of debate, there is evidence suggesting that chronic adolescent marijuana exposure may be associated with a higher risk for neuropsychiatric diseases, including schizophrenia (Rubino and Parolaro 2015; Renard et al. 2014). During adolescence, the endocannabinoid system is highly active, driving the central nervous system maturation. Thus,

T. Rubino
Department of Biotechnology and Life Sciences, University of Insubria, via Manara 7, 21052 Busto Arsizio, VA, Italy

D. Parolaro (✉)
Department of Biotechnology and Life Sciences, University of Insubria, via Manara 7, 21052 Busto Arsizio, VA, Italy

Fondazione Zardi Gori, via Curtatone 25, 20100 Milano, Italy
e-mail: daniela.parolaro@uninsubria.it

M. Melis (ed.), *Endocannabinoids and Lipid Mediators in Brain Functions*,
DOI 10.1007/978-3-319-57371-7_9

adolescent exposure to exogenous cannabinoids might interfere with normal cortical maturation, causing neurobiological changes that impair brain function. This chapter will describe the behavioral and cellular evidence of enhanced susceptibility to neuropsychiatric disorders after cannabinoid abuse in adolescence obtained using validated animal models. Moreover, a preliminary description of the maturational processes occurring in the endocannabinoid system during adolescence will be provided.

1 The Endocannabinoid System in Adolescence

Several physical, neural, and behavioral changes occur simultaneously during adolescence. Indeed, the brain of adolescents is subjected to extensive synaptic and neurochemical remodeling in areas involved in emotions, learning, decision-making, and reward-motivated behaviors. These complex changes may make the human brain more vulnerable, and transient imbalances, primarily among developmental trajectories of corticolimbic structures, can leave an individual susceptible to mental illness. The endocannabinoid system (eCB) is a lipid signaling system consisting of specific receptors (cannabinoid CB1 and CB2 receptors), endogenous ligands (mainly anandamide and 2-arachidonoylglycerol), and a battery of enzymes responsible for the synthesis and degradation of these ligands. The eCB system plays a major role in neurodevelopmental processes which are particularly active during adolescence, and several studies have highlighted its active and dynamic nature during this age.

Rodriguez de Fonseca et al. (1993) showed that CB1 receptor (CB1R) binding in the rat brain is highest just prior to the onset of adolescence (postnatal day (PND) 25–29), followed by a general linear reduction to adult levels within limbic, striatal, and cortical structures. Accordingly, recent findings from different laboratories demonstrated that prefrontal cortex (PFC) CB1R expression in male rats declines in pre- to early adolescence with a region-specific rate (Ellgren et al. 2008; Heng et al. 2011). Indeed there is a gradual decline of the CB1R expression in limbic/associative regions, whereas major changes in sensorimotor regions are not evident until mid- to late adolescence, with the functionality of these receptors following the same developmental pattern (Heng et al. 2011). Accordingly, Rubino et al. (2015) showed that in female rodents, CB1Rs increase from mid- to late adolescence (PND 60) and decline by adulthood (PND 75; Rubino et al. 2015). The efficacy of CB1 receptor coupling with G proteins through adolescence does not show significant alteration, at least in the PFC, implying that CB1Rs seem to be more efficient during adolescence (Rubino et al. 2015). Recently, an interesting paper was published by Schneider et al. (2015) in which the introduction of a gain-of-function mutation in the gene that encodes for the CB1R generated rats with sustaining features of adolescent behavior in adulthood. Indeed, adult mutant rats exhibited typical high risk/novelty seeking, increased peer interaction, enhanced impulsivity, and augmented reward sensitivity for drug and nondrug reward when compared to wild-type littermates. Partial inhibition of CB1R activity normalized mutant rats' behavior.

These observations in adult mutant rats together with the above cited reports on enhanced CB1R signaling in adolescence highlight that the activity state and functionality of the CB1R are critical for mediating adolescent behavior. Interestingly no differences were observed between the genotypes for the two main endocannabinoids, anandamide (AEA) and 2-arachidonoylglycerol (2-AG).

The picture regarding changes in the level of the two main endocannabinoids, AEA and 2-AG, during adolescence is still at the beginning, and the few data available are somewhat divergent, probably for the difference in the strain and sex of the used animals.

In male rats, a continuous increase in PFC AEA levels throughout the adolescent period was reported, anandamide being almost three times higher in later adolescence. However, 2-arachidonoylglycerol concentrations in the same brain area were highest very early in adolescence (PND 29), decreased by PND 38, and increased again in late adolescence (PND 50; Ellgren et al. 2008). Similarly, Rubino et al. (2015), in the PFC of female rats, observed that AEA levels increased from mid- to late adolescence and then decreased into adulthood, while 2-AG levels first decreased and subsequently increased. Interestingly, despite these dynamic changes, the activity of the two degrading enzymes, fatty acid amide hydrolase (FAAH) and monoacylglycerol lipase, did not show any variation throughout the developmental window, suggesting a more likely involvement of the synthetic enzymes in regulating endocannabinoid levels.

Concerning other brain areas, it was reported that hypothalamic AEA content in female rats increases immediately preceding vaginal opening (as a physical marker of pubertal onset; Wenger et al. 2002), whereas in male rats the pattern of the AEA adolescent fluctuation in the amygdala, hippocampus, and hypothalamus (PND 25–70) was similar as in the PFC (Lee et al. 2013). Moreover, two other related N-acylethanolamines hydrolyzed by FAAH, oleoylethanolamine and palmitoylethanolamine, exhibit the same temporal-specific pattern as AEA, indicating that the corticolimbic AEA fluctuations are at least partly due to corresponding changes in FAAH activity (Lee et al. 2013).

Concluding the eCB system during adolescence shows a high instability and participates actively to the final brain's refinement that occur in this period. External stimulation of this system can profoundly alter its physiological role triggering altered brain maturation.

2 Behavioral Consequences of Cannabinoid Abuse in Adolescence

2.1 *Emotional Aspects*

Chronic administration of cannabinoid compounds in adolescent animals has been reported to lead to altered emotional reactivity later in life.

The picture emerging from studies evaluating the effect on anxiety behavior is quite complex, since different results have been obtained depending on the specific component of anxiety under study and the tests used to monitor it. The most difficult interpretation regards data on general anxiety, monitored in the elevated plus maze test or the open field test. Indeed, adult animals that were exposed to cannabinoids during adolescence have been reported to exhibit no changes in their behavior (Rubino et al. 2008; Higuera-Matas et al. 2009; Bambico et al. 2010; O'Tuathaigh et al. 2010; Mateos et al. 2011; Bortolato et al. 2014; Cadoni et al. 2015), an anxiolytic-like response (Biscaia et al. 2003; Wegener and Koch 2009; Cadoni et al. 2015), or even anxiety (Llorente-Berzal et al. 2013; Stopponi et al. 2014; O'Tuathaigh et al. 2010; Abboussi et al. 2015; but see also Renard et al. 2016 who used the light–dark box test). This lack of consistency might be due to differences in several experimental parameters such as the cannabinoid compound used (synthetic versus natural), the precise developmental period of exposure, and the strain of animals used. For example, it is well known that synthetic cannabinoids are full CB1 receptor agonists, whereas the natural ingredient of *Cannabis*, delta-9 tetrahydrocannabinol (THC), is a partial agonist. This could greatly affect the adaptive changes occurring in the brain in response to chronic exposure. Also the strain of animals used in the experimentation could play a fundamental role in the response to cannabinoid treatment, due to a possible different genetic background. This is the case, for example, in Cadoni's study (Cadoni et al. 2015), where it was reported an anxiolytic-like effect in adult Lewis rats exposed to THC during adolescence and no changes in anxiety behavior in the Fisher344 ones. In the same line, O'Tuathaigh et al. (2010) reported that only mice KO for the catechol-O-methyltransferase (COMT) gene were able to develop an anxiety-like behavior at adulthood after adolescent exposure to THC, while wild-type ones did not.

These different experimental parameters seem to play a minor role when social anxiety has been investigated. All the researchers who monitored this aspect reported a decrease in social interaction in adult rats after adolescent cannabinoid exposure. This was demonstrated with both synthetic and natural cannabinoid agonists, in male and female animals (Leweke and Schneider 2011; O'Shea et al. 2004, 2006; Quinn et al. 2008; Realini et al. 2011; Zamberletti et al. 2014; Renard et al. 2016). The only exception is the Gleason's paper (2012) performed in mice, even if a trend to a decreased social interaction might be seen after the adolescent treatment. The most likely reason for the lack of the effect in Gleason's work may reside in the fact that the behavior was monitored long after the end of the treatment (nearly 3 months later), whereas in all the other studies it was checked within

2 months. This suggests that at least some of the altered behaviors induced by the adolescent cannabinoid exposure might be normalized with time. Although reduction in social interaction has been considered an anxiogenic behavior in rodents (File and Hyde 1978), it is also well known that a significant and pervasive impairment in social functioning is associated with major depressive disorder (Hirschfeld et al. 2000). According to this view, impaired social behavior might also be a sign present in a depressive-like phenotype. Besides reduced social behavior, two other features of depression-like reactivity in animals are behavioral despair/passive coping strategy (measured in the forced swim test) and anhedonia (assessed as sucrose preference or palatable food consumption). Accumulating literature provides evidence that adolescent exposure to THC or WIN 55,212-2 induces passive coping strategy in the forced swim test and a reduction in the sucrose preference or palatable food consumption (Rubino et al. 2008; Realini et al. 2011; Bambico et al. 2010; Zamberletti et al. 2014). This suggests the presence of dysfunction in motivational processes, and, intriguingly, these effects appear to be stronger in female rats (Rubino et al. 2008). In line with this, Chadwick et al. (2011) reported that adolescent exposure to CP-55,940 decreased sexual motivation in adult female rats.

As a whole, these findings suggest the presence of altered emotional reactivity and hedonic processes after adolescent cannabinoid exposure, especially in female animals.

2.2 Cognition

Most literature on adolescent cannabinoid exposure in animal models agrees on the presence of long-lasting impairments in learning and memory.

Impairments in episodic memory tested in the classical or spatial version of the novel object recognition test or in the social recognition test have been reported after adolescent exposure to synthetic or natural cannabinoid agonists in adult rats and mice and in both sexes (Abush and Akirav 2012; O'Shea et al. 2004, 2006; Quinn et al. 2008; Realini et al. 2011; Renard et al. 2013, 2016; Schneider and Koch 2003; Zamberletti et al. 2014; Abboussi et al. 2015; Lovelace et al. 2015). Few exceptions to this rule are represented by papers where adolescent exposure was performed for very short periods of time (3 days, Cadoni et al. 2015) or where the cognitive effect was monitored much longer than in other studies after the end of the treatment (Higuera-Matas et al. 2009). The former exception might suggest that a longer period of exposure is needed to elicit the cognitive deficit, whereas the latter suggests, again, the possibility that some signs might spontaneously return to control value with time. Spatial working memory deficits have been observed after adolescent THC exposure in adult rats (Rubino et al. 2009a, b, 2015), mice (O'Tuathaigh et al. 2010), and monkeys (Verrico et al. 2014), as well as impairments in cognitive flexibility assessed through the attentional set-shifting task (Gomes et al. 2015). When other forms of memory were considered, no lasting effects were observed. This was the case for aversive memory (Rubino et al. 2009a, b) or "pure" spatial learning in the Morris water maze (Abush and Akirav 2012; Cha et al. 2006, 2007; Higuera-Matas et al. 2009).

These data point toward a task-specific effect of adolescent cannabinoid exposure on cognition, suggesting that the exposure might more likely affect the forms of memory where both PFC and hippocampus play a role and the integrity of their interaction is needed to perform the task.

2.3 Psychosis

Epidemiological studies suggest that cannabis use during adolescence confers an increased risk for developing psychotic symptoms later in life (D'Souza et al. 2009; Evins et al. 2012).

Schizophrenia represents a complex disorder associated with various disturbances in motivational, social, emotional, and cognitive processing. Some of these features are uniquely human (i.e., hallucination and delusions) and cannot be reproduced in animals. Differently, several experimental protocols can reproduce impairment in cognition and altered emotionality present in schizophrenia, and the consequences of cannabis in adolescence on these signs has been already examined in the previous chapters. Thus, in the present chapter, only the effect of adolescent cannabis use on two other paradigms considered as valid translational models of schizophrenia will be reported: the pre-pulse inhibition of startle (PPI—a measure of sensorimotor gating, the ability of an organism to attain information and process it correctly) and drug-induced hyperactivity (Lodge and Grace 2009; Powell and Miyakawa 2006).

2.3.1 Pre-pulse Inhibition

Chronic adolescent treatment with the cannabinoid agonist WIN 55,212-2 provoked impairment in PPI in rats and mice that was still present long after the treatment withdrawal (Gleason et al. 2012; Schneider and Koch 2003; Wegener and Koch 2009). In contrast, Llorente-Berzal et al. (2011) and O'Tuathaigh et al. (2012) did not show alterations in this behavior after adolescent exposure to synthetic cannabinoid agonists (CP55940, 0.4 mg/kg, or WIN 55,212-2, 1 or 2.5 mg/kg, respectively). The reason for this discrepancy is quite difficult to find. Data for both evidence were obtained in mice and rats, using a synthetic cannabinoid agonist. The only speculation we may put forward is that the last two groups performed a longer treatment (15 or 21 days) with the same dose of agonist (thus triggering a deep state of tolerance), while the former ones performed a shorter treatment (10 days) or used an irregular protocol of injections (none, one, or two daily injections for 25 days), thus reducing the risk to develop profound tolerance. More recently, Renard et al. (2016) showed that adult rats pretreated with THC in adolescence displayed PPI deficits at both the pre-pulse intensity levels of 76 and 80 dB compared with control rats. Interestingly the same protocols in adult rats did not affect PPI.

2.3.2 Locomotor Hyperactivity

Finally, locomotor hyperactivity in rodents may have some face validity for certain components of the positive symptoms of schizophrenia, such as psychotic agitation (Van Den Buuse 2010). However, despite its ease of quantification in rodents, very few data are available about the long-term effect of adolescent cannabinoid exposure on this behavioral sign. A significant increase in locomotor activity in the open field was observed in adult rats exposed to WIN 55,212-2 during adolescence (Wegener and Koch 2009). Differently, other groups showed no significant alterations in the open field recordings (Biscaia et al. 2003; Rubino et al. 2008), and some others found that young adult rats chronically treated with CP55,940 showed reduced baseline locomotor activity 2 weeks after the treatment's cessation (Klug and Van Den Buuse 2013). Our group recently observed that adolescent exposure to THC increased the locomotor activating effect of phencyclidine (PCP) when this was administered acutely in adulthood (Zamberletti et al. 2014). In fact, the low dose of PCP injected (2.5 mg/kg) did not induced locomotor activation in vehicle-treated animals but significantly increased locomotion in THC-treated rats. Moreover, stereotyped behaviors were observed in vehicle-treated animals following acute PCP injection, but this effect was significantly enhanced in THC-treated rats.

2.3.3 Two-Hit Hypothesis

Schizophrenia is a multifactorial disease characterized by the combined action of multiple genes of small effect size (Owen et al. 2005) and a number of environmental risk factors (McGrath et al. 2003), which causes the development of this mental disorder (Mackay-Sim et al. 2004). This is conceptualized in the "two-hit hypothesis" of schizophrenia, which predicts that genetic and environmental risk factors interactively (G $\times$ E interaction) cause the development of the disorder (Bayer et al. 1999; Caspi and Moffitt 2006).

Genetic factors may confer vulnerability to psychosis outcomes following exposure to cannabis, i.e., a gene-environment interaction. In specific, the genes encoding for COMT, neuregulin (Nrg1), brain-derived neurotrophic factor (BDNF), and disrupted in schizophrenia 1 (DISC1) have been implicated in conferring such vulnerability.

COMT

In one of the first studies that drew attention to gene $\times$ environment interactions, Caspi et al. (2005) reported that the COMT gene moderated the risk of psychotic disorder with adolescent cannabis exposure. The enzyme COMT plays a critical role in the breakdown of dopamine (DA) in the PFC (Papaleo et al. 2008), in contrast to the striatum where DA is cleared by a transporter. The COMT gene has a common polymorphism in humans where valine (Val) is substituted for methionine at the158/108 locus, and this results in 40% higher enzymatic activity and thus more rapid degradation of DA. Lower cortical DA levels in individuals homozygous for the Val (158) polymorphism are associated with, among other things, poorer cognitive performance and inefficient pre-cortical functioning (Tunbridge et al. 2006). Accordingly, COMT knockout (KO) mice were more vulnerable than wild types (WT) to the disruptive effects of WIN55212 adolescent treatment on PPI. Moreover, acute pharmacological inhibition of COMT in mice modified acute cannabinoid effects on startle reactivity, as well as PPI. COMT KO mice also demonstrated differential effects of adolescent cannabinoid administration on sociability and anxiety-related behavior, confirming and extending earlier reports of COMT×cannabinoid effects on the expression of schizophrenia-related endophenotypes (O'Tuathaigh et al. 2012).

Neuregulin

Nrg1, a leading schizophrenia susceptibility gene, is relevant to several schizophrenia-related neurodevelopmental processes due to its involvement in axonal guidance, myelination, and GABAergic and glutamatergic neurotransmission (Mei and Xiong 2008). Heterozygous deletion of Nrg1 results in increased sensitivity of mice to schizophrenia-like symptoms induced by THC, especially under stressful conditions (Boucher et al. 2007). Long's lab team exposed adolescent WT and Nrg1 heterozygous (HET) mice to chronic THC during adolescence (Long et al. 2013). Surprisingly, Nrg1 mutants appeared less susceptible to THC-induced suppression of investigative social behaviors than control mice. However, adolescent THC exacerbated the hyperlocomotive phenotype characteristic for adult Nrg1 mutant mice (Karl et al. 2007; Long et al. 2013). Nrg1 deficiency also modulated the effects of adolescent THC on neurotransmitter systems involved in the pathophysiology of schizophrenia. Genotype-specific THC effects on CB1 expression in the substantia nigra were found: reduced CB1 level was present in Nrg1 HET drug-free mice, but its level increased in Nrg1 mice post THC challenge. Lower CB1 expression levels in the substantia nigra might be responsible for the observed decreased susceptibility of adolescent Nrg1 mutant mice. Nrg1 also conferred opposing effects of THC on serotonin 2A receptor expression in the insular cortex, and NMDA receptor binding was selectively increased in the hippocampus and cingulate cortex of Nrg1 HET mice (Long et al. 2013) (for a better mechanistic understanding of Nrg1-THC interaction on NMDA receptor expression in the hippocampus, see Spencer et al. 2013).

Concluding Nrg1 modulated the behavioral sensitivity of mice to cannabinoids during adolescence providing evidence for a role of Nrg1-cannabis interactions in schizophrenia.

BDNF

Reduced BDNF signaling has been shown in the frontal cortex and hippocampus in schizophrenia. Young BDNF HET mice and wild-type controls were chronically treated with the cannabinoid receptor agonist, CP55,940. Two weeks later, baseline PPI was lower but average startle was increased in BDNF HET compared to wild-type controls. Acute CP55,940 administration before the PPI session increased PPI only in male HET mice. In females, only small increases of PPI in all groups upon acute CP55,940 administration were observed. Acute CP55,940 administration furthermore reduced startle, and this effect was greater in HET mice irrespective of chronic CP55,940 pretreatment. Moreover, male "two-hit" mice, but not females, were hypersensitive to the effect of acute CP55,940 on sensorimotor gating. These effects may be related to a selective upregulation of CB1 receptor density in the nucleus accumbens (Klug and Van Den Buuse 2013).

DISC1

Recent research evaluated the interaction between cannabis in adolescence and another genetic risk factor, DISC1 (Brandon and Sawa 2011; Kamiya et al. 2012). Ballinger et al. (2015) demonstrated that a perturbation in DISC1 exacerbates the response to adolescent THC exposure in adult mice, such as deficits in fear-associated memory. Moreover, downregulation of the expression of CB1Rs in the PFC, hippocampus, and amygdala was induced by either expression of dominant-negative mutant of DISC1 (DN-DISC1) or adolescent THC treatment. A synergistic reduction of c-Fos expression induced by cue-dependent fear memory retrieval in DN-DISC1 with adolescent THC exposure was also found. These results suggest that alteration of CB1R-mediated signaling in DN-DISC1 mice may underlie susceptibility to detrimental effects of adolescent cannabis exposure on adult behaviors.

Environment × Environment Interaction

Some evidence has been recently accumulated highlighting the presence of environment × environment interaction in cannabis vulnerability in adolescence. Neonatal rodents subjected to prefrontal cortex lesioning showed impairment in various forms of social behavior and in object recognition memory after WIN 55,212-2 exposure in adolescence (Schneider and Koch 2007). Similarly, exposure to THC in adolescence produced a larger disruption of PPI in rats reared in social isolation (Malone and Taylor 2006), and a greater cognitive impairment was shown after THC in rats preexposed to PCP (Vigano et al. 2009). Interestingly stressful events early in life (maternal deprivation/separation) associated with adolescent exposure to natural or synthetic cannabinoids provoked different behavioral results when compared with non-stressed control animals. Indeed no effect, increased cannabinoid-induced effect, or even decreased cannabinoid-induced effect was observed, depending on the sex of the animals and the considered behavior (Llorente-Berzal et al. 2011; Zamberletti et al. 2012; Klug and Van Den Buuse 2012). Recently, Gomes et al. (2015) showed that WIN 55212-2 administration in adolescence did not exacerbate the behavioral and electrophysiological changes (increased locomotor response to amphetamine administration and increased number of spontaneously active dopamine neurons in the ventral tegmental area) present in the methylazoxymethanol acetate (MAM) developmental disruption model of schizophrenia. WIN 55212-2 treatment attenuated the locomotor response to amphetamine in MAM rats without affecting dopamine neuron activity.

To conclude, the interaction between a previous hit (genetic or environmental) and cannabinoid exposure in adolescence warrants further investigations, and the different outcomes seem to be dependent by the considered gene profile, the nature and the time of the environmental factors, and the sex.

3 The Gateway Hypothesis

The possibility that cannabis use during adolescence increases the vulnerability to drug abuse disorders later in life, the so-called gateway hypothesis, has been heavily debated over the last decades due to contrasting results obtained with epidemiological studies. The definitive causal link between cannabis use and subsequent abuse of other illicit drugs should come from the use of experimental animal models. However, despite the use of these models, no conclusive findings are available, due to the existence of contrasting results even under these controlled conditions.

The most consistent data have been collected in studies considering cannabinoid-opioid interaction. Male, but not female, rats presented significant increases in the acquisition of both morphine and heroin self-administration after adolescent exposure to natural or synthetic cannabinoid agonists (Ellgren et al. 2007; Biscaia et al. 2008). Similarly, chronic cannabinoids in adolescence increased the sensitivity to morphine or heroin conditioned place preference (CPP) (Morel et al. 2009; Cadoni et al. 2015). Conversely, the same authors under different conditions or other authors reported that adolescent THC exposure neither affected heroin CPP (Cadoni et al. 2015) nor heroin self-administration (Stopponi et al. 2014). However Stopponi et al. (2014) found that adolescent THC pretreatment enhanced subsequent vulnerability to relapse to heroin seeking caused by yohimbine administration (as a pharmacological stressor). These discrepancies might be explained by differences in experimental parameters such as the strain of animals under investigation and the different heroin doses used for self-administration.

More contrasting results have been found regarding the interaction with psychostimulant drugs. Increased acquisition of cocaine self-administration was reported in adult female, but not male, rats pretreated with cannabinoids in adolescence (Higuera-Matas et al. 2008). However, the locomotor activating effects of cocaine are enhanced after adolescent THC exposure (Dow-Edwards and Izenwasser 2012). Conversely, pretreatment with natural or synthetic cannabinoid agonists during early adolescence did not alter locomotor responses to amphetamine (Ellgren et al. 2004). Finally, exposure to cannabinoids during adolescence enhanced the acquisition and reinstatement of 3,4-methylenedioxymetamphetamine hydrochloride-induced CPP in mice (Rodriguez-Arias et al. 2010).

Very recently, data on cannabinoid–cannabinoid interaction have been reported (Scherma et al. 2015). These authors demonstrated that adolescent exposure to THC significantly increases WIN 55212-2 self-administration in adulthood. Indeed, THC-exposed rats acquired cannabinoid self-administration more rapidly than controls and showed higher rates of consumption when self-administration behavior reached asymptote.

In conclusion, the picture is still quite unclear regarding the role of Cannabis as a possible gateway drug to subsequent abuse of other illicit compounds. Maybe a more complex theory taking into account also influences of individual (personality traits) and environmental (substance availability, peer influence) characteristics will better describe this complex phenomenon.

4 Cellular Mechanisms Underlying Adolescent Brain Vulnerability

Despite their paucity, the available data seem to support the hypothesis that cannabis in adolescence might exert developmental interference on the eCB system. Accordingly, adolescent THC exposure in male and female rats induced lasting CB1 receptor downregulation and desensitization in different cerebral areas, females being more sensitive than males (Rubino et al. 2008, 2015; Burston et al. 2010). Moreover, in the PFC of THC-exposed female animals, the significant decrease of CB1R binding, still present in adulthood, was paralleled by a significant decrease of AEA levels (Rubino et al. 2015). Mice exposed to WIN 55,212-2 during adolescence exhibited at adulthood increased MGL and FAAH levels, suggesting increases in endocannabinoid degradation (Gleason et al. 2012). Recently, Ortega-Alvaro et al. (2015) showed that CB1R deletion induces a pre-attentional deficit and CB1 KO mice have significantly lower PPI values than WT mice. Moreover, treatment with the CB1R antagonist AM251 produces an impairment of PPI in WT mice, and typical and atypical antipsychotics did not ameliorate PPI deficit in CB1 KO mice. These data emphasize the important role of CB1R in sensorimotor gating modulation and suggest that the long-lasting downregulation of CB1R observed after cannabinoid in adolescence can play a role in the PPI impairment.

Collectively strong cannabis exposure in adolescence can profoundly affect the developmental trajectories of the different components of the eCB system such as receptors (CB1 and CB2) and endocannabinoid levels (AEA), thus provoking its dysregulation and indirectly interfering with the final brain connectivity. These data strengthen the hypothesis that the alteration in the dynamic changes present in the eCB system during adolescence provoked by cannabis could profoundly affect the neurodevelopmental processes in which this system is involved.

For example, working memory and decision-making are refined during adolescence and require the functional maturation of the PFC. Interestingly, the eCB tone seems to play a fundamental role in some maturational processes occurring in the adolescent PFC. Indeed, adolescent THC exposure induced alterations in the maturational fluctuations of NMDA and AMPA subunits in this brain area, leading to larger amounts of gluN2B and gluA1 at adulthood (Rubino et al. 2015). Moreover, due to the fact that NMDA receptors play a critical role in regulating adolescent maturation of GABAergic networks in the PFC (Thomases et al. 2013), adolescent cannabinoid exposure might also affect the GABAergic system. Accordingly, Cass et al. (2014) showed that WIN55,212-2 exposure during early or mid-adolescence, but not later in life, caused a functional downregulation of GABAergic transmission in the PFC. Similarly, Zamberletti et al. (2014) demonstrated that adolescent THC exposure resulted in reduced GAD67 and basal GABA levels in the same brain area. These data suggest that adolescent cannabinoid exposure impacts not only the endocannabinoid system but also the glutamatergic and the GABAergic ones. These three systems are important in shaping cortical oscillations, a neural network activity in the neocortex (Uhlhaas et al. 2009) that is implicated in cognitive and sensory processing (Buzsaki and Draguhn 2004; Wang 2010). Chronic exposure to cannabinoids during adolescence permanently suppresses pharmacologically evoked cortical oscillations (Raver et al. 2013), and this effect seems to be mediated partly by CB1 receptors, but there is also some evidence of the involvement of CB2 and other non-cannabinoid receptors (Raver and Keller 2014). Moreover, the functional downregulation of GABAergic transmission in the PFC induced by adolescent cannabinoid exposure might impact the functionality of PFC excitatory descending projections. According to this hypothesis, increased ventral tegmental area (VTA) DA neuronal firing frequencies and relative proportion of spikes firing in burst have been described in adult rats chronically exposed to THC during adolescence (Renard et al. 2016). Similarly, an increased number of spontaneously active VTA dopaminergic neurons were seen in adult rats treated with WIN55,212-2 during puberty (Gomes et al. 2015). This VTA hyper-DAergic activity impacts molecular and neuronal activity parameters in the VTA DAergic output targets, such as the PFC and NAc. Indeed, an increased dopamine turnover was observed in the ventral striatum of adult rats chronically treated with WIN55,212-2 during adolescence (Bortolato et al. 2014), as well as altered amounts of D1 and D2 receptors in PFC and NAc (Higuera-Matas et al. 2010; Zamberletti et al. 2012). Consistent with this state of hyper-DAergic function in the mesocorticolimbic system, acute administration of the D3 receptor antagonists U-99194A partially restored the cognitive performances in the novel object recognition test disrupted by chronic cannabinoid exposure during adolescence (Abboussi et al. 2015).

Besides the impact on the maturation of different neurotransmitter systems, the endocannabinoid system has been recently suggested to affect the process of synaptic pruning (Rubino et al. 2015). Alterations in this event can lead to structural changes in the adult brain that might play a part in the dysfunctionality triggered by adolescent cannabinoid exposure. In male rats, adolescent exposure to WIN55,212-2 significantly decreased spine density in the nucleus accumbens immediately after treatment (Carvalho et al. 2016). More interestingly, long after the end of treatment, adolescent cannabinoid exposure has been shown to reduce spine density in the dentate gyrus of the hippocampus at adulthood, paralleled by a significant decrease in dendrite length and number (Rubino et al. 2009b) as well as a significant decrease in the basal dendritic arborization of pyramidal neurons in layer II/III of the PFC (Rubino et al. 2015).

As the brain is composed of many different cell types, including, but not limited to, neurons, it is only logical that adolescent cannabinoid exposure could also affect the functionality of other cell types besides neurons. According to this view, adolescent WIN55,212-2 treatment has been shown to induce an increase in the survival of oligodendroglia precursors in the striatum and PFC immediately after treatment (Bortolato et al. 2014). The role of this alteration in cannabinoid-induced effects is still not known, but pave the way to the increasing importance for brain function of the interaction between glial cells and neurons. More recently, Zamberletti et al. (2015) reported that adolescent THC administration induces a persistent neuroinflammatory state specifically localized within the adult PFC, characterized by increased expression of pro-inflammatory markers (TNF-α, iNOS, and COX-2), and reduction of the anti-inflammatory cytokine IL-10. Moreover, this neuroinflammatory phenotype was associated with downregulation of CB1 receptor on neuronal cells and upregulation of CB2 on microglia cells. Interestingly, blocking microglia activation with ibudilast during THC treatment significantly attenuated short-term memory impairment in adulthood. This was paralleled by the prevention of the increases in TNF-α, iNOS, and COX-2 levels as well as of the upregulation of CB2 receptors on microglia cells.

5 Conclusions

In conclusion, this review shows that cannabinoid administration in adolescence can profoundly affect the maturation of the endocannabinoid system, thus triggering important dysregulation in the final remodeling of other neurotransmitter pathways (glutamate, GABA, and dopamine). These dysfunctions represent the neurobiological substrate of the behavioral abnormalities resembling psychiatric diseases that become evident at adulthood. Moreover, recent evidence suggests that besides neuron also glial cells are affected by cannabinoid in adolescence, thus underlying the presence of a more generalized dysfunctional picture involving all the cellular elements of the central nervous system. Finally, no conclusive findings are available for the gateway hypothesis due to the presence of limited and contrasting data mainly focused on opioids and psychostimulants.

References

Abboussi O, Said N, Fifel K, Lakehayli S, Tazi A, El Ganouni S (2015) Behavioral effects of D3 receptor inhibition and 5-HT4 receptor activation on animals undergoing chronic cannabinoid exposure during adolescence. Metab Brain Dis 31(2):321–327

Abush H, Akirav I (2012) Short- and long-term cognitive effects of chronic cannabinoids administration in late-adolescence rats. PLoS One 7:e31731

Ballinger MD, Saito A, Abazyan B, Taniguchi Y, Huang CH, Ito K, Zhu X, Segal H, Jaaro-Peled H, Sawa A, Mackie K, Pletnikov MV, Kamiya A (2015) Adolescent cannabis exposure interacts with mutant DISC1 to produce impaired adult emotional memory. Neurobiol Dis 82:176–184

Bambico FR, Nguyen NT, Katz N, Gobbi G (2010) Chronic exposure to cannabinoids during adolescence but not during adulthood impairs emotional behaviour and monoaminergic neurotransmission. Neurobiol Dis 37:641–655

Bayer TA, Falkai P, Maier W (1999) Genetic and non-genetic vulnerability factors in schizophrenia: the basis of the "two hit hypothesis". J Psychiatr Res 33:543–548

Biscaia M, Marin S, Fernandez B, Marco EM, Rubio M, Guaza C, Ambrosio E, Viveros MP (2003) Chronic treatment with CP 55,940 during the peri-adolescent period differentially affects the behavioural responses of male and female rats in adulthood. Psychopharmacology (Berl) 170:301–308

Biscaia M, Fernandez B, Higuera-Matas A, Miguens M, Viveros MP, Garcia-Lecumberri C, Ambrosio E (2008) Sex-dependent effects of periadolescent exposure to the cannabinoid agonist CP-55,940 on morphine self-administration behaviour and the endogenous opioid system. Neuropharmacology 54:863–873

Bortolato M, Bini V, Frau R, Devoto P, Pardu A, Fan Y, Solbrig MV (2014) Juvenile cannabinoid treatment induces frontostriatal gliogenesis in Lewis rats. Eur Neuropsychopharmacol 24:974–985

Boucher AA, Arnold JC, Duffy L, Schofield PR, Micheau J, Karl T (2007) Heterozygous neuregulin 1 mice are more sensitive to the behavioural effects of delta9-tetrahydrocannabinol. Psychopharmacology (Berl) 192:325–336

Brandon NJ, Sawa A (2011) Linking neurodevelopmental and synaptic theories of mental illness through DISC1. Nat Rev Neurosci 12:707–722

Burston JJ, Wiley JL, Craig AA, Selley DE, Sim-Selley LJ (2010) Regional enhancement of cannabinoid CB₁ receptor desensitization in female adolescent rats following repeated delta-tetrahydrocannabinol exposure. Br J Pharmacol 161:103–112

Buzsaki G, Draguhn A (2004) Neuronal oscillations in cortical networks. Science 304:1926–1929

Cadoni C, Simola N, Espa E, Fenu S, Di Chiara G (2015) Strain dependence of adolescent Cannabis influence on heroin reward and mesolimbic dopamine transmission in adult Lewis and Fischer 344 rats. Addict Biol 20:132–142

Carvalho AF, Reyes BA, Ramalhosa F, Sousa N, Van Bockstaele EJ (2016) Repeated administration of a synthetic cannabinoid receptor agonist differentially affects cortical and accumbal neuronal morphology in adolescent and adult rats. Brain Struct Funct 221:407–419

Caspi A, Moffitt TE (2006) Gene-environment interactions in psychiatry: joining forces with neuroscience. Nat Rev Neurosci 7:583–590

Caspi A, Moffitt TE, Cannon M, McClay J, Murray R, Harrington H, Taylor A, Arseneault L, Williams B, Braithwaite A, Poulton R, Craig IW (2005) Moderation of the effect of adolescent-onset cannabis use on adult psychosis by a functional polymorphism in the catechol-O-

methyltransferase gene: longitudinal evidence of a gene X environment interaction. Biol Psychiatry 57:1117–1127
Cass DK, Flores-Barrera E, Thomases DR, Vital WF, Caballero A, Tseng KY (2014) CB1 cannabinoid receptor stimulation during adolescence impairs the maturation of GABA function in the adult rat prefrontal cortex. Mol Psychiatry 19:536–543
Cha YM, White AM, Kuhn CM, Wilson WA, Swartzwelder HS (2006) Differential effects of delta9-THC on learning in adolescent and adult rats. Pharmacol Biochem Behav 83:448–455
Cha YM, Jones KH, Kuhn CM, Wilson WA, Swartzwelder HS (2007) Sex differences in the effects of delta9-tetrahydrocannabinol on spatial learning in adolescent and adult rats. Behav Pharmacol 18:563–569
Chadwick B, Saylor AJ, Lopez HH (2011) Adolescent cannabinoid exposure attenuates adult female sexual motivation but does not alter adulthood CB1R expression or estrous cyclicity. Pharmacol Biochem Behav 100:157–164
Dow-Edwards D, Izenwasser S (2012) Pretreatment with delta9-tetrahydrocannabinol (THC) increases cocaine-stimulated activity in adolescent but not adult male rats. Pharmacol Biochem Behav 100:587–591
D'Souza DC, Sewell RA, Ranganathan M (2009) Cannabis and psychosis/schizophrenia: human studies. Eur Arch Psychiatry Clin Neurosci 259:413–431
Ellgren M, Hurd YL, Franck J (2004) Amphetamine effects on dopamine levels and behavior following cannabinoid exposure during adolescence. Eur J Pharmacol 497:205–213
Ellgren M, Spano SM, Hurd YL (2007) Adolescent cannabis exposure alters opiate intake and opioid limbic neuronal populations in adult rats. Neuropsychopharmacology 32:607–615
Ellgren M, Artmann A, Tkalych O, Gupta A, Hansen HS, Hansen SH, Devi LA, Hurd YL (2008) Dynamic changes of the endogenous cannabinoid and opioid mesocorticolimbic systems during adolescence: THC effects. Eur Neuropsychopharmacol 18:826–834
Evins AE, Green AI, Kane JM, Murray RM (2012) The effect of marijuana use on the risk for schizophrenia. J Clin Psychiatry 73:1463–1468
File SE, Hyde JR (1978) Can social interaction be used to measure anxiety? Br J Pharmacol 62:19–24
Gleason KA, Birnbaum SG, Shukla A, Ghose S (2012) Susceptibility of the adolescent brain to cannabinoids: long-term hippocampal effects and relevance to schizophrenia. Transl Psychiatry 2:e199
Gomes FV, Guimaraes FS, Grace AA (2015) Effects of pubertal cannabinoid administration on attentional set-shifting and dopaminergic hyper-responsivity in a developmental disruption model of schizophrenia. Int J Neuropsychopharmacol 18(2):1–10
Heng L, Beverley JA, Steiner H, Tseng KY (2011) Differential developmental trajectories for CB1 cannabinoid receptor expression in limbic/associative and sensorimotor cortical areas. Synapse 65:278–286
Higuera-Matas A, Soto-Montenegro ML, Del Olmo N, Miguens M, Torres I, Vaquero JJ, Sanchez J, Garcia-Lecumberri C, Desco M, Ambrosio E (2008) Augmented acquisition of cocaine self-administration and altered brain glucose metabolism in adult female but not male rats exposed to a cannabinoid agonist during adolescence. Neuropsychopharmacology 33:806–813
Higuera-Matas A, Botreau F, Miguens M, Del Olmo N, Borcel E, Perez-Alvarez L, Garcia-Lecumberri C, Ambrosio E (2009) Chronic periadolescent cannabinoid treatment enhances adult hippocampal PSA-NCAM expression in male Wistar rats but only has marginal effects on anxiety, learning and memory. Pharmacol Biochem Behav 93:482–490
Higuera-Matas A, Botreau F, Del Olmo N, Miguéns M, Olías O, Montoya GL, García-Lecumberri C, Ambrosio E (2010) Periadolescent exposure to cannabinoids alters the striatal and hippocampal dopaminergic system in the adult rat brain. Eur Neuropsychopharmacol 20:895–906

Hirschfeld RM, Montgomery SA, Keller MB, Kasper S, Schatzberg AF, Moller HJ, Healy D, Baldwin D, Humble M, Versiani M, Montenegro R, Bourgeois M (2000) Social functioning in depression: a review. J Clin Psychiatry 61:268–275

Kamiya A, Sedlak TW, Pletnikov MV (2012) DISC1 pathway in brain development: exploring therapeutic targets for major psychiatric disorders. Front Psych 3:25

Karl T, Duffy L, Scimone A, Harvey RP, Schofield PR (2007) Altered motor activity, exploration and anxiety in heterozygous neuregulin 1 mutant mice: implications for understanding schizophrenia. Genes Brain Behav 6:677–687

Klug M, Van Den Buuse M (2012) Chronic cannabinoid treatment during young adulthood induces sex-specific behavioural deficits in maternally separated rats. Behav Brain Res 233:305–313

Klug M, Van Den Buuse M (2013) An investigation into "two hit" effects of BDNF deficiency and young-adult cannabinoid receptor stimulation on prepulse inhibition regulation and memory in mice. Front Behav Neurosci 7:149

Lee TT, Hill MN, Hillard CJ, Gorzalka BB (2013) Temporal changes in N-acylethanolamine content and metabolism throughout the peri-adolescent period. Synapse 67:4–10

Leweke FM, Schneider M (2011) Chronic pubertal cannabinoid treatment as a behavioural model for aspects of schizophrenia: effects of the atypical antipsychotic quetiapine. Int J Neuropsychopharmacol 14:43–51

Llorente-Berzal A, Fuentes S, Gagliano H, Lopez-Gallardo M, Armario A, Viveros MP, Nadal R (2011) Sex-dependent effects of maternal deprivation and adolescent cannabinoid treatment on adult rat behaviour. Addict Biol 16:624–637

Llorente-Berzal A, Puighermanal E, Burokas A, Ozaita A, Maldonado R, Marco EM, Viveros MP (2013) Sex-dependent psychoneuroendocrine effects of THC and MDMA in an animal model of adolescent drug consumption. PLoS One 8:e78386

Lodge DJ, Grace AA (2009) Gestational methylazoxymethanol acetate administration: a developmental disruption model of schizophrenia. Behav Brain Res 204:306–312

Long LE, Chesworth R, Huang XF, McGregor IS, Arnold JC, Karl T (2013) Transmembrane domain Nrg1 mutant mice show altered susceptibility to the neurobehavioural actions of repeated THC exposure in adolescence. Int J Neuropsychopharmacol 16:163–175

Lovelace JW, Corches A, Vieira PA, Hiroto AS, Mackie K, Korzus E (2015) An animal model of female adolescent cannabinoid exposure elicits a long-lasting deficit in presynaptic long-term plasticity. Neuropharmacology 99:242–255

Mackay-Sim A, Feron F, Eyles D, Burne T, McGrath J (2004) Schizophrenia, vitamin D, and brain development. Int Rev Neurobiol 59:351–380

Malone DT, Taylor DA (2006) The effect of delta9-tetrahydrocannabinol on sensorimotor gating in socially isolated rats. Behav Brain Res 166:101–109

Mateos B, Borcel E, Loriga R, Luesu W, Bini V, Llorente R, Castelli MP, Viveros MP (2011) Adolescent exposure to nicotine and/or the cannabinoid agonist CP 55,940 induces gender-dependent long-lasting memory impairments and changes in brain nicotinic and CB(1) cannabinoid receptors. J Psychopharmacol 25:1676–1690

McGrath JJ, Feron FP, Burne TH, Mackay-Sim A, Eyles DW (2003) The neurodevelopmental hypothesis of schizophrenia: a review of recent developments. Ann Med 35:86–93

Mei L, Xiong WC (2008) Neuregulin 1 in neural development, synaptic plasticity and schizophrenia. Nat Rev Neurosci 9:437–452

Morel LJ, Giros B, Dauge V (2009) Adolescent exposure to chronic delta-9-tetrahydrocannabinol blocks opiate dependence in maternally deprived rats. Neuropsychopharmacology 34:2469–2476

Ortega-Alvaro A, Navarrete F, Aracil-Fernandez A, Navarro D, Berbel P, Manzanares J (2015) Differential pharmacological regulation of sensorimotor gating deficit in CB1 knockout mice and associated neurochemical and histological alterations. Neuropsychopharmacology 40:2639–2647

O'Shea M, Singh ME, McGregor IS, Mallet PE (2004) Chronic cannabinoid exposure produces lasting memory impairment and increased anxiety in adolescent but not adult rats. J Psychopharmacol 18:502–508

O'Shea M, McGregor IS, Mallet PE (2006) Repeated cannabinoid exposure during perinatal, adolescent or early adult ages produces similar longlasting deficits in object recognition and reduced social interaction in rats. J Psychopharmacol 20:611–621

O'Tuathaigh CM, Hryniewiecka M, Behan A, Tighe O, Coughlan C, Desbonnet L, Cannon M, Karayiorgou M, Gogos JA, Cotter DR, Waddington JL (2010) Chronic adolescent exposure to delta-9-tetrahydrocannabinol in COMT mutant mice: impact on psychosis-related and other phenotypes. Neuropsychopharmacology 35:2262–2273

O'Tuathaigh CM, Clarke G, Walsh J, Desbonnet L, Petit E, O'Leary C, Tighe O, Clarke N, Karayiorgou M, Gogos JA, Dinan TG, Cryan JF, Waddington JL (2012) Genetic vs. pharmacological inactivation of COMT influences cannabinoid-induced expression of schizophrenia-related phenotypes. Int J Neuropsychopharmacol 15:1331–1342

Owen MJ, Craddock N, O'Donovan MC (2005) Schizophrenia: genes at last? Trends Genet 21:518–525

Papaleo F, Crawley JN, Song J, Lipska BK, Pickel J, Weinberger DR, Chen J (2008) Genetic dissection of the role of catechol-O-methyltransferase in cognition and stress reactivity in mice. J Neurosci 28:8709–8723

Powell CM, Miyakawa T (2006) Schizophrenia-relevant behavioral testing in rodent models: a uniquely human disorder? Biol Psychiatry 59:1198–1207

Quinn HR, Matsumoto I, Callaghan PD, Long LE, Arnold JC, Gunasekaran N, Thompson MR, Dawson B, Mallet PE, Kashem MA, Matsuda-Matsumoto H, Iwazaki T, Mcgregor IS (2008) Adolescent rats find repeated delta(9)-THC less aversive than adult rats but display greater residual cognitive deficits and changes in hippocampal protein expression following exposure. Neuropsychopharmacology 33:1113–1126

Raver SM, Keller A (2014) Permanent suppression of cortical oscillations in mice after adolescent exposure to cannabinoids: receptor mechanisms. Neuropharmacology 86:161–173

Raver SM, Haughwout SP, Keller A (2013) Adolescent cannabinoid exposure permanently suppresses cortical oscillations in adult mice. Neuropsychopharmacology 38:2338–2347

Realini N, Vigano D, Guidali C, Zamberletti E, Rubino T, Parolaro D (2011) Chronic URB597 treatment at adulthood reverted most depressive-like symptoms induced by adolescent exposure to THC in female rats. Neuropharmacology 60:235–243

Renard J, Krebs MO, Jay TM, Le Pen G (2013) Long-term cognitive impairments induced by chronic cannabinoid exposure during adolescence in rats: a strain comparison. Psychopharmacology (Berl) 225:781–790

Renard J, Krebs MO, Le Pen G, Jay TM (2014) Long-term consequences of adolescent cannabinoid exposure in adult psychopathology. Front Neurosci 8:361

Renard J, Vitalis T, Rame M, Krebs MO, Lenkei Z, Le Pen G, Jay TM (2016) Chronic cannabinoid exposure during adolescence leads to long-term structural and functional changes in the prefrontal cortex. Eur Neuropsychopharmacol 26:55–64

Rodriguez De Fonseca F, Ramos JA, Bonnin A, Fernandez-Ruiz JJ (1993) Presence of cannabinoid binding sites in the brain from early postnatal ages. Neuroreport 4:135–138

Rodriguez-Arias M, Manzanedo C, Roger-Sanchez C, Do Couto BR, Aguilar MA, Minarro J (2010) Effect of adolescent exposure to WIN 55212-2 on the acquisition and reinstatement of MDMA-induced conditioned place preference. Prog Neuropsychopharmacol Biol Psychiatry 34:166–171

Rubino T, Parolaro D (2015) The impact of exposure to cannabinoids in adolescence: insights from animal models. Biol Psychiatry 79(7):578–585

Rubino T, Vigano D, Realini N, Guidali C, Braida D, Capurro V, Castiglioni C, Cherubino F, Romualdi P, Candeletti S, Sala M, Parolaro D (2008) Chronic delta 9-tetrahydrocannabinol during adolescence provokes sex-dependent changes in the emotional profile in adult rats: behavioral and biochemical correlates. Neuropsychopharmacology 33:2760–2771

Rubino T, Realini N, Braida D, Guidi S, Capurro V, Vigano D, Guidali C, Pinter M, Sala M, Bartesaghi R, Parolaro D (2009a) Changes in hippocampal morphology and neuroplasticity induced by adolescent THC treatment are associated with cognitive impairment in adulthood. Hippocampus 19:763–772

Rubino T, Realini N, Braida D, Alberio T, Capurro V, Vigano D, Guidali C, Sala M, Fasano M, Parolaro D (2009b) The depressive phenotype induced in adult female rats by adolescent exposure to THC is associated with cognitive impairment and altered neuroplasticity in the prefrontal cortex. Neurotox Res 15:291–302

Rubino T, Prini P, Piscitelli F, Zamberletti E, Trusel M, Melis M, Sagheddu C, Ligresti A, Tonini R, Di Marzo V, Parolaro D (2015) Adolescent exposure to THC in female rats disrupts developmental changes in the prefrontal cortex. Neurobiol Dis 73:60–69

Scherma M, Dessi C, Muntoni AL, Lecca S, Satta V, Luchicchi A, Pistis M, Panlilio LV, Fattore L, Goldberg SR, Fratta W, Fadda P (2015) Adolescent delta-tetrahydrocannabinol exposure alters WIN55,212-2 self-administration in adult rats. Neuropsychopharmacology 41(5):1416–1426

Schneider M, Koch M (2003) Chronic pubertal, but not adult chronic cannabinoid treatment impairs sensorimotor gating, recognition memory, and the performance in a progressive ratio task in adult rats. Neuropsychopharmacology 28:1760–1769

Schneider M, Koch M (2007) The effect of chronic peripubertal cannabinoid treatment on deficient object recognition memory in rats after neonatal mPFC lesion. Eur Neuropsychopharmacol 17:180–186

Schneider M, Kasanetz F, Lynch DL, Friemel CM, Lassalle O, Hurst DP, Steindel F, Monory K, Schafer C, Miederer I, Leweke FM, Schreckenberger M, Lutz B, Reggio PH, Manzoni OJ, Spanagel R (2015) Enhanced functional activity of the cannabinoid type-1 receptor mediates adolescent behavior. J Neurosci 35:13975–13988

Spencer JR, Darbyshire KM, Boucher AA, Kashem MA, Long LE, McGregor IS, Karl T, Arnold JC (2013) Novel molecular changes induced by Nrg1 hypomorphism and Nrg1-cannabinoid interaction in adolescence: a hippocampal proteomic study in mice. Front Cell Neurosci 7:15

Stopponi S, Soverchia L, Ubaldi M, Cippitelli A, Serpelloni G, Ciccocioppo R (2014) Chronic THC during adolescence increases the vulnerability to stress-induced relapse to heroin seeking in adult rats. Eur Neuropsychopharmacol 24:1037–1045

Thomases DR, Cass DK, Tseng KY (2013) Periadolescent exposure to the NMDA receptor antagonist MK-801 impairs the functional maturation of local GABAergic circuits in the adult prefrontal cortex. J Neurosci 33:26–34

Tunbridge EM, Harrison PJ, Weinberger DR (2006) Catechol-o-methyltransferase, cognition, and psychosis: Val158Met and beyond. Biol Psychiatry 60:141–151

Uhlhaas PJ, Roux F, Singer W, Haenschel C, Sireteanu R, Rodriguez E (2009) The development of neural synchrony reflects late maturation and restructuring of functional networks in humans. Proc Natl Acad Sci USA 106:9866–9871

Van Den Buuse M (2010) Modeling the positive symptoms of schizophrenia in genetically modified mice: pharmacology and methodology aspects. Schizophr Bull 36:246–270

Verrico CD, Gu H, Peterson ML, Sampson AR, Lewis DA (2014) Repeated delta9-tetrahydrocannabinol exposure in adolescent monkeys: persistent effects selective for spatial working memory. Am J Psychiatry 171:416–425

Vigano D, Guidali C, Petrosino S, Realini N, Rubino T, Di Marzo V, Parolaro D (2009) Involvement of the endocannabinoid system in phencyclidine-induced cognitive deficits modelling schizophrenia. Int J Neuropsychopharmacol 12:599–614

Wang XJ (2010) Neurophysiological and computational principles of cortical rhythms in cognition. Physiol Rev 90:1195–1268

Wegener N, Koch M (2009) Behavioural disturbances and altered Fos protein expression in adult rats after chronic pubertal cannabinoid treatment. Brain Res 1253:81–91

Wenger T, Gerendai I, Fezza F, Gonzalez S, Bisogno T, Fernandez-Ruiz J, Di Marzo V (2002) The hypothalamic levels of the endocannabinoid, anandamide, peak immediately before the onset of puberty in female rats. Life Sci 70:1407–1414

Zamberletti E, Prini P, Speziali S, Gabaglio M, Solinas M, Parolaro D, Rubino T (2012) Gender-dependent behavioral and biochemical effects of adolescent delta-9-tetrahydrocannabinol in adult maternally deprived rats. Neuroscience 204:245–257

Zamberletti E, Beggiato S, Steardo L Jr, Prini P, Antonelli T, Ferraro L, Rubino T, Parolaro D (2014) Alterations of prefrontal cortex GABAergic transmission in the complex psychotic-like phenotype induced by adolescent delta-9-tetrahydrocannabinol exposure in rats. Neurobiol Dis 63:35–47

Zamberletti E, Gabaglio M, Prini P, Rubino T, Parolaro D (2015) Cortical neuroinflammation contributes to long-term cognitive dysfunctions following adolescent delta-9-tetrahydrocannabinol treatment in female rats. Eur Neuropsychopharmacol 25:2404–2415

Endocannabinoid Signaling in Reward and Addiction: From Homeostasis to Pathology

Sarah A. Laredo, William R. Marrs, and Loren H. Parsons

Abstract The endogenous cannabinoid system is an important regulatory system involved in physiological homeostasis. Endocannabinoid signaling is known to modulate neural development, immune function, metabolism, synaptic plasticity, and emotional state. Accumulating evidence also implicates brain endocannabinoid signaling in the processing of natural and drug-induced reward states and dysregulated endocannabinoid signaling in the etiology of aberrant reward function and drug addiction. In this chapter, we discuss the influence of endocannabinoid signaling on the rewarding and motivational effects of natural rewards such as food, sex, and social interaction, as well as evidence demonstrating an endocannabinoid influence in the rewarding effects of abused drugs. The effects of long-term drug consumption on endocannabinoid signaling are discussed, along with evidence that the resultant dysregulation of endocannabinoid function contributes to various aspects of drug dependence and addiction including physical symptoms of drug withdrawal, increased stress responsivity, negative affective states, dysregulated synaptic plasticity, dysregulated extinction of drug-related memories, relapse to drug taking, and impaired cognitive function. Lastly, consideration is given to the role for dysregulated endocannabinoid signaling in pathological food reward and eating disorders.

Present knowledge of the endogenous cannabinoid system (ECS) derives from decades of research demonstrating that the effects of *Cannabis* are mediated by cannabinoid receptors in the brain and the subsequent identification and characterization of the endogenous ligands for these receptors. Although our understanding of the nature and breadth of endocannabinoid (eCB) influences is constantly expanding, it is now acknowledged that the brain ECS plays a prominent role in modulating brain reward function and the maintenance of emotional homeostasis. This chapter examines the evidence for an eCB influence in the positive reinforcing effects of natural rewards and drugs of abuse, as well as dysregulation of ECS function that may influence or contribute to aberrant synaptic plasticity, negative

S.A. Laredo (✉) • W.R. Marrs • L.H. Parsons
Committee on the Neurobiology of Addictive Disorders, The Scripps Research Institute, 10550 N. Torrey Pines Road, La Jolla, CA, USA
e-mail: salaredo@gmail.com; lparsons@scripps.edu

M. Melis (ed.), *Endocannabinoids and Lipid Mediators in Brain Functions*,
DOI 10.1007/978-3-319-57371-7_10

emotional states, and impaired learning and memory processes resulting from long-term drug exposure that sustains compulsive drug consumption characteristic of the addicted state.

1 Neurobiology of Reward

The brain reward system plays an important role in our survival as a species. In addition to homeostatic factors that drive behavior, the hedonic or pleasurable effects produced by eating, exercise, and sexual activity provide important motivational effects that increase the likelihood of future engagement in these critical activities (known as positive reinforcement). In a similar manner, though less commonly considered, the reward system participates in negative hedonic responses in which aversive or unpleasant events (sickness, bodily harm) increase the likelihood of behaviors that will relieve or avoid these negative states (negative reinforcement). Accordingly, proper reward system function is critical for the maintenance of behaviors that promote survival.

Our understanding of the neurobiological substrates for reward and reinforcement is derived from studies in the 1950s by Olds and Milner who demonstrated that rodents will work to receive electrical stimulation in various brain regions (Olds and Milner 1954). Although stimulation of many brain sites supports operant behavior, the most sensitive sites involve the trajectory of the medial forebrain bundle that connects the ventral tegmental area (VTA) to the basal forebrain, including the nucleus accumbens (NAc) in the ventral striatum. Substantial subsequent evidence has demonstrated a critical involvement of mesocorticolimbic dopamine (DA) signaling in the mediation of reward and reinforcement, suggesting that activation of this system induces general behavioral activation (Le Moal and Simon 1991), promotes goal-directed behavior (Salamone et al. 2007), and increases incentive salience to environmental stimuli (Robinson and Berridge 1993). Functional neuroimaging studies in humans demonstrate that the mesocorticolimbic system is activated by natural rewards including the sight of appetizing food and orgasm (Komisaruk et al. 2004; Volkow et al. 2011), and rodent studies demonstrate that food consumption, sex, and exercise each increase NAc DA signaling (Bassareo and Di Chiara 1999a; Fiorino et al. 1997; Freed and Yamamoto 1985; Hattori et al. 1994). PET studies in humans have shown that psychostimulants, nicotine, alcohol, and marijuana each increase DA in the dorsal and ventral striatum and that this is associated with the subjective rewarding effects of the drugs (Tomasi and Volkow 2013). Similarly, rodent studies demonstrate that all addictive drugs possess the ability to increase mesocorticolimbic DA and that this contributes to the positive reinforcing effects of these substances (particularly psychostimulants and nicotine) (Di Chiara and Bassareo 2007; Koob and Volkow 2010). Substantial evidence suggests that DA-independent signaling in the mesolimbic system also substantially impacts reward processing, including cholinergic, opioid, glutamatergic, and GABAergic signaling (Koob 1992; Nestler 2005).

The NAc receives input from the amygdala, frontal cortex, and hippocampus that may be converted to motivational actions through its connections with the extra-pyramidal motor system. In particular, the central nucleus of the amygdala (CeA) appears to play a role in mediating/modulating the positive reinforcing effects of abused drugs (Koob and Volkow 2010), and the ventral pallidum/substantia innominata, which receives a major innervation from the NAc, is also critically involved in modulating the positive reinforcing effects of both natural and drug rewards (Koob and Volkow 2010).

In general, NAc DA levels are decreased by many aversive conditions such as unavoidable shock, medial hypothalamic stimulation, chronic pain, and certain patterns of over- or undereating (Umberg and Pothos 2011). Numerous studies have also demonstrated deficiencies in NAc DA during withdrawal from addictive drugs (Umberg and Pothos 2011; Koob and Volkow 2010). Further, dysregulation of key neurochemical elements involved in stress responses mediated by the CeA, bed nucleus of the stria terminalis (BNST), frontal cortex, and medial shell of the NAc plays a major role in the negative reinforcing effects of withdrawal from highly palatable food and drugs of abuse (Koob and Volkow 2010; Koob et al. 2014; Blasio et al. 2013; Iemolo et al. 2013). Signaling systems involved in these processes include dysregulation of both pro-stress (corticotropin-releasing factor (CRF), dynorphin) and anti-stress (NPY, nociceptin) systems.

Thus, reward processing is mediated in large part through an interconnected network of structures that includes the VTA, NAc, ventral pallidum, CeA, BNST, and prefrontal cortex (PFC). In addition to the well-known involvement of mesocorticolimbic DA, reward processing is also heavily influenced by many other systems including glutamate, GABA, opioid peptides, and stress-related signaling systems.

2 The Endogenous Cannabinoid System (ECS)

Endocannabinoids are neuroactive lipids that participate in a range of physiological processes including reward, motivation, emotional homeostasis, pain processing, neuroprotection, and synaptic plasticity contributing to learning and memory. For the purposes of this chapter, a superficial overview of the main ECS components is provided below, and readers are referred to more in-depth coverage of this topic provided by several excellent recently published reviews (Lu and Mackie 2015; Hillard 2015; Mechoulam and Parker 2013).

In addition to the endocannabinoid lipid moieties themselves, the ECS is comprised of G protein-coupled receptors and metabolic enzymes for the synthesis and degradation of the ligands. Two major types of cannabinoid receptor have been characterized and cloned: CB_1 and CB_2. CB_1 receptors (CB_1Rs) are the most abundant G protein-coupled receptor expressed in the adult brain, with particularly dense expression in each of the interconnected structures involved in reward (Glass et al. 1997; Wang et al. 2003b; Herkenham et al. 1991) where they exert widespread modulatory influences on excitatory and inhibitory signaling in a manner that influences reward processing (Sidhpura and Parsons 2011; Panagis et al. 2014). In particular, eCBs play a prominent role in fine-tuning the activity of the VTA-NAc DA projection and its influence on approach and avoidance behaviors that govern reward acquisition (Hernandez and Cheer 2015). CB_2 receptors (CB_2Rs) are mainly expressed by immune cells with recent evidence also suggesting CB_2R expression in neurons, glia, and endothelial cells in the brain (Atwood and Mackie 2010). CB_1R and CB_2R are coupled to similar transduction systems primarily through G_i or G_o proteins. CB_1Rs directly inhibit the release of GABA, glutamate, and acetylcholine that produce widespread effects on neural signaling across many neurotransmitter systems.

At present the best characterized endocannabinoid ligands are N-arachidonylethanolamide (anandamide, AEA) and 2-arachidonoylglycerol (2-AG). Due to their lipid nature, AEA and 2-AG are not stored in vesicles but are synthesized on an "on demand" basis by cleavage from membrane precursors and immediate release through Ca^{2+}-dependent mechanisms. AEA is derived from the phospholipid precursor N-arachidonoyl phosphatidylethanolamine (NAPE), and while the precise mechanisms for AEA formation are not known, a very likely candidate is through N-acyl-phosphatidylethanolamine phospholipase D (NAPE-PLD). 2-AG derives primarily from the hydrolytic metabolism of 1,2-diacylglycerol (DAG) by sn-1-selective DAG lipases (DAGLα, DAGLβ). AEA is primarily catabolized through fatty acid amide hydrolase 1 (FAAH1), and 2-AG is catabolized through monoacylglycerol lipase (MAGL) and to a lesser extent α,β-hydrolase 6 (ABHD6), cyclooxygenase-2 (COX-2), and FAAH1. The eCB catabolic enzymes have distinct cellular anatomical locations with MAGL localized predominantly in presynaptic terminals and FAAH1 to the postsynaptic domain of neurons. AEA and 2-AG each exert agonist activity at CB_1R and CB_2R. AEA binds with slightly higher affinity to CB_1R vs. CB_2R, and like Δ9-tetrahydrocannabinol (Δ9-THC, the main psychoactive component of the *Cannabis* plant), AEA exhibits low efficacy as an agonist at both receptors, producing submaximal signaling upon binding. 2-AG binds with essentially equal affinity at CB_1R and CB_2R and exhibits greater agonist efficacy than AEA. AEA and 2-AG also exhibit agonist properties at several secondary receptors including peroxisome proliferator-activated receptors (PPARs), GPR55, and GPR119, and AEA exerts potent agonist effects at transient receptor potential ion channels including TRPV1.

3 Rewarding Effects of Altered eCB Signaling

Both exogenous AEA and exogenous 2-AG increase extracellular DA levels in the NAc in a CB_1-dependent manner (Solinas et al. 2006), and substantial evidence demonstrates a strong eCB influence in fine-tuning the activity of midbrain DA cells (Melis and Pistis 2012). Given this influence, and the rewarding effects produced by natural and synthetic CB_1 agonists, there has been considerable interest in the influence of eCB signaling in the modulation of brain reward function.

Several preclinical studies have investigated whether enhanced eCB signaling itself produces rewarding effects, and much of this has compared eCB-induced behaviors with those produced by exogenous CB_1 agonists. The metabolically stable AEA analogs (R)-methanandamide, O-1812, and AM1346 each fully generalize to Δ9-THC in drug discrimination tests (Burkey and Nation 1997; Jarbe et al. 2006; Alici and Appel 2004; Solinas et al. 2007), and exogenous AEA itself also produces Δ9-THC-like discriminative effects when administered following FAAH inhibition (Solinas et al. 2007; Vann et al. 2009; Wiley et al. 2014). Because neither AEA nor FAAH inhibition produces Δ9-THC-like responding in their own right (Burkey and Nation 1997; Wiley et al. 1997, 2014; Gobbi et al. 2005; Solinas et al. 2007), these findings suggest that administration of exogenous AEA produces Δ9-THC-like interoceptive effects only under conditions of reduced hydrolytic clearance. In the absence of MAGL inhibition, exogenous 2-AG also does not substitute for Δ9-THC in discrimination tests (Wiley et al. 2014), though interestingly the MAGL inhibitor JZL184 produces partial generalization to Δ9-THC even without exogenously administered 2-AG (Wiley et al. 2014; Long et al. 2009b). However, these effects of JZL184 may result from concurrent MAGL and FAAH inhibition based on evidence that dual inhibition of MAGL and FAAH fully substitutes for Δ9-THC in discrimination tests (Wiley et al. 2014; Long et al. 2009b), and the more selective MAGL inhibitor KML29 does not generalize to Δ9-THC (Ignatowska-Jankowska et al. 2014). The Δ9-THC-like discriminative properties of enhanced eCB signaling are all blocked by CB_1 antagonism. Collectively, these findings indicate that systemic administration of eCBs produces CB_1-dependent Δ9-THC-like discriminative stimulus effects, though combined enhancement of AEA and 2-AG signaling is most effective for producing Δ9-THC-like effects. Neither exogenously administered AEA nor selective FAAH, MAGL, or dual FAAH/MAGL inhibition produces rewarding effects in the conditioned place preference model (Mallet and Beninger 1998; Gobbi et al. 2005; Gamage et al. 2015), and FAAH inhibitors and putative eCB transport inhibitors do not alter brain stimulation reward thresholds (Vlachou et al. 2006). Drugs that inhibit eCB hydrolysis do not support operant self-administration behavior in rodents, though exogenous AEA, the metabolically stable AEA analog methanandamide, and exogenous 2-AG do support operant self-administration behavior by squirrel monkeys through a CB_1R-reliant mechanism (Justinova et al. 2005, 2008a, 2011). Moreover, the eCB clearance inhibitors URB597, AM404, and VDM11 support operant self-administration behavior by squirrel monkeys with a prior history of nicotine, cocaine, or AEA self-administration (Justinova et al. 2015; Schindler et al. 2016). However, unlike drugs with prominent abuse liability, these eCB clearance inhibitors do not increase mesolimbic DA release (Solinas et al. 2007; Murillo-Rodriguez et al. 2013; Justinova et al. 2015), and it remains to be determined whether these compounds will be self-administered by drug-naïve monkeys or other species. Collectively, these findings suggest that administration of exogenous eCBs can produce positive reinforcing effects in a variety of paradigms indexing brain reward, though most evidence suggests that pharmacological compounds that enhance eCB tone (e.g., selective FAAH or MAGL clearance inhibitors) do not produce substantial rewarding effects.

4 Endocannabinoid Influence on Natural Rewards

4.1 Food Reward

It is well known that palatable foods can lead people to eat, even when satiated. As such, eating can be motivated not only by hunger or homeostatic processes but also by hedonics. The ECS plays a prominent homeostatic role in controlling metabolic functions such as energy balance and food intake (Silvestri and Di Marzo 2013), and growing evidence also demonstrates an important ECS influence in the hedonic or rewarding properties of food intake. Both exogenous and endogenous CB_1 agonists reduce the latency to feed in pre-satiated or free-feeding animals, increase the amount of food consumed, and increase an animal's' willingness to work for food reward (Kirkham and Williams 2001; Farrimond et al. 2011; Gallate and McGregor 1999; Kirkham et al. 2002; Williams and Kirkham 1999; Martinez-Gonzalez et al. 2004; Solinas and Goldberg 2005). Manipulations of CB_1 tone appear to preferentially influence the motivation for highly palatable rewards compared with standard isocaloric chow or neutral or mildly aversive rewards (Guegan et al. 2013; Dipatrizio and Simansky 2008; Shinohara et al. 2009). For example, studies on the microstructure of fluid consumption and taste reactivity demonstrate that CB_1 receptor agonists increase the hedonic effects of palatable rewards such as sucrose (Higgs et al. 2003; Jarrett et al. 2005, 2007; Mahler et al. 2007). Conversely, CB_1 receptor antagonism preferentially reduces the consumption of palatable and/or high-fat foods vs. standard chow (Simiand et al. 1998; South et al. 2007; Mathes et al. 2008), reduces the conditioned reinforcing effects produced by palatable substances such as sucrose or chocolate (Chaperon et al. 1998), and reduces the motivation for palatable foods indexed under progressive ratio schedules of reinforcement (Gallate and McGregor 1999; Gallate et al. 1999; Solinas and Goldberg 2005).

The neural pathways contributing to the hedonic aspects of feeding include the mesocorticolimbic system (Kalivas and Volkow 2005; Nestler 2005). Food-related visual or olfactory cues increase mesolimbic DA signaling in both humans (Macht and Mueller 2007) and laboratory animals (Hajnal et al. 2004; Rada et al. 2005; Sclafani et al. 1998) in part through CB_1-reliant mechanisms (Melis et al. 2007). The mesocorticolimbic system influences feeding behavior through interconnections with hypothalamic nuclei (Berridge et al. 2010), and many of the hypothalamic signaling molecules that modulate homeostatic feeding behavior are also present in mesocorticolimbic structures such as the NAc, VTA, and PFC (Horvath and Diano 2004; Kelley et al. 2005). In fact, an influence of feeding-related hormones on eCB production (Di Marzo et al. 2001; Bermudez-Silva et al. 2012) and signaling (Edwards and Abizaid 2016) has been described, and this mechanism may play a prominent role in the regulation of consummatory behaviors. Fasting increases AEA and 2-AG levels in the limbic forebrain and hypothalamus of rodents, an effect that is ameliorated following feeding (Kirkham et al. 2002; Hanus et al. 2003). Doses of CB_1 agonists that induce food consumption in satiated rodents activate the corticostriatal-hypothalamic pathway (including the NAc) (Dodd et al. 2009), and food consumption is increased by direct infusion of eCBs or FAAH inhibitors into the hypothalamic nuclei (Anderson-Baker et al. 1979; Jamshidi and Taylor 2001; Verty et al. 2005) or NAc shell (Kirkham et al. 2002; Soria-Gomez et al. 2007). In particular, the NAc shell has been identified as a critical locus for the eCB modulation of the hedonic properties of food (Berridge et al. 2010; Mahler et al. 2007), and intra-NAc eCB infusions increase neural activity in hypothalamic nuclei through a CB_1-dependent mechanism (Soria-Gomez et al. 2007). Furthermore, transgenic mice overexpressing MAGL in the forebrain (MGL-Tg mice) do not form a conditioned place preference for high-fat foods, indicating that reductions in 2-AG may impede high-fat food reward (Wei et al. 2016). Collectively, these findings demonstrate that eCB signaling in the CNS plays a role in the motivation for and hedonic response to food consumption. Interestingly, there is emerging evidence that in addition to sweet taste receptors (T1R2/T1R3), sweet-sensitive taste cells in the peripheral system also express CB_1 receptors (Jyotaki et al. 2010), and both AEA and 2-AG increase gustatory nerve responses to sweet solutions (Yoshida et al. 2010). Thus, the ECS modulates food palatability via both central and peripheral pathways.

4.2 *Sexual Reward*

CB_1 receptors are densely expressed in the neural circuits that regulate sexual behavior, and substantial evidence indicates that both exogenous and endogenous cannabinoids can alter sexual behavior, though the effects are often highly divergent in males and females (Gorzalka et al. 2010; Lopez 2010). With regard to the hedonic aspects of sexual behavior, the majority of studies indicate a facilitatory effect of *Cannabis* on subjective indices of sexual arousal in women including desire, orgasmic function, and pleasure/enjoyment/satisfaction (Koff 1974; Dawley et al. 1979; Halikas et al. 1982; Klein et al. 2012). Similarly, low-dose *Cannabis* exposure increases sexual desire in males though simultaneously hindering erectile functioning. Rodent studies have revealed both facilitatory and inhibitory effects of cannabinoid agonists and eCB clearance inhibition (e.g., FAAH inhibitors) on female sexual receptivity and proceptivity and a general inhibitory effect on male sexual performance (Gorzalka et al. 2010; Lopez 2010; Canseco-Alba and Rodriguez-Manzo 2016), though the relative influence of these manipulations on the motivational vs. performance aspects of sexual behavior is not clear. In the most direct evaluation of an eCB influence on sexual motivation, Klein and colleagues observed that physiological indices of sexual arousal in women are associated with decreased serum AEA levels, and subjective indices of arousal are associated with decreases in serum levels of both AEA and 2-AG (Klein et al. 2012). However, while these indices reflect an ECS involvement in the anticipatory/motivational effects produced by erotic visual stimuli, it remains unclear whether endocannabinoid signaling is altered by sexual activity itself and, if so, whether this results in a facilitation or attenuation of brain reward processing. Endocannabinoid signaling may also influence sexual reward indirectly through modulation of emotional state and/or sensation perception, though these possibilities have not been explored.

4.3 Social Interaction

Cooperative playing and social interaction are rewarding activities in both humans and rodents (Tabibnia and Lieberman 2007; Douglas et al. 2004). In rodents, social play is enhanced following inhibition of FAAH or putative eCB transporters (with VDM11) but diminished following administration of exogenous CB_1 agonists such as WIN 55,212-2 (Trezza and Vanderschuren 2008a, b, 2009). This highlights the distinct effects produced by selectively enhancing the effects of eCB signaling in selected synapses, wherein eCB formation is evoked (e.g., eCB clearance inhibition) vs. widespread CB_1 activation by exogenous agonists that engage CB_1 signaling in circuits not normally activated by a given behavior (Trezza et al. 2010). The hedonic aspects of social play are mediated through both opioid (mu opioid in particular) and CB_1 mechanisms, and the CB_1 influence appears to involve subsequent increases in DA signaling (Trezza et al. 2010). Interestingly, social reward is influenced by opioid and cannabinoid interactions in a manner similar to drug and food reward (Fattore et al. 2004; Solinas and Goldberg 2005).

5 eCB Influence on the Acute Rewarding Effects of Drugs of Abuse

There is substantial evidence implicating the ECS in the motivational effects produced by several classes of abused drugs including ethanol, nicotine, opiates, and psychostimulants. Much of this is based on the influence of CB_1 receptor signaling on drug-induced behaviors, though some studies have evaluated manipulations of eCB tone on drug reinforcement.

In general, CB_1 receptors exert a facilitatory influence on the rewarding effects of several classes of abused drugs (Table 1). Rodent studies employing the conditioned place preference and operant self-administration paradigms demonstrate that CB_1 agonists increase the motivational and rewarding effects of alcohol (Colombo

Table 1 Summary of CB_1R influence on motivation-related effects of abused drugs

Genetic or pharmacological manipulation	Ethanol	Nicotine	Opiates	Psychostimulants
CB_1R Knockout	↓ CPP[a] ↓ Operant SA ↓ EtOH-induced NAc DA	↓ CPP[b] ↓ Operant SA	↓ CPP[c] ↓ Operant SA	**No change** in CPP[d] **No change** in operant SA
CB_1R Antagonist	↓ Preference[e] ↓ Operant SA ↓ EtOH-induced NAc DA	↓ CPP[f] ↓ Operant SA ↓ Nic-induced NAc DA	↓ CPP[g] ↓ Operant SA	↓↑ Operant SA[h] ↓↑ Cocaine effects on ICSS
CB_1R Agonist	↑ Operant SA[i] ↑ Motivation for EtOH	↑ CPP[j]	↑ CPP[k] ↑ Motivation for heroin	↓ Operant SA[l] ↓ Cocaine effects on ICSS

[a]Houchi et al. (2005), Thanos et al. (2005), Hungund et al. (2003), Naassila et al. (2004), Wang et al. (2003a)
[b]Cossu et al. (2001)
[c]Martin et al. (2000), Ledent et al. (1999), Cossu et al. (2001)
[d]Martin et al. (2000), Houchi et al. (2005), Cossu et al. (2001), Soria et al. (2005), Li et al. (2009)
[e]Arnone et al. (1997), Freedland et al. (2001), Gallate and McGregor (1999), Colombo et al. (1998), Gessa et al. (2005), Malinen and Hyytia (2008), Alvarez-Jaimes and Parsons (2009), Caille et al. (2007), Wang et al. (2003a), Perra et al. (2005), Cheer et al. (2007)
[f]Cohen et al. (2002), Shoaib (2008), Simonnet et al. (2013), Cheer et al. (2007)
[g]Caille and Parsons (2003, 2006), De Vries et al. (2003), Solinas et al. (2003), Singh et al. (2004), Chaperon et al. (1998), Navarro et al. (2001)
[h]Filip et al. (2006), Caille and Parsons (2006), Xi et al. (2008), Vinklerova et al. (2002), Chaperon et al. (1998), Cheer et al. (2007), Soria et al. (2005), Vlachou et al. (2003), Fattore et al. (1999), Tanda et al. (2000), De Vries et al. (2001), Caille et al. (2007), Lesscher et al. (2005)
[i]Colombo et al. (2002), Wang et al. (2003a), Gallate et al. (1999), Malinen and Hyytia (2008), Getachew et al. (2011)
[j]Valjent et al. (2002), Gamaleddin et al. (2012b)
[k]Manzanedo et al. (2004), Solinas et al. (2005)
[l]Vlachou et al. (2003), Fattore et al. (1999)

et al. 2002; Wang et al. 2003a; Gallate et al. 1999; Malinen and Hyytia 2008; Getachew et al. 2011), nicotine (Valjent et al. 2002; Gamaleddin et al. 2012b), and opiates (such as morphine and heroin) (Manzanedo et al. 2004; Solinas et al. 2005). In contrast, diminished CB_1 receptor signaling (either genetic receptor deletion or pharmacological inhibition) attenuates the motivational and rewarding effects of alcohol (Houchi et al. 2005; Thanos et al. 2005; Hungund et al. 2003; Naassila et al. 2004; Arnone et al. 1997; Wang et al. 2003a; Freedland et al. 2001; Gallate and McGregor 1999; Colombo et al. 1998; Gessa et al. 2005), nicotine (Cohen et al. 2002; Shoaib 2008; Cossu et al. 2001; Simonnet et al. 2013), and opiates (Chaperon et al. 1998; Navarro et al. 2001; Singh et al. 2004; Martin et al. 2000; Ledent et al. 1999; Cossu et al. 2001; Caille and Parsons 2003; De Vries et al. 2003; Solinas et al. 2003). The CB_1 influence on alcohol and nicotine reward is mediated in part, through modulation of the mesolimbic DA response to these drugs as CB_1 receptor antagonism blocks the excitation of VTA DA cells produced by alcohol and nicotine, thereby attenuating the effects of these drugs on NAc DA release (Cohen et al. 2002; Hungund et al. 2003; Perra et al. 2005; Cheer et al. 2007). Blockade of CB_1 receptors in the VTA decreases both alcohol and nicotine self-administration (Malinen and Hyytia 2008; Alvarez-Jaimes and Parsons 2009; Caille et al. 2007; Simonnet et al. 2013), and antagonism of NAc CB_1 receptors similarly reduces alcohol consumption (Alvarez-Jaimes and Parsons 2009; Caille et al. 2007). However, while nicotine reward is critically dependent on the mesolimbic DA system (D'Souza and Markou 2011), the motivational and rewarding effects of alcohol and opiates are less DA dependent (Koob 2013; Shippenberg and Elmer 1998; Pettit et al. 1984; Gerrits and Van Ree 1996; Platt et al. 2001) with CB_1R modulation of the rewarding effects of these drugs likely involving non-dopaminergic mechanisms. Indeed, although CB_1 receptor deletion results in attenuated opiate-induced increases in NAc DA (Mascia et al. 1999), acute CB_1 receptor blockade does not alter opiate-induced increases in NAc DA (Caille and Parsons 2003, 2006; Tanda et al. 1997). Rather, CB_1 receptor antagonism may modulate opiate reward through DA-independent attenuation of opiate-induced reductions in ventral pallidal GABA signaling (Caille and Parsons 2006).

The effects of CB_1 receptor manipulations on psychostimulant reward are less consistent. In contrast to the facilitation of ethanol, nicotine, and opiate reward, CB_1 agonists reduce both cocaine-induced facilitation of brain stimulation reward and cocaine self-administration by rats (Vlachou et al. 2003; Fattore et al. 1999). Further, CB_1 receptor inactivation does not affect psychostimulant-induced place conditioning (Martin et al. 2000; Houchi et al. 2005), psychostimulant self-administration (Cossu et al. 2001; Tanda et al. 2000; Fattore et al. 1999; De Vries et al. 2001; Caille and Parsons 2006; Filip et al. 2006; Caille et al. 2007; Lesscher et al. 2005), cocaine-induced increases in nucleus accumbens dopamine (Caille and Parsons 2003; Soria et al. 2005), or cocaine-induced enhancement of brain stimulation reward (Vlachou et al. 2003; Xi et al. 2008). However, some reports indicate that CB_1 inactivation attenuates cocaine and amphetamine self-administration (Soria et al. 2005; Xi et al. 2008; Vinklerova et al. 2002), blocks cocaine-induced increases in NAc DA signaling (Li et al. 2009; Cheer et al. 2007), and decreases cocaine-induced enhancement of brain stimulation reward sensitivity (Xi et al. 2008). In general, the inconsistent and generally subtle effects reported suggest that CB_1 receptors exert a relatively modest influence over psychostimulant-induced reward.

Recent evidence in mice also implicates CB_2Rs in the modulation of drug reward. CB_2R agonists reduce cocaine-induced CPP and cocaine self-administration in wild-type mice but not CB_2R knockout mice (Xi et al. 2011; Ignatowska-Jankowska et al. 2013; Zhang et al. 2014a), and cocaine self-administration is also reduced by CB_2R overexpression in the brain (Aracil-Fernandez et al. 2012). CB_2R agonists also reduce alcohol-induced CPP and alcohol consumption (Al Mansouri et al. 2014). Mouse VTA DA neurons express CB_2R mRNA and receptor immunostaining, and CB_2R agonists decrease the activity of these neurons in wild-type but not CB_2R knockout mice (Zhang et al. 2014b). Surprisingly, CB_2R agonists facilitate nicotine-induced CPP (Ignatowska-Jankowska et al. 2013), and CB_2R inactivation reduces both nicotine-induced CPP and nicotine self-administration in mice (Navarrete et al. 2013; Ignatowska-Jankowska et al. 2013). In contrast, CB_2R antagonism does not alter cocaine or nicotine self-administration in rats but attenuates cocaine-induced and conditioned locomotion and reinstatement of cocaine-seeking behavior (without altering nicotine-induced drug seeking) (Blanco-Calvo et al. 2014; Adamczyk et al. 2012b; Gamaleddin et al. 2012b). Further, a recent study reported the CB_2R agonism reduces cocaine consumption by rats, but not mice, and produces opposite effects on the motivation for cocaine in these species (Zhang et al. 2014a). These distinctions may result from species differences in the splicing and expression of CB_2R genes, possibly conferring distinct CB_2R structure, function, or pharmacology (Zhang et al. 2014a).

6 Evidence that Drugs of Abuse Increase Brain eCB Levels

The inhibitory influence of CB_1 receptor antagonism on drug reward has led to the hypothesis that abused drugs increase brain eCB formation possibly resulting in aberrant eCB signaling following long-term drug exposure. Several observations support these hypotheses.

Early studies reported that chronic alcohol exposure increases 2-AG and AEA formation in human neuroblastoma cells and cultured rodent neurons (Basavarajappa and Hungund 1999; Basavarajappa et al. 2000, 2003). Subsequent evaluations of postmortem brain tissue eCB content clearly demonstrate alcohol-induced alterations in 2-AG and AEA production, though substantial inconsistencies among studies make it difficult to draw clear conclusions on the direction of change and regional nature of the effects (Vinod et al. 2006, 2012; Gonzalez et al. 2002, 2004b; Malinen et al. 2009; Rubio et al. 2007). In vivo microdialysis studies in rats demonstrate that voluntary alcohol consumption robustly increases nucleus accumbens 2-AG but not AEA levels (Caille et al. 2007), while forced alcohol administration decreases AEA levels and induces more modest increases in 2-AG (Ferrer et al. 2007; Alvarez-Jaimes et al. 2009). This provides initial evidence that the volitional nature of drug exposure (e.g., voluntary vs. forced administration) may differentially impact eCB responses. The effects of alcohol on brain eCB production may also be region specific as alcohol-induced disruptions of extracellular eCB levels and mRNA expression of their associated enzymes are consistently observed in striatal regions (Caille et al. 2007; Ceccarini et al. 2013; Ferrer et al. 2007; Alvarez-Jaimes et al. 2009; Henricks et al. 2016) but not frontal cortical areas (Alvarez-Jaimes and Parsons 2009), and ethanol self-administration is reduced by localized antagonism of CB_1 receptors in the VTA and NAc shell but not PFC of outbred rats (Malinen and Hyytia 2008; Alvarez-Jaimes and Parsons 2009; Caille et al. 2007) (though disruptions in eCB processing have been observed in cortical regions of rats selectively bred for high alcohol consumption (Hansson et al. 2007) and in mice following long-term alcohol exposure (Vinod et al. 2006)).

Δ9-THC and synthetic CB_1 agonists also induce region-specific alterations in brain tissue eCB content. For example, chronic Δ9-THC increases AEA (but not 2-AG) in limbic forebrain tissue and increases 2-AG (but not AEA) in the hippocampus, brainstem, and cerebellar tissue (Di Marzo et al. 2000; Gonzalez et al. 2004a; Castelli et al. 2007). In vivo microdialysis studies demonstrate that acute CB_1 agonist exposure increases 2-AG but decreases AEA levels in the rat hypothalamus (Bequet et al. 2007), and preliminary evidence from human clinical research studies also points to cannabinoid-induced increases in 2-AG and decrements in AEA production (Leweke et al. 2007; Morgan et al. 2013).

Chronic nicotine exposure also induces region-specific disruptions in brain tissue eCB content, with increased AEA (but not 2-AG) levels evident in limbic forebrain and caudate but decreased AEA and 2-AG levels in the cortex (Gonzalez et al. 2002). In vivo microdialysis studies demonstrate nicotine-induced increases in both 2-AG and AEA in the VTA and a sensitization of nicotine-induced 2-AG, but not AEA, formation results from chronic nicotine exposure (Buczynski et al. 2013). This enhancement of nicotine-induced VTA 2-AG formation results in diminished inhibitory constraint of VTA cell excitability, thereby enhancing nicotine-induced elevations in mesolimbic DA levels (Buczynski et al. 2016). Interestingly, while VTA 2-AG levels are increased by both voluntary self-administration and response-independent forced nicotine administration, VTA AEA levels are increased only by voluntary self-administration. Together with evidence of distinct eCB responses to self-administered vs. forced alcohol exposure (Caille et al. 2007; Ferrer et al. 2007; Alvarez-Jaimes et al. 2009), these observations suggest that brain eCB production can be influenced by activity of neural systems involved in the motivation for drug consumption, in addition to drug-related pharmacological effects.

Opiates also induce differential effects on brain 2-AG and AEA levels. Chronic morphine dose dependently reduces tissue 2-AG content in the striatum, cortex, hippocampus, hypothalamus, and limbic forebrain, while increased AEA content is observed in these regions (Vigano et al. 2003, 2004). Consistently, in vivo microdialysis studies demonstrate that heroin self-administration significantly increases interstitial AEA levels but decreases interstitial 2-AG levels in the nucleus accumbens (Caille et al. 2007).

In general, psychostimulants induce more subtle alterations in brain eCB levels as compared with the drugs described above (Zlebnik and Cheer 2016). Early studies reported that acute high-dose cocaine administration induces a subtle but significant increase in forebrain 2-AG content in mice (Patel et al. 2003), while chronic cocaine induces comparably subtle decreases in forebrain 2-AG content in rats (Gonzalez et al. 2002), though neither 2-AG nor AEA content is altered in other regions following acute or chronic cocaine exposure (Patel et al. 2003; Gonzalez et al. 2002). Similarly, cocaine self-administration does not alter either 2-AG or AEA content in nucleus accumbens microdialysates (Caille et al. 2007). However, recent studies provide some evidence of psychostimulant-induced alterations in eCB levels and processing. For example, cocaine self-administration is reported to induce increased cerebellar AEA content; decreased 2-AG content in the frontal cortex, hippocampus, and cerebellum; and an increased ratio of DAGLa/MAGL expression in the hippocampus (Bystrowska et al. 2014; Rivera et al. 2013). Further, sensitization of the motor-activating effects of cocaine is associated with altered eCB metabolic enzyme expression in the cerebellum (Palomino et al. 2014), and detoxified cocaine addicts present significant increases in plasma AEA and decreases in plasma 2-AG content (Pavon et al. 2013).

These collective findings indicate that cannabinoids, ethanol, nicotine, and opiates can alter brain eCB content, consistent with substantial evidence of a CB_1 receptor influence on the behavioral effects produced by these drugs. Although there is relatively less information on psychostimulant-induced alterations in brain eCB content, the modest effects that have been reported are consistent with the subtle disruptions in psychostimulant-induced behaviors observed following CB_1 receptor antagonism.

eCBs are rapidly degraded, and thus strategies that reduce eCB clearance have been employed as a means to further investigate the eCB influence on drug reward. Most investigations have focused on the effects of FAAH inhibition because selective tools for inhibiting MAGL and other eCB clearance enzymes were not available until recently. Such studies have shed light on important species differences that confound the overall conclusions that can be made from existing data. For example, FAAH inhibition in mice increases nicotine reward in the CPP paradigm (Merritt et al. 2008; Muldoon et al. 2013), though in rats, FAAH inhibition prevents nicotine-induced CPP, diminishes nicotine self-administration, and blunts nicotine-induced increases in NAc DA release (Scherma et al. 2008). The potentiation of nicotine reward in mice by FAAH inhibition is CB_1R mediated, whereas the reduction in nicotine reward in rats results from activation of PPAR-α by non-cannabinoid lipids such as oleoylethanolamide (OEA) and palmitoylethanolamide (PEA) that are hydrolytically cleared by FAAH (Melis and Pistis 2014). FAAH inhibition also produces distinct species-related alterations in alcohol consumption, with increased intake observed in mice but not rats (Blednov et al. 2007; Vinod et al. 2008; Hansson et al. 2007; Cippitelli et al. 2008). The mechanisms underlying these differences are not understood. Brain region-specific disruptions in FAAH activity may be an important factor in regard to alcohol reward as inhibition of FAAH activity specifically in the PFC results in increased alcohol consumption, and rats selectively bred for high alcohol intake and preference are characterized by reduced FAAH activity specifically in the PFC (Hansson et al. 2007; Cippitelli et al. 2008). The effects of FAAH inhibition on opiate and psychostimulant reward have primarily been studied in rats. FAAH inhibition does not alter morphine- or cocaine-induced disruptions in VTA DA cell firing or the self-administration of either drug (Justinova et al. 2008a; Luchicchi et al. 2010). However, FAAH inhibition diminishes cocaine-induced alterations in NAc medium-spiny neuron activity (Luchicchi et al. 2010), and this may contribute to enhanced sensitization of both cocaine-induced motor activity and mesolimbic dopamine responses following repeated cocaine exposure (Mereu et al. 2013). Other studies have investigated the effects of putative eCB transport inhibitors such as AM404 and VDM11, and the findings thus far suggest that these compounds produce subtle and inconsistent effects on nicotine and cocaine reward (Scherma et al. 2012; Gamaleddin et al. 2011, 2013; Vlachou et al. 2008).

While growing evidence implicates ECS influences in the modulation of acute drug reward, additional efforts are needed to further clarify the nature of eCB disruptions caused by different classes of abused drugs and the neural mechanisms through which these eCB influences are mediated. Selective inhibitors of 2-AG clearance have recently been developed, but studies are still in their infancy, so there are presently no published reports on the effects of enhanced 2-AG tone on drug reward and related physiological events. As such there remains a substantial gap of knowledge given the prominent 2-AG influence on neural signaling and plasticity related to both drug and natural rewards. Nevertheless, the role of the CB_1R in drug reward is unequivocal. Although there is evident complexity related to the effects produced by eCB clearance inhibition (producing discrete modulation of eCB tone in specific synapses/circuits as compared with broad CB_1R activation by exogenous CB_1R agonists), the extant evidence strongly supports an eCB influence on the sensitivity to and motivation for several drugs of abuse.

7 Overview of Drug Dependence, Addiction, and Withdrawal

The transition from intermittent and controlled drug use to compulsive forms of drug seeking and drug taking that characterize addiction is influenced by a number of factors. Substantial evidence implicates genetic influences in the development of substance use disorders and pathological forms of eating, sexual behavior, and gambling (Ducci and Goldman 2012). A crucial role for epigenetic mechanisms driving lasting changes in addiction-related gene expression is also increasingly recognized (Nestler 2014). Additionally, long-term drug exposure induces lasting neuroadaptations that alter the motivational mechanisms that propel drug seeking and use. Although initial drug use is motivated by the hedonic effects of drug consumption, prolonged drug exposure results in progressive blunting of reward system function that can motivate consumption of increasingly larger amounts of drug. Evidence suggests this results in part from decreased function of the mesolimbic DA system and its output to the ventral pallidum. Escalated frequency and amount of drug consumption lead to a dependent state wherein negative affective symptoms emerge during drug abstinence (e.g., dysphoria, anxiety, irritability). These negative emotional states arise from the recruitment of stress-signaling systems within the extended amygdala (such as corticotropin-releasing factor and dynorphin) and dysregulation of systems that constrains these responses (such as neuropeptide Y and nociceptin) (Koob and Volkow 2010). Renewed drug consumption alleviates these negative affective states, and this is conceptualized to contribute to compulsive drug use through negative reinforcement mechanisms

(Koob and Volkow 2010). Superimposed on these processes is a dysregulation of corticostriatal mechanisms contributing to stimulus response learning, conditioned reinforcement, and incentive motivation resulting in a narrowed focus on drug seeking at the expense of natural rewards (Kalivas and Volkow 2005; Everitt et al. 2008). eCBs exert prominent modulatory influence in the extended amygdala and corticostriatal circuits implicated in the etiology of addiction, and as described below, increasing evidence suggests that preexisting genetic influences on the ECS and/or drug-induced dysregulation of eCB function may participate in the development and maintenance of addiction. The following sections consider the consequences of chronic drug exposure on eCB signaling within the reward circuitry and related disruptions in synaptic plasticity, affective state, and learning and memory mechanisms related to extinction and relapse.

8 Evidence of Altered ECS Function Following Chronic Drug Exposure

It stands to reason that chronic *Cannabis* use induces dysregulation of brain cannabinoid receptors' availability and function. In vivo PET imaging studies in humans have revealed downregulation of brain CB_1R binding that correlates with the duration of prior *Cannabis* use, with particularly robust decrements observed in the temporal lobe, anterior and posterior cingulate cortex, and nucleus accumbens (Hirvonen et al. 2013; Ceccarini et al. 2015). These decrements in CB_1R availability appear to progressively recover with prolonged *Cannabis* abstinence. A similar profile of decreased brain CB_1R function and posttreatment recovery has been consistently reported in rodents given chronic CB_1R agonist exposure (Breivogel et al. 1999; Sim et al. 1996). Recent experiments employing stochastic optical reconstruction microscopy (STORM) demonstrate that chronic exposure to clinically relevant doses of Δ9-THC results in a startling loss of CB_1Rs on terminals of perisomatically projecting GABA interneurons in the mouse hippocampus and internalization of the remaining CB_1Rs (Dudok et al. 2015). The resulting deficits in inhibitory CB_1R control over hippocampal GABA release persisted during several weeks of Δ9-THC abstinence, and this may underlie the enduring loss of hippocampal LTP in rodents and memory deficits in humans evident following chronic cannabinoid exposure (Puighermanal et al. 2012). Surprisingly little is known of the effect of chronic cannabinoid exposure on other facets of ECS function. Chronic cannabinoid exposure increases enzymatic clearance of AEA and reduces brain tissue AEA content in rodents (Di Marzo et al. 2000; Schlosburg et al. 2009), and frequent *Cannabis* smokers present decreased AEA and increased 2-AG levels in the blood (Leweke et al. 2007; Morgan et al. 2013), though increased serum AEA levels are evident following at least 6 months of *Cannabis* abstinence (Muhl et al. 2014). The contribution of these disruptions to *Cannabis* use disorder and related physiological and behavioral disruptions is presently unexplored. However, as discussed below, eCBs provide important homeostatic constraint over emotional state (Lutz 2009) and sleep function (Murillo-Rodriguez et al. 2011), and it's conceivable that Δ9-THC-induced impairment of eCB signaling contributes to negative emotional states and sleep disturbances present during protracted *Cannabis* abstinence (Budney et al. 2001; Gates et al. 2015).

Chronic exposure to non-cannabinoid drugs also disrupts eCB signaling and processing. Chronic alcohol exposure in rodents alters eCB-related gene expression in a manner sensitive to the intermittent nature of alcohol exposure and post-alcohol abstinence period (Serrano et al. 2012) and downregulates CB_1R expression and function (Mitrirattanakul et al. 2007; Ceccarini et al. 2013). Postmortem studies of alcohol-dependent humans also demonstrate disrupted CB_1R expression in the ventral striatum and cortical regions (Vinod et al. 2010), and in vivo imaging studies demonstrate decreased CB_1R availability in heavy drinking alcoholics that persist for at least 1 month of abstinence (Hirvonen et al. 2013; Ceccarini et al. 2014) (but see Neumeister et al. (2012)). Although a potential contribution of *CNR1* gene variants to these observations cannot be excluded, a common interpretation based on animal studies is that these CB_1R adaptations in alcoholic humans are a consequence of prolonged alcohol-induced increases in brain eCB levels. This is supported by evidence of transient recovery (and perhaps eventual upregulation) of CB_1R function in humans during protracted alcohol abstinence (Vinod et al. 2006; Mitrirattanakul et al. 2007). In rodents, chronic nicotine exposure induces distinct age-related disruptions in CB_1R binding, with increased levels evident in the PFC, VTA, and hippocampus of adolescent but not adult rats (Werling et al. 2009), and increased hippocampal and decreased striatal CB_1R binding in adult rats during protracted nicotine abstinence (Marco et al. 2007). Few studies have investigated altered CB_1R binding following chronic opiate or psychostimulant exposure, but findings in rodents implicate impaired CB_1R function in the development and expression of opiate dependence (Rubino et al. 1997; Cichewicz et al. 2001) and demonstrate that chronic cocaine increases CB_1R binding in dorsal striatum, NAc, and cortical areas (Adamczyk et al. 2012a). Interestingly, detoxified cocaine addicts present significant increases in plasma AEA and decreases in plasma 2-AG content (Pavon et al. 2013), but the functional consequence of these disturbances is not known. Overall, accruing data suggests that long-term exposure to a variety of drug classes compromises eCB processing and CB_1R expression and function. As discussed below, these perturbations may contribute to aberrant neural signaling during acute and protracted drug abstinence.

9 eCB Influence on Physical Withdrawal Symptoms

In dependent individuals, drug abstinence is associated with transient physical or somatic symptoms that can persist for several days. The intensity of these symptoms varies between drug classes, with severe somatic withdrawal signs evident during abstinence from opiates and alcohol and substantially less severe symptoms during abstinence from nicotine, cocaine, and cannabinoids (West and Gossop 1994). Early conceptualizations of addiction are focused on the relief from somatic withdrawal symptoms as the motivational basis for drug use by dependent individuals (Wikler 1948; Dole 1965; Dole et al. 1966), though more recent evidence suggests that alleviation of somatic withdrawal is not a major factor contributing to relapse (Heilig et al. 2010; Hershon 1977). In fact the greatest susceptibility to drug relapse typically occurs well after the abatement of somatic withdrawal symptoms. Nonetheless, there is value in understanding the mechanisms contributing to somatic withdrawal symptoms for the development of palliative therapies for dependent individuals and avoidance of medically serious conditions associated with acute drug detoxification.

Chronic alcohol exposure results in significant downregulation of CB_1 receptor expression and function in many brain regions (Basavarajappa et al. 1998; Basavarajappa and Hungund 1999; Vinod et al. 2006, 2010; Moranta et al. 2006; Mitrirattanakul et al. 2007; Ceccarini et al. 2013, 2014; Hirvonen et al. 2013; Neumeister et al. 2012) and leads to disruptions in brain tissue AEA and 2-AG content (Vinod et al. 2006, 2012; Gonzalez et al. 2002, 2004b; Malinen et al. 2009; Rubio et al. 2007). Clinical research has revealed a correlation between the severity of alcohol withdrawal symptoms and polymorphisms in the gene encoding CB_1 receptors (Schmidt et al. 2002), and alcohol withdrawal severity is increased in CB_1 receptor knockout mice (Naassila et al. 2004). Accordingly, it is possible that impaired CB_1 signaling contributes to the intensity of physical symptoms of alcohol withdrawal. In this regard, it is notable that FAAH knockout mice exhibit significantly diminished alcohol withdrawal-related seizures relative to wild-type mice (Vinod et al. 2008; Blednov et al. 2007). The somatic symptoms of alcohol withdrawal, including seizure activity, result primarily from excessive excitatory glutamate signaling (De Witte et al. 2003). Interestingly, AEA modulates NMDA receptor function (Hampson et al. 1998) and reduces NMDA- and electroshock-induced seizures (Hayase et al. 2001; Wallace et al. 2002). Therefore, enhancement of AEA tone may provide protection against alcohol withdrawal-related seizures, though more research is necessary to support this possibility.

Chronic nicotine exposure does not appear to alter brain CB_1 receptor expression (Gonzalez et al. 2002), and precipitated somatic withdrawal signs in nicotine-dependent mice are not altered by CB_1 receptor deletion or moderate doses of the CB_1 receptor antagonist SR141716A (Castane et al. 2002, 2005; Cossu et al. 2001; Balerio et al. 2004; Merritt et al. 2008). In contrast, the Manzanares group observed that somatic indices of nicotine withdrawal are absent in CB_2 knockout mice and are attenuated in WT mice by treatment with the CB_2 antagonist AM630 (Navarrete et al. 2013), though no genotypic differences in somatic withdrawal were evident following a shorter duration of nicotine exposure (7 days vs. 14 days) (Ignatowska-Jankowska et al. 2013). This suggests a potential CB_2 influence in the somatic symptoms of withdrawal following long-term nicotine exposure. Acute Δ9-THC administration decreases somatic symptoms of nicotine withdrawal (Balerio et al. 2004, 2006). Moreover, nicotine-dependent MAGL knockout mice exhibit diminished precipitated withdrawal signs, and the MAGL inhibitor JZL184 dose-dependently reduces somatic and aversive signs of precipitated nicotine withdrawal in dependent mice through a CB_1 receptor-dependent mechanism (Muldoon et al. 2015). This same study also points to an association between MAGL gene polymorphisms and the severity of nicotine withdrawal symptoms in humans. In contrast, acute FAAH inhibition does not diminish somatic nicotine withdrawal symptoms in rats (Cippitelli et al. 2011), while somatic symptoms of withdrawal are worsened in mice by FAAH inhibition, though this may occur through non-CB_1 mechanisms (Merritt et al. 2008). Collectively these findings suggest that enhanced CB_1 receptor signaling resulting from MAGL inhibition may provide therapeutic benefit for the somatic symptoms of nicotine withdrawal.

Chronic opiate exposure alters CB_1 receptor expression and function (Rubino et al. 1997; Cichewicz et al. 2001; Smith et al. 2007), and this appears to play a role in the development and expression of opiate dependence. For example, the somatic symptoms of opiate withdrawal are significantly reduced in CB_1 knockout mice or WT mice receiving chronic CB_1 antagonist treatment concurrent with chronic morphine administration (Ledent et al. 1999; Lichtman et al. 2001; Mas-Nieto et al. 2001; Rubino et al. 2000), and it has long been recognized that Δ9-THC effectively reduces the intensity of somatic symptoms of opiate withdrawal (Bhargava 1976; Hine et al. 1975). These findings suggest that chronic opiate exposure dysregulates CB_1 receptor or eCB function in a manner that contributes to elicitation of somatic symptoms during drug withdrawal. Consistent with this hypothesis, administration of Δ9-THC, exogenous AEA, or exogenous 2-AG attenuates the intensity of somatic withdrawal symptoms (Vela et al. 1995; Yamaguchi et al. 2001; Gamage et al. 2015), and high doses of the selective MAGL inhibitor JZL184 significantly reduce all indices of somatic withdrawal in opiate-dependent mice, while the selective FAAH inhibitor PF-3845 reduces only a subset of withdrawal responses (Ramesh et al. 2011, 2013; Gamage et al. 2015). Perhaps more importantly, a novel dual inhibitor exhibiting >100-fold greater potency at FAAH vs. MAGL (SA-57) (Niphakis et al. 2012) significantly reduces somatic indices of precipitated withdrawal (Gamage et al. 2015). This is an important observation given that high-dose JZL184 exposure elicits some cannabimimetic effects (Long et al. 2009a), and repeated high-dose JZL184 administration induces functional CB_1 receptor tolerance and results in cannabinoid dependence (Schlosburg et al. 2010); these effects may limit the clinical viability of selective MAGL inhibition for treatment of opiate withdrawal. In contrast, SA-57 doses that efficaciously reduce opiate withdrawal do not produce cannabimimetic effects (Ramesh et al. 2013), and the moderate MAGL inhibition induced by this compound is not likely to induce CB_1 receptor downregulation or cannabinoid dependence (Kinsey et al. 2013; Ghosh et al. 2013). However, a caveat to these findings is that while SA-57 attenuates somatic aspects of withdrawal, its administration does not prevent the aversive aspects of withdrawal as measured by the conditioned place aversion assay in mice (Gamage et al. 2015). Nonetheless, these collective findings suggest that modest pharmacological enhancement of eCB signaling may provide palliative effects for the physical symptoms of opiate withdrawal.

10 Stress Responsivity

Substantial clinical and preclinical data indicate that various forms of stress are involved in the etiology and maintenance of addiction. High levels of stress often precede the development of substance use disorders (Jose et al. 2000; Richman et al. 1996; Rospenda et al. 2000), and stress-related increases in glucocorticoid signaling increases the acquisition of drug self-administration by rodents (Goeders 2002; Goeders and Guerin 1996; Koob and Kreek 2007; Vengeliene et al. 2003). Consistently, evidence suggests that both stress- and drug-induced activation of the hypothalamic-pituitary-adrenal (HPA) axis sensitizes the function of reward pathways through glucocorticoid mechanisms (Piazza and Le Moal 1998; Rouge-Pont et al. 1995, 1998; Tidey and Miczek 1997). However, long-term drug use results in a blunting of reward system function, increased influence of CNS stress systems, and dysregulation of the HPA axis that is theorized to induce an allostatic shift in motivational influence that propels an escalation of drug consumption (Koob and Kreek 2007; Le Moal 2009; Koob and Le Moal 1997).

It is now well established that eCB signaling serves a homeostatic role in the constraint of HPA axis activation (Cota 2008; Gorzalka et al. 2008; Morena et al. 2016; also see chapter "Endocannabinoids, Stress, and Negative Affect" that details this function in greater depth). Stress-induced glucocorticoid secretion increases eCB production in several regions implicated in drug dependence and addiction including the amygdala, hippocampus, PFC, hypothalamic nuclei, and periaqueductal gray (Di et al. 2003, 2005, 2009; Hill et al. 2010a; Hohmann et al. 2005; Patel et al. 2004, 2005). Strong evidence demonstrates that eCB-mediated CB_1 activation in these and other regions constrains stress-induced HPA axis activation and resultant increases in glucocorticoid secretion (Barna et al. 2004; Cota et al. 2007; Di et al. 2003; Manzanares et al. 1999; Patel et al. 2004; Uriguen et al. 2004; Wade et al. 2006). In contrast, stress-induced reductions in AEA content are consistently observed in the amygdala, PFC, and hippocampus (Hill et al. 2008, 2009; Patel et al. 2005; Rademacher and Hillard 2007), and this is theorized to contribute to stress responsivity through a release of tonic HPA axis inhibition (Patel et al. 2004). Regardless of the specific mechanism that prevails, it is clear that eCB signaling participates in the constraint of HPA axis activation and may contribute to the termination of stress responses.

In this regard, it is conceivable that disrupted eCB-mediated plasticity induced by long-term drug exposure contributes to a persistent dysregulation of HPA axis function and increases stress sensitivity associated with addiction (Koob and Kreek 2007; Le Moal 2009; Koob and Le Moal 1997). Because stress exposure is implicated in the development of addiction (Goeders 2002; Goeders and Guerin 1996; Jose et al. 2000; Koob and Kreek 2007; Richman et al. 1996; Rospenda et al. 2000; Vengeliene et al. 2003), it is notable that stress can alter eCB-mediated plasticity in addiction-related brain regions (Natividad et al. 2017). For example, prolonged stress exposure impairs eCB-mediated DSI, LTD, and fEPSPs in the NAc (Wang et al. 2010), attenuates eCB-mediated DSE and DSI in the PVN

(Wamsteeker et al. 2010), and enhances 2-AG-mediated DSI in the basolateral amygdala (Patel et al. 2009). Similarly, establishment of conditioned fear increases the efficacy of DSE and DSI in the central nucleus of the amygdala (Kamprath et al. 2011).

11 Affective Disruptions

In addition to transient somatic symptoms (West and Gossop 1994), withdrawal from most drugs of abuse is associated with increased negative affective symptoms such as anxiety and depression (Alling et al. 1982; Coffey et al. 2000; Janiri et al. 2005; Nunes and Levin 2004; Nunes et al. 2004). Withdrawal-related negative affective states can persist for months and in some cases years during protracted abstinence, and the severity of these symptoms is closely associated with susceptibility to relapse (Annis et al. 1998; Miller and Harris 2000). Moreover, there is a prevalent comorbidity between affective disorders and substance use disorders, and affective disruption may be an antecedent to addiction (Bruijnzeel et al. 2004; Castle 2008; Conway et al. 2006; Pani et al. 2010; Schuckit 2006). As discussed in detail in chapter "Lipid Mediators in the Regulation of Emotions, Memory and Cognitive Functions" of this book, substantial evidence implicates the ECS in the regulation of affective state, and dysfunctional eCB signaling is associated with increased anxiety and depression. In light of the evidence for drug-induced changes in brain eCB levels and eCB-mediated synaptic regulation, it is conceivable that dysregulated eCB function contributes to affective disturbances associated with drug dependence and protracted withdrawal. Several studies have begun to investigate an involvement of eCB signaling in withdrawal-related affective disruptions (or the efficacy of eCB manipulations as treatments for these disruptions), and as described below, these studies have predominantly focused on anxiety-like behavior that accompanies withdrawal from chronic exposure to drugs of abuse and highly palatable food.

In tobacco-dependent individuals, smoking cessation leads to withdrawal symptoms that persist for several months (Hughes 2007; Hughes et al. 1991; Markou 2008). Symptoms such as dysphoria/aversion, anxiety, and irritability contribute to negative reinforcement mechanisms that perpetuate nicotine addiction (Hughes et al. 1991; Koob and Volkow 2010; Piper et al. 2011; Allen et al. 2008; Rose et al. 2010), and the severity of these symptoms is a predictor of increased relapse risk (Piasecki et al. 2003a, b, c; al'Absi et al. 2004). Because eCB signaling is implicated in the initiation and maintenance of nicotine self-administration, it is possible that persistent disruptions in eCB signaling contribute to the negative affective states of nicotine withdrawal. However, few studies have investigated this possibility.

The Maldonado group reported that CB_1 knockout mice exhibit more robust anxiety-like behavior during nicotine withdrawal than similarly treated wild-type mice do (Bura et al. 2010). However, nicotine-naïve CB_1 knockout mice also exhibit greater anxiety-like behavior than wild types (Bura et al. 2010; Haller et al. 2002; Martin et al. 2002; Uriguen et al. 2004), and this innate phenotype clouds the determination as to whether or not withdrawal-related anxiety-like behavior is exacerbated by CB_1 receptor deletion. In rats, significant increases in anxiety-like behavior are evident after 36 h of spontaneous nicotine withdrawal, and this is dose-dependently reversed by treatment with the FAAH inhibitor URB597 at doses that do not alter anxiety-like behavior in nicotine-naïve rats (Cippitelli et al. 2011). This is temporally correlated with significant increases in AEA content in the amygdala, hypothalamus, and prefrontal cortical tissue, though no significant alterations in tissue 2-AG content are evident regardless of region analyzed (Cippitelli et al. 2011). It is somewhat surprising that FAAH inhibition ameliorates anxiety-like behavior at a time when AEA levels appear to be increased in brain regions known to modulate anxiety-like behaviors. However, while increased AEA content in these regions may reflect a compensatory response to withdrawal, it is possible the effect magnitude is insufficient to fully constrain withdrawal-related disturbances underlying increased anxiety. Thus, the bolstering of this response through FAAH inhibition provides more efficacious reversal of withdrawal-related neural dysfunction (Cippitelli et al. 2011). It is worth noting, however, that disruptions in postmortem tissue AEA content may not provide a precise index of signaling-competent AEA levels in vivo (Buczynski and Parsons 2010), and a more clear in vivo view of withdrawal-related disruptions in regional eCBs may require further study. Nonetheless, these findings provide compelling evidence that acute FAAH inhibition ameliorates increased anxiety-like states during protracted nicotine withdrawal.

The putative eCB uptake inhibitor AM404 has been found to attenuate nicotine withdrawal-related increases in immobility in the Porsolt forced swim test (Mannucci et al. 2011). Increased immobility in this test is often interpreted as a form of despair or hopelessness, and this paradigm is commonly used as a screen for antidepressant efficacy. Accordingly, withdrawal-related increases in immobility in the Porsolt test may reflect a form of depressive-like behavior, and the effects of AM404 may thus reflect a beneficial effect of enhanced eCB signaling on withdrawal-related depressive-like states (Mannucci et al. 2011). However, AM404 also dose-dependently reduced immobility in nicotine-naïve mice in this study, and as such a preferential influence of this eCB clearance inhibitor on withdrawal-related behaviors is unclear. Moreover, modeling depressive-like behavior in rodents is best accomplished through the use of several paradigms (Abelaira et al. 2013; Yan et al. 2010), and further work is clearly needed to evaluate possible eCB influences on nicotine withdrawal-related depression-like behavior.

Post-traumatic stress disorder (PTSD) is an anxiety disorder with particularly high prevalence among individuals with alcohol use disorders (Kessler et al. 1995). The fear-potentiated startle (FPS) paradigm is often used to model PTSD-like behaviors (Grillon 2002; Hijzen et al. 1995). Rodent lines selectively bred for high alcohol consumption exhibit significantly greater FPS than corollary lines bred for low alcohol consumption (McKinzie et al. 2000; Barrenha and Chester

2007), and these rodent lines may thus provide models for characterizing mechanisms contributing to the anxiety, alcohol use disorders, and intersection of these pathologies. In this regard, the nonselective FAAH inhibitor LY2183240 selectively reduces FPS in high alcohol-preferring mice but not low alcohol-preferring mice (Powers et al. 2010). However, LY2183240 effects on FPS were evident only following repeated FPS testing and not upon first exposure to the FPS paradigm, suggesting that FAAH inhibition facilitates the extinction of learned fear responses in high alcohol-preferring mice, consistent with substantial evidence that FAAH inhibition generally accelerates extinction of aversive memories (Gunduz-Cinar et al. 2013). LY2183240 also enhanced the expression of alcohol-induced conditioned place preference without altering alcohol consumption itself. This suggests that FAAH inhibition influences memory-related processes regulating the expression of conditioned fear and conditioned alcohol reward in animals genetically predisposed toward high alcohol consumption. Recent evidence also points to the efficacy of the more selective FAAH inhibitor URB597 for attenuating anxiety-like behavior evident in rats following acute, high-dose alcohol administration (Cippitelli et al. 2008). In this study, URB597 did not alter alcohol self-administration, cue-induced alcohol-seeking, or stress-induced alcohol-seeking behavior. Collectively, these findings suggest that FAAH inhibition may be beneficial for reducing alcohol-related anxiety-like behavior, though further investigation is warranted, particularly evaluations of persistent anxiety-like behavior associated with protracted withdrawal in alcohol-dependent subjects.

In humans, the chronic use of the NMDA glutamate receptor antagonist phencyclidine (PCP) evokes both positive and negative symptoms of schizophrenia and many of the characteristic cognitive deficits associated with the disease (Murray 2002). In rats, withdrawal from sub-chronic PCP exposure is associated with reduced social interaction (Lee et al. 2005; Seillier and Giuffrida 2009; Seillier et al. 2013), a phenotype often interpreted as an anxiety-like behavior (Lapiz-Bluhm et al. 2008) and reminiscent of negative symptoms of schizophrenia. PCP withdrawal is associated with reduced AEA content in the amygdala and prefrontal cortex (Seillier and Giuffrida 2009; Seillier et al. 2013), and withdrawal-related reductions in social interaction are reversed by the FAAH inhibitor URB597 through a CB_1-reliant mechanism (Seillier and Giuffrida 2009; Seillier et al. 2013).

Similar to drug addiction, eating disorders and pathological obesity are conceptualized as chronic relapsing conditions with alternating periods of abstinence (e.g., dieting) and relapse (compulsive overeating) (Corwin and Grigson 2009; Cottone et al. 2009a; Epstein and Shaham 2010; Johnson and Kenny 2010; Parylak et al. 2011). Rats given scheduled intermittent access to highly palatable sucrose food exhibit a withdrawal state upon forced abstinence from the palatable food, including increased anxiety-like behavior and decreased motivation for standard lab food (Cottone et al. 2008, 2009a, b). These effects are most evident during early abstinence and abate within 2–3 days. Moreover, these animals exhibit robust excessive food consumption of the palatable food when given renewed access. The negative affective state during abstinence and the excessive consumption upon renewed access to palatable food are each mediated by increased CRF1 signaling in the amygdala (Iemolo et al. 2013), similar to withdrawal-related negative affective states and compulsive drug consumption evident in drug-dependent subjects (Koob et al. 2014; Koob 2010). This suggests that dysregulation of amygdalar signaling is a common factor contributing to pathological motivation for and consumption of food and drugs. Interestingly, 2-AG content is significantly increased in the CeA following 4 days' abstinence from palatable food, and this occurs in conjunction with increased CeA CB_1 mRNA and CB_1 protein expression (Blasio et al. 2013). Moreover, either systemic or intra-CeA administration of the CB_1 antagonist rimonabant precipitates an anxiogenic-like state in animals with a history of palatable food consumption but not in rats given only with standard lab chow. This suggests that increased 2-AG-mediated CB_1 signaling in CeA counteracts a withdrawal-like state in rats during abstinence from palatable food (Blasio et al. 2013).

12 Synaptic Plasticity

Substantial evidence demonstrates that exposure to most drugs of abuse dysregulates synaptic plasticity mechanisms in a variety of brain regions involved in the evolution of addiction (Hyman et al. 2006; Luscher and Malenka 2011). Acute exposure to alcohol, nicotine, morphine, or cocaine induces transient loss of LTP at inhibitory GABAergic synapses in the VTA (Niehaus et al. 2010; Liu et al. 2005; Melis et al. 2002; Nugent et al. 2007) and enhanced excitatory strength onto VTA DA neurons (Borgland et al. 2004; Faleiro et al. 2004; Saal et al. 2003). This tandem potentiation of excitatory transmission and loss of inhibitory LTP at VTA synapses serves to heighten midbrain DA cell excitability, increasing dopaminergic signaling throughout the mesocorticolimbic system.

As described in detail in the chapter titled "Endocannabinoid-Dependent Synaptic Plasticity in Striatum", eCBs and CB_1 receptors are implicated in several forms of synaptic plasticity involving presynaptic expression mechanisms (Lovinger 2008; Heifets and Castillo 2009). Both short- and long-term forms of eCB-mediated synaptic plasticity occur in many brain regions critically involved in various stages of the addiction cycle including the NAc, VTA, amygdala, prefrontal cortex, hippocampus, and dorsal striatum. Accordingly, drug-induced disruption of these mechanisms may contribute to aberrations in learning, memory, and affective state that propel addiction. Although a full description of these processes is provided in "Endocannabinoid-Dependent Synaptic Plasticity in Striatum", it is worth noting here that exposure to several classes of abused drugs dysregulates eCB-mediated forms of synaptic plasticity. For example, eCB-mediated LTD of excitatory and inhibitory signaling in the rodent NAc and hippocampus is abolished following a 7-day treatment with either Δ9-THC or the synthetic CB_1 agonist WIN 55,212-2 (Hoffman et al. 2003; Mato et al. 2004), correlating with decreased sensitivity to CB_1 inhibition of both excitatory and inhibitory signaling.

Both acute and chronic alcohol exposure substantially reduces a CB_1-dependent form of plasticity that results in long-lasting disinhibition of striatal output neurons (Clarke and Adermark 2010; Adermark et al. 2011) and reduces eCB-mediated LTD at inhibitory striatal synapses (though alcohol-induced disruption of eCB-LTD at excitatory synapses has not been observed) (Clarke and Adermark 2010). Because the dorsal striatum is involved in reward-guided learning and habitual behavior (Volkow et al. 2007; Yin et al. 2008), it is possible that these alcohol-induced disruptions in eCB-LTD contribute to maladaptive habitual behavior that contributes to addiction. Consistent with this hypothesis, recent work by the Lovinger lab demonstrates that chronic and intermittent alcohol exposure increases 2-AG levels in the dorsolateral striatum (DLS), leading to a compensatory downregulation of CB_1 receptor signaling and loss of CB_1-mediated LTD at excitatory synapses (DePoy et al. 2013). This was associated with enhancement of DLS-mediated learning processes that may contribute to habitual behaviors.

Cocaine exposure results in the loss of eCB-LTD of evoked excitatory transmission in the NAc (Fourgeaud et al. 2004) and facilitated eCB-LTD of GABAergic signaling at VTA DA synapses (Pan et al. 2008; Fourgeaud et al. 2004; Liu et al. 2005). Collectively, these cocaine-induced disruptions in eCB-LTD result in imbalanced mesolimbic DA function characterized by diminished inhibitory control over VTA DA cell bodies and heightened excitatory signaling in the NAc. Chronic cocaine exposure also results in disruption of eCB-LTD of excitatory transmission in the bed nucleus of the stria terminalis (BNST) (Grueter et al. 2006), a stress-responsive structure wherein excitatory transmission is critically involved in mediating stress-reward interactions and anxiety-like behavior (Delfs et al. 2000; McElligott and Winder 2009). The BNST sends substantial projections to the VTA, and accordingly disruption of BNST plasticity likely influences motivational responses to stress.

Little is known regarding the potential influence of opiate or nicotine exposure on eCB-mediated forms of synaptic plasticity. However, a recent report demonstrated that a history of nicotine self-administration induces CB_1-dependent LTP at synapses of infralimbic cortex afferents to the BNST (Reisiger et al. 2014). This perturbation may have relevance to many consequences of chronic nicotine exposure, and clear evidence is provided that this neuroplastic change contributes to cue-induced reinstatement of nicotine-seeking behavior (in an animal model of relapse).

As noted above, stress is implicated in the development of addiction (Koob and Kreek 2007), and it is therefore notable that stress can alter eCB-mediated plasticity in addiction-related brain regions. Prolonged exposure to unpredictable stress impairs eCB-mediated DSI, LTD, and fEPSPs in the NAc (Wang et al. 2010), while repeated restraint stress reduces eCB-mediated DSE and DSI in the rat PVN (Wamsteeker et al. 2010) and increases 2-AG-mediated DSI in the mouse basolateral amygdala (Patel et al. 2009). Similarly, establishment of conditioned fear increases the efficacy of eCB-mediated DSE and DSI in the central nucleus of the amygdala (Kamprath et al. 2011). Further, it has recently been shown that acute restraint stress induces a switch from eCB-mediated LTD to eCB-mediated LTP at synapses of mPFC afferents to the BNST (Glangetas et al. 2013). As previously discussed, the ECS participates in a negative feedback system that limits the expression of anxiety under stressful circumstances and contributes to the suppression of aversive memories. These processes are mediated in part through eCB-mediated synaptic plasticity in the amygdalar nuclei (Kamprath et al. 2011; Lafenetre et al. 2007; Viveros et al. 2007), and disruptions of this eCB-mediated plasticity may contribute to its dysregulated affect (including increased anxiety) associated with protracted drug abstinence.

13 Extinction and Relapse

The ECS plays a prominent role in learning and memory processes (Hashimotodani et al. 2007; Heifets and Castillo 2009; Marsicano and Lafenetre 2009), and because eCB signaling is disrupted by most drugs of abuse (see above sections), a role for the ECS in learning and memory components of addiction may be hypothesized. As previously reviewed, CB_1 receptors play an important role in the conditioned rewarding effects of alcohol (Houchi et al. 2005; Hungund et al. 2003; Thanos et al. 2005), opiates (Chaperon et al. 1998; Martin et al. 2000; Navarro et al. 2001; Singh et al. 2004), and nicotine (Castane et al. 2002; Forget et al. 2005, 2006; Le Foll and Goldberg 2004; Merritt et al. 2008), with a lesser CB_1 influence reported for the conditioning effects of psychostimulants (Houchi et al. 2005; Martin et al. 2000). Although these behaviors are generally interpreted in the context of drug reward, a CB_1 receptor influence on the associative learning aspects of drug exposure is also likely, which as discussed below may have relevance to the persistent reactivity to drug-related memories that characterize addiction.

13.1 Drug Seeking (Relapse)

Drug exposure produces powerful interoceptive effects, and memory of these effects increases the likelihood of continued drug use. With continued drug use, these interoceptive effects become associated with environmental cues to the extent where these drug-associated cues alone can induce craving and thereby propel drug use. This form of drug-related memory is also causal in relapse to drug use following periods of abstinence (Carter and Tiffany 1999; McLellan et al. 2000). Several factors are believed to be causal in drug relapse including craving induced by environments or situations previously associated with drug use (e.g., conditioning factors), acute exposure to the drug itself or a pharmacologically related agent during abstinence (e.g., drug priming), and stressful events.

Recently developed animal models of drug seeking demonstrate an important influence of cannabinoid signaling in the reinstatement of extinguished drug-seeking and drug-taking behaviors. For example, Δ9-THC and various synthetic

CB_1 agonists reinstate drug seeking for cannabinoids (Justinova et al. 2008b; Spano et al. 2004), opioids (De Vries et al. 2003; Fattore et al. 2003, 2005), ethanol (Lopez-Moreno et al. 2004; McGregor et al. 2005), nicotine (Biala and Budzynska 2008; Gamaleddin et al. 2012a), and cocaine (De Vries et al. 2001). Conversely, CB_1 receptor antagonism attenuates drug-seeking behavior associated with a variety of abused substances. Rimonabant (SR141716A) significantly reduces operant responding for conditioned cues previously associated with Δ9-THC (Justinova et al. 2008b), heroin (De Vries et al. 2003; Fattore et al. 2003; 2005), nicotine (Cohen et al. 2005; De Vries et al. 2005; Diergaarde et al. 2008), and ethanol (Cippitelli et al. 2005; Economidou et al. 2006). CB_1 antagonism also reduces excessive ethanol intake exhibited following periods of abstinence (an animal model of relapse-like alcohol consumption) (Gessa et al. 2005), and the novel CB_1 antagonist SLV330 reduces both drug- and cue-induced reinstatement of ethanol- and nicotine-seeking behavior (de Bruin et al. 2011). Studies employing site-specific antagonist infusions have uncovered important contributions of CB_1 receptors in the PFC and NAc shell in cue-induced reinstatement of both heroin- and nicotine-seeking behavior, and CB_1 receptors in the BLA appear to contribute to cue-induced nicotine-seeking, but not heroin-seeking, behavior (Alvarez-Jaimes et al. 2008; Kodas et al. 2007). Although CB_1 receptor inactivation produces subtle and inconsistent effects on psychostimulant self-administration as compared with other drugs of abuse, substantial evidence indicates a strong CB_1 receptor influence on the reinstatement of psychostimulant-seeking behavior. CB_1 antagonism attenuates both drug-primed and cue-induced reinstatement of cocaine-seeking (De Vries et al. 2001; Filip et al. 2006; Xi et al. 2006; Adamczyk et al. 2012b) and methamphetamine-seeking behavior in rats (Anggadiredja et al. 2004; Boctor et al. 2007). Further, both drug- and cue-induced reinstatement of cocaine and methamphetamine seeking is attenuated by the negative allosteric CB_1 modulator ORG27569 (Jing et al. 2014). Interestingly, although footshock-induced reinstatement of cocaine seeking is not blocked by either rimonabant or the CB_1 antagonist AM251 (De Vries et al. 2001; Kupferschmidt et al. 2012a), AM251 blocks the reinstatement of cocaine seeking induced by forced swim stress and exogenous CRF administration (Vaughn et al. 2012; Kupferschmidt et al. 2012a), suggesting a CB_1 influence on some forms of stress-induced reinstatement of drug seeking. These latter findings may relate specifically to an AM251-induced modulation of the effects produced by the stress peptide CRF, given that AM251 also attenuates the anxiogenic effects of both cocaine withdrawal (an effect mediated by increased CRF signaling (DeVries and Pert 1998)) and exogenously administered CRF (Kupferschmidt et al. 2012b). However, the mechanisms for this putative interaction are not clear, and these findings are at odds with evidence that CB_1 signaling generally suppresses the physiological and behavioral responses to stress (see discussion below).

Thus, CB_1 receptor signaling is implicated in drug-seeking behavior for a variety of abused substances that differ substantially in pharmacodynamic mechanisms including cannabinoids, opioids, alcohol, nicotine, and psychostimulants. Moreover, CB_1 receptor antagonism blocks both cue- and "prime"-induced reinstatements of seeking behavior for nondrug rewards such as sucrose and corn oil ((De Vries et al. 2005; Ward et al. 2009), but see Xi et al. (2006)). Accordingly, CB_1 receptor signaling appears to participate in the modulation of conditioned reward in general.

Few studies have characterized the influence of eCB signaling on drug-seeking behavior through more direct manipulations of eCB processing such as the inhibition of hydrolytic clearance mechanisms. Three studies have found that both drug-primed and cue-induced nicotine- and cocaine-seeking behaviors are reduced by URB597 and the less selective FAAH inhibitor PMSF (Forget et al. 2009; Scherma et al. 2008; Adamczyk et al. 2009). This is somewhat surprising given that CB_1 receptor stimulation enhances both nicotine- and cocaine-seeking behavior (Biala and Budzynska 2008; De Vries et al. 2001). However, because tonic eCB production is believed to be low, the inhibition of eCB clearance is anticipated to preferentially amplify eCB signaling in circuits/synapses activated by a given stimulus (in this case, drug-seeking behavior), rather than producing more widespread indiscriminate CB_1 activation as is produced by exogenous CB_1 agonists. Moreover, FAAH hydrolyzes a large variety of fatty acid moieties (Ahn et al. 2008), and it is conceivable that the effects of FAAH inhibition on drug seeking are mediated through actions of non-cannabinoid lipids. In this regard, it's notable that the putative eCB uptake inhibitor VDM11 also attenuates both nicotine- and cue-induced nicotine-seeking behavior (Gamaleddin et al. 2011). This latter finding is of interest as unlike URB597, VDM11 elevates AEA levels with reduced effects on non-cannabinoid FAAH substrates that may influence behavior through PPARα signaling (De Petrocellis et al. 2000; van der Stelt et al. 2006). Similar to the effects of VDM11, the putative eCB uptake inhibitor AM404 dose-dependently attenuates both nicotine- and cue-induced nicotine-seeking behavior, though AM404 did not alter nicotine self-administration itself (Gamaleddin et al. 2013). In contrast to the effects on nicotine- and cocaine-seeking behavior, neither URB597 nor AM404 alters either cue- or stress-induced reinstatement of alcohol-seeking behavior (Cippitelli et al. 2007; 2008).

Recent data demonstrate that detoxified alcoholics present significantly lower baseline plasma AEA levels as compared with nondependent social drinkers (Mangieri et al. 2009). Further, a significant negative relationship between resting plasma AEA levels and the severity of cue-induced craving was evident in social drinkers but not alcoholics. In social drinkers, alcohol-related cues significantly increased both craving and plasma AEA levels, and the relative magnitude of cue-induced increases in AEA was significantly correlated with the severity of craving. However, cue-induced changes in AEA were indexed by a percentage change from baseline, and thus this index is substantially influenced by baseline AEA levels (e.g., the same absolute change in AEA content will be reflected as a greater percentage change in subjects with lower vs. higher baseline AEA levels). Accordingly, the positive correlation between cue-induced changes in plasma AEA and craving intensity further supports the importance of baseline AEA levels in cue-induced craving in social drinkers. Interestingly, although alcohol-related cues elicited more intense craving in alcoholics vs. social drinkers, alcoholics did not present significant cue-induced increases in plasma AEA. Based on these observations, it was concluded that cue-induced increases in plasma AEA contributes to interoceptive signaling that moderates the desire for alcohol in social drinkers. It may also be concluded that deficits in plasma AEA levels confer increased susceptibility to cue-induced alcohol craving. While cue-induced elevations in plasma AEA may contribute to craving responses, it is curious that elevations in AEA were not evident in alcoholics who also presented the greatest level of cue-induced craving. It is conceivable that cue-induced increases in plasma AEA reflect an adaptive response, and its absence in alcoholics somehow contributes to excessive craving responses to alcohol-related cues. Further evaluations in humans are clearly needed to investigate these issues.

13.2 Extinction Learning

The potent motivational effects of drug-related cues create substantial difficulties during periods of attempted drug abstinence and are causal in the reinstatement of drug intake (e.g., relapse) (Carter and Tiffany 1999). One approach for lessening the motivational impact of drug-associated cues is through extinction learning, wherein a subject learns that a drug-associated cue no longer has predictive value. However, extinction therapy for addiction is generally ineffective for reducing relapse in both humans (Conklin and Tiffany 2002) and rodents (Crombag and Shaham 2002), and it is conceivable that this results from disruptions in the learning mechanisms that override the original association memory (Bouton 2004; Rescorla 1996). The ECS plays a prominent role in memory extinction, particularly the extinction of aversive memory. Substantial evidence demonstrates that CB_1 signaling blockade results in impaired extinction of cued fear memory, contextual fear memory, and fear-potentiated startle and spatial memory under aversive mildly stressful conditions (Marsicano et al. 2002; Suzuki et al. 2004; Chhatwal et al. 2005, 2009; Niyuhire et al. 2007; Varvel et al. 2005; Varvel and Lichtman 2002; Pamplona et al. 2006, 2008). Moreover, recent findings demonstrate that a line of mice selectively bred for high alcohol consumption exhibits excessive fear-potentiated startle responses and that the nonselective FAAH inhibitor LY2183240 facilitates extinction of fear-potentiated startle in these mice but not mice bred for low alcohol consumption (Powers et al. 2010). This suggests that FAAH inhibition influences memory-related processes regulating the expression of conditioned fear in animals genetically predisposed toward high alcohol consumption. However, it has consistently been reported that CB_1 inactivation does not alter the extinction of behaviors motivated by appetitive and non-aversive memories (Holter et al. 2005; Niyuhire et al. 2007; Ward et al. 2007; Harloe et al. 2008; Hernandez and Cheer 2011). Because aversive memory may be involved in relapse to drug taking (Kaplan et al. 2011; Quirk and Gehlert 2003; Stewart and Kushner 2001), deficient eCB signaling following long-term drug exposure may contribute to the limited efficacy of extinction therapy for addiction.

13.3 Amygdala: Cortical Communication as a Possible Mechanism for eCB Modulation of Drug-Seeking Behavior

Considerable evidence highlights the importance of functional communication between the basolateral amygdala and prefrontal cortex (BLA->PFC) as a mediator of emotionally salient learning, memory, and synaptic plasticity. Distinct subpopulations of neurons in the PFC control the processing of aversive fear-related memory (Laviolette et al. 2005; Lauzon et al. 2009; Laviolette and Grace 2006) and rewarding and appetitive memory (Bishop et al. 2011; Sun et al. 2011), and both of these subpopulations are influenced by CB_1 modulation of BLA inputs (Tan et al. 2014). Growing evidence demonstrates that excessive activation of CB_1 receptors in either the BLA or PFC leads to amplification of aversive emotional information, while deficient CB_1 signaling in these areas blunts the salience of normally aversive events while enhancing the salience of sub-reward threshold stimuli (Tan et al. 2010, 2014; Laviolette and Grace 2006). These processes appear to be mediated through direct interactions with subcortical DAergic motivational systems arising from the VTA, and through these interactions, cortical CB_1 receptors are capable of modulating reward processing, even the switching of emotional valence from normally rewarding conditioned stimuli to aversive stimuli (Ahmad et al. 2013). The specific contribution of these mechanisms in the pathologically elevated motivational salience to drug-predictive cues and/or motivational effects of aversive memory contributing to drug seeking and relapse remains to be characterized (Tan et al. 2014). However, the potential involvement of these mechanisms not only in addiction but also in abnormal emotional learning and memory in other neuropsychiatric conditions is worth noting.

14 Cognitive Function

Impairments in cognitive processing are hallmarks of addiction. The progression from initial, intermittent, controlled drug use to compulsive drug seeking and consumption in the addicted state is conceptualized as a progressive deterioration of executive control over behavior (primarily reliant on frontal-cortical and hippocampal function) and the emergence of habitual or compulsive behaviors that drive drug seeking and drug taking (mediated in large part through dorsal striatal signaling) (Everitt and Robbins 2005; Jentsch and Taylor 1999). Endocannabinoids exert prominent control of frontal-cortical, hippocampal, and striatal synaptic signaling, and growing evidence implicates dysregulated eCB signaling in both aberrant executive function and habit-based behaviors driven by dorsal striatal signaling.

Endocannabinoid signaling regulates neural activity, structural plasticity, and functional output of the medial prefrontal cortex (McLaughlin et al. 2014) and is implicated in several PFC-mediated cognitive functions including attentional set shifting (Klugmann et al. 2011) and reversal learning in the Morris water maze (Varvel and Lichtman 2002; Lee et al. 2014). CB_1 receptor antagonism or deletion results in shorter and less complex dendrites of neurons in layer II/III of the prelimbic cortex (Hill et al. 2011; Lee et al. 2014) and impairments in cognitive flexibility that resemble behaviors in mPFC-lesioned animals (Lacroix et al. 2002; Lee et al. 2014). Conversely, genetic deletion or pharmacological inhibition of FAAH enhances the acquisition of the Barnes maze task (Wise et al. 2009), enhances memory acquisition in a passive avoidance task (Mazzola et al. 2009), and facilitates acquisition and extinction of a spatial memory task (Varvel et al. 2007). Physical activity increases circulating AEA levels in both humans and rodents (Sparling et al. 2003; Hill et al. 2010b; Raichlen et al. 2013; Heyman et al. 2012), and this may contribute to the beneficial effects of exercise on hippocampal cell proliferation, neurogenesis, and synaptic plasticity (Hill et al. 2010b; Wolf et al. 2010; Madronal et al. 2012). Treadmill running increases spatial working memory through a CB_1-reliant mechanism, and similar working memory enhancement is observed in sedentary animals treated with URB597 (Ferreira-Vieira et al. 2014).

Despite growing evidence for an eCB influence in cognitive functions mediated by the mPFC and hippocampus, few studies have investigated disruptions in these mechanisms following long-term drug exposure. Drug-induced impairments in recognition memory may result from altered hippocampal CB_1 receptor function, and blockade of CB_1 receptors during drug exposure has been shown to prevent deficits in recognition memory produced by chronic morphine, MDMA, and nicotine exposure (Nawata et al. 2010; Vaseghi et al. 2012, 2013; Saravia et al. 2016). Regarding drug-induced disruptions in frontal cortical function, the Woodward lab recently demonstrated that alcohol withdrawal is associated with deficient CB_1-mediated inhibition of GABAergic signaling in layers II/III of the PFC and possible increases in CB_1-mediated inhibition of GABAergic signaling in layers V/VI (Pava and Woodward 2014), suggesting a role for dysregulated CB_1 signaling in altered cortical network activity that may underlie impaired cognitive function in detoxified alcoholics.

15 Food Addiction and Eating Disorders

Food addiction is defined as "a loss of control over food intake," and growing evidence demonstrates that food and drug addiction have overlapping neuroadaptations in the mesolimbic system (Lutter and Nestler 2009). Indeed, eating disorders such as binge eating disorder (BED), anorexia nervosa (AN), and bulimia nervosa (BN) are hypothesized to derive in part from aberrant brain reward function (Kaye 2008; Marco et al. 2012; Scherma et al. 2014; Stoving et al. 2009). Similar to drug addiction, food addiction is defined in part by compulsive eating, continued aberrant feeding behavior despite negative consequences, and unsuccessful attempts to "normalize" dysfunctional eating (Cassin and von Ranson 2007;

Table 2 Summary of ECS involvement in food intake and eating disorders

Pharmacological manipulation	Healthy animals	Binge eating disorder	Anorexia nervosa
Baseline ECS	↑↓ Energy balance[a] ↑↓ Food intake ↑↓ Hedonic aspects of feeding	↑ AEA levels in blood[b]	↑ AEA levels in blood[c] ↑ CB_1 mRNA in blood ↑ CB_1 binding in brain ↓ FAAH activity
CB_1R antagonist	↓ Motivation for food[d] ↓ Preference for palatable foods ↓ Hedonic effects of food	↓ Compulsive eating[e] ↑ Weight loss	
CB_1R agonist	↑ Motivation for food[f] ↑ Preference for palatable foods ↑ Hedonic effects of food		↑ Weight gain[g]

[a]Silvestri and Di Marzo (2013), Melis et al. (2007), Di Marzo et al. (2001), Bermudez-Silva et al. (2012), Kirkham et al. (2002), Hanus et al. (2003), Berridge et al. (2010), Jyotaki et al. (2010), Yoshida et al. (2010), Cervino et al. (2009)
[b]Monteleone et al. (2005)
[c]Monteleone et al. (2005, 2009), Frieling et al. (2009), Gerard et al. (2011)
[d]Simiand et al. (1998), South et al. (2007), Mathes et al. (2008), Chaperon et al. (1998), Gallate and McGregor (1999), Gallate et al. (1999), Solinas and Goldberg (2005), Jarrett et al. (2007), Melis et al. (2007)
[e]Scherma et al. (2013), Pataky et al. (2013)
[f]Kirkham and Williams (2001), Farrimond et al. (2011), Kirkham et al. (2002), Williams and Kirkham (1999), Martinez-Gonzalez et al. (2004), Solinas and Goldberg (2005), Guegan et al. (2013), Dipatrizio and Simansky (2008), Shinohara et al. (2009), Higgs et al. (2003), Jarrett et al. (2005, 2007), Mahler et al. (2007), Gallate et al. (1999), Dodd et al. (2009), Anderson-Baker et al. (1979), Jamshidi and Taylor (2001), Verty et al. (2005), Soria-Gomez et al. (2007)
[g]Andries et al. (2014)

Gearhardt et al. 2012). The influence of cultural pressures and stress in these disorders is well documented, though there are also strong genetic contributions to these diseases (Kaye 2008; Klump et al. 2001). Furthermore, it is estimated that more than 50% of individuals with BED meet the criteria for food addiction (Cassin and von Ranson 2007; Gearhardt et al. 2012).

Several findings suggest an ECS involvement in eating disorders, including aberrant reward processing associated with these conditions (Table 2). For example, abnormal eating behaviors like self-starvation (in AN) or binge eating (in BN or BED) become rewarding in their own right, and this may be influenced by disrupted eCB function (Monteleone et al. 2012; Cervino et al. 2009). Patients with AN or BED (but not BN) present with increased blood AEA levels as compared with controls, though no changes in blood 2-AG levels are associated with these disorders (Monteleone et al. 2005). Moreover, blood AEA levels are significantly and inversely correlated with plasma leptin levels in women with AN, consistent with the inhibitory influence of leptin on AEA synthesis (Di Marzo et al. 2001) and leptin deficiency in these eating disorders. It should be noted, however, that although plasma eCB levels provide a potential index of disrupted eCB processing, they do not directly reflect dynamic changes in eCB signaling in the brain or other tissues. A *CNR1* polymorphism is associated with vulnerability to AN (Siegfried et al. 2004) (but see Muller et al. (2008)), CB_1 mRNA levels are increased in the blood of AN and BN patients (Frieling et al. 2009), and increased brain CB_1 binding has been observed in patients with AN and BN (Gerard et al. 2011). Specifically, AN and BN patients exhibit increased CB_1 availability in the bilateral insular cortex (a region critically involved in reward and emotional processing as well as interoception), and increased CB_1 availability is evident in the inferior frontal and temporal cortices of AN patients (regions involved in emotional processing and executive function). An association between the *C385A* polymorphism in the gene encoding FAAH (leading to diminished FAAH activity) and the *CNR1* polymorphism rs1049353 has been observed in higher frequency in patients with AN and BN (Monteleone et al. 2009) with a synergistic effect of these polymorphisms evident in patients with AN. The *C385A* polymorphism is also significantly associated with obesity (Monteleone et al. 2008; Sipe et al. 2005), though the relative contributions of possible disruptions in metabolism and hedonic mechanisms have not been determined. An association between a nonsynonymous *CNR2* polymorphism and both AN and BN has also recently been reported (Ishiguro et al. 2010).

Limited research has been conducted on the effects of CB manipulations in animal models of eating disorders. Sub-chronic Δ9-THC treatment (0.5–2 mg/kg/day) was found to increase food intake and reduce body weight loss in an activity-based rat model of anorexia nervosa (Verty et al. 2005), though these effects were not evident in similar studies conducted in mice (Lewis and Brett 2010). Conversely, CB_1 receptor antagonism selectively reduces binge-like vs. normal patterns of consumption for highly palatable sweet and high-fat foods (Scherma et al. 2013). It is possible this latter finding involves a mesolimbic DA mechanism in light of evidence that binge-like consumption of palatable or high-fat foods elicits increased NAc DA release upon repeated bouts of consumption (Rada et al. 2005; Liang et al. 2006), while the DAergic response to normal patterns of consumption abates upon repeated exposure to palatable foods (Bassareo and Di Chiara 1999b). Regarding clinical studies in humans, two small initial trials assessing the efficacy of Δ9-THC treatments for AN failed to demonstrate significant benefits on weight gain, though some beneficial effects on depression and perfectionism were evident (for review see Stoving et al. 2009). However, a more recent and larger trial demonstrated that a 4-week treatment with dronabinol induced a small but significant weight gain in women suffering from severe-enduring AN (Andries et al. 2014). Further, a recent randomized, placebo-controlled, double-blind trial in 289 obese subjects with binge eating disorders demonstrated that a 6-month treatment with rimonabant significantly reduced binge eating and led to significant weight loss with modest presentation of adverse psychiatric side effects (Pataky et al. 2013).

16 Conclusions and Future Directions

Endocannabinoid signaling participates in the mediation and modulation of both natural reward (such as food consumption, exercise, sex, and social interaction) and drug-induced reward. Most natural rewards and drugs of abuse alter brain eCB levels, and there is a robust CB_1R influence on the motivation for both natural and drug rewards. Chronic drug use results in a neuroadaptive downregulation of CB_1R and/or CB_2R availability and function and may also lead to disruptions in eCB biosynthesis and clearance. The resulting deficit in eCB signaling may contribute to increased stress responsivity, increased negative affect, impaired extinction of drug-related memories, and drug seeking/craving that are known contributors to relapse. Recent animal studies suggest that eCB clearance inhibitors may have therapeutic potential for ameliorating these behavioral abnormalities related to addiction disorders. This therapeutic approach may present fewer unwanted behavioral effects than that produced by exogenous cannabinoids, given that eCBs are generally produced in a synapse-specific manner, and as such, eCB clearance inhibition may preferentially facilitate eCB signaling in specific circuits engaged by distinct stimuli (e.g., stress, drug-associated cues, etc.).

There are a number of notable gaps in our understanding of the eCB influence on reward and addiction. The ECS plays a prominent role in neuronal guidance and brain development (Maccarrone et al. 2014), and as such, disruptions in eCB function at an early age likely have substantial consequences for adult brain function. This is underscored by increasing evidence of the long-term consequences of prenatal or adolescent cannabinoid exposure (Hurd et al. 2014; Calvigioni et al. 2014). Though the effects of early-life exposure to non-cannabinoid drugs are well studied, the specific contributions of persistent drug-induced disruptions in eCB signaling on adult neural function and behavior are not understood. Robust bidirectional interactions between the ECS and sex hormones are now recognized (Gorzalka and Dang 2012), but few studies have characterized possible sex differences in the eCB influence on reward function, addiction, stress, and cognitive processing. There are also substantial limitations in the interpretation and replication of genetic analyses of the eCB influence in addiction due to heterogeneity of the populations studied, drug class, polysubstance use, and even drug use phenotypes examined. Large-scale future studies across different populations and drug classes will be critical to understand the relative impact and causal nature of ECS-related genetic mutations in the vulnerability to addictive disorders. Filling these gaps of knowledge is critical given the important need for scientific data to help guide current discussions and changes being made in marijuana legalization policies.

Acknowledgments This work was supported by grants from the US National Institutes of Health (NIH): AA020404, AA006420, AA022249 and AA017447 to LHP, and T32 AA 7456-33 and F32 AA025257-01 to SAL. This is manuscript number 29294 from The Scripps Research Institute. This chapter is dedicated in loving memory of Dr. Loren H Parsons. He was an exceptional mentor and compassionate friend who contributed heavily to endocannabinoid research. He has left a strong legacy for the scientific community to build on and will be sorely missed.

References

al'Absi M, Hatsukami D, Davis GL, Wittmers LE (2004) Prospective examination of effects of smoking abstinence on cortisol and withdrawal symptoms as predictors of early smoking relapse. Drug Alcohol Depend 73(3):267–278. doi:10.1016/j.drugalcdep.2003.10.014

Abelaira HM, Reus GZ, Quevedo J (2013) Animal models as tools to study the pathophysiology of depression. Revista brasileira de psiquiatria 35(Suppl 2):S112–S120. doi:10.1590/1516-4446-2013-1098

Adamczyk P, McCreary AC, Przegalinski E, Mierzejewski P, Bienkowski P, Filip M (2009) The effects of fatty acid amide hydrolase inhibitors on maintenance of cocaine and food self-administration and on reinstatement of cocaine-seeking and food-taking behavior in rats. J Physiol Pharmacol 60(3):119–125

Adamczyk P, Faron-Gorecka A, Kusmider M, Dziedzicka-Wasylewska M, Papp M, Filip M (2012a) Long-lasting increase in [(3)H]CP55,940 binding to CB1 receptors following cocaine self-administration and its withdrawal in rats. Brain Res 1451:34–43. doi:10.1016/j.brainres.2012.02.052

Adamczyk P, Miszkiel J, McCreary AC, Filip M, Papp M, Przegalinski E (2012b) The effects of cannabinoid CB1, CB2 and vanilloid TRPV1 receptor antagonists on cocaine addictive behavior in rats. Brain Res 1444:45–54. doi:10.1016/j.brainres.2012.01.030

Adermark L, Jonsson S, Ericson M, Soderpalm B (2011) Intermittent ethanol consumption depresses endocannabinoid-signaling in the dorsolateral striatum of rat. Neuropharmacology 61(7):1160–1165. doi:10.1016/j.neuropharm.2011.01.014

Ahmad T, Lauzon NM, de Jaeger X, Laviolette SR (2013) Cannabinoid transmission in the prelimbic cortex bidirectionally controls opiate reward and aversion signaling through dissociable kappa versus mu-opiate receptor dependent mechanisms. J Neurosci 33 (39):15642–15651. doi:10.1523/JNEUROSCI.1686-13.2013

Ahn K, McKinney MK, Cravatt BF (2008) Enzymatic pathways that regulate endocannabinoid signaling in the nervous system. Chem Rev 108(5):1687–1707. doi:10.1021/cr0782067

Al Mansouri S, Ojha S, Al Maamari E, Al Ameri M, Nurulain SM, Bahi A (2014) The cannabinoid receptor 2 agonist, beta-caryophyllene, reduced voluntary alcohol intake and attenuated ethanol-induced place preference and sensitivity in mice. Pharmacol Biochem Behav 124:260–268. doi:10.1016/j.pbb.2014.06.025

Alici T, Appel JB (2004) Increasing the selectivity of the discriminative stimulus effects of delta 9-tetrahydrocannabinol: complete substitution with methanandamide. Pharmacol Biochem Behav 79(3):431–437. doi:10.1016/j.pbb.2004.08.020

Allen SS, Bade T, Hatsukami D, Center B (2008) Craving, withdrawal, and smoking urges on days immediately prior to smoking relapse. Nicotine Tob Res 10(1):35–45. doi:10.1080/14622200701705076

Alling C, Balldin J, Bokstrom K, Gottfries CG, Karlsson I, Langstrom G (1982) Studies on duration of a late recovery period after chronic abuse of ethanol. A cross-sectional study of biochemical and psychiatric indicators. Acta Psychiatr Scand 66(5):384–397

Alvarez-Jaimes L, Parsons LH (2009) Regional influence of CB1 receptor signaling on ethanol self-administration by rats. Open Neuropsychopharmacol 2:77–85

Alvarez-Jaimes L, Polis I, Parsons LH (2008) Attenuation of cue-induced heroin-seeking behavior by cannabinoid CB1 antagonist infusions into the nucleus accumbens core and prefrontal cortex, but not basolateral amygdala. Neuropsychopharmacology 33(10):2483–2493. doi:10.1038/sj.npp.1301630

Alvarez-Jaimes L, Stouffer DG, Parsons LH (2009) Chronic ethanol treatment potentiates ethanol-induced increases in interstitial nucleus accumbens endocannabinoid levels in rats. J Neurochem 111(1):37-48. doi:10.1111/j.1471-4159.2009.06301.x

Anderson-Baker WC, McLaughlin CL, Baile CA (1979) Oral and hypothalamic injections of barbiturates, benzodiazepines and cannabinoids and food intake in rats. Pharmacol Biochem Behav 11(5):487–491

Andries A, Frystyk J, Flyvbjerg A, Stoving RK (2014) Dronabinol in severe, enduring anorexia nervosa: a randomized controlled trial. Int J Eat Disord 47(1):18–23. doi:10.1002/eat.22173

Anggadiredja K, Nakamichi M, Hiranita T, Tanaka H, Shoyama Y, Watanabe S, Yamamoto T (2004) Endocannabinoid system modulates relapse to methamphetamine seeking: possible mediation by the arachidonic acid cascade. Neuropsychopharmacology 29(8):1470–1478. doi:10.1038/sj.npp.1300454

Annis HM, Sklar SM, Moser AE (1998) Gender in relation to relapse crisis situations, coping, and outcome among treated alcoholics. Addict Behav 23(1):127–131

Aracil-Fernandez A, Trigo JM, Garcia-Gutierrez MS, Ortega-Alvaro A, Ternianov A, Navarro D, Robledo P, Berbel P, Maldonado R, Manzanares J (2012) Decreased cocaine motor sensitization and self-administration in mice overexpressing cannabinoid CB(2) receptors. Neuropsychopharmacology 37(7):1749–1763. doi:10.1038/npp.2012.22

Arnone M, Maruani J, Chaperon F, Thiebot MH, Poncelet M, Soubrie P, Le Fur G (1997) Selective inhibition of sucrose and ethanol intake by SR 141716, an antagonist of central cannabinoid (CB1) receptors. Psychopharmacology 132(1):104–106

Atwood BK, Mackie K (2010) CB2: a cannabinoid receptor with an identity crisis. Br J Pharmacol 160(3):467–479. doi:10.1111/j.1476-5381.2010.00729.x

Balerio GN, Aso E, Berrendero F, Murtra P, Maldonado R (2004) Delta9-tetrahydrocannabinol decreases somatic and motivational manifestations of nicotine withdrawal in mice. Eur J Neurosci 20(10):2737–2748. doi:10.1111/j.1460-9568.2004.03714.x

Balerio GN, Aso E, Maldonado R (2006) Role of the cannabinoid system in the effects induced by nicotine on anxiety-like behaviour in mice. Psychopharmacology 184(3-4):504–513. doi:10.1007/s00213-005-0251-9

Barna I, Zelena D, Arszovszki AC, Ledent C (2004) The role of endogenous cannabinoids in the hypothalamo-pituitary-adrenal axis regulation: in vivo and in vitro studies in CB1 receptor knockout mice. Life Sci 75(24):2959–2970. doi:10.1016/j.lfs.2004.06.006

Barrenha GD, Chester JA (2007) Genetic correlation between innate alcohol preference and fear-potentiated startle in selected mouse lines. Alcohol Clin Exp Res 31(7):1081–1088. doi:10.1111/j.1530-0277.2007.00396.x

Basavarajappa BS, Hungund BL (1999) Chronic ethanol increases the cannabinoid receptor agonist anandamide and its precursor N-arachidonoyl phosphatidylethanolamine in SK-N-SH cells. J Neurochem 72(2):522–528

Basavarajappa BS, Cooper TB, Hungund BL (1998) Chronic ethanol administration down-regulates cannabinoid receptors in mouse brain synaptic plasma membrane. Brain Res 793 (1-2):212–218

Basavarajappa BS, Saito M, Cooper TB, Hungund BL (2000) Stimulation of cannabinoid receptor agonist 2-arachidonylglycerol by chronic ethanol and its modulation by specific neuromodulators in cerebellar granule neurons. Biochim Biophys Acta 1535(1):78–86

Basavarajappa BS, Saito M, Cooper TB, Hungund BL (2003) Chronic ethanol inhibits the anandamide transport and increases extracellular anandamide levels in cerebellar granule neurons. Eur J Pharmacol 466(1-2):73–83

Bassareo V, Di Chiara G (1999a) Differential responsiveness of dopamine transmission to food-stimuli in nucleus accumbens shell/core compartments. Neuroscience 89(3):637–641

Bassareo V, Di Chiara G (1999b) Modulation of feeding-induced activation of mesolimbic dopamine transmission by appetitive stimuli and its relation to motivational state. Eur J Neurosci 11(12):4389–4397

Bequet F, Uzabiaga F, Desbazeille M, Ludwiczak P, Maftouh M, Picard C, Scatton B, Le Fur G (2007) CB1 receptor-mediated control of the release of endocannabinoids (as assessed by microdialysis coupled with LC/MS) in the rat hypothalamus. Eur J Neurosci 26 (12):3458–3464. doi:10.1111/j.1460-9568.2007.05900.x

Bermudez-Silva FJ, Cardinal P, Cota D (2012) The role of the endocannabinoid system in the neuroendocrine regulation of energy balance. J Psychopharmacol 26(1):114–124. doi:10.1177/0269881111408458

Berridge KC, Ho CY, Richard JM, DiFeliceantonio AG (2010) The tempted brain eats: pleasure and desire circuits in obesity and eating disorders. Brain Res 1350:43–64. doi:10.1016/j.brainres.2010.04.003

Bhargava HN (1976) Effect of some cannabinoids on naloxone-precipitated abstinence in morphine-dependent mice. Psychopharmacology 49(3):267–270

Biala G, Budzynska B (2008) Calcium-dependent mechanisms of the reinstatement of nicotine-conditioned place preference by drug priming in rats. Pharmacol Biochem Behav 89 (1):116–125. doi:10.1016/j.pbb.2007.12.005

Bishop SF, Lauzon NM, Bechard M, Gholizadeh S, Laviolette SR (2011) NMDA receptor hypofunction in the prelimbic cortex increases sensitivity to the rewarding properties of opiates via dopaminergic and amygdalar substrates. Cereb Cortex 21(1):68–80. doi:10.1093/cercor/bhq060

Blanco-Calvo E, Rivera P, Arrabal S, Vargas A, Pavon FJ, Serrano A, Castilla-Ortega E, Galeano P, Rubio L, Suarez J, Rodriguez de Fonseca F (2014) Pharmacological blockade of either cannabinoid CB1 or CB2 receptors prevents both cocaine-induced conditioned locomotion and cocaine-induced reduction of cell proliferation in the hippocampus of adult male rat. Front Integr Neurosci 7:106. doi:10.3389/fnint.2013.00106

Blasio A, Iemolo A, Sabino V, Petrosino S, Steardo L, Rice KC, Orlando P, Iannotti FA, Di Marzo V, Zorrilla EP, Cottone P (2013) Rimonabant precipitates anxiety in rats withdrawn from palatable food: role of the central amygdala. Neuropsychopharmacology 38 (12):2498–2507. doi:10.1038/npp.2013.153

Blednov YA, Cravatt BF, Boehm SL 2nd, Walker D, Harris RA (2007) Role of endocannabinoids in alcohol consumption and intoxication: studies of mice lacking fatty acid amide hydrolase. Neuropsychopharmacology 32(7):1570–1582. doi:10.1038/sj.npp.1301274

Boctor SY, Martinez JL Jr, Koek W, France CP (2007) The cannabinoid CB1 receptor antagonist AM251 does not modify methamphetamine reinstatement of responding. Eur J Pharmacol 571 (1):39–43. doi:10.1016/j.ejphar.2007.06.004

Borgland SL, Malenka RC, Bonci A (2004) Acute and chronic cocaine-induced potentiation of synaptic strength in the ventral tegmental area: electrophysiological and behavioral correlates in individual rats. J Neurosci 24(34):7482–7490. doi:10.1523/JNEUROSCI.1312-04.2004

Bouton ME (2004) Context and behavioral processes in extinction. Learn Mem 11(5):485–494. doi:10.1101/lm.78804

Breivogel CS, Childers SR, Deadwyler SA, Hampson RE, Vogt LJ, Sim-Selley LJ (1999) Chronic delta9-tetrahydrocannabinol treatment produces a time-dependent loss of cannabinoid receptors and cannabinoid receptor-activated G proteins in rat brain. J Neurochem 73(6):2447–2459

Bruijnzeel AW, Repetto M, Gold MS (2004) Neurobiological mechanisms in addictive and psychiatric disorders. Psychiatr Clin North Am 27(4):661–674. doi:10.1016/j.psc.2004.06.005

Buczynski MW, Parsons LH (2010) Quantification of brain endocannabinoid levels: methods, interpretations and pitfalls. Br J Pharmacol 160(3):423–442. doi:10.1111/j.1476-5381.2010.00787.x

Buczynski MW, Polis IY, Parsons LH (2013) The volitional nature of nicotine exposure alters anandamide and oleoylethanolamide levels in the ventral tegmental area. Neuropsychopharmacology 38(4):574–584. doi:10.1038/npp.2012.210

Buczynski MW, Herman MA, Hsu KL, Natividad LA, Irimia C, Polis IY, Pugh H, Chang JW, Niphakis MJ, Cravatt BF, Roberto M, Parsons LH (2016) Diacylglycerol lipase disinhibits

VTA dopamine neurons during chronic nicotine exposure. Proc Natl Acad Sci USA 113 (4):1086–1091. doi:10.1073/pnas.1522672113

Budney AJ, Hughes JR, Moore BA, Novy PL (2001) Marijuana abstinence effects in marijuana smokers maintained in their home environment. Arch Gen Psychiatry 58(10):917–924

Bura SA, Burokas A, Martin-Garcia E, Maldonado R (2010) Effects of chronic nicotine on food intake and anxiety-like behaviour in CB(1) knockout mice. Eur Neuropsychopharmacol 20 (6):369–378. doi:10.1016/j.euroneuro.2010.02.003

Burkey RT, Nation JR (1997) (R)-methanandamide, but not anandamide, substitutes for delta 9-THC in a drug-discrimination procedure. Exp Clin Psychopharmacol 5(3):195–202

Bystrowska B, Smaga I, Frankowska M, Filip M (2014) Changes in endocannabinoid and N-acylethanolamine levels in rat brain structures following cocaine self-administration and extinction training. Prog Neuropsychopharmacol Biol Psychiatry 50:1–10. doi:10.1016/j.pnpbp.2013.12.002

Caille S, Parsons LH (2003) SR141716A reduces the reinforcing properties of heroin but not heroin-induced increases in nucleus accumbens dopamine in rats. Eur J Neurosci 18 (11):3145–3149

Caille S, Parsons LH (2006) Cannabinoid modulation of opiate reinforcement through the ventral striatopallidal pathway. Neuropsychopharmacology 31(4):804–813. doi:10.1038/sj.npp.1300848

Caille S, Alvarez-Jaimes L, Polis I, Stouffer DG, Parsons LH (2007) Specific alterations of extracellular endocannabinoid levels in the nucleus accumbens by ethanol, heroin, and cocaine self-administration. J Neurosci 27(14):3695–3702. doi:10.1523/JNEUROSCI.4403-06.2007

Calvigioni D, Hurd YL, Harkany T, Keimpema E (2014) Neuronal substrates and functional consequences of prenatal cannabis exposure. Eur Child Adolesc Psychiatry 23(10):931–941. doi:10.1007/s00787-014-0550-y

Canseco-Alba A, Rodriguez-Manzo G (2016) Intra-VTA anandamide infusion produces dose-based biphasic effects on male rat sexual behavior expression. Pharmacol Biochem Behav 150-151:182–189. doi:10.1016/j.pbb.2016.11.004

Carter BL, Tiffany ST (1999) Cue-reactivity and the future of addiction research. Addiction 94 (3):349–351

Cassin SE, von Ranson KM (2007) Is binge eating experienced as an addiction? Appetite 49 (3):687–690. doi:10.1016/j.appet.2007.06.012

Castane A, Valjent E, Ledent C, Parmentier M, Maldonado R, Valverde O (2002) Lack of CB1 cannabinoid receptors modifies nicotine behavioural responses, but not nicotine abstinence. Neuropharmacology 43(5):857–867

Castane A, Berrendero F, Maldonado R (2005) The role of the cannabinoid system in nicotine addiction. Pharmacol Biochem Behav 81(2):381–386. doi:10.1016/j.pbb.2005.01.025

Castelli MP, Paola Piras A, D'Agostino A, Pibiri F, Perra S, Gessa GL, Maccarrone M, Pistis M (2007) Dysregulation of the endogenous cannabinoid system in adult rats prenatally treated with the cannabinoid agonist WIN 55,212-2. Eur J Pharmacol 573(1-3):11–19. doi:10.1016/j.ejphar.2007.06.047

Castle DJ (2008) Anxiety and substance use: layers of complexity. Expert Rev Neurother 8 (3):493–501. doi:10.1586/14737175.8.3.493

Ceccarini J, Casteels C, Koole M, Bormans G, Van Laere K (2013) Transient changes in the endocannabinoid system after acute and chronic ethanol exposure and abstinence in the rat: a combined PET and microdialysis study. Eur J Nucl Med Mol Imaging 40(10):1582–1594. doi:10.1007/s00259-013-2456-1

Ceccarini J, Hompes T, Verhaeghen A, Casteels C, Peuskens H, Bormans G, Claes S, Van Laere K (2014) Changes in cerebral CB1 receptor availability after acute and chronic alcohol abuse and monitored abstinence. J Neurosci 34(8):2822–2831. doi:10.1523/JNEUROSCI.0849-13.2014

Ceccarini J, Kuepper R, Kemels D, van Os J, Henquet C, Van Laere K (2015) [18F]MK-9470 PET measurement of cannabinoid CB1 receptor availability in chronic cannabis users. Addict Biol 20(2):357–367. doi:10.1111/adb.12116

Cervino C, Vicennati V, Pasquali R, Pagotto U (2009) Feeding disorders and obesity. Curr Top Behav Neurosci 1:373–385. doi:10.1007/978-3-540-88955-7_15

Chaperon F, Soubrie P, Puech AJ, Thiebot MH (1998) Involvement of central cannabinoid (CB1) receptors in the establishment of place conditioning in rats. Psychopharmacology (Berl) 135 (4):324–332

Cheer JF, Wassum KM, Sombers LA, Heien ML, Ariansen JL, Aragona BJ, Phillips PE, Wightman RM (2007) Phasic dopamine release evoked by abused substances requires cannabinoid receptor activation. J Neurosci 27(4):791–795. doi:10.1523/JNEUROSCI.4152-06.2007

Chhatwal JP, Davis M, Maguschak KA, Ressler KJ (2005) Enhancing cannabinoid neurotransmission augments the extinction of conditioned fear. Neuropsychopharmacology 30 (3):516–524. doi:10.1038/sj.npp.1300655

Chhatwal JP, Gutman AR, Maguschak KA, Bowser ME, Yang Y, Davis M, Ressler KJ (2009) Functional interactions between endocannabinoid and CCK neurotransmitter systems may be critical for extinction learning. Neuropsychopharmacology 34(2):509–521. doi:10.1038/npp.2008.97

Cichewicz DL, Haller VL, Welch SP (2001) Changes in opioid and cannabinoid receptor protein following short-term combination treatment with delta(9)-tetrahydrocannabinol and morphine. J Pharmacol Exp Ther 297(1):121–127

Cippitelli A, Bilbao A, Hansson AC, del Arco I, Sommer W, Heilig M, Massi M, Bermudez-Silva FJ, Navarro M, Ciccocioppo R, de Fonseca FR (2005) Cannabinoid CB1 receptor antagonism reduces conditioned reinstatement of ethanol-seeking behavior in rats. Eur J Neurosci 21 (8):2243–2251. doi:10.1111/j.1460-9568.2005.04056.x

Cippitelli A, Bilbao A, Gorriti MA, Navarro M, Massi M, Piomelli D, Ciccocioppo R, Rodriguez de Fonseca F (2007) The anandamide transport inhibitor AM404 reduces ethanol self-administration. Eur J Neurosci 26(2):476–486. doi:10.1111/j.1460-9568.2007.05665.x

Cippitelli A, Cannella N, Braconi S, Duranti A, Tontini A, Bilbao A, Defonseca FR, Piomelli D, Ciccocioppo R (2008) Increase of brain endocannabinoid anandamide levels by FAAH inhibition and alcohol abuse behaviours in the rat. Psychopharmacology (Berl) 198(4):449–460. doi:10.1007/s00213-008-1104-0

Cippitelli A, Astarita G, Duranti A, Caprioli G, Ubaldi M, Stopponi S, Kallupi M, Sagratini G, Rodriguez de Fonseca F, Piomelli D, Ciccocioppo R (2011) Endocannabinoid regulation of acute and protracted nicotine withdrawal: effect of FAAH inhibition. PLoS One 6(11):e28142. doi:10.1371/journal.pone.0028142

Clarke RB, Adermark L (2010) Acute ethanol treatment prevents endocannabinoid-mediated long-lasting disinhibition of striatal output. Neuropharmacology 58(4–5):799–805. doi:10.1016/j.neuropharm.2009.12.006

Coffey SF, Dansky BS, Carrigan MH, Brady KT (2000) Acute and protracted cocaine abstinence in an outpatient population: a prospective study of mood, sleep and withdrawal symptoms. Drug Alcohol Depend 59(3):277–286

Cohen C, Perrault G, Voltz C, Steinberg R, Soubrie P (2002) SR141716, a central cannabinoid (CB (1)) receptor antagonist, blocks the motivational and dopamine-releasing effects of nicotine in rats. Behav Pharmacol 13(5–6):451–463

Cohen C, Perrault G, Griebel G, Soubrie P (2005) Nicotine-associated cues maintain nicotine-seeking behavior in rats several weeks after nicotine withdrawal: reversal by the cannabinoid (CB1) receptor antagonist, rimonabant (SR141716). Neuropsychopharmacology 30 (1):145–155. doi:10.1038/sj.npp.1300541

Colombo G, Agabio R, Fa M, Guano L, Lobina C, Loche A, Reali R, Gessa GL (1998) Reduction of voluntary ethanol intake in ethanol-preferring sP rats by the cannabinoid antagonist SR-141716. Alcohol Alcohol 33(2):126–130

Colombo G, Serra S, Brunetti G, Gomez R, Melis S, Vacca G, Carai MM, Gessa L (2002) Stimulation of voluntary ethanol intake by cannabinoid receptor agonists in ethanol-preferring sP rats. Psychopharmacology 159(2):181–187. doi:10.1007/s002130100887

Conklin CA, Tiffany ST (2002) Applying extinction research and theory to cue-exposure addiction treatments. Addiction 97(2):155–167

Conway KP, Compton W, Stinson FS, Grant BF (2006) Lifetime comorbidity of DSM-IV mood and anxiety disorders and specific drug use disorders: results from the National Epidemiologic Survey on Alcohol and Related Conditions. J Clin Psychiatry 67(2):247–257

Corwin RL, Grigson PS (2009) Symposium overview--Food addiction: fact or fiction? J Nutr 139 (3):617–619. doi:10.3945/jn.108.097691

Cossu G, Ledent C, Fattore L, Imperato A, Bohme GA, Parmentier M, Fratta W (2001) Cannabinoid CB1 receptor knockout mice fail to self-administer morphine but not other drugs of abuse. Behav Brain Res 118(1):61–65

Cota D (2008) The role of the endocannabinoid system in the regulation of hypothalamic-pituitary-adrenal axis activity. J Neuroendocrinol 20(Suppl 1):35–38. doi:10.1111/j.1365-2826.2008.01673.x

Cota D, Steiner MA, Marsicano G, Cervino C, Herman JP, Grubler Y, Stalla J, Pasquali R, Lutz B, Stalla GK, Pagotto U (2007) Requirement of cannabinoid receptor type 1 for the basal modulation of hypothalamic-pituitary-adrenal axis function. Endocrinology 148 (4):1574–1581. doi:10.1210/en.2005-1649

Cottone P, Sabino V, Steardo L, Zorrilla EP (2008) Intermittent access to preferred food reduces the reinforcing efficacy of chow in rats. Am J Physiol Regul Integr Comp Physiol 295(4): R1066–R1076. doi:10.1152/ajpregu.90309.2008

Cottone P, Sabino V, Roberto M, Bajo M, Pockros L, Frihauf JB, Fekete EM, Steardo L, Rice KC, Grigoriadis DE, Conti B, Koob GF, Zorrilla EP (2009a) CRF system recruitment mediates dark side of compulsive eating. Proc Natl Acad Sci USA 106(47):20016–20020. doi:10.1073/pnas.0908789106

Cottone P, Sabino V, Steardo L, Zorrilla EP (2009b) Consummatory, anxiety-related and metabolic adaptations in female rats with alternating access to preferred food. Psychoneuroendocrinology 34(1):38–49. doi:10.1016/j.psyneuen.2008.08.010

Crombag HS, Shaham Y (2002) Renewal of drug seeking by contextual cues after prolonged extinction in rats. Behav Neurosci 116(1):169–173

Dawley HH Jr, Winstead DK, Baxter AS, Gay JR (1979) An attitude survey of the effects of marijuana on sexual enjoyment. J Clin Psychol 35(1):212–217

de Bruin NM, Lange JH, Kruse CG, Herremans AH, Schoffelmeer AN, van Drimmelen M, De Vries TJ (2011) SLV330, a cannabinoid CB(1) receptor antagonist, attenuates ethanol and nicotine seeking and improves inhibitory response control in rats. Behav Brain Res 217 (2):408–415. doi:10.1016/j.bbr.2010.11.013

De Petrocellis L, Bisogno T, Davis JB, Pertwee RG, Di Marzo V (2000) Overlap between the ligand recognition properties of the anandamide transporter and the VR1 vanilloid receptor: inhibitors of anandamide uptake with negligible capsaicin-like activity. FEBS Lett 483 (1):52–56

De Vries TJ, Shaham Y, Homberg JR, Crombag H, Schuurman K, Dieben J, Vanderschuren LJ, Schoffelmeer AN (2001) A cannabinoid mechanism in relapse to cocaine seeking. Nat Med 7 (10):1151–1154. doi:10.1038/nm1001-1151

De Vries TJ, Homberg JR, Binnekade R, Raaso H, Schoffelmeer AN (2003) Cannabinoid modulation of the reinforcing and motivational properties of heroin and heroin-associated cues in rats. Psychopharmacology 168(1-2):164–169. doi:10.1007/s00213-003-1422-1

De Vries TJ, de Vries W, Janssen MC, Schoffelmeer AN (2005) Suppression of conditioned nicotine and sucrose seeking by the cannabinoid-1 receptor antagonist SR141716A. Behav Brain Res 161(1):164–168. doi:10.1016/j.bbr.2005.02.021

De Witte P, Pinto E, Ansseau M, Verbanck P (2003) Alcohol and withdrawal: from animal research to clinical issues. Neurosci Biobehav Rev 27(3):189–197

Delfs JM, Zhu Y, Druhan JP, Aston-Jones G (2000) Noradrenaline in the ventral forebrain is critical for opiate withdrawal-induced aversion. Nature 403(6768):430–434. doi:10.1038/35000212

DePoy L, Daut R, Brigman JL, MacPherson K, Crowley N, Gunduz-Cinar O, Pickens CL, Cinar R, Saksida LM, Kunos G, Lovinger DM, Bussey TJ, Camp MC, Holmes A (2013) Chronic alcohol produces neuroadaptations to prime dorsal striatal learning. Proc Natl Acad Sci USA 110(36):14783–14788. doi:10.1073/pnas.1308198110

DeVries AC, Pert A (1998) Conditioned increases in anxiogenic-like behavior following exposure to contextual stimuli associated with cocaine are mediated by corticotropin-releasing factor. Psychopharmacology 137(4):333–340

Di S, Malcher-Lopes R, Halmos KC, Tasker JG (2003) Nongenomic glucocorticoid inhibition via endocannabinoid release in the hypothalamus: a fast feedback mechanism. J Neurosci 23 (12):4850–4857

Di S, Boudaba C, Popescu IR, Weng FJ, Harris C, Marcheselli VL, Bazan NG, Tasker JG (2005) Activity-dependent release and actions of endocannabinoids in the rat hypothalamic supraoptic nucleus. J Physiol 569(Pt 3):751–760. doi:10.1113/jphysiol.2005.097477

Di S, Maxson MM, Franco A, Tasker JG (2009) Glucocorticoids regulate glutamate and GABA synapse-specific retrograde transmission via divergent nongenomic signaling pathways. J Neurosci 29(2):393–401. doi:10.1523/JNEUROSCI.4546-08.2009

Di Chiara G, Bassareo V (2007) Reward system and addiction: what dopamine does and doesn't do. Curr Opin Pharmacol 7(1):69–76. doi:10.1016/j.coph.2006.11.003

Di Marzo V, Berrendero F, Bisogno T, Gonzalez S, Cavaliere P, Romero J, Cebeira M, Ramos JA, Fernandez-Ruiz JJ (2000) Enhancement of anandamide formation in the limbic forebrain and reduction of endocannabinoid contents in the striatum of delta9-tetrahydrocannabinol-tolerant rats. J Neurochem 74(4):1627–1635

Di Marzo V, Goparaju SK, Wang L, Liu J, Batkai S, Jarai Z, Fezza F, Miura GI, Palmiter RD, Sugiura T, Kunos G (2001) Leptin-regulated endocannabinoids are involved in maintaining food intake. Nature 410(6830):822–825. doi:10.1038/35071088

Diergaarde L, de Vries W, Raaso H, Schoffelmeer AN, De Vries TJ (2008) Contextual renewal of nicotine seeking in rats and its suppression by the cannabinoid-1 receptor antagonist Rimonabant (SR141716A). Neuropharmacology 55(5):712–716. doi:10.1016/j.neuropharm.2008.06.003

Dipatrizio NV, Simansky KJ (2008) Inhibiting parabrachial fatty acid amide hydrolase activity selectively increases the intake of palatable food via cannabinoid CB1 receptors. Am J Physiol Regul Integr Comp Physiol 295(5):R1409–R1414. doi:10.1152/ajpregu.90484.2008

Dodd GT, Stark JA, McKie S, Williams SR, Luckman SM (2009) Central cannabinoid signaling mediating food intake: a pharmacological-challenge magnetic resonance imaging and functional histology study in rat. Neuroscience 163(4):1192–1200. doi:10.1016/j.neuroscience.2009.07.022

Dole VP (1965) Thoughts on narcotics addiction. Bull NY Acad Med 41:211–213

Dole VP, Nyswander ME, Kreek MJ (1966) Narcotic blockade. Arch Intern Med 118(4):304–309

Douglas LA, Varlinskaya EI, Spear LP (2004) Rewarding properties of social interactions in adolescent and adult male and female rats: impact of social versus isolate housing of subjects and partners. Dev Psychobiol 45(3):153–162. doi:10.1002/dev.20025

D'Souza MS, Markou A (2011) Neuronal mechanisms underlying development of nicotine dependence: implications for novel smoking-cessation treatments. Addict Sci Clin Pract 6 (1):4–16

Ducci F, Goldman D (2012) The genetic basis of addictive disorders. Psychiatr Clin North Am 35 (2):495–519. doi:10.1016/j.psc.2012.03.010

Dudok B, Barna L, Ledri M, Szabo SI, Szabadits E, Pinter B, Woodhams SG, Henstridge CM, Balla GY, Nyilas R, Varga C, Lee SH, Matolcsi M, Cervenak J, Kacskovics I, Watanabe M, Sagheddu C, Melis M, Pistis M, Soltesz I, Katona I (2015) Cell-specific STORM super-resolution imaging reveals nanoscale organization of cannabinoid signaling. Nat Neurosci 18 (1):75–86. doi:10.1038/nn.3892

Economidou D, Mattioli L, Cifani C, Perfumi M, Massi M, Cuomo V, Trabace L, Ciccocioppo R (2006) Effect of the cannabinoid CB1 receptor antagonist SR-141716A on ethanol self-administration and ethanol-seeking behaviour in rats. Psychopharmacology 183(4):394–403. doi:10.1007/s00213-005-0199-9

Edwards A, Abizaid A (2016) Driving the need to feed: insight into the collaborative interaction between ghrelin and endocannabinoid systems in modulating brain reward systems. Neurosci Biobehav Rev 66:33–53. doi:10.1016/j.neubiorev.2016.03.032

Epstein DH, Shaham Y (2010) Cheesecake-eating rats and the question of food addiction. Nat Neurosci 13(5):529–531. doi:10.1038/nn0510-529

Everitt BJ, Robbins TW (2005) Neural systems of reinforcement for drug addiction: from actions to habits to compulsion. Nat Neurosci 8(11):1481–1489. doi:10.1038/nn1579

Everitt BJ, Belin D, Economidou D, Pelloux Y, Dalley JW, Robbins TW (2008) Review. Neural mechanisms underlying the vulnerability to develop compulsive drug-seeking habits and addiction. Philos Trans R Soc Lond B Biol Sci 363(1507):3125–3135. doi:10.1098/rstb.2008.0089

Faleiro LJ, Jones S, Kauer JA (2004) Rapid synaptic plasticity of glutamatergic synapses on dopamine neurons in the ventral tegmental area in response to acute amphetamine injection. Neuropsychopharmacology 29(12):2115–2125. doi:10.1038/sj.npp.1300495

Farrimond JA, Mercier MS, Whalley BJ, Williams CM (2011) Cannabis sativa and the endogenous cannabinoid system: therapeutic potential for appetite regulation. Phytother Res 25 (2):170–188. doi:10.1002/ptr.3375

Fattore L, Martellotta MC, Cossu G, Mascia MS, Fratta W (1999) CB1 cannabinoid receptor agonist WIN 55,212-2 decreases intravenous cocaine self-administration in rats. Behav Brain Res 104(1-2):141–146

Fattore L, Spano MS, Cossu G, Deiana S, Fratta W (2003) Cannabinoid mechanism in reinstatement of heroin-seeking after a long period of abstinence in rats. Eur J Neurosci 17 (8):1723–1726

Fattore L, Cossu G, Spano MS, Deiana S, Fadda P, Scherma M, Fratta W (2004) Cannabinoids and reward: interactions with the opioid system. Crit Rev Neurobiol 16(1-2):147–158

Fattore L, Deiana S, Spano SM, Cossu G, Fadda P, Scherma M, Fratta W (2005) Endocannabinoid system and opioid addiction: behavioural aspects. Pharmacol Biochem Behav 81(2):343–359. doi:10.1016/j.pbb.2005.01.031

Ferreira-Vieira TH, Bastos CP, Pereira GS, Moreira FA, Massensini AR (2014) A role for the endocannabinoid system in exercise-induced spatial memory enhancement in mice. Hippocampus 24(1):79–88. doi:10.1002/hipo.22206

Ferrer B, Bermudez-Silva FJ, Bilbao A, Alvarez-Jaimes L, Sanchez-Vera I, Giuffrida A, Serrano A, Baixeras E, Khaturia S, Navarro M, Parsons LH, Piomelli D, Rodriguez de Fonseca F (2007) Regulation of brain anandamide by acute administration of ethanol. Biochem J 404 (1):97–104. doi:10.1042/BJ20061898

Filip M, Golda A, Zaniewska M, McCreary AC, Nowak E, Kolasiewicz W, Przegalinski E (2006) Involvement of cannabinoid CB1 receptors in drug addiction: effects of rimonabant on behavioral responses induced by cocaine. Pharmacol Rep 58(6):806–819

Fiorino DF, Coury A, Phillips AG (1997) Dynamic changes in nucleus accumbens dopamine efflux during the Coolidge effect in male rats. J Neurosci 17(12):4849–4855

Forget B, Hamon M, Thiebot MH (2005) Cannabinoid CB1 receptors are involved in motivational effects of nicotine in rats. Psychopharmacology 181(4):722–734. doi:10.1007/s00213-005-0015-6

Forget B, Barthelemy S, Saurini F, Hamon M, Thiebot MH (2006) Differential involvement of the endocannabinoid system in short- and long-term expression of incentive learning supported by nicotine in rats. Psychopharmacology 189(1):59–69. doi:10.1007/s00213-006-0525-x

Forget B, Coen KM, Le Foll B (2009) Inhibition of fatty acid amide hydrolase reduces reinstatement of nicotine seeking but not break point for nicotine self-administration--comparison with CB(1) receptor blockade. Psychopharmacology (Berl) 205(4):613–624. doi:10.1007/s00213-009-1569-5

Fourgeaud L, Mato S, Bouchet D, Hemar A, Worley PF, Manzoni OJ (2004) A single in vivo exposure to cocaine abolishes endocannabinoid-mediated long-term depression in the nucleus accumbens. J Neurosci 24(31):6939–6945. doi:10.1523/JNEUROSCI.0671-04.2004

Freed CR, Yamamoto BK (1985) Regional brain dopamine metabolism: a marker for the speed, direction, and posture of moving animals. Science 229(4708):62–65

Freedland CS, Sharpe AL, Samson HH, Porrino LJ (2001) Effects of SR141716A on ethanol and sucrose self-administration. Alcohol Clin Exp Res 25(2):277–282

Frieling H, Albrecht H, Jedtberg S, Gozner A, Lenz B, Wilhelm J, Hillemacher T, de Zwaan M, Kornhuber J, Bleich S (2009) Elevated cannabinoid 1 receptor mRNA is linked to eating disorder related behavior and attitudes in females with eating disorders. Psychoneuroendocrinology 34(4):620–624. doi:10.1016/j.psyneuen.2008.10.014

Gallate JE, McGregor IS (1999) The motivation for beer in rats: effects of ritanserin, naloxone and SR 141716. Psychopharmacology (Berl) 142(3):302–308

Gallate JE, Saharov T, Mallet PE, McGregor IS (1999) Increased motivation for beer in rats following administration of a cannabinoid CB1 receptor agonist. Eur J Pharmacol 370 (3):233–240

Gamage TF, Ignatowska-Jankowska BM, Muldoon PP, Cravatt BF, Damaj MI, Lichtman AH (2015) Differential effects of endocannabinoid catabolic inhibitors on morphine withdrawal in mice. Drug Alcohol Depend 146:7–16. doi:10.1016/j.drugalcdep.2014.11.015

Gamaleddin I, Guranda M, Goldberg SR, Le Foll B (2011) The selective anandamide transport inhibitor VDM11 attenuates reinstatement of nicotine seeking behaviour, but does not affect nicotine intake. Br J Pharmacol 164(6):1652–1660. doi:10.1111/j.1476-5381.2011.01440.x

Gamaleddin I, Wertheim C, Zhu AZ, Coen KM, Vemuri K, Makryannis A, Goldberg SR, Le Foll B (2012a) Cannabinoid receptor stimulation increases motivation for nicotine and nicotine seeking. Addict Biol 17(1):47–61. doi:10.1111/j.1369-1600.2011.00314.x

Gamaleddin I, Zvonok A, Makriyannis A, Goldberg SR, Le Foll B (2012b) Effects of a selective cannabinoid CB2 agonist and antagonist on intravenous nicotine self administration and reinstatement of nicotine seeking. PLoS One 7(1):e29900. doi:10.1371/journal.pone.0029900

Gamaleddin I, Guranda M, Scherma M, Fratta W, Makriyannis A, Vadivel SK, Goldberg SR, Le Foll B (2013) AM404 attenuates reinstatement of nicotine seeking induced by nicotine-associated cues and nicotine priming but does not affect nicotine- and food-taking. J Psychopharmacol 27(6):564–571. doi:10.1177/0269881113477710

Gates P, Albertella L, Copeland J (2015) Cannabis withdrawal and sleep: a systematic review of human studies. Subst Abus. doi:10.1080/08897077.2015.1023484

Gearhardt AN, White MA, Masheb RM, Morgan PT, Crosby RD, Grilo CM (2012) An examination of the food addiction construct in obese patients with binge eating disorder. Int J Eat Disord 45(5):657–663. doi:10.1002/eat.20957

Gerard N, Pieters G, Goffin K, Bormans G, Van Laere K (2011) Brain type 1 cannabinoid receptor availability in patients with anorexia and bulimia nervosa. Biol Psychiatry 70(8):777–784. doi:10.1016/j.biopsych.2011.05.010

Gerrits MA, Van Ree JM (1996) Effect of nucleus accumbens dopamine depletion on motivational aspects involved in initiation of cocaine and heroin self-administration in rats. Brain Res 713 (1-2):114–124

Gessa GL, Serra S, Vacca G, Carai MA, Colombo G (2005) Suppressing effect of the cannabinoid CB1 receptor antagonist, SR147778, on alcohol intake and motivational properties of alcohol in alcohol-preferring sP rats. Alcohol Alcohol 40(1):46–53. doi:10.1093/alcalc/agh114

Getachew B, Hauser SR, Dhaher R, Katner SN, Bell RL, Oster SM, McBride WJ, Rodd ZA (2011) CB1 receptors regulate alcohol-seeking behavior and alcohol self-administration of alcohol-preferring (P) rats. Pharmacol Biochem Behav 97(4):669–675. doi:10.1016/j.pbb.2010.11.006

Ghosh S, Wise LE, Chen Y, Gujjar R, Mahadevan A, Cravatt BF, Lichtman AH (2013) The monoacylglycerol lipase inhibitor JZL184 suppresses inflammatory pain in the mouse carrageenan model. Life Sci 92(8–9):498–505. doi:10.1016/j.lfs.2012.06.020

Glangetas C, Girard D, Groc L, Marsicano G, Chaouloff F, Georges F (2013) Stress switches cannabinoid type-1 (CB1) receptor-dependent plasticity from LTD to LTP in the bed nucleus of the stria terminalis. J Neurosci 33(50):19657–19663. doi:10.1523/JNEUROSCI.3175-13.2013

Glass M, Dragunow M, Faull RL (1997) Cannabinoid receptors in the human brain: a detailed anatomical and quantitative autoradiographic study in the fetal, neonatal and adult human brain. Neuroscience 77(2):299–318

Gobbi G, Bambico FR, Mangieri R, Bortolato M, Campolongo P, Solinas M, Cassano T, Morgese MG, Debonnel G, Duranti A, Tontini A, Tarzia G, Mor M, Trezza V, Goldberg SR, Cuomo V, Piomelli D (2005) Antidepressant-like activity and modulation of brain monoaminergic transmission by blockade of anandamide hydrolysis. Proc Natl Acad Sci USA 102 (51):18620–18625. doi:10.1073/pnas.0509591102

Goeders NE (2002) Stress and cocaine addiction. J Pharmacol Exp Ther 301(3):785–789

Goeders NE, Guerin GF (1996) Role of corticosterone in intravenous cocaine self-administration in rats. Neuroendocrinology 64(5):337–348

Gonzalez S, Cascio MG, Fernandez-Ruiz J, Fezza F, Di Marzo V, Ramos JA (2002) Changes in endocannabinoid contents in the brain of rats chronically exposed to nicotine, ethanol or cocaine. Brain Res 954(1):73–81

Gonzalez S, Fernandez-Ruiz J, Di Marzo V, Hernandez M, Arevalo C, Nicanor C, Cascio MG, Ambrosio E, Ramos JA (2004a) Behavioral and molecular changes elicited by acute administration of SR141716 to Delta9-tetrahydrocannabinol-tolerant rats: an experimental model of cannabinoid abstinence. Drug Alcohol Depend 74(2):159–170. doi:10.1016/j.drugalcdep.2003.12.011

Gonzalez S, Valenti M, de Miguel R, Fezza F, Fernandez-Ruiz J, Di Marzo V, Ramos JA (2004b) Changes in endocannabinoid contents in reward-related brain regions of alcohol-exposed rats, and their possible relevance to alcohol relapse. Br J Pharmacol 143(4):455–464. doi:10.1038/sj.bjp.0705963

Gorzalka BB, Dang SS (2012) Minireview: Endocannabinoids and gonadal hormones: bidirectional interactions in physiology and behavior. Endocrinology 153(3):1016–1024. doi:10.1210/en.2011-1643

Gorzalka BB, Hill MN, Hillard CJ (2008) Regulation of endocannabinoid signaling by stress: implications for stress-related affective disorders. Neurosci Biobehav Rev 32(6):1152–1160. doi:10.1016/j.neubiorev.2008.03.004

Gorzalka BB, Hill MN, Chang SC (2010) Male-female differences in the effects of cannabinoids on sexual behavior and gonadal hormone function. Horm Behav 58(1):91–99. doi:10.1016/j.yhbeh.2009.08.009

Grillon C (2002) Startle reactivity and anxiety disorders: aversive conditioning, context, and neurobiology. Biol Psychiatry 52(10):958–975

Grueter BA, Gosnell HB, Olsen CM, Schramm-Sapyta NL, Nekrasova T, Landreth GE, Winder DG (2006) Extracellular-signal regulated kinase 1-dependent metabotropic glutamate receptor 5-induced long-term depression in the bed nucleus of the stria terminalis is disrupted by cocaine administration. J Neurosci 26(12):3210–3219. doi:10.1523/JNEUROSCI.0170-06.2006

Guegan T, Cutando L, Ayuso E, Santini E, Fisone G, Bosch F, Martinez A, Valjent E, Maldonado R, Martin M (2013) Operant behavior to obtain palatable food modifies neuronal plasticity in the brain reward circuit. Eur Neuropsychopharmacol 23(2):146–159. doi:10.1016/j.euroneuro.2012.04.004

Gunduz-Cinar O, Hill MN, McEwen BS, Holmes A (2013) Amygdala FAAH and anandamide: mediating protection and recovery from stress. Trends Pharmacol Sci 34(11):637–644. doi:10.1016/j.tips.2013.08.008

Hajnal A, Smith GP, Norgren R (2004) Oral sucrose stimulation increases accumbens dopamine in the rat. Am J Physiol Regul Integr Comp Physiol 286(1):R31–R37. doi:10.1152/ajpregu.00282.2003

Halikas J, Weller R, Morse C (1982) Effects of regular marijuana use on sexual performance. J Psychoactive Drugs 14(1-2):59–70. doi:10.1080/02791072.1982.10471911

Haller J, Bakos N, Szirmay M, Ledent C, Freund TF (2002) The effects of genetic and pharmacological blockade of the CB1 cannabinoid receptor on anxiety. Eur J Neurosci 16 (7):1395–1398

Hampson AJ, Bornheim LM, Scanziani M, Yost CS, Gray AT, Hansen BM, Leonoudakis DJ, Bickler PE (1998) Dual effects of anandamide on NMDA receptor-mediated responses and neurotransmission. J Neurochem 70(2):671–676

Hansson AC, Bermudez-Silva FJ, Malinen H, Hyytia P, Sanchez-Vera I, Rimondini R, Rodriguez de Fonseca F, Kunos G, Sommer WH, Heilig M (2007) Genetic impairment of frontocortical endocannabinoid degradation and high alcohol preference. Neuropsychopharmacology 32 (1):117–126. doi:10.1038/sj.npp.1301034

Hanus L, Avraham Y, Ben-Shushan D, Zolotarev O, Berry EM, Mechoulam R (2003) Short-term fasting and prolonged semistarvation have opposite effects on 2-AG levels in mouse brain. Brain Res 983(1-2):144–151

Harloe JP, Thorpe AJ, Lichtman AH (2008) Differential endocannabinoid regulation of extinction in appetitive and aversive Barnes maze tasks. Learn Mem 15(11):806–809. doi:10.1101/lm. 1113008

Hashimotodani Y, Ohno-Shosaku T, Kano M (2007) Endocannabinoids and synaptic function in the CNS. Neuroscientist 13(2):127–137. doi:10.1177/1073858406296716

Hattori S, Naoi M, Nishino H (1994) Striatal dopamine turnover during treadmill running in the rat: relation to the speed of running. Brain Res Bull 35(1):41–49

Hayase T, Yamamoto Y, Yamamoto K (2001) Protective effects of cannabinoid receptor agonists against cocaine and other convulsant-induced toxic behavioural symptoms. J Pharmacy Pharmacol 53(11):1525–1532

Heifets BD, Castillo PE (2009) Endocannabinoid signaling and long-term synaptic plasticity. Annu Rev Physiol 71:283–306. doi:10.1146/annurev.physiol.010908.163149

Heilig M, Egli M, Crabbe JC, Becker HC (2010) Acute withdrawal, protracted abstinence and negative affect in alcoholism: are they linked? Addict Biol 15(2):169–184. doi:10.1111/j.1369-1600.2009.00194.x

Henricks AM, Berger AL, Lugo JM, Baxter-Potter LN, Bieniasz KV, Craft RM, McLaughlin RJ (2016) Sex differences in alcohol consumption and alterations in nucleus accumbens endocannabinoid mRNA in alcohol-dependent rats. Neuroscience 335:195–206. doi:10.1016/j.neuroscience.2016.08.032

Herkenham M, Lynn AB, Johnson MR, Melvin LS, de Costa BR, Rice KC (1991) Characterization and localization of cannabinoid receptors in rat brain: a quantitative in vitro autoradiographic study. J Neurosci 11(2):563–583

Hernandez G, Cheer JF (2011) Extinction learning of rewards in the rat: is there a role for CB1 receptors? Psychopharmacology 217(2):189–197. doi:10.1007/s00213-011-2275-7

Hernandez G, Cheer JF (2015) To act or not to act: endocannabinoid/dopamine interactions in decision-making. Front Behav Neurosci 9:336. doi:10.3389/fnbeh.2015.00336

Hershon HI (1977) Alcohol withdrawal symptoms and drinking behavior. J Stud Alcohol 38 (5):953–971

Heyman E, Gamelin FX, Goekint M, Piscitelli F, Roelands B, Leclair E, Di Marzo V, Meeusen R (2012) Intense exercise increases circulating endocannabinoid and BDNF levels in humans--possible implications for reward and depression. Psychoneuroendocrinology 37(6):844–851. doi:10.1016/j.psyneuen.2011.09.017

Higgs S, Williams CM, Kirkham TC (2003) Cannabinoid influences on palatability: microstructural analysis of sucrose drinking after delta(9)-tetrahydrocannabinol, anandamide, 2-arachidonoyl glycerol and SR141716. Psychopharmacology (Berl) 165(4):370–377. doi:10.1007/s00213-002-1263-3

Hijzen TH, Houtzager SW, Joordens RJ, Olivier B, Slangen JL (1995) Predictive validity of the potentiated startle response as a behavioral model for anxiolytic drugs. Psychopharmacology (Berl) 118(2):150–154
Hill MN, Carrier EJ, McLaughlin RJ, Morrish AC, Meier SE, Hillard CJ, Gorzalka BB (2008) Regional alterations in the endocannabinoid system in an animal model of depression: effects of concurrent antidepressant treatment. J Neurochem 106(6):2322–2336. doi:10.1111/j.1471-4159.2008.05567.x
Hill MN, McLaughlin RJ, Morrish AC, Viau V, Floresco SB, Hillard CJ, Gorzalka BB (2009) Suppression of amygdalar endocannabinoid signaling by stress contributes to activation of the hypothalamic-pituitary-adrenal axis. Neuropsychopharmacology 34(13):2733–2745. doi:10.1038/npp.2009.114
Hill MN, Karatsoreos IN, Hillard CJ, McEwen BS (2010a) Rapid elevations in limbic endocannabinoid content by glucocorticoid hormones in vivo. Psychoneuroendocrinology 35(9):1333–1338. doi:10.1016/j.psyneuen.2010.03.005
Hill MN, Titterness AK, Morrish AC, Carrier EJ, Lee TT, Gil-Mohapel J, Gorzalka BB, Hillard CJ, Christie BR (2010b) Endogenous cannabinoid signaling is required for voluntary exercise-induced enhancement of progenitor cell proliferation in the hippocampus. Hippocampus 20(4):513–523. doi:10.1002/hipo.20647
Hill MN, Hillard CJ, McEwen BS (2011) Alterations in corticolimbic dendritic morphology and emotional behavior in cannabinoid CB1 receptor-deficient mice parallel the effects of chronic stress. Cereb Cortex 21(9):2056–2064. doi:10.1093/cercor/bhq280
Hillard CJ (2015) The endocannabinoid signaling system in the CNS: a primer. Int Rev Neurobiol 125:1–47. doi:10.1016/bs.irn.2015.10.001
Hine B, Friedman E, Torrelio M, Gershon S (1975) Morphine-dependent rats: blockade of precipitated abstinence by tetrahydrocannabinol. Science 187(4175):443–445
Hirvonen J, Zanotti-Fregonara P, Umhau JC, George DT, Rallis-Frutos D, Lyoo CH, Li CT, Hines CS, Sun H, Terry GE, Morse C, Zoghbi SS, Pike VW, Innis RB, Heilig M (2013) Reduced cannabinoid CB1 receptor binding in alcohol dependence measured with positron emission tomography. Mol Psychiatry 18(8):916–921. doi:10.1038/mp.2012.100
Hoffman AF, Oz M, Caulder T, Lupica CR (2003) Functional tolerance and blockade of long-term depression at synapses in the nucleus accumbens after chronic cannabinoid exposure. J Neurosci 23(12):4815–4820
Hohmann AG, Suplita RL, Bolton NM, Neely MH, Fegley D, Mangieri R, Krey JF, Walker JM, Holmes PV, Crystal JD, Duranti A, Tontini A, Mor M, Tarzia G, Piomelli D (2005) An endocannabinoid mechanism for stress-induced analgesia. Nature 435(7045):1108–1112. doi:10.1038/nature03658
Holter SM, Kallnik M, Wurst W, Marsicano G, Lutz B, Wotjak CT (2005) Cannabinoid CB1 receptor is dispensable for memory extinction in an appetitively-motivated learning task. Eur J Pharmacol 510(1-2):69–74. doi:10.1016/j.ejphar.2005.01.008
Horvath TL, Diano S (2004) The floating blueprint of hypothalamic feeding circuits. Nat Rev Neurosci 5(8):662–667. doi:10.1038/nrn1479
Houchi H, Babovic D, Pierrefiche O, Ledent C, Daoust M, Naassila M (2005) CB1 receptor knockout mice display reduced ethanol-induced conditioned place preference and increased striatal dopamine D2 receptors. Neuropsychopharmacology 30(2):339–349. doi:10.1038/sj.npp.1300568
Hughes JR (2007) Effects of abstinence from tobacco: valid symptoms and time course. Nicotine Tob Res 9(3):315–327. doi:10.1080/14622200701188919
Hughes JR, Gust SW, Skoog K, Keenan RM, Fenwick JW (1991) Symptoms of tobacco withdrawal. A replication and extension. Arch Gen Psychiatry 48(1):52–59
Hungund BL, Szakall I, Adam A, Basavarajappa BS, Vadasz C (2003) Cannabinoid CB1 receptor knockout mice exhibit markedly reduced voluntary alcohol consumption and lack alcohol-induced dopamine release in the nucleus accumbens. J Neurochem 84(4):698–704

Hurd YL, Michaelides M, Miller ML, Jutras-Aswad D (2014) Trajectory of adolescent cannabis use on addiction vulnerability. Neuropharmacology 76(Pt B):416–424. doi:10.1016/j.neuropharm.2013.07.028

Hyman SE, Malenka RC, Nestler EJ (2006) Neural mechanisms of addiction: the role of reward-related learning and memory. Annu Rev Neurosci 29:565–598. doi:10.1146/annurev.neuro.29.051605.113009

Iemolo A, Blasio A, St Cyr SA, Jiang F, Rice KC, Sabino V, Cottone P (2013) CRF-CRF1 receptor system in the central and basolateral nuclei of the amygdala differentially mediates excessive eating of palatable food. Neuropsychopharmacology 38(12):2456–2466. doi:10.1038/npp.2013.147

Ignatowska-Jankowska BM, Muldoon PP, Lichtman AH, Damaj MI (2013) The cannabinoid CB2 receptor is necessary for nicotine-conditioned place preference, but not other behavioral effects of nicotine in mice. Psychopharmacology (Berl) 229(4):591–601. doi:10.1007/s00213-013-3117-6

Ignatowska-Jankowska BM, Ghosh S, Crowe MS, Kinsey SG, Niphakis MJ, Abdullah RA, Tao Q, ST ON, Walentiny DM, Wiley JL, Cravatt BF, Lichtman AH (2014) In vivo characterization of the highly selective monoacylglycerol lipase inhibitor KML29: antinociceptive activity without cannabimimetic side effects. Br J Pharmacol 171 (6):1392-1407. doi:10.1111/bph.12298

Ishiguro H, Carpio O, Horiuchi Y, Shu A, Higuchi S, Schanz N, Benno R, Arinami T, Onaivi ES (2010) A nonsynonymous polymorphism in cannabinoid CB2 receptor gene is associated with eating disorders in humans and food intake is modified in mice by its ligands. Synapse 64 (1):92–96. doi:10.1002/syn.20714

Jamshidi N, Taylor DA (2001) Anandamide administration into the ventromedial hypothalamus stimulates appetite in rats. Br J Pharmacol 134(6):1151–1154. doi:10.1038/sj.bjp.0704379

Janiri L, Martinotti G, Dario T, Reina D, Paparello F, Pozzi G, Addolorato G, Di Giannantonio M, De Risio S (2005) Anhedonia and substance-related symptoms in detoxified substance-dependent subjects: a correlation study. Neuropsychobiology 52(1):37–44. doi:10.1159/000086176

Jarbe TU, Lamb RJ, Liu Q, Makriyannis A (2006) Discriminative stimulus functions of AM-1346, a CB1R selective anandamide analog in rats trained with Delta9-THC or (R)-methanandamide (AM-356). Psychopharmacology (Berl) 188(3):315–323. doi:10.1007/s00213-006-0517-x

Jarrett MM, Limebeer CL, Parker LA (2005) Effect of Delta9-tetrahydrocannabinol on sucrose palatability as measured by the taste reactivity test. Physiol Behav 86(4):475–479. doi:10.1016/j.physbeh.2005.08.033

Jarrett MM, Scantlebury J, Parker LA (2007) Effect of delta9-tetrahydrocannabinol on quinine palatability and AM251 on sucrose and quinine palatability using the taste reactivity test. Physiol Behav 90(2-3):425–430. doi:10.1016/j.physbeh.2006.10.003

Jentsch JD, Taylor JR (1999) Impulsivity resulting from frontostriatal dysfunction in drug abuse: implications for the control of behavior by reward-related stimuli. Psychopharmacology 146 (4):373–390

Jing L, Qiu Y, Zhang Y, Li JX (2014) Effects of the cannabinoid CB(1) receptor allosteric modulator ORG 27569 on reinstatement of cocaine- and methamphetamine-seeking behavior in rats. Drug Alcohol Depend 143:251–256. doi:10.1016/j.drugalcdep.2014.08.004

Johnson PM, Kenny PJ (2010) Dopamine D2 receptors in addiction-like reward dysfunction and compulsive eating in obese rats. Nat Neurosci 13(5):635–641. doi:10.1038/nn.2519

Jose BS, van Oers HA, van de Mheen HD, Garretsen HF, Mackenbach JP (2000) Stressors and alcohol consumption. Alcohol Alcohol 35(3):307–312

Justinova Z, Solinas M, Tanda G, Redhi GH, Goldberg SR (2005) The endogenous cannabinoid anandamide and its synthetic analog R(+)-methanandamide are intravenously self-administered by squirrel monkeys. J Neurosci 25(23):5645–5650. doi:10.1523/JNEUROSCI.0951-05.2005

Justinova Z, Mangieri RA, Bortolato M, Chefer SI, Mukhin AG, Clapper JR, King AR, Redhi GH, Yasar S, Piomelli D, Goldberg SR (2008a) Fatty acid amide hydrolase inhibition heightens

anandamide signaling without producing reinforcing effects in primates. Biol Psychiatry 64 (11):930–937. doi:10.1016/j.biopsych.2008.08.008
Justinova Z, Munzar P, Panlilio LV, Yasar S, Redhi GH, Tanda G, Goldberg SR (2008b) Blockade of THC-seeking behavior and relapse in monkeys by the cannabinoid CB(1)-receptor antagonist rimonabant. Neuropsychopharmacology 33(12):2870–2877. doi:10.1038/npp.2008.21
Justinova Z, Yasar S, Redhi GH, Goldberg SR (2011) The endogenous cannabinoid 2-arachidonoylglycerol is intravenously self-administered by squirrel monkeys. J Neurosci 31(19):7043–7048. doi:10.1523/JNEUROSCI.6058-10.2011
Justinova Z, Panlilio LV, Moreno-Sanz G, Redhi GH, Auber A, Secci ME, Mascia P, Bandiera T, Armirotti A, Bertorelli R, Chefer SI, Barnes C, Yasar S, Piomelli D, Goldberg SR (2015) Effects of fatty acid amide hydrolase (FAAH) inhibitors in non-human primate models of nicotine reward and relapse. Neuropsychopharmacology. doi:10.1038/npp.2015.62
Jyotaki M, Shigemura N, Ninomiya Y (2010) Modulation of sweet taste sensitivity by orexigenic and anorexigenic factors. Endocr J 57(6):467–475
Kalivas PW, Volkow ND (2005) The neural basis of addiction: a pathology of motivation and choice. Am J Psychiatry 162(8):1403–1413. doi:10.1176/appi.ajp.162.8.1403
Kamprath K, Romo-Parra H, Haring M, Gaburro S, Doengi M, Lutz B, Pape HC (2011) Short-term adaptation of conditioned fear responses through endocannabinoid signaling in the central amygdala. Neuropsychopharmacology 36(3):652–663. doi:10.1038/npp.2010.196
Kaplan GB, Heinrichs SC, Carey RJ (2011) Treatment of addiction and anxiety using extinction approaches: neural mechanisms and their treatment implications. Pharmacol Biochem Behav 97(3):619–625. doi:10.1016/j.pbb.2010.08.004
Kaye W (2008) Neurobiology of anorexia and bulimia nervosa. Physiol Behav 94(1):121–135. doi:10.1016/j.physbeh.2007.11.037
Kelley AE, Baldo BA, Pratt WE, Will MJ (2005) Corticostriatal-hypothalamic circuitry and food motivation: integration of energy, action and reward. Physiol Behav 86(5):773–795. doi:10.1016/j.physbeh.2005.08.066
Kessler RC, Sonnega A, Bromet E, Hughes M, Nelson CB (1995) Posttraumatic stress disorder in the National Comorbidity Survey. Arch Gen Psychiatry 52(12):1048–1060
Kinsey SG, Wise LE, Ramesh D, Abdullah R, Selley DE, Cravatt BF, Lichtman AH (2013) Repeated low-dose administration of the monoacylglycerol lipase inhibitor JZL184 retains cannabinoid receptor type 1-mediated antinociceptive and gastroprotective effects. J Pharmacol Exp Ther 345(3):492–501. doi:10.1124/jpet.112.201426
Kirkham TC, Williams CM (2001) Endogenous cannabinoids and appetite. Nutr Res Rev 14 (1):65–86. doi:10.1079/NRR200118
Kirkham TC, Williams CM, Fezza F, Di Marzo V (2002) Endocannabinoid levels in rat limbic forebrain and hypothalamus in relation to fasting, feeding and satiation: stimulation of eating by 2-arachidonoyl glycerol. Br J Pharmacol 136(4):550–557. doi:10.1038/sj.bjp.0704767
Klein C, Hill MN, Chang SC, Hillard CJ, Gorzalka BB (2012) Circulating endocannabinoid concentrations and sexual arousal in women. J Sex Med 9(6):1588–1601. doi:10.1111/j.1743-6109.2012.02708.x
Klugmann M, Goepfrich A, Friemel CM, Schneider M (2011) AAV-mediated overexpression of the CB1 receptor in the mPFC of adult rats alters cognitive flexibility, social behavior, and emotional reactivity. Front Behav Neurosci 5:37. doi:10.3389/fnbeh.2011.00037
Klump KL, Miller KB, Keel PK, McGue M, Iacono WG (2001) Genetic and environmental influences on anorexia nervosa syndromes in a population-based twin sample. Psychol Med 31(4):737–740
Kodas E, Cohen C, Louis C, Griebel G (2007) Cortico-limbic circuitry for conditioned nicotine-seeking behavior in rats involves endocannabinoid signaling. Psychopharmacology (Berl) 194 (2):161–171. doi:10.1007/s00213-007-0813-0
Koff WC (1974) Marijuana and sexual activity. J Sex Res 10(3):194–204. doi:10.1080/00224497409550850

Komisaruk BR, Whipple B, Crawford A, Liu WC, Kalnin A, Mosier K (2004) Brain activation during vaginocervical self-stimulation and orgasm in women with complete spinal cord injury: fMRI evidence of mediation by the vagus nerves. Brain Res 1024(1–2):77–88. doi:10.1016/j.brainres.2004.07.029

Koob GF (1992) Drugs of abuse: anatomy, pharmacology and function of reward pathways. Trends Pharmacol Sci 13(5):177–184

Koob GF (2010) The role of CRF and CRF-related peptides in the dark side of addiction. Brain Res 1314:3–14. doi:10.1016/j.brainres.2009.11.008

Koob GF (2013) Theoretical frameworks and mechanistic aspects of alcohol addiction: alcohol addiction as a reward deficit disorder. Curr Top Behav Neurosci 13:3–30. doi:10.1007/7854_2011_129

Koob G, Kreek MJ (2007) Stress, dysregulation of drug reward pathways, and the transition to drug dependence. Am J Psychiatry 164(8):1149–1159. doi:10.1176/appi.ajp.2007.05030503

Koob GF, Le Moal M (1997) Drug abuse: hedonic homeostatic dysregulation. Science 278 (5335):52–58

Koob GF, Volkow ND (2010) Neurocircuitry of addiction. Neuropsychopharmacology 35 (1):217–238. doi:10.1038/npp.2009.110

Koob GF, Buck CL, Cohen A, Edwards S, Park PE, Schlosburg JE, Schmeichel B, Vendruscolo LF, Wade CL, Whitfield TW Jr, George O (2014) Addiction as a stress surfeit disorder. Neuropharmacology 76(Pt B):370–382. doi:10.1016/j.neuropharm.2013.05.024

Kupferschmidt DA, Klas PG, Erb S (2012a) Cannabinoid CB1 receptors mediate the effects of corticotropin-releasing factor on the reinstatement of cocaine seeking and expression of cocaine-induced behavioural sensitization. Br J Pharmacol 167(1):196–206. doi:10.1111/j.1476-5381.2012.01983.x

Kupferschmidt DA, Newman AE, Boonstra R, Erb S (2012b) Antagonism of cannabinoid 1 receptors reverses the anxiety-like behavior induced by central injections of corticotropin-releasing factor and cocaine withdrawal. Neuroscience 204:125–133. doi:10.1016/j.neuroscience.2011.07.022

Lacroix L, White I, Feldon J (2002) Effect of excitotoxic lesions of rat medial prefrontal cortex on spatial memory. Behav Brain Res 133(1):69–81

Lafenetre P, Chaouloff F, Marsicano G (2007) The endocannabinoid system in the processing of anxiety and fear and how CB1 receptors may modulate fear extinction. Pharmacol Res 56 (5):367–381. doi:10.1016/j.phrs.2007.09.006

Lapiz-Bluhm MD, Bondi CO, Doyen J, Rodriguez GA, Bedard-Arana T, Morilak DA (2008) Behavioural assays to model cognitive and affective dimensions of depression and anxiety in rats. J Neuroendocrinol 20(10):1115–1137. doi:10.1111/j.1365-2826.2008.01772.x

Lauzon NM, Bishop SF, Laviolette SR (2009) Dopamine D1 versus D4 receptors differentially modulate the encoding of salient versus nonsalient emotional information in the medial prefrontal cortex. J Neurosci 29(15):4836–4845. doi:10.1523/JNEUROSCI.0178-09.2009

Laviolette SR, Grace AA (2006) Cannabinoids potentiate emotional learning plasticity in neurons of the medial prefrontal cortex through basolateral amygdala inputs. J Neurosci 26 (24):6458–6468. doi:10.1523/JNEUROSCI.0707-06.2006

Laviolette SR, Lipski WJ, Grace AA (2005) A subpopulation of neurons in the medial prefrontal cortex encodes emotional learning with burst and frequency codes through a dopamine D4 receptor-dependent basolateral amygdala input. J Neurosci 25(26):6066–6075. doi:10.1523/JNEUROSCI.1168-05.2005

Le Foll B, Goldberg SR (2004) Rimonabant, a CB1 antagonist, blocks nicotine-conditioned place preferences. Neuroreport 15(13):2139–2143

Le Moal M (2009) Drug abuse: vulnerability and transition to addiction. Pharmacopsychiatry 42 (Suppl 1):S42–S55. doi:10.1055/s-0029-1216355

Le Moal M, Simon H (1991) Mesocorticolimbic dopaminergic network: functional and regulatory roles. Physiol Rev 71(1):155–234

Ledent C, Valverde O, Cossu G, Petitet F, Aubert JF, Beslot F, Bohme GA, Imperato A, Pedrazzini T, Roques BP, Vassart G, Fratta W, Parmentier M (1999) Unresponsiveness to cannabinoids and reduced addictive effects of opiates in CB1 receptor knockout mice. Science 283(5400):401–404

Lee PR, Brady DL, Shapiro RA, Dorsa DM, Koenig JI (2005) Social interaction deficits caused by chronic phencyclidine administration are reversed by oxytocin. Neuropsychopharmacology 30 (10):1883–1894. doi:10.1038/sj.npp.1300722

Lee TT, Filipski SB, Hill MN, McEwen BS (2014) Morphological and behavioral evidence for impaired prefrontal cortical function in female CB1 receptor deficient mice. Behav Brain Res 271:106–110. doi:10.1016/j.bbr.2014.05.064

Lesscher HM, Hoogveld E, Burbach JP, van Ree JM, Gerrits MA (2005) Endogenous cannabinoids are not involved in cocaine reinforcement and development of cocaine-induced behavioural sensitization. Eur Neuropsychopharmacol 15(1):31–37. doi:10.1016/j.euroneuro.2004.04.003

Leweke FM, Giuffrida A, Koethe D, Schreiber D, Nolden BM, Kranaster L, Neatby MA, Schneider M, Gerth CW, Hellmich M, Klosterkotter J, Piomelli D (2007) Anandamide levels in cerebrospinal fluid of first-episode schizophrenic patients: impact of cannabis use. Schizophr Res 94(1–3):29–36. doi:10.1016/j.schres.2007.04.025

Lewis DY, Brett RR (2010) Activity-based anorexia in C57/BL6 mice: effects of the phytocannabinoid, Delta9-tetrahydrocannabinol (THC) and the anandamide analogue, OMDM-2. Eur Neuropsychopharmacol 20(9):622–631. doi:10.1016/j.euroneuro.2010.04.002

Li X, Hoffman AF, Peng XQ, Lupica CR, Gardner EL, Xi ZX (2009) Attenuation of basal and cocaine-enhanced locomotion and nucleus accumbens dopamine in cannabinoid CB1-receptor-knockout mice. Psychopharmacology (Berl) 204(1):1–11. doi:10.1007/s00213-008-1432-0

Liang NC, Hajnal A, Norgren R (2006) Sham feeding corn oil increases accumbens dopamine in the rat. Am J Physiol Regul Integr Comp Physiol 291(5):R1236–R1239. doi:10.1152/ajpregu.00226.2006

Lichtman AH, Sheikh SM, Loh HH, Martin BR (2001) Opioid and cannabinoid modulation of precipitated withdrawal in delta(9)-tetrahydrocannabinol and morphine-dependent mice. J Pharmacol Exp Ther 298(3):1007–1014

Liu QS, Pu L, Poo MM (2005) Repeated cocaine exposure in vivo facilitates LTP induction in midbrain dopamine neurons. Nature 437(7061):1027–1031. doi:10.1038/nature04050

Long JZ, Li W, Booker L, Burston JJ, Kinsey SG, Schlosburg JE, Pavon FJ, Serrano AM, Selley DE, Parsons LH, Lichtman AH, Cravatt BF (2009a) Selective blockade of 2-arachidonoylglycerol hydrolysis produces cannabinoid behavioral effects. Nat Chem Biol 5(1):37–44. doi:10.1038/nchembio.129

Long JZ, Nomura DK, Vann RE, Walentiny DM, Booker L, Jin X, Burston JJ, Sim-Selley LJ, Lichtman AH, Wiley JL, Cravatt BF (2009b) Dual blockade of FAAH and MAGL identifies behavioral processes regulated by endocannabinoid crosstalk in vivo. Proc Natl Acad Sci USA 106(48):20270–20275. doi:10.1073/pnas.0909411106

Lopez HH (2010) Cannabinoid-hormone interactions in the regulation of motivational processes. Horm Behav 58(1):100–110. doi:10.1016/j.yhbeh.2009.10.005

Lopez-Moreno JA, Gonzalez-Cuevas G, Rodriguez de Fonseca F, Navarro M (2004) Long-lasting increase of alcohol relapse by the cannabinoid receptor agonist WIN 55,212-2 during alcohol deprivation. J Neurosci 24(38):8245–8252. doi:10.1523/JNEUROSCI.2179-04.2004

Lovinger DM (2008) Presynaptic modulation by endocannabinoids. Handb Exp Pharmacol 184:435–477. doi:10.1007/978-3-540-74805-2_14

Lu HC, Mackie K (2015) An introduction to the endogenous cannabinoid system. Biol Psychiatry. doi:10.1016/j.biopsych.2015.07.028

Luchicchi A, Lecca S, Carta S, Pillolla G, Muntoni AL, Yasar S, Goldberg SR, Pistis M (2010) Effects of fatty acid amide hydrolase inhibition on neuronal responses to nicotine, cocaine and

morphine in the nucleus accumbens shell and ventral tegmental area: involvement of PPAR-alpha nuclear receptors. Addict Biol 15(3):277–288. doi:10.1111/j.1369-1600.2010.00222.x
Luscher C, Malenka RC (2011) Drug-evoked synaptic plasticity in addiction: from molecular changes to circuit remodeling. Neuron 69(4):650–663. doi:10.1016/j.neuron.2011.01.017
Lutter M, Nestler EJ (2009) Homeostatic and hedonic signals interact in the regulation of food intake. J Nutr 139(3):629–632. doi:10.3945/jn.108.097618
Lutz B (2009) Endocannabinoid signals in the control of emotion. Curr Opin Pharmacol 9 (1):46–52. doi:10.1016/j.coph.2008.12.001
Maccarrone M, Guzman M, Mackie K, Doherty P, Harkany T (2014) Programming of neural cells by (endo)cannabinoids: from physiological rules to emerging therapies. Nat Rev Neurosci 15 (12):786–801. doi:10.1038/nrn3846
Macht M, Mueller J (2007) Immediate effects of chocolate on experimentally induced mood states. Appetite 49(3):667–674. doi:10.1016/j.appet.2007.05.004
Madronal N, Gruart A, Valverde O, Espadas I, Moratalla R, Delgado-Garcia JM (2012) Involvement of cannabinoid CB1 receptor in associative learning and in hippocampal CA3-CA1 synaptic plasticity. Cereb Cortex 22(3):550–566. doi:10.1093/cercor/bhr103
Mahler SV, Smith KS, Berridge KC (2007) Endocannabinoid hedonic hotspot for sensory pleasure: anandamide in nucleus accumbens shell enhances 'liking' of a sweet reward. Neuropsychopharmacology 32(11):2267–2278. doi:10.1038/sj.npp.1301376
Malinen H, Hyytia P (2008) Ethanol self-administration is regulated by CB1 receptors in the nucleus accumbens and ventral tegmental area in alcohol-preferring AA rats. Alcohol Clin Exp Res 32(11):1976–1983. doi:10.1111/j.1530-0277.2008.00786.x
Malinen H, Lehtonen M, Hyytia P (2009) Modulation of brain endocannabinoid levels by voluntary alcohol consumption in alcohol-preferring AA rats. Alcohol Clin Exp Res 33 (10):1711–1720. doi:10.1111/j.1530-0277.2009.01008.x
Mallet PE, Beninger RJ (1998) Delta9-tetrahydrocannabinol, but not the endogenous cannabinoid receptor ligand anandamide, produces conditioned place avoidance. Life Sci 62 (26):2431–2439
Mangieri RA, Hong KI, Piomelli D, Sinha R (2009) An endocannabinoid signal associated with desire for alcohol is suppressed in recently abstinent alcoholics. Psychopharmacology (Berl) 205(1):63–72. doi:10.1007/s00213-009-1518-3
Mannucci C, Navarra M, Pieratti A, Russo GA, Caputi AP, Calapai G (2011) Interactions between endocannabinoid and serotonergic systems in mood disorders caused by nicotine withdrawal. Nicotine Tob Res 13(4):239–247. doi:10.1093/ntr/ntq242
Manzanares J, Corchero J, Fuentes JA (1999) Opioid and cannabinoid receptor-mediated regulation of the increase in adrenocorticotropin hormone and corticosterone plasma concentrations induced by central administration of delta(9)-tetrahydrocannabinol in rats. Brain Res 839 (1):173–179
Manzanedo C, Aguilar MA, Rodriguez-Arias M, Navarro M, Minarro J (2004) Cannabinoid agonist-induced sensitisation to morphine place preference in mice. Neuroreport 15 (8):1373–1377
Marco EM, Granstrem O, Moreno E, Llorente R, Adriani W, Laviola G, Viveros MP (2007) Subchronic nicotine exposure in adolescence induces long-term effects on hippocampal and striatal cannabinoid-CB1 and mu-opioid receptors in rats. Eur J Pharmacol 557(1):37–43. doi:10.1016/j.ejphar.2006.11.013
Marco EM, Romero-Zerbo SY, Viveros MP, Bermudez-Silva FJ (2012) The role of the endocannabinoid system in eating disorders: pharmacological implications. Behav Pharmacol 23(5-6):526–536. doi:10.1097/FBP.0b013e328356c3c9
Markou A (2008) Review. Neurobiology of nicotine dependence. Philos Trans R Soc Lond B Biol Sci 363(1507):3159–3168. doi:10.1098/rstb.2008.0095
Marsicano G, Lafenetre P (2009) Roles of the endocannabinoid system in learning and memory. Curr Top Behav Neurosci 1:201–230. doi:10.1007/978-3-540-88955-7_8

Marsicano G, Wotjak CT, Azad SC, Bisogno T, Rammes G, Cascio MG, Hermann H, Tang J, Hofmann C, Zieglgansberger W, Di Marzo V, Lutz B (2002) The endogenous cannabinoid system controls extinction of aversive memories. Nature 418(6897):530–534. doi:10.1038/nature00839

Martin M, Ledent C, Parmentier M, Maldonado R, Valverde O (2000) Cocaine, but not morphine, induces conditioned place preference and sensitization to locomotor responses in CB1 knockout mice. Eur J Neurosci 12(11):4038–4046

Martin M, Ledent C, Parmentier M, Maldonado R, Valverde O (2002) Involvement of CB1 cannabinoid receptors in emotional behaviour. Psychopharmacology 159(4):379–387. doi:10.1007/s00213-001-0946-5

Martinez-Gonzalez D, Bonilla-Jaime H, Morales-Otal A, Henriksen SJ, Velazquez-Moctezuma J, Prospero-Garcia O (2004) Oleamide and anandamide effects on food intake and sexual behavior of rats. Neurosci Lett 364(1):1–6. doi:10.1016/j.neulet.2004.03.080

Mascia MS, Obinu MC, Ledent C, Parmentier M, Bohme GA, Imperato A, Fratta W (1999) Lack of morphine-induced dopamine release in the nucleus accumbens of cannabinoid CB(1) receptor knockout mice. Eur J Pharmacol 383(3):R1–R2

Mas-Nieto M, Pommier B, Tzavara ET, Caneparo A, Da Nascimento S, Le Fur G, Roques BP, Noble F (2001) Reduction of opioid dependence by the CB(1) antagonist SR141716A in mice: evaluation of the interest in pharmacotherapy of opioid addiction. Br J Pharmacol 132(8):1809–1816. doi:10.1038/sj.bjp.0703990

Mathes CM, Ferrara M, Rowland NE (2008) Cannabinoid-1 receptor antagonists reduce caloric intake by decreasing palatable diet selection in a novel dessert protocol in female rats. Am J Physiol Regul Integr Comp Physiol 295(1):R67–R75. doi:10.1152/ajpregu.00150.2008

Mato S, Chevaleyre V, Robbe D, Pazos A, Castillo PE, Manzoni OJ (2004) A single in-vivo exposure to delta 9THC blocks endocannabinoid-mediated synaptic plasticity. Nat Neurosci 7(6):585–586. doi:10.1038/nn1251

Mazzola C, Medalie J, Scherma M, Panlilio LV, Solinas M, Tanda G, Drago F, Cadet JL, Goldberg SR, Yasar S (2009) Fatty acid amide hydrolase (FAAH) inhibition enhances memory acquisition through activation of PPAR-alpha nuclear receptors. Learn Mem 16(5):332–337. doi:10.1101/lm.1145209

McElligott ZA, Winder DG (2009) Modulation of glutamatergic synaptic transmission in the bed nucleus of the stria terminalis. Prog Neuropsychopharmacol Biol Psychiatry 33(8):1329–1335. doi:10.1016/j.pnpbp.2009.05.022

McGregor IS, Dam KD, Mallet PE, Gallate JE (2005) Delta9-THC reinstates beer- and sucrose-seeking behaviour in abstinent rats: comparison with midazolam, food deprivation and predator odour. Alcohol Alcohol 40(1):35–45. doi:10.1093/alcalc/agh113

McKinzie DL, Sajdyk TJ, McBride WJ, Murphy JM, Lumeng L, Li TK, Shekhar A (2000) Acoustic startle and fear-potentiated startle in alcohol-preferring (P) and -nonpreferring (NP) lines of rats. Pharmacol Biochem Behav 65(4):691–696

McLaughlin RJ, Hill MN, Gorzalka BB (2014) A critical role for prefrontocortical endocannabinoid signaling in the regulation of stress and emotional behavior. Neurosci Biobehav Rev 42:116–131. doi:10.1016/j.neubiorev.2014.02.006

McLellan AT, Lewis DC, O'Brien CP, Kleber HD (2000) Drug dependence, a chronic medical illness: implications for treatment, insurance, and outcomes evaluation. JAMA 284(13):1689–1695

Mechoulam R, Parker LA (2013) The endocannabinoid system and the brain. Annu Rev Psychol 64:21–47. doi:10.1146/annurev-psych-113011-143739

Melis M, Pistis M (2012a) Hub and switches: endocannabinoid signalling in midbrain dopamine neurons. Philos Trans R Soc Lond B Biol Sci 367(1607):3276–3285. doi:10.1098/rstb.2011.0383

Melis M, Pistis M (2012b) Hub and switches: endocannabinoid signalling in midbrain dopamine neurons. Philos Trans R Soc Lond B Biol Sci 367(1607):3276–3285. doi:10.1098/rstb.2011.0383

Melis M, Pistis M (2014) Targeting the interaction between fatty acid ethanolamides and nicotinic receptors: therapeutic perspectives. Pharmacol Res 86:42–49. doi:10.1016/j.phrs.2014.03.009

Melis M, Camarini R, Ungless MA, Bonci A (2002) Long-lasting potentiation of GABAergic synapses in dopamine neurons after a single in vivo ethanol exposure. J Neurosci 22 (6):2074–2082

Melis T, Succu S, Sanna F, Boi A, Argiolas A, Melis MR (2007) The cannabinoid antagonist SR 141716A (Rimonabant) reduces the increase of extra-cellular dopamine release in the rat nucleus accumbens induced by a novel high palatable food. Neurosci Lett 419(3):231–235. doi:10.1016/j.neulet.2007.04.012

Mereu M, Tronci V, Chun LE, Thomas AM, Green JL, Katz JL, Tanda G (2013) Cocaine-induced endocannabinoid release modulates behavioral and neurochemical sensitization in mice. Addict Biol 20(4):91–103. doi:10.1111/adb.12080

Merritt LL, Martin BR, Walters C, Lichtman AH, Damaj MI (2008) The endogenous cannabinoid system modulates nicotine reward and dependence. J Pharmacol Exp Ther 326(2):483–492. doi:10.1124/jpet.108.138321

Miller WR, Harris RJ (2000) A simple scale of Gorski's warning signs for relapse. J Stud Alcohol 61(5):759–765

Mitrirattanakul S, Lopez-Valdes HE, Liang J, Matsuka Y, Mackie K, Faull KF, Spigelman I (2007) Bidirectional alterations of hippocampal cannabinoid 1 receptors and their endogenous ligands in a rat model of alcohol withdrawal and dependence. Alcohol Clin Exp Res 31(5):855–867. doi:10.1111/j.1530-0277.2007.00366.x

Monteleone P, Matias I, Martiadis V, De Petrocellis L, Maj M, Di Marzo V (2005) Blood levels of the endocannabinoid anandamide are increased in anorexia nervosa and in binge-eating disorder, but not in bulimia nervosa. Neuropsychopharmacology 30(6):1216–1221. doi:10.1038/sj.npp.1300695

Monteleone P, Tortorella A, Martiadis V, Di Filippo C, Canestrelli B, Maj M (2008) The cDNA 385C to A missense polymorphism of the endocannabinoid degrading enzyme fatty acid amide hydrolase (FAAH) is associated with overweight/obesity but not with binge eating disorder in overweight/obese women. Psychoneuroendocrinology 33(4):546–550. doi:10.1016/j.psyneuen.2008.01.004

Monteleone P, Bifulco M, Di Filippo C, Gazzerro P, Canestrelli B, Monteleone F, Proto MC, Di Genio M, Grimaldi C, Maj M (2009) Association of CNR1 and FAAH endocannabinoid gene polymorphisms with anorexia nervosa and bulimia nervosa: evidence for synergistic effects. Genes Brain Behav 8(7):728–732. doi:10.1111/j.1601-183X.2009.00518.x

Monteleone P, Piscitelli F, Scognamiglio P, Monteleone AM, Canestrelli B, Di Marzo V, Maj M (2012) Hedonic eating is associated with increased peripheral levels of ghrelin and the endocannabinoid 2-arachidonoyl-glycerol in healthy humans: a pilot study. J Clin Endocrinol Metab 97(6):E917–E924. doi:10.1210/jc.2011-3018

Moranta D, Esteban S, Garcia-Sevilla JA (2006) Ethanol desensitizes cannabinoid CB1 receptors modulating monoamine synthesis in the rat brain in vivo. Neurosci Lett 392(1–2):58–61. doi:10.1016/j.neulet.2005.08.061

Morena M, Patel S, Bains JS, Hill MN (2016) Neurobiological interactions between stress and the endocannabinoid system. Neuropsychopharmacology 41(1):80–102. doi:10.1038/npp.2015.166

Morgan CJ, Page E, Schaefer C, Chatten K, Manocha A, Gulati S, Curran HV, Brandner B, Leweke FM (2013) Cerebrospinal fluid anandamide levels, cannabis use and psychotic-like symptoms. Br J Psychiatry 202(5):381–382. doi:10.1192/bjp.bp.112.121178

Muhl D, Kathmann M, Hoyer C, Kranaster L, Hellmich M, Gerth CW, Faulhaber J, Schlicker E, Leweke FM (2014) Increased CB2 mRNA and anandamide in human blood after cessation of cannabis abuse. Naunyn Schmiedebergs Arch Pharmacol 387(7):691–695. doi:10.1007/s00210-014-0984-2

Muldoon PP, Lichtman AH, Parsons LH, Damaj MI (2013) The role of fatty acid amide hydrolase inhibition in nicotine reward and dependence. Life Sci 92(8–9):458–462. doi:10.1016/j.lfs.2012.05.015

Muldoon PP, Chen J, Harenza JL, Abdullah RA, Sim-Selley LJ, Cravatt BF, Miles MF, Chen X, Lichtman AH, Damaj MI (2015) Inhibition of monoacylglycerol lipase reduces nicotine withdrawal. Br J Pharmacol 172(3):869–882. doi:10.1111/bph.12948

Muller TD, Reichwald K, Bronner G, Kirschner J, Nguyen TT, Scherag A, Herzog W, Herpertz-Dahlmann B, Lichtner P, Meitinger T, Platzer M, Schafer H, Hebebrand J, Hinney A (2008) Lack of association of genetic variants in genes of the endocannabinoid system with anorexia nervosa. Child Adolesc Psychiatr Ment Health 2(1):33. doi:10.1186/1753-2000-2-33

Murillo-Rodriguez E, Poot-Ake A, Arias-Carrion O, Pacheco-Pantoja E, Fuente-Ortegon Ade L, Arankowsky-Sandoval G (2011) The emerging role of the endocannabinoid system in the sleep-wake cycle modulation. Cent Nerv Syst Agents Med Chem 11(3):189–196

Murillo-Rodriguez E, Palomero-Rivero M, Millan-Aldaco D, Di Marzo V (2013) The administration of endocannabinoid uptake inhibitors OMDM-2 or VDM-11 promotes sleep and decreases extracellular levels of dopamine in rats. Physiol Behav 109:88–95. doi:10.1016/j.physbeh.2012.11.007

Murray JB (2002) Phencyclidine (PCP): a dangerous drug, but useful in schizophrenia research. J Psychol 136(3):319–327. doi:10.1080/00223980209604159

Naassila M, Pierrefiche O, Ledent C, Daoust M (2004) Decreased alcohol self-administration and increased alcohol sensitivity and withdrawal in CB1 receptor knockout mice. Neuropharmacology 46(2):243–253

Natividad LA, Buczynski MW, Herman MA, Kirson D, Oleata CS, Irimia C, Polis I, Ciccocioppo R, Roberto M, Parsons LH (2017) Constitutive increases in amygdalar corticotropin-releasing factor and fatty acid amide hydrolase drive an anxious phenotype. Biol Psychiatry. doi:10.1016/j.biopsych.2017.01.005

Navarrete F, Rodriguez-Arias M, Martin-Garcia E, Navarro D, Garcia-Gutierrez MS, Aguilar MA, Aracil-Fernandez A, Berbel P, Minarro J, Maldonado R, Manzanares J (2013) Role of CB2 cannabinoid receptors in the rewarding, reinforcing, and physical effects of nicotine. Neuropsychopharmacology 38(12):2515–2524. doi:10.1038/npp.2013.157

Navarro M, Carrera MR, Fratta W, Valverde O, Cossu G, Fattore L, Chowen JA, Gomez R, del Arco I, Villanua MA, Maldonado R, Koob GF, Rodriguez de Fonseca F (2001) Functional interaction between opioid and cannabinoid receptors in drug self-administration. J Neurosci 21(14):5344–5350

Nawata Y, Hiranita T, Yamamoto T (2010) A cannabinoid CB(1) receptor antagonist ameliorates impairment of recognition memory on withdrawal from MDMA (Ecstasy). Neuropsychopharmacology 35(2):515–520. doi:10.1038/npp.2009.158

Nestler EJ (2005) Is there a common molecular pathway for addiction? Nat Neurosci 8 (11):1445–1449. doi:10.1038/nn1578

Nestler EJ (2014) Epigenetic mechanisms of drug addiction. Neuropharmacology 76 (Pt B):259–268. doi:10.1016/j.neuropharm.2013.04.004

Neumeister A, Normandin MD, Murrough JW, Henry S, Bailey CR, Luckenbaugh DA, Tuit K, Zheng MQ, Galatzer-Levy IR, Sinha R, Carson RE, Potenza MN, Huang Y (2012) Positron emission tomography shows elevated cannabinoid CB1 receptor binding in men with alcohol dependence. Alcohol Clin Exp Res 36(12):2104–2109. doi:10.1111/j.1530-0277.2012.01815.x

Niehaus JL, Murali M, Kauer JA (2010) Drugs of abuse and stress impair LTP at inhibitory synapses in the ventral tegmental area. Eur J Neurosci 32(1):108–117. doi:10.1111/j.1460-9568.2010.07256.x

Niphakis MJ, Johnson DS, Ballard TE, Stiff C, Cravatt BF (2012) O-hydroxyacetamide carbamates as a highly potent and selective class of endocannabinoid hydrolase inhibitors. ACS Chem Neurosci 3(5):418–426. doi:10.1021/cn200089j

Niyuhire F, Varvel SA, Thorpe AJ, Stokes RJ, Wiley JL, Lichtman AH (2007) The disruptive effects of the CB1 receptor antagonist rimonabant on extinction learning in mice are task-specific. Psychopharmacology 191(2):223–231. doi:10.1007/s00213-006-0650-6

Nugent FS, Penick EC, Kauer JA (2007) Opioids block long-term potentiation of inhibitory synapses. Nature 446(7139):1086–1090. doi:10.1038/nature05726

Nunes EV, Levin FR (2004) Treatment of depression in patients with alcohol or other drug dependence: a meta-analysis. JAMA 291(15):1887–1896. doi:10.1001/jama.291.15.1887

Nunes EV, Sullivan MA, Levin FR (2004) Treatment of depression in patients with opiate dependence. Biol Psychiatry 56(10):793–802. doi:10.1016/j.biopsych.2004.06.037

Olds J, Milner P (1954) Positive reinforcement produced by electrical stimulation of septal area and other regions of rat brain. J Comp Physiol Psychol 47(6):419–427

Palomino A, Pavon FJ, Blanco-Calvo E, Serrano A, Arrabal S, Rivera P, Alen F, Vargas A, Bilbao A, Rubio L, Rodriguez de Fonseca F, Suarez J (2014) Effects of acute versus repeated cocaine exposure on the expression of endocannabinoid signaling-related proteins in the mouse cerebellum. Front Integr Neurosci 8:22. doi:10.3389/fnint.2014.00022

Pamplona FA, Prediger RD, Pandolfo P, Takahashi RN (2006) The cannabinoid receptor agonist WIN 55,212-2 facilitates the extinction of contextual fear memory and spatial memory in rats. Psychopharmacology 188(4):641–649. doi:10.1007/s00213-006-0514-0

Pamplona FA, Bitencourt RM, Takahashi RN (2008) Short- and long-term effects of cannabinoids on the extinction of contextual fear memory in rats. Neurobiol Learn Mem 90(1):290–293. doi:10.1016/j.nlm.2008.04.003

Pan B, Hillard CJ, Liu QS (2008) Endocannabinoid signaling mediates cocaine-induced inhibitory synaptic plasticity in midbrain dopamine neurons. J Neurosci 28(6):1385–1397. doi:10.1523/JNEUROSCI.4033-07.2008

Panagis G, Mackey B, Vlachou S (2014) Cannabinoid regulation of brain reward processing with an emphasis on the role of CB1 receptors: a step back into the future. Front Psychiatry 5:92. doi:10.3389/fpsyt.2014.00092

Pani PP, Maremmani I, Trogu E, Gessa GL, Ruiz P, Akiskal HS (2010) Delineating the psychic structure of substance abuse and addictions: should anxiety, mood and impulse-control dysregulation be included? J Affect Disord 122(3):185–197. doi:10.1016/j.jad.2009.06.012

Parylak SL, Koob GF, Zorrilla EP (2011) The dark side of food addiction. Physiol Behav 104 (1):149–156. doi:10.1016/j.physbeh.2011.04.063

Pataky Z, Gasteyger C, Ziegler O, Rissanen A, Hanotin C, Golay A (2013) Efficacy of rimonabant in obese patients with binge eating disorder. Exp Clin Endocrinol Diabetes 121(1):20–26. doi:10.1055/s-0032-1329957

Patel S, Rademacher DJ, Hillard CJ (2003) Differential regulation of the endocannabinoids anandamide and 2-arachidonylglycerol within the limbic forebrain by dopamine receptor activity. J Pharmacol Exp Ther 306(3):880–888. doi:10.1124/jpet.103.054270

Patel S, Roelke CT, Rademacher DJ, Cullinan WE, Hillard CJ (2004) Endocannabinoid signaling negatively modulates stress-induced activation of the hypothalamic-pituitary-adrenal axis. Endocrinology 145(12):5431–5438. doi:10.1210/en.2004-0638

Patel S, Cravatt BF, Hillard CJ (2005) Synergistic interactions between cannabinoids and environmental stress in the activation of the central amygdala. Neuropsychopharmacology 30 (3):497–507. doi:10.1038/sj.npp.1300535

Patel S, Kingsley PJ, Mackie K, Marnett LJ, Winder DG (2009) Repeated homotypic stress elevates 2-arachidonoylglycerol levels and enhances short-term endocannabinoid signaling at inhibitory synapses in basolateral amygdala. Neuropsychopharmacology 34(13):2699–2709. doi:10.1038/npp.2009.101

Pava MJ, Woodward JJ (2014) Chronic ethanol alters network activity and endocannabinoid signaling in the prefrontal cortex. Front Integr Neurosci 8:58. doi:10.3389/fnint.2014.00058

Pavon FJ, Araos P, Pastor A, Calado M, Pedraz M, Campos-Cloute R, Ruiz JJ, Serrano A, Blanco E, Rivera P, Suarez J, Romero-Cuevas M, Pujadas M, Vergara-Moragues E, Gornemann I, Torrens M, de la Torre R, Rodriguez de Fonseca F (2013) Evaluation of plasma-free endocannabinoids and their congeners in abstinent cocaine addicts seeking outpatient treatment: impact of psychiatric co-morbidity. Addict Biol 18(6):955–969. doi:10.1111/adb.12107

Perra S, Pillolla G, Melis M, Muntoni AL, Gessa GL, Pistis M (2005) Involvement of the endogenous cannabinoid system in the effects of alcohol in the mesolimbic reward circuit: electrophysiological evidence in vivo. Psychopharmacology (Berl) 183(3):368–377. doi:10.1007/s00213-005-0195-0

Pettit HO, Ettenberg A, Bloom FE, Koob GF (1984) Destruction of dopamine in the nucleus accumbens selectively attenuates cocaine but not heroin self-administration in rats. Psychopharmacology (Berl) 84(2):167–173

Piasecki TM, Jorenby DE, Smith SS, Fiore MC, Baker TB (2003a) Smoking withdrawal dynamics: I. Abstinence distress in lapsers and abstainers. J Abnorm Psychol 112(1):3–13

Piasecki TM, Jorenby DE, Smith SS, Fiore MC, Baker TB (2003b) Smoking withdrawal dynamics: II. Improved tests of withdrawal-relapse relations. J Abnorm Psychol 112(1):14–27

Piasecki TM, Jorenby DE, Smith SS, Fiore MC, Baker TB (2003c) Smoking withdrawal dynamics: III. Correlates of withdrawal heterogeneity. Exp Clin Psychopharmacol 11(4):276–285. doi:10.1037/1064-1297.11.4.276

Piazza PV, Le Moal M (1998) The role of stress in drug self-administration. Trends Pharmacol Sci 19(2):67–74

Piper ME, Cook JW, Schlam TR, Jorenby DE, Baker TB (2011) Anxiety diagnoses in smokers seeking cessation treatment: relations with tobacco dependence, withdrawal, outcome and response to treatment. Addiction 106(2):418–427. doi:10.1111/j.1360-0443.2010.03173.x

Platt DM, Rowlett JK, Spealman RD (2001) Discriminative stimulus effects of intravenous heroin and its metabolites in rhesus monkeys: opioid and dopaminergic mechanisms. J Pharmacol Exp Ther 299(2):760–767

Powers MS, Barrenha GD, Mlinac NS, Barker EL, Chester JA (2010) Effects of the novel endocannabinoid uptake inhibitor, LY2183240, on fear-potentiated startle and alcohol-seeking behaviors in mice selectively bred for high alcohol preference. Psychopharmacology (Berl) 212(4):571–583. doi:10.1007/s00213-010-1997-2

Puighermanal E, Busquets-Garcia A, Maldonado R, Ozaita A (2012) Cellular and intracellular mechanisms involved in the cognitive impairment of cannabinoids. Philos Trans R Soc Lond B Biol Sci 367(1607):3254–3263. doi:10.1098/rstb.2011.0384

Quirk GJ, Gehlert DR (2003) Inhibition of the amygdala: key to pathological states? Ann NY Acad Sci 985:263–272

Rada P, Avena NM, Hoebel BG (2005) Daily bingeing on sugar repeatedly releases dopamine in the accumbens shell. Neuroscience 134(3):737–744. doi:10.1016/j.neuroscience.2005.04.043

Rademacher DJ, Hillard CJ (2007) Interactions between endocannabinoids and stress-induced decreased sensitivity to natural reward. Prog Neuropsychopharmacol Biol Psychiatry 31 (3):633–641. doi:10.1016/j.pnpbp.2006.12.013

Raichlen DA, Foster AD, Seillier A, Giuffrida A, Gerdeman GL (2013) Exercise-induced endocannabinoid signaling is modulated by intensity. Eur J Appl Physiol 113(4):869–875. doi:10.1007/s00421-012-2495-5

Ramesh D, Ross GR, Schlosburg JE, Owens RA, Abdullah RA, Kinsey SG, Long JZ, Nomura DK, Sim-Selley LJ, Cravatt BF, Akbarali HI, Lichtman AH (2011) Blockade of endocannabinoid hydrolytic enzymes attenuates precipitated opioid withdrawal symptoms in mice. J Pharmacol Exp Ther 339(1):173–185. doi:10.1124/jpet.111.181370

Ramesh D, Gamage TF, Vanuytsel T, Owens RA, Abdullah RA, Niphakis MJ, Shea-Donohue T, Cravatt BF, Lichtman AH (2013) Dual inhibition of endocannabinoid catabolic enzymes produces enhanced antiwithdrawal effects in morphine-dependent mice. Neuropsychopharmacology 38(6):1039–1049. doi:10.1038/npp.2012.269

Reisiger AR, Kaufling J, Manzoni O, Cador M, Georges F, Caille S (2014) Nicotine self-administration induces CB1-dependent LTP in the bed nucleus of the stria terminalis. J Neurosci 34(12):4285–4292. doi:10.1523/JNEUROSCI.3149-13.2014

Rescorla RA (1996) Preservation of Pavlovian associations through extinction. Q J Exp Psychol B Comp Physiol Psychol 49(3):245–258
Richman JA, Flaherty JA, Rospenda KM (1996) Perceived workplace harassment experiences and problem drinking among physicians: broadening the stress/alienation paradigm. Addiction 91 (3):391–403
Rivera P, Miguens M, Coria SM, Rubio L, Higuera-Matas A, Bermudez-Silva FJ, de Fonseca FR, Suarez J, Ambrosio E (2013) Cocaine self-administration differentially modulates the expression of endogenous cannabinoid system-related proteins in the hippocampus of Lewis vs. Fischer 344 rats. Int J Neuropsychopharmacol 16(6):1277–1293. doi:10.1017/S1461145712001186
Robinson TE, Berridge KC (1993) The neural basis of drug craving: an incentive-sensitization theory of addiction. Brain research Brain research reviews 18(3):247–291
Rose JE, Salley A, Behm FM, Bates JE, Westman EC (2010) Reinforcing effects of nicotine and non-nicotine components of cigarette smoke. Psychopharmacology 210(1):1–12. doi:10.1007/s00213-010-1810-2
Rospenda KM, Richman JA, Wislar JS, Flaherty JA (2000) Chronicity of sexual harassment and generalized work-place abuse: effects on drinking outcomes. Addiction 95(12):1805–1820. doi:10.1080/09652140020011117
Rouge-Pont F, Marinelli M, Le Moal M, Simon H, Piazza PV (1995) Stress-induced sensitization and glucocorticoids. II. Sensitization of the increase in extracellular dopamine induced by cocaine depends on stress-induced corticosterone secretion. J Neurosci 15(11):7189–7195
Rouge-Pont F, Deroche V, Le Moal M, Piazza PV (1998) Individual differences in stress-induced dopamine release in the nucleus accumbens are influenced by corticosterone. Eur J Neurosci 10 (12):3903–3907
Rubino T, Tizzoni L, Vigano D, Massi P, Parolaro D (1997) Modulation of rat brain cannabinoid receptors after chronic morphine treatment. Neuroreport 8(15):3219–3223
Rubino T, Massi P, Vigano D, Fuzio D, Parolaro D (2000) Long-term treatment with SR141716A, the CB1 receptor antagonist, influences morphine withdrawal syndrome. Life Sci 66 (22):2213–2219
Rubio M, McHugh D, Fernandez-Ruiz J, Bradshaw H, Walker JM (2007) Short-term exposure to alcohol in rats affects brain levels of anandamide, other N-acylethanolamines and 2-arachidonoyl-glycerol. Neurosci Lett 421(3):270–274. doi:10.1016/j.neulet.2007.05.052
Saal D, Dong Y, Bonci A, Malenka RC (2003) Drugs of abuse and stress trigger a common synaptic adaptation in dopamine neurons. Neuron 37(4):577–582
Salamone JD, Correa M, Farrar A, Mingote SM (2007) Effort-related functions of nucleus accumbens dopamine and associated forebrain circuits. Psychopharmacology 191 (3):461–482. doi:10.1007/s00213-006-0668-9
Saravia R, Flores A, Plaza-Zabala A, Busquets-Garcia A, Pastor A, de la Torre R, Di Marzo V, Marsicano G, Ozaita A, Maldonado R, Berrendero F (2016) cb1 cannabinoid receptors mediate cognitive deficits and structural plasticity changes during nicotine withdrawal. Biol Psychiatry. doi:10.1016/j.biopsych.2016.07.007
Scherma M, Panlilio LV, Fadda P, Fattore L, Gamaleddin I, Le Foll B, Justinova Z, Mikics E, Haller J, Medalie J, Stroik J, Barnes C, Yasar S, Tanda G, Piomelli D, Fratta W, Goldberg SR (2008) Inhibition of anandamide hydrolysis by cyclohexyl carbamic acid 3′-carbamoyl-3-yl ester (URB597) reverses abuse-related behavioral and neurochemical effects of nicotine in rats. J Pharmacol Exp Ther 327 (2):482–490. doi:10.1124/jpet.108.142224
Scherma M, Justinova Z, Zanettini C, Panlilio LV, Mascia P, Fadda P, Fratta W, Makriyannis A, Vadivel SK, Gamaleddin I, Le Foll B, Goldberg SR (2012) The anandamide transport inhibitor AM404 reduces the rewarding effects of nicotine and nicotine-induced dopamine elevations in the nucleus accumbens shell in rats. Br J Pharmacol 165(8):2539–2548. doi:10.1111/j.1476-5381.2011.01467.x

Scherma M, Fattore L, Satta V, Businco F, Pigliacampo B, Goldberg SR, Dessi C, Fratta W, Fadda P (2013) Pharmacological modulation of the endocannabinoid signalling alters binge-type eating behaviour in female rats. Br J Pharmacol 169(4):820–833. doi:10.1111/bph.12014

Scherma M, Fattore L, Castelli MP, Fratta W, Fadda P (2014) The role of the endocannabinoid system in eating disorders: neurochemical and behavioural preclinical evidence. Curr Pharm Des 20(13):2089–2099

Schindler CW, Scherma M, Redhi GH, Vadivel SK, Makriyannis A, Goldberg SR, Justinova Z (2016) Self-administration of the anandamide transport inhibitor AM404 by squirrel monkeys. Psychopharmacology. doi:10.1007/s00213-016-4211-3

Schlosburg JE, Carlson BL, Ramesh D, Abdullah RA, Long JZ, Cravatt BF, Lichtman AH (2009) Inhibitors of endocannabinoid-metabolizing enzymes reduce precipitated withdrawal responses in THC-dependent mice. AAPS J 11(2):342–352. doi:10.1208/s12248-009-9110-7

Schlosburg JE, Blankman JL, Long JZ, Nomura DK, Pan B, Kinsey SG, Nguyen PT, Ramesh D, Booker L, Burston JJ, Thomas EA, Selley DE, Sim-Selley LJ, Liu QS, Lichtman AH, Cravatt BF (2010) Chronic monoacylglycerol lipase blockade causes functional antagonism of the endocannabinoid system. Nat Neurosci 13(9):1113–1119. doi:10.1038/nn.2616

Schmidt LG, Samochowiec J, Finckh U, Fiszer-Piosik E, Horodnicki J, Wendel B, Rommelspacher H, Hoehe MR (2002) Association of a CB1 cannabinoid receptor gene (CNR1) polymorphism with severe alcohol dependence. Drug Alcohol Depend 65(3):221–224

Schuckit MA (2006) Comorbidity between substance use disorders and psychiatric conditions. Addiction 101(Suppl 1):76–88. doi:10.1111/j.1360-0443.2006.01592.x

Sclafani A, Bodnar RJ, Delamater AR (1998) Pharmacology of food conditioned preferences. Appetite 31(3):406. doi:10.1006/appe.1998.0211

Seillier A, Giuffrida A (2009) Evaluation of NMDA receptor models of schizophrenia: divergences in the behavioral effects of sub-chronic PCP and MK-801. Behav Brain Res 204 (2):410–415. doi:10.1016/j.bbr.2009.02.007

Seillier A, Martinez AA, Giuffrida A (2013) Phencyclidine-induced social withdrawal results from deficient stimulation of cannabinoid CB(1) receptors: implications for schizophrenia. Neuropsychopharmacology 38(9):1816–1824. doi:10.1038/npp.2013.81

Serrano A, Rivera P, Pavon FJ, Decara J, Suarez J, Rodriguez de Fonseca F, Parsons LH (2012) Differential effects of single versus repeated alcohol withdrawal on the expression of endocannabinoid system-related genes in the rat amygdala. Alcohol Clin Exp Res 36 (6):984–994. doi:10.1111/j.1530-0277.2011.01686.x

Shinohara Y, Inui T, Yamamoto T, Shimura T (2009) Cannabinoid in the nucleus accumbens enhances the intake of palatable solution. Neuroreport 20(15):1382–1385. doi:10.1097/WNR.0b013e3283318010

Shippenberg TS, Elmer GI (1998) The neurobiology of opiate reinforcement. Crit Rev Neurobiol 12(4):267–303

Shoaib M (2008) The cannabinoid antagonist AM251 attenuates nicotine self-administration and nicotine-seeking behaviour in rats. Neuropharmacology 54(2):438–444. doi:10.1016/j.neuropharm.2007.10.011

Sidhpura N, Parsons LH (2011) Endocannabinoid-mediated synaptic plasticity and addiction-related behavior. Neuropharmacology 61(7):1070–1087. doi:10.1016/j.neuropharm.2011.05.034

Siegfried Z, Kanyas K, Latzer Y, Karni O, Bloch M, Lerer B, Berry EM (2004) Association study of cannabinoid receptor gene (CNR1) alleles and anorexia nervosa: differences between restricting and binging/purging subtypes. Am J Med Genet B Neuropsychiatr Genet 125B (1):126–130. doi:10.1002/ajmg.b.20089

Silvestri C, Di Marzo V (2013) The endocannabinoid system in energy homeostasis and the etiopathology of metabolic disorders. Cell Metab 17(4):475–490. doi:10.1016/j.cmet.2013.03.001

Sim LJ, Hampson RE, Deadwyler SA, Childers SR (1996) Effects of chronic treatment with delta9-tetrahydrocannabinol on cannabinoid-stimulated [35S]GTPgammaS autoradiography in rat brain. J Neurosci 16(24):8057–8066

Simiand J, Keane M, Keane PE, Soubrie P (1998) SR 141716, a CB1 cannabinoid receptor antagonist, selectively reduces sweet food intake in marmoset. Behav Pharmacol 9(2):179–181

Simonnet A, Cador M, Caille S (2013) Nicotine reinforcement is reduced by cannabinoid CB1 receptor blockade in the ventral tegmental area. Addict Biol 18(6):930–936. doi:10.1111/j.1369-1600.2012.00476.x

Singh ME, Verty AN, McGregor IS, Mallet PE (2004) A cannabinoid receptor antagonist attenuates conditioned place preference but not behavioural sensitization to morphine. Brain Res 1026(2):244–253. doi:10.1016/j.brainres.2004.08.027

Sipe JC, Waalen J, Gerber A, Beutler E (2005) Overweight and obesity associated with a missense polymorphism in fatty acid amide hydrolase (FAAH). Int J Obes (Lond) 29(7):755–759. doi:10.1038/sj.ijo.0802954

Smith PA, Selley DE, Sim-Selley LJ, Welch SP (2007) Low dose combination of morphine and delta9-tetrahydrocannabinol circumvents antinociceptive tolerance and apparent desensitization of receptors. Eur J Pharmacol 571(2-3):129–137. doi:10.1016/j.ejphar.2007.06.001

Solinas M, Goldberg SR (2005) Motivational effects of cannabinoids and opioids on food reinforcement depend on simultaneous activation of cannabinoid and opioid systems. Neuropsychopharmacology 30(11):2035–2045. doi:10.1038/sj.npp.1300720

Solinas M, Panlilio LV, Antoniou K, Pappas LA, Goldberg SR (2003) The cannabinoid CB1 antagonist N-piperidinyl-5-(4-chlorophenyl)-1-(2,4-dichlorophenyl) -4-methylpyrazole-3-carboxamide (SR-141716A) differentially alters the reinforcing effects of heroin under continuous reinforcement, fixed ratio, and progressive ratio schedules of drug self-administration in rats. J Pharmacol Exp Ther 306(1):93–102. doi:10.1124/jpet.102.047928

Solinas M, Panlilio LV, Tanda G, Makriyannis A, Matthews SA, Goldberg SR (2005) Cannabinoid agonists but not inhibitors of endogenous cannabinoid transport or metabolism enhance the reinforcing efficacy of heroin in rats. Neuropsychopharmacology 30(11):2046–2057. doi:10.1038/sj.npp.1300754

Solinas M, Justinova Z, Goldberg SR, Tanda G (2006) Anandamide administration alone and after inhibition of fatty acid amide hydrolase (FAAH) increases dopamine levels in the nucleus accumbens shell in rats. J Neurochem 98(2):408–419. doi:10.1111/j.1471-4159.2006.03880.x

Solinas M, Tanda G, Justinova Z, Wertheim CE, Yasar S, Piomelli D, Vadivel SK, Makriyannis A, Goldberg SR (2007) The endogenous cannabinoid anandamide produces delta-9-tetrahydrocannabinol-like discriminative and neurochemical effects that are enhanced by inhibition of fatty acid amide hydrolase but not by inhibition of anandamide transport. J Pharmacol Exp Ther 321(1):370–380. doi:10.1124/jpet.106.114124

Soria G, Mendizabal V, Tourino C, Robledo P, Ledent C, Parmentier M, Maldonado R, Valverde O (2005) Lack of CB1 cannabinoid receptor impairs cocaine self-administration. Neuropsychopharmacology 30(9):1670–1680. doi:10.1038/sj.npp.1300707

Soria-Gomez E, Matias I, Rueda-Orozco PE, Cisneros M, Petrosino S, Navarro L, Di Marzo V, Prospero-Garcia O (2007) Pharmacological enhancement of the endocannabinoid system in the nucleus accumbens shell stimulates food intake and increases c-Fos expression in the hypothalamus. Br J Pharmacol 151(7):1109–1116. doi:10.1038/sj.bjp.0707313

South T, Deng C, Huang XF (2007) AM 251 and beta-Funaltrexamine reduce fat intake in a fat-preferring strain of mouse. Behav Brain Res 181(1):153–157. doi:10.1016/j.bbr.2007.03.028

Spano MS, Fattore L, Cossu G, Deiana S, Fadda P, Fratta W (2004) CB1 receptor agonist and heroin, but not cocaine, reinstate cannabinoid-seeking behaviour in the rat. Br J Pharmacol 143(3):343–350. doi:10.1038/sj.bjp.0705932

Sparling PB, Giuffrida A, Piomelli D, Rosskopf L, Dietrich A (2003) Exercise activates the endocannabinoid system. Neuroreport 14(17):2209–2211. doi:10.1097/01.wnr.0000097048.56589.47

Stewart SH, Kushner MG (2001) Introduction to the special issue on "Anxiety sensitivity and addictive behaviors". Addict Behav 26(6):775–785

Stoving RK, Andries A, Brixen K, Flyvbjerg A, Horder K, Frystyk J (2009) Leptin, ghrelin, and endocannabinoids: potential therapeutic targets in anorexia nervosa. J Psychiatr Res 43 (7):671–679. doi:10.1016/j.jpsychires.2008.09.007

Sun N, Chi N, Lauzon N, Bishop S, Tan H, Laviolette SR (2011) Acquisition, extinction, and recall of opiate reward memory are signaled by dynamic neuronal activity patterns in the prefrontal cortex. Cereb Cortex 21(12):2665–2680. doi:10.1093/cercor/bhr031

Suzuki A, Josselyn SA, Frankland PW, Masushige S, Silva AJ, Kida S (2004) Memory reconsolidation and extinction have distinct temporal and biochemical signatures. J Neurosci 24(20):4787–4795. doi:10.1523/JNEUROSCI.5491-03.2004

Tabibnia G, Lieberman MD (2007) Fairness and cooperation are rewarding: evidence from social cognitive neuroscience. Ann NY Acad Sci 1118:90–101. doi:10.1196/annals.1412.001

Tan H, Lauzon NM, Bishop SF, Bechard MA, Laviolette SR (2010) Integrated cannabinoid CB1 receptor transmission within the amygdala-prefrontal cortical pathway modulates neuronal plasticity and emotional memory encoding. Cereb Cortex 20(6):1486–1496. doi:10.1093/cercor/bhp210

Tan H, Ahmad T, Loureiro M, Zunder J, Laviolette SR (2014) The role of cannabinoid transmission in emotional memory formation: implications for addiction and schizophrenia. Front Psychiatry 5:73. doi:10.3389/fpsyt.2014.00073

Tanda G, Pontieri FE, Di Chiara G (1997) Cannabinoid and heroin activation of mesolimbic dopamine transmission by a common mu1 opioid receptor mechanism. Science 276 (5321):2048–2050

Tanda G, Munzar P, Goldberg SR (2000) Self-administration behavior is maintained by the psychoactive ingredient of marijuana in squirrel monkeys. Nat Neurosci 3(11):1073–1074. doi:10.1038/80577

Thanos PK, Dimitrakakis ES, Rice O, Gifford A, Volkow ND (2005) Ethanol self-administration and ethanol conditioned place preference are reduced in mice lacking cannabinoid CB1 receptors. Behav Brain Res 164(2):206–213. doi:10.1016/j.bbr.2005.06.021

Tidey JW, Miczek KA (1997) Acquisition of cocaine self-administration after social stress: role of accumbens dopamine. Psychopharmacology 130(3):203–212

Tomasi D, Volkow ND (2013) Striatocortical pathway dysfunction in addiction and obesity: differences and similarities. Crit Rev Biochem Mol Biol 48(1):1–19. doi:10.3109/10409238.2012.735642

Trezza V, Vanderschuren LJ (2008a) Bidirectional cannabinoid modulation of social behavior in adolescent rats. Psychopharmacology (Berl) 197(2):217–227. doi:10.1007/s00213-007-1025-3

Trezza V, Vanderschuren LJ (2008b) Cannabinoid and opioid modulation of social play behavior in adolescent rats: differential behavioral mechanisms. Eur Neuropsychopharmacol 18 (7):519–530. doi:10.1016/j.euroneuro.2008.03.001

Trezza V, Vanderschuren LJ (2009) Divergent effects of anandamide transporter inhibitors with different target selectivity on social play behavior in adolescent rats. J Pharmacol Exp Ther 328 (1):343–350. doi:10.1124/jpet.108.141069

Trezza V, Baarendse PJ, Vanderschuren LJ (2010) The pleasures of play: pharmacological insights into social reward mechanisms. Trends Pharmacol Sci 31(10):463–469. doi:10.1016/j.tips.2010.06.008

Umberg EN, Pothos EN (2011) Neurobiology of aversive states. Physiol Behav 104(1):69–75. doi:10.1016/j.physbeh.2011.04.045

Uriguen L, Perez-Rial S, Ledent C, Palomo T, Manzanares J (2004) Impaired action of anxiolytic drugs in mice deficient in cannabinoid CB1 receptors. Neuropharmacology 46(7):966–973. doi:10.1016/j.neuropharm.2004.01.003

Valjent E, Mitchell JM, Besson MJ, Caboche J, Maldonado R (2002) Behavioural and biochemical evidence for interactions between delta 9-tetrahydrocannabinol and nicotine. Br J Pharmacol 135(2):564–578. doi:10.1038/sj.bjp.0704479 van der Stelt M, Mazzola C, Esposito G, Matias I, Petrosino S, De Filippis D, Micale V, Steardo L, Drago F, Iuvone T, Di Marzo V (2006) Endocannabinoids and beta-amyloid-induced neurotoxicity in vivo: effect of pharmacological elevation of endocannabinoid levels. Cell Mol Life Sci 63(12):1410–1424. doi:10.1007/s00018-006-6037-3

Vann RE, Warner JA, Bushell K, Huffman JW, Martin BR, Wiley JL (2009) Discriminative stimulus properties of delta9-tetrahydrocannabinol (THC) in C57Bl/6J mice. Eur J Pharmacol 615(1-3):102–107. doi:10.1016/j.ejphar.2009.05.010

Varvel SA, Lichtman AH (2002) Evaluation of CB1 receptor knockout mice in the Morris water maze. J Pharmacol Exp Ther 301(3):915–924

Varvel SA, Anum EA, Lichtman AH (2005) Disruption of CB(1) receptor signaling impairs extinction of spatial memory in mice. Psychopharmacology 179(4):863–872. doi:10.1007/s00213-004-2121-2

Varvel SA, Wise LE, Niyuhire F, Cravatt BF, Lichtman AH (2007) Inhibition of fatty-acid amide hydrolase accelerates acquisition and extinction rates in a spatial memory task. Neuropsychopharmacology 32(5):1032–1041. doi:10.1038/sj.npp.1301224

Vaseghi G, Rabbani M, Hajhashemi V (2012) The CB(1) receptor antagonist, AM281, improves recognition loss induced by naloxone in morphine withdrawal mice. Basic Clin Pharmacol Toxicol 111(3):161–165. doi:10.1111/j.1742-7843.2012.00881.x

Vaseghi G, Rabbani M, Hajhashemi V (2013) The effect of AM281, a cannabinoid antagonist, on memory performance during spontaneous morphine withdrawal in mice. Res Pharm Sci 8 (1):59–64

Vaughn LK, Mantsch JR, Vranjkovic O, Stroh G, Lacourt M, Kreutter M, Hillard CJ (2012) Cannabinoid receptor involvement in stress-induced cocaine reinstatement: potential interaction with noradrenergic pathways. Neuroscience 204:117–124. doi:10.1016/j.neuroscience.2011.08.021

Vela G, Ruiz-Gayo M, Fuentes JA (1995) Anandamide decreases naloxone-precipitated withdrawal signs in mice chronically treated with morphine. Neuropharmacology 34(6):665–668

Vengeliene V, Siegmund S, Singer MV, Sinclair JD, Li TK, Spanagel R (2003) A comparative study on alcohol-preferring rat lines: effects of deprivation and stress phases on voluntary alcohol intake. Alcohol Clin Exp Res 27(7):1048–1054. doi:10.1097/01.ALC.0000075829.81211.0C

Verty AN, McGregor IS, Mallet PE (2005) Paraventricular hypothalamic CB(1) cannabinoid receptors are involved in the feeding stimulatory effects of Delta(9)-tetrahydrocannabinol. Neuropharmacology 49(8):1101–1109. doi:10.1016/j.neuropharm.2005.03.025

Vigano D, Grazia Cascio M, Rubino T, Fezza F, Vaccani A, Di Marzo V, Parolaro D (2003) Chronic morphine modulates the contents of the endocannabinoid, 2-arachidonoyl glycerol, in rat brain. Neuropsychopharmacology 28(6):1160–1167. doi:10.1038/sj.npp.1300117

Vigano D, Valenti M, Cascio MG, Di Marzo V, Parolaro D, Rubino T (2004) Changes in endocannabinoid levels in a rat model of behavioural sensitization to morphine. Eur J Neurosci 20(7):1849–1857. doi:10.1111/j.1460-9568.2004.03645.x

Vinklerova J, Novakova J, Sulcova A (2002) Inhibition of methamphetamine self-administration in rats by cannabinoid receptor antagonist AM251. J Psychopharmacol 16(2):139–143

Vinod KY, Yalamanchili R, Xie S, Cooper TB, Hungund BL (2006) Effect of chronic ethanol exposure and its withdrawal on the endocannabinoid system. Neurochem Int 49(6):619–625. doi:10.1016/j.neuint.2006.05.002

Vinod KY, Sanguino E, Yalamanchili R, Manzanares J, Hungund BL (2008) Manipulation of fatty acid amide hydrolase functional activity alters sensitivity and dependence to ethanol. J Neurochem 104(1):233–243. doi:10.1111/j.1471-4159.2007.04956.x

Vinod KY, Kassir SA, Hungund BL, Cooper TB, Mann JJ, Arango V (2010) Selective alterations of the CB1 receptors and the fatty acid amide hydrolase in the ventral striatum of alcoholics and suicides. J Psychiatr Res 44(9):591–597. doi:10.1016/j.jpsychires.2009.11.013

Vinod KY, Maccioni P, Garcia-Gutierrez MS, Femenia T, Xie S, Carai MA, Manzanares J, Cooper TB, Hungund BL, Colombo G (2012) Innate difference in the endocannabinoid signaling and its modulation by alcohol consumption in alcohol-preferring sP rats. Addict Biol 17(1):62–75. doi:10.1111/j.1369-1600.2010.00299.x

Viveros MP, Marco EM, Llorente R, Lopez-Gallardo M (2007) Endocannabinoid system and synaptic plasticity: implications for emotional responses. Neural Plast 2007:52908. doi:10.1155/2007/52908

Vlachou S, Nomikos GG, Panagis G (2003) WIN 55,212-2 decreases the reinforcing actions of cocaine through CB1 cannabinoid receptor stimulation. Behav Brain Res 141(2):215–222

Vlachou S, Nomikos GG, Panagis G (2006) Effects of endocannabinoid neurotransmission modulators on brain stimulation reward. Psychopharmacology (Berl) 188(3):293–305. doi:10.1007/s00213-006-0506-0

Vlachou S, Stamatopoulou F, Nomikos GG, Panagis G (2008) Enhancement of endocannabinoid neurotransmission through CB1 cannabinoid receptors counteracts the reinforcing and psychostimulant effects of cocaine. Int J Neuropsychopharmacol 11(7):905–923. doi:10.1017/S1461145708008717

Volkow ND, Fowler JS, Wang GJ, Swanson JM, Telang F (2007) Dopamine in drug abuse and addiction: results of imaging studies and treatment implications. Arch Neurol 64 (11):1575–1579. doi:10.1001/archneur.64.11.1575

Volkow ND, Wang GJ, Baler RD (2011) Reward, dopamine and the control of food intake: implications for obesity. Trends Cogn Sci 15(1):37–46. doi:10.1016/j.tics.2010.11.001

Wade MR, Degroot A, Nomikos GG (2006) Cannabinoid CB1 receptor antagonism modulates plasma corticosterone in rodents. Eur J Pharmacol 551(1–3):162–167. doi:10.1016/j.ejphar.2006.08.083

Wallace MJ, Martin BR, DeLorenzo RJ (2002) Evidence for a physiological role of endocannabinoids in the modulation of seizure threshold and severity. Eur J Pharmacol 452 (3):295–301

Wamsteeker JI, Kuzmiski JB, Bains JS (2010) Repeated stress impairs endocannabinoid signaling in the paraventricular nucleus of the hypothalamus. J Neurosci 30(33):11188–11196. doi:10.1523/JNEUROSCI.1046-10.2010

Wang L, Liu J, Harvey-White J, Zimmer A, Kunos G (2003a) Endocannabinoid signaling via cannabinoid receptor 1 is involved in ethanol preference and its age-dependent decline in mice. Proc Natl Acad Sci USA 100(3):1393–1398. doi:10.1073/pnas.0336351100

Wang X, Dow-Edwards D, Keller E, Hurd YL (2003b) Preferential limbic expression of the cannabinoid receptor mRNA in the human fetal brain. Neuroscience 118(3):681–694

Wang W, Sun D, Pan B, Roberts CJ, Sun X, Hillard CJ, Liu QS (2010) Deficiency in endocannabinoid signaling in the nucleus accumbens induced by chronic unpredictable stress. Neuropsychopharmacology 35(11):2249–2261. doi:10.1038/npp.2010.99

Ward SJ, Walker EA, Dykstra LA (2007) Effect of cannabinoid CB1 receptor antagonist SR141716A and CB1 receptor knockout on cue-induced reinstatement of Ensure and corn-oil seeking in mice. Neuropsychopharmacology 32(12):2592–2600. doi:10.1038/sj.npp.1301384

Ward SJ, Rosenberg M, Dykstra LA, Walker EA (2009) The CB1 antagonist rimonabant (SR141716) blocks cue-induced reinstatement of cocaine seeking and other context and extinction phenomena predictive of relapse. Drug Alcohol Depend 105(3):248–255. doi:10.1016/j.drugalcdep.2009.07.002

Wei D, Lee D, Li D, Daglian J, Jung KM, Piomelli D (2016) A role for the endocannabinoid 2-arachidonoyl-sn-glycerol for social and high-fat food reward in male mice. Psychopharmacology (Berl) 233(10):1911–1919. doi:10.1007/s00213-016-4222-0

Werling LL, Reed SC, Wade D, Izenwasser S (2009) Chronic nicotine alters cannabinoid-mediated locomotor activity and receptor density in periadolescent but not adult male rats. Int J Dev Neurosci 27(3):263–269. doi:10.1016/j.ijdevneu.2008.12.008

West R, Gossop M (1994) Overview: A comparison of withdrawal symptoms from different drug classes. Addiction 89(11):1483–1489

Wikler A (1948) Recent progress in research on the neurophysiologic basis of morphine addiction. Am J Psychiatry 105(5):329–338. doi:10.1176/ajp.105.5.329

Wiley JL, Golden KM, Ryan WJ, Balster RL, Razdan RK, Martin BR (1997) Evaluation of cannabimimetic discriminative stimulus effects of anandamide and methylated fluoroanandamide in rhesus monkeys. Pharmacol Biochem Behav 58(4):1139–1143

Wiley JL, Walentiny DM, Wright MJ Jr, Beardsley PM, Burston JJ, Poklis JL, Lichtman AH, Vann RE (2014) Endocannabinoid contribution to Delta(9)-tetrahydrocannabinol discrimination in rodents. Eur J Pharmacol 737:97–105. doi:10.1016/j.ejphar.2014.05.013

Williams CM, Kirkham TC (1999) Anandamide induces overeating: mediation by central cannabinoid (CB1) receptors. Psychopharmacology (Berl) 143(3):315–317

Wise LE, Harloe JP, Lichtman AH (2009) Fatty acid amide hydrolase (FAAH) knockout mice exhibit enhanced acquisition of an aversive, but not of an appetitive, Barnes maze task. Neurobiol Learn Mem 92(4):597–601. doi:10.1016/j.nlm.2009.06.001

Wolf SA, Bick-Sander A, Fabel K, Leal-Galicia P, Tauber S, Ramirez-Rodriguez G, Muller A, Melnik A, Waltinger TP, Ullrich O, Kempermann G (2010) Cannabinoid receptor CB1 mediates baseline and activity-induced survival of new neurons in adult hippocampal neurogenesis. Cell Commun Signal 8:12. doi:10.1186/1478-811X-8-12

Xi ZX, Gilbert JG, Peng XQ, Pak AC, Li X, Gardner EL (2006) Cannabinoid CB1 receptor antagonist AM251 inhibits cocaine-primed relapse in rats: role of glutamate in the nucleus accumbens. J Neurosci 26(33):8531–8536. doi:10.1523/JNEUROSCI.0726-06.2006

Xi ZX, Spiller K, Pak AC, Gilbert J, Dillon C, Li X, Peng XQ, Gardner EL (2008) Cannabinoid CB1 receptor antagonists attenuate cocaine's rewarding effects: experiments with self-administration and brain-stimulation reward in rats. Neuropsychopharmacology 33 (7):1735–1745. doi:10.1038/sj.npp.1301552

Xi ZX, Peng XQ, Li X, Song R, Zhang HY, Liu QR, Yang HJ, Bi GH, Li J, Gardner EL (2011) Brain cannabinoid CB(2) receptors modulate cocaine's actions in mice. Nat Neurosci 14 (9):1160–1166. doi:10.1038/nn.2874

Yamaguchi T, Hagiwara Y, Tanaka H, Sugiura T, Waku K, Shoyama Y, Watanabe S, Yamamoto T (2001) Endogenous cannabinoid, 2-arachidonoylglycerol, attenuates naloxone-precipitated withdrawal signs in morphine-dependent mice. Brain Res 909(1-2):121–126

Yan HC, Cao X, Das M, Zhu XH, Gao TM (2010) Behavioral animal models of depression. Neurosci Bull 26(4):327–337. doi:10.1007/s12264-010-0323-7

Yin HH, Ostlund SB, Balleine BW (2008) Reward-guided learning beyond dopamine in the nucleus accumbens: the integrative functions of cortico-basal ganglia networks. Eur J Neurosci 28(8):1437–1448. doi:10.1111/j.1460-9568.2008.06422.x

Yoshida R, Ohkuri T, Jyotaki M, Yasuo T, Horio N, Yasumatsu K, Sanematsu K, Shigemura N, Yamamoto T, Margolskee RF, Ninomiya Y (2010) Endocannabinoids selectively enhance sweet taste. Proc Natl Acad Sci USA 107(2):935–939. doi:10.1073/pnas.0912048107

Zhang HY, Bi GH, Li X, Li J, Qu H, Zhang SJ, Li CY, Onaivi ES, Gardner EL, Xi ZX, Liu QR (2014a) Species differences in cannabinoid receptor 2 and receptor responses to cocaine self-administration in mice and rats. Neuropsychopharmacology 40(4):1037–1051. doi:10.1038/npp.2014.297

Zhang HY, Gao M, Liu QR, Bi GH, Li X, Yang HJ, Gardner EL, Wu J, Xi ZX (2014b) Cannabinoid CB2 receptors modulate midbrain dopamine neuronal activity and dopamine-related behavior in mice. Proc Natl Acad Sci USA 111(46):E5007–E5015. doi:10.1073/pnas.1413210111

Zlebnik NE, Cheer JF (2016) Drug-induced alterations of endocannabinoid-mediated plasticity in brain reward regions. J Neurosci 36(40):10230–10238. doi:10.1523/JNEUROSCI.1712-16.2016

Roles of *N*-Acylethanolamines in Brain Functions and Neuropsychiatric Diseases

Marco Pistis and Anna Lisa Muntoni

Abstract *N*-acylethanolamines (NAEs) are bioactive lipids, structural analogues to the endocannabinoid arachidonoylethanolamide (anandamide), whose functions and properties are being elucidated in recent years. By activating their receptors, specifically peroxisome proliferator-activated receptors (PPARs), these molecules exert a variety of physiological effects via genomic and rapid non-genomic mechanisms. Regulation of lipid metabolism, energy homeostasis, and anti-inflammation are among the best-characterized effects of PPAR activation. NAEs are abundant in the CNS and their receptors are widely expressed both in neurons and in glial cells, where they modulate brain functions and are involved in the pathophysiology of neurological and psychiatric disorders. In the brain, they participate in the regulation of feeding behavior, cognitive functions, mood, reward, and sleep-wake cycles, and evidence suggests that they might be therapeutically exploited as neuroprotective agents, "anti-addictive" medications, anticonvulsant, and antidepressant.

In this chapter, we will review the state of the art on these neuromodulators and their receptors in the brain and will discuss new hypotheses on their physiological and pathophysiological roles.

1 Introduction

Fatty acid ethanolamides, generally referred as *N*-acylethanolamines (NAEs), are a group of endogenous lipid molecules with long-chain fatty acids (Schmid et al. 1990; Hansen et al. 2000). It was long known that these molecules are ubiquitously found in animal tissues (Bachur et al. 1965), in both the periphery and the brain, and

M. Pistis (✉)
Department of Biomedical Sciences, Division of Neuroscience and Clinical Pharmacology, University of Cagliari, 09042 Cagliari, Italy

Neuroscience Institute, National Research Council of Italy, Section of Cagliari, Cagliari, Italy
e-mail: mpistis@unica.it

A.L. Muntoni
Neuroscience Institute, National Research Council of Italy, Section of Cagliari, Cagliari, Italy

M. Melis (ed.), *Endocannabinoids and Lipid Mediators in Brain Functions*,
DOI 10.1007/978-3-319-57371-7_11

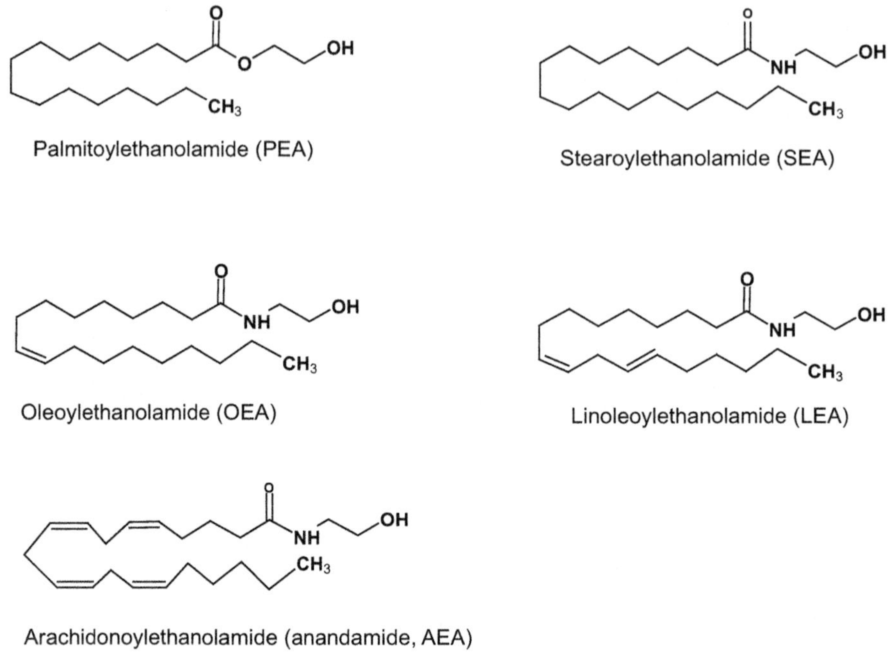

Fig. 1 Chemical structures of representative NAEs

have attracted attention in recent years as they possess a variety of biological activities.

Among NAEs, the most studied are arachidonoylethanolamide (anandamide, AEA), palmitoylethanolamide (PEA), and oleoylethanolamide (OEA); other congeners are stearoylethanolamide (SEA) and linoleoylethanolamide (LEA) (Fig. 1) (Hansen 2010; Rahman et al. 2014). Anandamide is the first endocannabinoid to be discovered (Devane et al. 1992) and the only NAE to bind to cannabinoid type 1 (CB1) and type 2 (CB2) receptors. Quantitatively, anandamide is a minor component in most animal tissues when compared with other NAEs such as PEA, SEA, OEA, and LEA (Hansen and Diep 2009). NAEs, although belonging to the same extended endocannabinoid-like family as the cannabinoid agonist AEA, exert a variety of biological effects through several other receptors, in particular peroxisome proliferator-activated receptor-α (PPARα) (Hansen 2010; Lo Verme et al. 2005; Petrosino et al. 2010), but also G-protein-coupled receptors GPR55 (Baker et al. 2006) and GPR119 and transient receptor potential vanilloid type 1 (TRPV1) (Piomelli 2013).

Although these molecules were known from many years, the role of PEA and OEA, as well as of other NAEs, in the CNS has been elucidated only recently, when the discovery of AEA (Devane et al. 1992) fueled a renewed interest in these lipid messengers.

Specifically, functions and properties of PEA, a saturated fatty acid (palmitic acid) derivative, were discovered first in 1957, when this lipid (from soybeans, peanuts, and egg yolk) was found to exert anti-inflammatory activity in guinea pigs (Kuehl et al. 1957). High levels of PEA were measured in mammalian tissues, especially in the brain (Bachur et al. 1965). The finding that PEA is the most abundant NAE in the rat brain has been more recently substantiated by the use of lipidomics techniques (Kilaru et al. 2010).

The monounsaturated OEA is a potent anorectic lipid mediator (Fu et al. 2003, 2005; Rodriguez de Fonseca et al. 2001; Piomelli 2013) and shares this property with PEA, LEA (Hansen 2014; Hansen and Diep 2009), and SEA (Terrazzino et al. 2004). As anorexiant, however, PEA is significantly less potent than OEA in reducing food intake, and LEA is similar in potency to OEA (Rodriguez de Fonseca et al. 2001; Diep et al. 2011).

Besides their anti-inflammatory activity and their physiological functions as modulators of feeding behavior, NAEs, and specifically PEA and OEA, have been recently involved in pathophysiology of neurological and psychiatric disorders, ranging from addiction, neurodegenerative diseases, epilepsy, and mood disorders (Pistis and Melis 2010; Scherma et al. 2016; Melis and Pistis 2014). In this review, we will focus on physiology of non-cannabinoid NAEs in the CNS and on their relevance in neuropsychiatric disorders.

2 Synthesis and Catabolism

NAEs including AEA and the non-cannabinoid PEA and OEA are not stored in vesicles but produced "on demand". Their endogenous levels are strictly regulated by enzymes responsible for their formation and degradation (Ueda et al. 2010a; Rahman et al. 2014). In fact, they are synthesized from glycerophospholipids via their corresponding *N*-acylphosphatidylethanolamines (NAPEs) (Okamoto et al. 2004; Rahman et al. 2014; Ueda et al. 2010b; Hansen 2010; Hansen et al. 2000). The classical pathway for NAE synthesis is a two-step process (Fig. 2). The first step is the generation of NAPE by a Ca^{2+}-dependent *N*-acyltransferase (NAT), which transfers the Sn-1 fatty acid to phosphatidylethanolamine from a donor phospholipid (Hansen et al. 2000; Hansen and Diep 2009). *N*-Acyltransferase is not specific for any fatty acids, since it catalyzes the transfer of any acyl group from the Sn-1 position of donor phospholipids. NAPE is then hydrolyzed by NAPE-hydrolyzing PLD (NAPE-PLD) with the generation of NAEs (Rahman et al. 2014). Human, mouse, and rat NAPE-PLD cloning allowed the functional characterization of NAE biosynthetic pathways (Okamoto et al. 2004). NAPE-PLD−/− mice (Leung et al. 2006) show a highly reduced Ca^{2+}-dependent conversion of NAPE to long-chain saturated and unsaturated NAEs in brain tissue. These mice display decreased levels of OEA and PEA, but similar levels of AEA compared with wild-type littermates (Leung et al. 2006). This indicates that AEA is formed by other

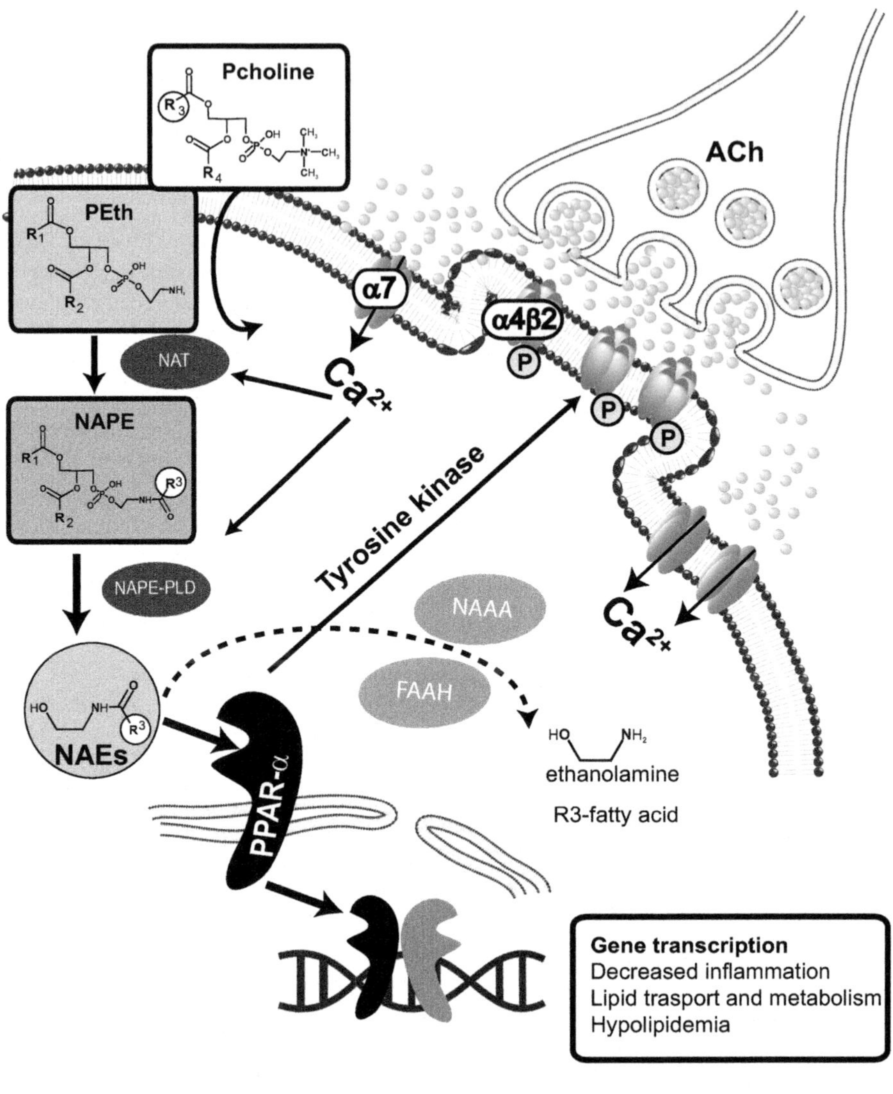

Fig. 2 Schematic diagram illustrating the canonical biosynthetic and catabolic pathways for NAE formation and their cellular mechanisms of actions through their receptor of PPARα. Phosphatidylcholine (Pcholine) donates a fatty acid moiety from the sn-1 position (R3) to a phosphatidylethanolamine (PEth). This reaction is catalyzed by N-acyltransferase (NAT). The resulting N-acylphosphatidylethanolamine (NAPE) is hydrolyzed by NAPE-PLD to the corresponding N-acylethanolamine (NAE). Activation of PPARα by NAEs results in genomic effects (gene transcription) and in non-genomic actions, such as activation of a tyrosine kinase and phosphorylation of β2*nAChRs (i.e., α4β2). Ca^{2+} entry mediated by α7-nAChRs activates NAS synthesis through the Ca^{2+}-dependent NAT and NAPE-PLD. Fatty acid amide hydrolase (FAAH) and NAE-hydrolyzing acid amidase (NAAA) are the major inactivating enzymes for OEA, PEA, and AEA which convert them into ethanolamine and corresponding fatty acids (oleic, palmitic, and arachidonic acids, respectively)

parallel pathways (Hansen and Diep 2009) and strongly suggests the existence of other NAE-forming enzymes within NAPE-PLD-independent pathways (Tsuboi et al. 2011; Simon and Cravatt 2010; Leung et al. 2006; Rahman et al. 2014). No gross phenotype in NAPE-PLD-deficient mice has been reported, whereas in humans a polymorphism of NAPE-PLD was associated with obesity in a Norwegian population (Wangensteen et al. 2010).

NAPE-PLD is expressed in many tissues, including the brain (Egertova et al. 2008; Morishita et al. 2005; Cristino et al. 2008; Suarez et al. 2008). The cellular localization of NAPE-PLD in the brain is informative on the functional significance of NAEs in the CNS. NAPE-PLD mRNA and immunoreactivity are located both presynaptically and postsynaptically and are intense in axons of the vomeronasal nerve projecting to the accessory olfactory bulb (Egertova et al. 2008) and in the hippocampus (Cristino et al. 2008; Egertova et al. 2008; Nyilas et al. 2008). Other brain areas with NAPE-PLD immunoreactivity are the cortex, thalamus, hypothalamus (Reguero et al. 2014; Egertova et al. 2008), and cerebellum (Suarez et al. 2008). The postsynaptic localization of NAPE-PLD indicates that NAEs may act as an autocrine or paracrine signal at receptors expressed in the same or neighboring cells. On the other hand, the axonal localization suggests that synthesized NAEs may be released to target postsynaptic neurons and regulate synaptic signaling.

The generated NAEs are then degraded to their corresponding free fatty acids and ethanolamine (Cravatt et al. 1996; Deutsch et al. 2002) (Fig. 2). This hydrolysis is catalyzed mainly by fatty acid amide hydrolase (FAAH) (Cravatt et al. 1996) and NAE-hydrolyzing acid amidase (NAAA) (Tsuboi et al. 2005). FAAH hydrolyzes all NAEs with high efficiency, and it is expressed in many different tissues and cell types, including in the brain.

NAAA has no sequence homology with FAAH and is expressed in lysosomes. NAAA displays same specificity toward unsaturated NAEs such as PEA to much greater extent than AEA (Tsuboi et al. 2007a). In the rat, NAAA is highly expressed in the lung, spleen, thymus, and intestine (Tsuboi et al. 2007a), very high in alveolar macrophages (Tsuboi et al. 2007b), but low in the brain.

Alternatively, AEA and other polyunsaturated NAEs are oxygenated in their polyunsaturated fatty acyl moieties and converted to prostamides by cyclooxygenase-2 (COX-2) (Yu et al. 1997; Kozak et al. 2002) or to other oxygenated metabolites by lipoxygenases (Hampson et al. 1995; Ueda et al. 1995) or by cytochrome P-450 (Bornheim et al. 1993). In addition to these enzymes, more recent studies revealed the involvement of new players in NAE metabolism (see Rahman et al. 2014 for a comprehensive review).

3 Receptors for NAEs

3.1 Peroxisome Proliferator-Activated Receptor-α (PPARα)

PPARs belong to the large superfamily of transcription factors, accounting 48 members in the human genome (Germain et al. 2006).The PPAR subfamily, specifically, is composed of three isotypes: PPARα, PPARγ, and PPARβ/δ.

The denomination of these receptors dates back to the 1960s, when the lipid-lowering drug clofibrate was found to induce proliferation of peroxisomes in the liver and hepatomegaly in rodents (Thorp and Waring 1962; Hess et al. 1965). The receptor activated by clofibrate, other fibrates, and synthetic ligands was cloned and denominated peroxisome proliferator-activated receptor-α (Issemann and Green 1990). Although the other members of this subfamily (β/δ and γ) are structural homologs of PPARα, they do not induce peroxisome proliferation (Graves et al. 1992; Dreyer et al. 1992; Schmidt et al. 1992). Nevertheless, the name has remained.

PPARs, like other nuclear receptors, are composed of four domains (Fidaleo et al. 2014), among which the DNA-binding domain and the ligand-binding domain, which are the most conserved portions of the protein across the different isotypes (Desvergne and Wahli 1999; Escriva et al. 1998; Laudet et al. 1992). The ligand-binding domain, in particular, is strikingly large in comparison with other nuclear receptors (Desvergne and Wahli 1999). This explains why PPARs are activated by a large number of diverse endogenous ligands as well as by synthetic agonists: saturated and unsaturated fatty acids (palmitic acid, oleic acid, linoleic acid, and arachidonic acid), NAEs, and eicosanoids (Desvergne and Wahli 1999).

Considerable evidence suggests that NAEs display most of their CNS activity through PPARα. In HeLa cells stably expressing a luciferase reporter gene together with the ligand-binding domain of human PPARα, OEA activates PPARα with an EC_{50} of 120 nM, whereas PEA displays lower potency (EC_{50} of 3.1 μM) (Lo Verme et al. 2005; Fu et al. 2003). Under identical conditions, SEA was reported to be ineffective (Lo Verme et al. 2005).

PPARα has been cloned and characterized in several species, including humans (Sher et al. 1993). PPARα expression is enriched in tissues with high fatty acid oxidation rates such as the liver, heart, skeletal muscle, brown adipose tissue, and kidney. It is also expressed in other tissues and cells including the intestine, vascular endothelium, smooth muscle, and immune cells such as monocytes, macrophages, and lymphocytes (Lefebvre et al. 2006; Chinetti et al. 1998) and in the CNS (Mandard et al. 2004; Galan-Rodriguez et al. 2009; Moreno et al. 2004; Braissant et al. 1996; Auboeuf et al. 1997). The highest PPARα expression was found in the basal ganglia; some thalamic, mesencephalic, and cranial motor nuclei; the reticular formation; and the large motoneurons of the spinal cord (Fidaleo et al. 2014; Moreno et al. 2004). Besides neuronal localization, a specific distribution of PPARα in ependymal and astroglial cells, but not in oligodendrocytes, was reported (Moreno et al. 2004).

Several structurally different classes of compounds bind to PPARα, including the hypolipidemic fibrates. Saturated or monounsaturated NAEs are high-affinity ligands for PPARα, particularly OEA which may be considered an endogenous ligand at concentrations normally achieved under physiological conditions (Fu et al. 2003).

3.2 GPR119 and GPR55

NAEs bind also to G-protein-coupled receptors 119 (GPR119) and 55 (GPR55). GPR119 is involved in insulin secretion, and, for this reason, it is emerging as a potential therapeutic target for type 2 diabetes, with beneficial effects on glucose homeostasis (Moran et al. 2014). There is no evidence of GPR119 expression in the CNS, therefore a detailed discussion of its functions is not within the aims of this chapter.

GPR55 is expressed in the CNS (Sawzdargo et al. 1999; Ryberg et al. 2007), as well as in peripheral tissues (intestine, bone marrow, immune and endothelial cells, spleen, and platelets) (Pietr et al. 2009; Balenga et al. 2011; Rowley et al. 2011; Cherif et al. 2015). GPR55 is a 319-amino acid protein that was cloned in 1999 and mapped to chromosome 2q37 in humans (Sawzdargo et al. 1999). GPR55 shows low amino acid homology with CB1 (13.5%) or CB2 (14.4%) (Ryberg et al. 2007; Kapur et al. 2009; Baker et al. 2006). Similarly to PPARα, GPR55 is a receptor for small lipid mediators, and it is also activated by some synthetic cannabinoids and related molecules. The lipid lysophosphatidylinositol (LPI), which activates GPR55 but has no affinity for CB1 or CB2 receptors, was the first endogenous ligand identified for this receptor (Oka et al. 2007). Among lipids and NAEs, PEA binds with high affinity at GPR55 ($EC_{50} = 1$–20 nM), whereas OEA is less potent ($EC_{50} = 440$ nM) (Baker et al. 2006; Ryberg et al. 2007). Recent evidence suggests that GPR55 is involved in the regulation of axonal growth during development (Cherif et al. 2015). GPR55 participates also in central processing of neuropathic and inflammatory pain (Deliu et al. 2015) as it has been reported to play a pronociceptive role in the periaqueductal gray. Consistently, GPR55$^{-/-}$ (knockout) mice have been shown to be protected in models of inflammatory and neuropathic pain. This suggests that GPR55 antagonists may have therapeutic potential as analgesics for both these pain types (Staton et al. 2008). It is not clear how the anti-inflammatory and anti-neuropathic effects of PEA can be reconciled with the functional role of GPR55 in pain and inflammation. As little is known on the physiological roles of GPR55, particularly in the CNS, further studies are needed to shed some lights into this receptor and its endogenous agonists.

4 Physiological Role of NAEs and PPAR

PPARα is a transcriptional regulator of genes involved in peroxisomal and mitochondrial β-oxidation, fatty acid transport, and, in rodents, hepatic glucose production (Xu et al. 2002). PPARα negatively regulates pro-inflammatory and acute phase response signaling pathways, as seen in rodent models of systemic inflammation (Gervois et al. 2004; Bensinger and Tontonoz 2008; Glass and Ogawa 2006). Consistently, PPARα$^{-/-}$ mice display longer inflammatory responses (Devchand et al. 1996), increased susceptibility to experimental colitis, and experimental autoimmune encephalitis (an experimental model of multiple sclerosis) (Straus and Glass 2007). PPARα agonists reduce peripheral inflammation in a PPARα-dependent manner (Sheu et al. 2002; Lo Verme et al. 2005).

The canonical mechanism of action downstream to PPARα activation is regulation of gene transcription (Ferre 2004; Berger and Moller 2002; Moreno et al. 2004). Specifically, ligand binding promotes dissociation of corepressor proteins, association of coactivators, and coactivator proteins and heterodimerization with the retinoid X receptor (RXR). This dimer is then translocated into the nucleus and binds to specific regions on DNA termed peroxisome proliferator response elements (PPRE). The result is either an increase or a decrease of gene transcription, depending on the target gene. Besides this typical transduction mechanism, non-transcriptional actions by PPARα have also been observed (Melis et al. 2008; Ropero et al. 2009; Gardner et al. 2005). These effects are rapid in onset (2–5 min). This short onset time rules out the possibility of a genomic mechanism of action and involves signaling pathways similar to those described for many other nuclear receptor ligands (Losel and Wehling 2003; Losel et al. 2003; Moraes et al. 2007; Ropero et al. 2009). Consistent with this scenario, PPARα activation induced production of cytosolic effectors, such as reactive oxygen species (Ropero et al. 2009; Melis et al. 2008).

These non-transcriptional effects take place also in the CNS and have the potential to regulate neuronal functions on a short timescale. In fact, our recent studies show that in ventral tegmental area (VTA), dopamine neurons OEA and PEA bind to PPARα within the cytosol and trigger a rapid non-genomic mechanism leading to increased endogenous hydrogen peroxide and consequent activation of tyrosine kinase(s) (Melis et al. 2008, 2010). In turn, these tyrosine kinases phosphorylate the β2 subunits of the nicotinic acetylcholine receptors (nAChRs) (Melis et al. 2013b). Phosphorylation of nAChRs is an efficient mechanism to control receptor functions and results in a faster desensitization rate or a downregulation (Huganir and Greengard 1990). As cholinergic afferents control firing rate and burst firing of midbrain dopamine cells, the functional regulation of β2-containing nAChRs (β2*nAChR, the asterisk indicates the presence of other subunits) alters dopaminergic activity. Thus, NAEs as PPARα ligands act as intrinsic modulators of cholinergic transmission and modify dopamine cell excitability, contributing to acetylcholine effects on dopamine system. Moreover, our studies revealed that PPARα-induced regulation of β2*nAChR abolished electrophysiological, neurochemical, and behavioral effects of nicotine (Scherma et al. 2008; Melis and Pistis 2014; Melis et al. 2013a, b; Panlilio et al. 2012). Therefore, enhancing brain levels of PPARα endogenous agonists or activating this receptor with synthetic ligands represents a novel strategy for anti-smoking medications (Melis and Pistis 2014) (see next section).

Noteworthy, this cellular mechanism of action explains why the effects of PPARα activation are specific to nicotine, as it does not attenuate cocaine or cannabinoid-induced electrophysiological or behavioral effects (Justinova et al. 2008, 2015; Luchicchi et al. 2010).

As mentioned above, PPARα activation can be achieved by preventing degradation of endogenous ligands or by promoting their synthesis (Melis et al. 2013a, b). NAE's synthesis is increased by elevating intracellular Ca^{2+}, as NAPE-PLD is a Ca^{2+}-dependent enzyme (Ueda et al. 2001). In fact, activation of α7-nAChRs, a low-affinity Ca^{2+}-permeable nAChR subunit combination, enhanced midbrain OEA and PEA levels and prevented nicotine-induced excitation of dopamine neurons both in vitro and in vivo (Melis et al. 2013b). Thus, these observations indicate that α7-nAChRs function as sensors in dopamine neurons for excessive cholinergic transmission, which is attenuated by subsequent PPARα activation by endogenous agonists and phosphorylation of β2*nAChRs.

5 Role of NAEs and PPARα in Neuropsychiatric Diseases

5.1 Addiction

In the past two decades, research advances have progressively sustained the idea of addiction as a brain disease (Volkow et al. 2016). In particular, addiction can be defined as a chronic, relapsing disorder of brain reward, motivation, cognition, affective functioning, memory, and related circuitry (American Society of Addiction Medicine, ASAM, 2011). Dysfunction in these pathways leads to a behavioral pathology characterized by compulsive drug seeking and use with progressive loss of control over consumption despite the emergence of significant negative or disadvantageous consequences (see Volkow et al. 2016 and references therein). In the fifth edition of the *Diagnostic and Statistical Manual of Mental Disorders* (DSM-5) (2013), the term addiction is synonymous with the most severe stage of substance use disorder (see Volkow et al. 2016). Drug addiction represents the most prevalent neuropsychiatric disorder affecting modern society (United Nations Office on Drugs and Crime, 2015). The global health burden attributable to alcohol and illicit drug use amounts to 5.4% of the total burden of diseases (WHO 2003).

A better knowledge of the neurobiological basis underlying addiction is certainly crucial for developing more effective pharmacological and behavioral interventions to counteract harmful consequences of drug use disorders. In this respect, the evidence that NAEs such as OEA and PEA block nicotine-induced excitation of midbrain dopamine neurons by acting on PPARα (Melis et al. 2008) pinpointed their key role in the modulation of the brain reward system and, therefore, in the pathophysiology of nicotine addiction ((Melis et al.2010); reviewed in (Pistis and Melis 2010)). In fact, since dopamine neurons projecting from the VTA to the nucleus accumbens (NAc) act as “reward sensors,” these results offered new perspectives in both understanding and treating tobacco dependence. Accordingly, a study by Scherma et al. (2008) unveiled that boosting brain levels of NAEs reverses abuse-related behavioral and neurochemical effects of nicotine in rats, including those predictive of relapse liability (Scherma et al. 2008). But how does URB597 exert its anti-addictive and anti-rewarding actions? Both the receptor and the mechanism responsible for the effect played by enhanced NAE levels against nicotine were identified soon after (Melis et al. 2008). In particular, it was demonstrated that URB597 counteracts nicotine-induced stimulation of the dopamine system by selectively enhancing intracellular levels of OEA and/or PEA within this circuit and by activating PPARα (Luchicchi et al. 2010; Melis et al. 2008; Mascia et al. 2011). However, even though PPARα was recognized as the receptor accounting for the “anti-nicotine” effect played by NAEs (Melis et al. 2008), a contribution of CB1 receptors activated by anandamide cannot be completely ruled out (Luchicchi et al. 2010; Melis et al. 2008; Mascia et al. 2011).

This wealth of experimental data revealed a novel property of endocannabinoid-like NAEs and suggested that increasing brain levels of OEA and PEA, or activating their receptors (PPARα) with synthetic ligands, might represent a novel and effective approach for achieving smoking cessation (Melis et al. 2013a, b; Melis and Pistis 2014; Scherma et al. 2008). Hence, a new frontier for the treatment of tobacco addiction, the number one preventable cause of death in the developed world, was opened. In keeping with these observations, synthetic PPARα ligands such as fibrates, long-standing clinically available hypolipidemic drugs, were found to suppress (i) nicotine-induced stimulation of dopamine neurons and increased extracellular dopamine levels in the shell of the NAc, which are key features of its acute rewarding properties (Panlilio et al. 2012), and (ii) nicotine intake and reinstatement of nicotine-seeking behavior after a period of prolonged abstinence (Panlilio et al. 2012). Thus, all this evidence has prompted the idea that targeting PPARα represents a promising therapeutic strategy in nicotine relapse prevention in humans and, in the end, for quitting smoking.

So far, anti-addictive actions of PPARα agonists seem to be almost exclusively limited to nicotine in laboratory animals (Justinova et al. 2008, 2015; Luchicchi et al. 2010). The fact that in dopamine neurons PPARα affects the phosphorylation status of β2*nAChRs by activating tyrosine kinase(s) (Melis et al. 2010, 2013b), therefore leading to a reduced excitation of dopamine neurons by nicotinic agonists, might provide a possible explanation for this specificity. Hence, it is not surprising that URB597 does not influence either self-administration of the main psychoactive ingredient of cannabis (i.e., Δ9-tetrahydrocannabinol, Δ9-THC) or cocaine in nonhuman primates (Justinova et al. 2008). Moreover, URB597 is unable to prevent dopamine neuronal responses to morphine or cocaine in rats (Luchicchi et al. 2010). Accordingly, in humans, it has been reported that a natural genetic variation in

FAAH, +385 A/A (P129T), is associated with an increased risk of illicit, but not licit (i.e., nicotine, alcohol), drug use problems (Sipe et al. 2002; Tyndale et al. 2007). Notably, this FAAH variant displays a reduced expression and activity in humans, thereby supporting a potential link between enhanced NAE circulating levels and illicit drug abuse and dependence (Chiang et al. 2004).

Most recently, a link of PPARα (and γ) with both alcohol consumption in rodents and withdrawal and dependence in humans emerged (Haile and Kosten 2017; Blednov et al. 2015). Specifically, Blednov et al. (2015) investigated the effects of different classes of PPAR agonists on chronic alcohol intake and preference in mice with a genetic predisposition for high alcohol consumption and then examined human genome-wide association data for polymorphisms in PPAR genes in alcohol-dependent subjects. According to the authors, the observed reduction of alcohol intake in mice and the genetic association between alcohol dependence and withdrawal (according to DSM-5 criteria) in humans underscore the potential for reconsidering clinically approved PPARα or PPARγ agonists for the treatment of alcohol use disorders and alcoholism (Blednov et al. 2015). Consistently, administration of either OEA or synthetic PPARα agonists was lately shown to block cue-induced reinstatement of alcohol-seeking behavior and reduce the severity of somatic withdrawal symptoms in alcohol-dependent rats (Bilbao et al. 2015; Haile and Kosten 2017; Blednov et al. 2016), thus supporting the intriguing possibility of using PPARα as novel therapeutic tool also for alcoholism.

5.2 Epilepsy and Other Neurological Disorders

Epilepsy is a common and prevalent neurological disorder affecting 1–2% of the population worldwide (Thurman et al. 2011). As of 2015, nearly 50 million of people suffer from this disease, characterized by recurrent spontaneous seizures which negatively impact quality of life and respond to medication in about 70% of cases (see for a recent review Varvel et al. 2015). In addition, currently available antiepileptic drugs are mainly symptomatic and have numerous side effects. Thus, an urgent need exists to identify and exploit new, effective therapies. Unraveling the neurobiological mechanisms that subserve the generation of epilepsy will help the development of drugs to modify the disease outcome and, potentially, to prevent epileptogenesis. Intriguingly, the interplay between PPARα and nAChRs shows relevance also for epilepsy and can be exploited as an innovative therapeutic target. Indeed, nAChRs are involved in the pathogenetic mechanisms underlying seizures and epilepsy. In particular, converging evidence from genetic studies in epileptic patients and animal models of seizures demonstrated that nAChR activity is increased in some types of epilepsy, including nocturnal frontal lobe epilepsy (De Fusco et al. 2000; Steinlein et al. 1997; Steinlein 2004; Sutor and Zolles 2001). From this standpoint, negative modulation of nAChRs exerted by endogenous NAEs, or by exogenous ligands of their receptors (PPARα), might represent a potential disease-modifying strategy for epilepsies where nAChRs contribute to neuronal excitation and synchronization. This is particularly relevant considering that, as above mentioned, current therapies, besides from being ineffective in the 30% of patients, are mainly symptomatic and do not necessarily affect the epileptogenic process or the disease progression.

Accordingly, several preclinical studies have shown that selective agonists of PPARα raise seizure thresholds, thus supporting PPARs as potential drug targets for seizure control (Auvin 2012). PEA, for example, proved to have antiepileptic effects in kindled rats (Sheerin et al. 2004) and showed anticonvulsant activity in mice (Lambert et al. 2001). Similarly, fenofibrate, a synthetic PPARα ligand long used as hypolipidemic agent in clinical practice, has been reported to be effective as anticonvulsant in pentylenetetrazole (PTZ)-induced seizures and on latencies to the onset of status epilepticus induced by lithium-pilocarpine in adult rats (Porta et al. 2009). Likewise, acute and chronic PPARα gonists (e.g., fenofibrate) are protective against nicotine-induced seizures and abolished nicotine-induced generation of ictal activity and synchronization in the frontal cortex (Puligheddu et al. 2013). Remarkably, these latter results were paralleled by an increased ratio of phosphorylated/dephosphorylated β2 nAChR subunits in the frontal cortex following acute PPARα ligand treatment, whereas the chronic regimen induced a threefold augment in OEA levels in the same brain region (Puligheddu et al., 2013).

Notably, these observations can also be applied to other models of epilepsy, since PEA was shown to display antiepileptic actions, though via direct PPARα and indirect CB1 activation, in a rat genetic model of absence epilepsy (Citraro et al. 2013) and in DBA/2 mice (Citraro et al. 2016). In addition, a recent study by Saha et al. (2014) reported the anti-kindling effect of the PPARα agonist bezafibrate in PTZ-induced kindling seizure model. Importantly, and not surprisingly, bezafibrate also reduced the neuronal damage and apoptosis in hippocampal areas of the rat brain (Saha et al. 2014).

In this scenario, it is worth to mention that ketogenic diet (KD) proved as efficacious as fenofibrate in exerting anticonvulsive properties in animal models of epilepsy (Porta et al. 2009). The KD is a high-fat, adequate-protein, and low-carbohydrate diet long used as beneficial therapy for refractory epilepsy, especially in children (Bough and Rho 2007). It has been proposed that KD, by increasing fast inhibition in the rat hippocampus dentate gyrus, might protect these neurons from excitotoxicity and, consequently, caused anticonvulsant and antiepileptogenic effects (Bough et al. 2003). However, even though the KD mechanism of action is still unclear and under investigation (Porta et al. 2009; Bough and Rho 2007), it is tempting to hypothesize the existence of a direct, causal relationship between augmented levels of NAEs and activation of PPARα following KD. In fact, both brain and peripheral tissue levels of NAEs change with high-fat diets (see Pistis and Melis, 2010 and references therein), thus suggesting that these lipid molecules would behave like PPARα agonists (Cullingford 2008) and dampen hippocampal neuronal excitability (Bough et al. 2003). Moreover, circulating levels of different PPARα endogenous ligands (e.g., fatty acids, oxysterols, hormones) might increase following KD and exert themselves the anticonvulsive properties ascribed to this diet (Pistis and Melis 2010).

Whether or not NAEs, as well as KD, have anticonvulsants and anti-epileptogenic effects via activation of PPARα, it is undeniable the physiological role played by these receptors in neuroinflammation and protection from excitotoxicity and, therefore, a link between these phenomena (Pistis and Melis 2010; Melis and Pistis 2014). Hence, considering their multifaceted pharmacological properties (neurotrophic/neuroprotective and anti-inflammatory), PPAR pathways are under evaluation as potential therapeutic targets as well as vulnerability factors in Parkinson's disease (PD), Alzheimer's disease (AD), and multiple sclerosis (MS) (Pistis and Melis 2010; Fidaleo et al. 2014). For example, NAEs seem to be directly implicated in the pathogenesis of AD (Pazos et al. 2004), the most common cause of dementia and one of the leading sources of morbidity and mortality in the aging population (Reitz et al. 2011). Accordingly, OEA, PEA, and the NAE-enhancing FAAH inhibitor URB597 increase memory acquisition and consolidation in naïve rats in a PPARα-dependent manner (Mazzola et al. 2009; Campolongo et al. 2009), while PPARγ improve both learning and memory not only in an experimental model of AD (Heneka et al. 2005) but also in AD patients (Risner et al. 2006; Watson et al. 2005; Landreth et al. 2008; Jiang et al. 2008). Importantly, activation of PPARγ is also able to ameliorate AD-related signs by reducing both microglial activation and Aβ plaques (Heneka et al. 2005). In line with these observations, Scuderi et al. (2011, 2012) demonstrated that PEA anti-inflammatory properties are responsible for counteracting Aβ-induced astrogliosis and helped to identify the molecular apparatus by which PEA, via PPARα activation, contributes to downregulate astroglial reaction, pro-inflammatory signal overproduction, and neuronal loss. Next, an elegant in vivo study by D'Agostino et al. (2012) substantiated the fundamental role of PPARα for the neuroprotective and memory-rescuing effects of PEA in an AD animal model (see (Scuderi and Steardo 2013; Mattace Raso et al. 2014) for recent reviews).

On the other hand, MS subjects show reduced levels of NAEs (Di Filippo et al. 2008), thus suggesting the involvement of endocannabinoid system imbalances also in the pathogenesis of MS (Shohami and Mechoulam 2006), the most widespread disabling neurological condition of young adults around the world (see for a recent review Ascherio and Munger 2016). However, it should be pointed out that during MS clinical exacerbations, levels of OEA, PEA, and AEA increased, although being still lower than in controls. This evidence consolidates the hypothesis that NAEs can have a significant impact in reducing inflammation and, therefore, act as endogenous neuroprotective molecules (Di Filippo et al. 2008). In accordance with this, an increase of CNS PEA levels has been observed in a mouse chronic model of MS (Loria et al. 2008), which might be interpreted as an adaptive mechanism to preserve function and integrity during the development of neurodegenerative damage (Mattace Raso et al. 2014). Noteworthy, in the same MS chronic model, PEA administration induced, together with an anti-inflammatory effect, a reduction of motor disability (Loria et al. 2008). Concerning PD, which is among the most prevalent neurodegenerative conditions (Pringsheim et al. 2014), a protective/neurotrophic (and, therefore, potentially therapeutic) role of NAEs has also been demonstrated. In particular, it has been shown that OEA significantly decreased behavioral PD symptoms and exerted neuroprotective effects on the nigrostriatal circuit in an experimental model of the disease through a PPARα-dependent mechanism (Galan-Rodriguez et al. 2009; Gonzalez-Aparicio and Moratalla 2014; Avagliano et al. 2016). Similarly, systemic administration of PEA reduced 1-methyl-4-phenyl-1,2,3,6-tetrahydropyridine (MPTP)-induced glial cell activation and protected against MPTP-related loss of tyrosine hydroxylase-positive neurons in the substantia nigra pars compacta, two effects that were blunted in PPARα−/− mice. Moreover, chronic administration of PEA reversed MPTP-associated motor deficits, as revealed by the analysis of forepaw step width and percentage of faults (Esposito et al. 2012). The fact that PEA proved to be neuroprotective even when administered after the insult is of particular importance, as the lack of PD biomarkers and the difficulties of early diagnosis make the pharmacotherapy of PD possible only when dopamine neuronal loss is advanced and the first symptoms have appeared (Esposito et al. 2012). Likewise, PPARγ ligands have been reported to arrest PD progression in preclinical settings (Randy and Guoying 2007; Carta 2013; Schintu et al. 2009; Pinto et al. 2016). However, it should be mentioned that a recent study published in *The Lancet Neurology* (NINDS Exploratory Trials in Parkinson Disease (NET-PD) FS-ZONE Investigators, 2015), which assessed the PPARγ agonist pioglitazone for its disease-modifying potential in early-stage PD, does not support initiation of further trials with this drug (see also Brundin and Wyse 2015).

5.3 *Mood Disorders and Schizophrenia*

The magnitude, suffering, and burden of mental health disorders in terms of disability and costs for individuals, families, and societies are astonishing. According to WHO's Global Burden of Disease reports, four of the six primary causes of years lived with disability are due to neuropsychiatric disorders (depression, alcohol use disorders, schizophrenia, and bipolar disorder). In particular, unipolar depression alone leads to 12.15% of years lived with disability and ranks as the third most important contributor to the global burden of diseases (Whiteford et al. 2013 and references therein). Depression is indeed a very common illness worldwide, with an estimated 350 million people affected, and is one of the priority conditions covered by WHO's Mental Health Gap Action Programme (Reed et al. 2015). Current therapies for this disorder are ineffective in many patients or cause intolerable side effects, thus highlighting the importance of novel, improved therapeutic options. Mesolimbic dopamine system dysfunctions have been implicated in numerous brain disorders, including depression (reviewed in Nestler and Carlezon 2006; Russo and Nestler 2013). Apart from specific anti-inflammatory and neuroprotective mechanisms, it is has been proposed that PPARs may affect mood by interfering with its neurobiological bases (Rolland et al. 2013). Thus, NAEs, as endogenous PPARα ligands and modulators of midbrain dopamine neuronal activity, are well suited to regulate several dopamine-dependent brain physiological and pathological processes. Specifically, NAEs regulate cholinergic input onto dopamine cells and dampen hypercholinergic drive (Fig. 2) and the consequent aberrant excitation of dopamine neurons. This intrinsic negative feedback mechanism might be exploited as a therapeutic strategy in depression, which is characterized also by an unbalance between dopamine and acetylcholine systems (Melis et al. 2013b). Accordingly, experimental evidence suggests that abnormal nicotinic signaling may contribute to depressive symptoms (Mineur and Picciotto 2010; Tizabi et al. 2000) and raises the possibility that β2*nAChR availability could be a biomarker of depression, correlated with severity of the symptoms (see for a review Picciotto et al. 2015). Consistently, a recent human imaging study has revealed that ACh levels are elevated in patients who are actively depressed, as measured by occupancy of nAChRs throughout the brain, and remain high in patients who have a history of depression (Saricicek et al. 2012). In this context, the PPARα agonist fenofibrate (Jiang et al. 2017; Scheggi et al. 2016) or WY-14643 (Yang et al. 2017), PEA (Yu et al. 2011), and FAAH inhibitors show antidepressant-like activity in rodents (Gobbi et al. 2005; Bortolato et al. 2007; Adamczyk et al. 2008; Umathe et al. 2011). In addition, OEA was also lately proven to be effective in ameliorating animal depression-like behaviors, even though through complex mechanisms (Yu et al. 2015; Jin et al. 2015). Similarly, PPARγ agonists (i.e., pioglitazone, rosiglitazone) have displayed some antidepressant-like properties in preclinical settings (Eissa Ahmed and Al-Rasheed 2009; Sadaghiani et al. 2011). Even though it is still unclear whether the PPARγ beneficial effect lays on the NO pathways (Sadaghiani et al. 2011) or whether it is NMDAR-related (Salehi-Sadaghiani et al. 2012), these results encouraged clinical research. In humans, few studies have been carried out so far, the first of which reporting a moderate improvement in two different depression scales in depressed (and insulin-resistant) patients treated with rosiglitazone (Rasgon et al. 2010). Subsequently, randomized controlled trials tested pioglitazone as adjunctive therapy (Sepanjnia et al. 2012) or as monotherapy (Colle et al. 2017) on subjects with a major depressive disorder and demonstrated a significant amelioration in early response, remission, and symptom attenuation for the pioglitazone-treated group. More recently, open-label administration of this PPARγ agonist was found to be associated with improvement of depressive symptoms in bipolar disorder patients (Kemp et al. 2014). Remarkably, a positive correlation has been observed between depressive symptomatology score and change in interleukin (IL)-6, thus indicating that reduction in inflammation might contribute to the mechanism by which pioglitazone modulates mood (Kemp et al. 2014).

Noteworthy, NAEs can also play a role in the etiopathogenesis of schizophrenia (see for a review Rolland et al. 2013), in which substantial dopamine transmission dysfunctions are strongly involved. Briefly, schizophrenia is a severe mental disorder afflicting approximately 1% of the population around the world (which corresponds to more than 21 million people) (see for a recent review Owen et al. 2016). Clinically, schizophrenia is characterized by profound disruptions in thinking that affect language, perception, and the sense of self. The so-called psychotic experiences (or positive symptoms), such as hearing voices or delusions, are the main target of current antipsychotic drugs which all block dopamine D2 receptors (Leucht et al. 2009). In line with this, and with the above discussed observations, PPARα may exert antipsychotic effects by dampening dopamine neuron activity. Data from preclinical and clinical studies seem to support this hypothesis. For example, the PPARα agonist fenofibrate reduces prepulse inhibition disruption in a neurodevelopmental model of schizophrenia, an effect that might be explained by a direct action of fenofibrate on dopaminergic transmission (Rolland et al. 2012). Consistently, PPARα or α7-nAChRs agonists have been demonstrated to beneficial in animal models of schizophrenia (Zanaletti et al. 2012; Rolland et al. 2012; Kucinski et al. 2012; Pichat et al. 2007; Thomsen et al. 2012). Moreover, since the idea is emerging that schizophrenia displays a pro-inflammatory phenotype (e.g., pro-inflammatory cytokines IL-6 and TNF-α are upregulated in obesity, inflammation, and schizophrenia), a recent human genetic study measured expression of prototypic obesogenic and immunogenic genes in peripheral blood cells from controls and schizophrenic patients (Chase et al. 2015). Intriguingly, Chase et al. reported a profound dysregulation of genes relating to metabolic inflammation, including a significant decrease in PPARα mRNA levels and increases in IL-6 and TNF-α, with additional BMI interactions. Finally, a recent (Croatian) population study aimed at determining whether a functional L162V polymorphism in PPARα gene, extensively investigated in the etiology of abnormal lipid and glucose metabolism, was also linked to schizophrenia risk (Nadalin et al. 2014). While the PPARα-L162V polymorphism did not show an association with schizophrenia risk, it impacts clinical expression of the illness and plasma lipid concentrations in female patients (Nadalin et al. 2014). According to the authors, these polymorphisms could have a possible protective effect toward negative symptom severity in female schizophrenic patients.

6 Concluding Remarks

The explosion of research in the field of the endocannabinoid system in the last two decades has fueled an increased interest in lipid messengers. NAEs have many similarities with endocannabinoids: they share synthetic and catabolic pathways but also the logic in signaling mechanism, as they are produced "on demand" and their diffusion and targets are not restricted to the prototypical anterograde signaling that characterizes classical neurotransmitters. Yet, peculiarities are emerging that might render NAEs and their receptors a new Pandora's box for pharmacologists. NAE signaling in the CNS is engaged under physiological conditions but, more importantly, under several pathological circumstances. Thus, exploiting these targets with novel pharmacological tools will help us to better understand pathophysiological mechanisms in neuropsychiatric disorders and will pave new roads to therapeutic intervention.

References

Adamczyk P, Golda A, McCreary AC, Filip M, Przegalinski E (2008) Activation of endocannabinoid transmission induces antidepressant-like effects in rats. J Physiol Pharmacol 59(2):217–228

American Psychiatric Association (2013) Diagnostic and statistical manual of mental disorders (DSM-5®). American Psychiatric Pub

Ascherio A, Munger KL (2016) Epidemiology of multiple sclerosis: from risk factors to prevention-an update. Semin Neurol 36(2):103–114. doi:10.1055/s-0036-1579693

Auboeuf D, Rieusset J, Fajas L, Vallier P, Frering V, Riou JP, Staels B, Auwerx J, Laville M, Vidal H (1997) Tissue distribution and quantification of the expression of mRNAs of peroxisome proliferator-activated receptors and liver X receptor-alpha in humans: no alteration in adipose tissue of obese and NIDDM patients. Diabetes 46(8):1319–1327

Auvin S (2012) Fatty acid oxidation and epilepsy. Epilepsy Res 100(3):224–228. doi:10.1016/j.eplepsyres.2011.05.022

Avagliano C, Russo R, De Caro C, Cristiano C, La Rana G, Piegari G, Paciello O, Citraro R, Russo E, De Sarro G, Meli R, Mattace Raso G, Calignano A (2016) Palmitoylethanolamide protects mice against 6-OHDA-induced neurotoxicity and endoplasmic reticulum stress: In vivo and in vitro evidence. Pharmacol Res 113(Pt A):276–289. doi:10.1016/j.phrs.2016.09.004

Bachur NR, Masek K, Melmon KL, Udenfriend S (1965) Fatty acid amides of ethanolamine in mammalian tissues. J Biol Chem 240:1019–1024

Baker D, Pryce G, Davies WL, Hiley CR (2006) In silico patent searching reveals a new cannabinoid receptor. Trends Pharmacol Sci 27(1):1–4. doi:10.1016/j.tips.2005.11.003

Balenga NA, Aflaki E, Kargl J, Platzer W, Schroder R, Blattermann S, Kostenis E, Brown AJ, Heinemann A, Waldhoer M (2011) GPR55 regulates cannabinoid 2 receptor-mediated responses in human neutrophils. Cell Res 21(10):1452–1469. doi:10.1038/cr.2011.60

Bensinger SJ, Tontonoz P (2008) Integration of metabolism and inflammation by lipid-activated nuclear receptors. Nature 454(7203):470–477. doi:10.1038/nature07202

Berger J, Moller DE (2002) The mechanisms of action of PPARs. Annu Rev Med 53:409–435

Bilbao A, Serrano A, Cippitelli A, Pavon FJ, Giuffrida A, Suarez J, Garcia-Marchena N, Baixeras E, Gomez de Heras R, Orio L, Alen F, Ciccocioppo R, Cravatt BF, Parsons LH, Piomelli D, Rodriguez de Fonseca F (2015) Role of the satiety factor oleoylethanolamide in alcoholism. Addict Biol. doi:10.1111/adb.12276

Blednov YA, Benavidez JM, Black M, Ferguson LB, Schoenhard GL, Goate AM, Edenberg HJ, Wetherill L, Hesselbrock V, Foroud T, Harris RA (2015) Peroxisome proliferator-activated receptors alpha and gamma are linked with alcohol consumption in mice and withdrawal and dependence in humans. Alcohol Clin Exp Res 39(1):136–145. doi:10.1111/acer.12610

Blednov YA, Black M, Benavidez JM, Stamatakis EE, Harris RA (2016) PPAR Agonists: II. Fenofibrate and Tesaglitazar Alter behaviors related to voluntary alcohol consumption. Alcohol Clin Exp Res 40(3):563–571. doi:10.1111/acer.12972

Bornheim LM, Kim KY, Chen B, Correia MA (1993) The effect of cannabidiol on mouse hepatic microsomal cytochrome P450-dependent anandamide metabolism. Biochem Biophys Res Commun 197(2):740–746

Bortolato M, Mangieri RA, Fu J, Kim JH, Arguello O, Duranti A, Tontini A, Mor M, Tarzia G, Piomelli D (2007) Antidepressant-like activity of the fatty acid amide hydrolase inhibitor URB597 in a rat model of chronic mild stress. Biol Psychiatry 62(10):1103–1110. doi:10.1016/j.biopsych.2006.12.001

Bough KJ, Rho JM (2007) Anticonvulsant mechanisms of the ketogenic diet. Epilepsia 48 (1):43–58. doi:10.1111/j.1528-1167.2007.00915.x

Bough KJ, Schwartzkroin PA, Rho JM (2003) Calorie restriction and ketogenic diet diminish neuronal excitability in rat dentate gyrus in vivo. Epilepsia 44(6):752–760. doi:55502 [pii]

Braissant O, Foufelle F, Scotto C, Dauca M, Wahli W (1996) Differential expression of peroxisome proliferator-activated receptors (PPARs): tissue distribution of PPAR-alpha, -beta, and -gamma in the adult rat. Endocrinology 137(1):354–366

Brundin P, Wyse R (2015) Parkinson disease: laying the foundations for disease-modifying therapies in PD. Nat Rev Neurol 11(10):553–555. doi:10.1038/nrneurol.2015.150

Campolongo P, Roozendaal B, Trezza V, Cuomo V, Astarita G, Fu J, McGaugh JL, Piomelli D (2009) Fat-induced satiety factor oleoylethanolamide enhances memory consolidation. Proc Natl Acad Sci U S A 106(19):8027–8031. doi:10.1073/pnas.0903038106

Carta AR (2013) PPAR-gamma: therapeutic prospects in Parkinson's disease. Curr Drug Targets 14(7):743–751

Chase KA, Rosen C, Gin H, Bjorkquist O, Feiner B, Marvin R, Conrin S, Sharma RP (2015) Metabolic and inflammatory genes in schizophrenia. Psychiatry Res 225(1–2):208–211. doi:10.1016/j.psychres.2014.11.007

Cherif H, Argaw A, Cecyre B, Bouchard A, Gagnon J, Javadi P, Desgent S, Mackie K, Bouchard JF (2015) Role of GPR55 during Axon growth and target innervation(1,2,3). eNeuro 2(5). doi:10.1523/ENEURO.0011-15.2015

Chiang KP, Gerber AL, Sipe JC, Cravatt BF (2004) Reduced cellular expression and activity of the P129T mutant of human fatty acid amide hydrolase: evidence for a link between defects in the endocannabinoid system and problem drug use. Hum Mol Genet 13(18):2113–2119. doi:10.1093/hmg/ddh216

Chinetti G, Griglio S, Antonucci M, Torra IP, Delerive P, Majd Z, Fruchart J-C, Chapman J, Najib J, Staels B (1998) Activation of proliferator-activated receptors alpha and gamma induces apoptosis of human monocyte-derived macrophages. J Biol Chem 273 (40):25573–25580. doi:10.1074/jbc.273.40.25573

Citraro R, Russo E, Scicchitano F, van Rijn CM, Cosco D, Avagliano C, Russo R, D'Agostino G, Petrosino S, Guida F, Gatta L, van Luijtelaar G, Maione S, Di Marzo V, Calignano A, De Sarro G (2013) Antiepileptic action of N-palmitoylethanolamine through CB1 and PPAR-alpha receptor activation in a genetic model of absence epilepsy. Neuropharmacology 69:115–126. doi:10.1016/j.neuropharm.2012.11.017

Citraro R, Russo E, Leo A, Russo R, Avagliano C, Navarra M, Calignano A, De Sarro G (2016) Pharmacokinetic-pharmacodynamic influence of N-palmitoylethanolamine, arachidonyl-2'-chloroethylamide and WIN 55,212-2 on the anticonvulsant activity of antiepileptic drugs against audiogenic seizures in DBA/2 mice. Eur J Pharmacol 791:523–534. doi:10.1016/j.ejphar.2016.09.029

Colle R, de Larminat D, Rotenberg S, Hozer F, Hardy P, Verstuyft C, Feve B, Corruble E (2017) Pioglitazone could induce remission in major depression: a meta-analysis. Neuropsychiatr Dis Treat 13:9–16. doi:10.2147/ndt.s121149

Cravatt BF, Giang DK, Mayfield SP, Boger DL, Lerner RA, Gilula NB (1996) Molecular characterization of an enzyme that degrades neuromodulatory fatty-acid amides. Nature 384 (6604):83–87

Cristino L, Starowicz K, De Petrocellis L, Morishita J, Ueda N, Guglielmotti V, Di Marzo V (2008) Immunohistochemical localization of anabolic and catabolic enzymes for anandamide and other putative endovanilloids in the hippocampus and cerebellar cortex of the mouse brain. Neuroscience 151(4):955–968. doi:10.1016/j.neuroscience.2007.11.047

Cullingford T (2008) Peroxisome proliferator-activated receptor alpha and the ketogenic diet. Epilepsia 49(Suppl 8):70–72. doi:10.1111/j.1528-1167.2008.01840.x

D'Agostino G, Russo R, Avagliano C, Cristiano C, Meli R, Calignano A (2012) Palmitoylethanolamide protects against the amyloid-beta25-35-induced learning and memory impairment in mice, an experimental model of Alzheimer disease. Neuropsychopharmacology 37(7):1784–1792. doi:10.1038/npp.2012.25

De Fusco M, Becchetti A, Patrignani A, Annesi G, Gambardella A, Quattrone A, Ballabio A, Wanke E, Casari G (2000) The nicotinic receptor beta 2 subunit is mutant in nocturnal frontal lobe epilepsy. Nat Genet 26 (3):275-276. doi:10.1038/81566

Deliu E, Sperow M, Console-Bram L, Carter RL, Tilley DG, Kalamarides DJ, Kirby LG, Brailoiu GC, Brailoiu E, Benamar K, Abood ME (2015) The lysophosphatidylinositol receptor GPR55 modulates pain perception in the periaqueductal Gray. Mol Pharmacol 88(2):265–272. doi:10.1124/mol.115.099333

Desvergne B, Wahli W (1999) Peroxisome proliferator-activated receptors: nuclear control of metabolism. Endocr Rev 20(5):649–688. doi:10.1210/er.20.5.649

Deutsch DG, Ueda N, Yamamoto S (2002) The fatty acid amide hydrolase (FAAH). Prostaglandins Leukot Essent Fatty Acids 66(2-3):201–210. doi:10.1054/plef.2001.0358

Devane WA, Hanus L, Breuer A, Pertwee RG, Stevenson LA, Griffin G, Gibson D, Mandelbaum A, Etinger A, Mechoulam R (1992) Isolation and structure of a brain constituent that binds to the cannabinoid receptor. Science 258(5090):1946–1949

Devchand PR, Keller H, Peters JM, Vazquez M, Gonzalez FJ, Wahli W (1996) The PPARalpha-leukotriene B4 pathway to inflammation control. Nature 384(6604):39–43. doi:10.1038/384039a0

Di Filippo M, Pini LA, Pelliccioli GP, Calabresi P, Sarchielli P (2008) Abnormalities in the cerebrospinal fluid levels of endocannabinoids in multiple sclerosis. J Neurol Neurosurg Psychiatry 79(11):1224–1229. doi:10.1136/jnnp.2007.139071

Diep TA, Madsen AN, Holst B, Kristiansen MM, Wellner N, Hansen SH, Hansen HS (2011) Dietary fat decreases intestinal levels of the anorectic lipids through a fat sensor. FASEB J 25 (2):765–774. doi:10.1096/fj.10-166595

Dreyer C, Krey G, Keller H, Givel F, Helftenbein G, Wahli W (1992) Control of the peroxisomal beta-oxidation pathway by a novel family of nuclear hormone receptors. Cell 68(5):879–887. 0092-8674(92)90031-7 [pii]

Egertova M, Simon GM, Cravatt BF, Elphick MR (2008) Localization of N-acyl phosphatidylethanolamine phospholipase D (NAPE-PLD) expression in mouse brain: a new perspective on N-acylethanolamines as neural signaling molecules. J Comp Neurol 506(4):604–615. doi:10.1002/cne.21568

Eissa Ahmed AA, Al-Rasheed NM (2009) Antidepressant-like effects of rosiglitazone, a PPARgamma agonist, in the rat forced swim and mouse tail suspension tests. Behav Pharmacol 20(7):635–642. doi:10.1097/FBP.0b013e328331b9bf

Escriva H, Langlois MC, Mendonca RL, Pierce R, Laudet V (1998) Evolution and diversification of the nuclear receptor superfamily. Ann N Y Acad Sci 839:143–146

Esposito E, Impellizzeri D, Mazzon E, Paterniti I, Cuzzocrea S (2012) Neuroprotective activities of palmitoylethanolamide in an animal model of Parkinson's disease. PloS One 7(8):e41880. doi:10.1371/journal.pone.0041880

Ferre P (2004) The biology of peroxisome proliferator-activated receptors: relationship with lipid metabolism and insulin sensitivity. Diabetes 53(Suppl 1):S43–S50

Fidaleo M, Fanelli F, Ceru MP, Moreno S (2014) Neuroprotective properties of peroxisome proliferator-activated receptor alpha (PPARalpha) and its lipid ligands. Curr Med Chem 21 (24):2803–2821

Fu J, Gaetani S, Oveisi F, Lo Verme J, Serrano A, Rodriguez De Fonseca F, Rosengarth A, Luecke H, Di Giacomo B, Tarzia G, Piomelli D (2003) Oleylethanolamide regulates feeding and body weight through activation of the nuclear receptor PPAR-alpha. Nature 425 (6953):90–93

Fu J, Oveisi F, Gaetani S, Lin E, Piomelli D (2005) Oleoylethanolamide, an endogenous PPAR-alpha agonist, lowers body weight and hyperlipidemia in obese rats. Neuropharmacology 48 (8):1147–1153

Galan-Rodriguez B, Suarez J, Gonzalez-Aparicio R, Bermudez-Silva FJ, Maldonado R, Robledo P, Rodriguez de Fonseca F, Fernandez-Espejo E (2009) Oleoylethanolamide exerts partial and dose-dependent neuroprotection of substantia nigra dopamine neurons. Neuropharmacology 56(3):653–664

Gardner OS, Dewar BJ, Graves LM (2005) Activation of mitogen-activated protein kinases by peroxisome proliferator-activated receptor ligands: an example of nongenomic signaling. Mol Pharmacol 68(4):933–941

Germain P, Staels B, Dacquet C, Spedding M, Laudet V (2006) Overview of nomenclature of nuclear receptors. Pharmacol Rev 58(4):685–704. doi:10.1124/pr.58.4.2

Gervois P, Kleemann R, Pilon A, Percevault F, Koenig W, Staels B, Kooistra T (2004) Global suppression of IL-6-induced acute phase response gene expression after chronic in vivo treatment with the peroxisome proliferator-activated receptor-alpha activator fenofibrate. J Biol Chem 279(16):16154–16160. doi:10.1074/jbc.M400346200

Glass CK, Ogawa S (2006) Combinatorial roles of nuclear receptors in inflammation and immunity. Nat Rev Immunol 6(1):44–55. doi:10.1038/nri1748

Gobbi G, Bambico FR, Mangieri R, Bortolato M, Campolongo P, Solinas M, Cassano T, Morgese MG, Debonnel G, Duranti A, Tontini A, Tarzia G, Mor M, Trezza V, Goldberg SR, Cuomo V, Piomelli D (2005) Antidepressant-like activity and modulation of brain monoaminergic transmission by blockade of anandamide hydrolysis. PNAS 102(51):18620–18625

Gonzalez-Aparicio R, Moratalla R (2014) Oleoylethanolamide reduces L-DOPA-induced dyskinesia via TRPV1 receptor in a mouse model of Parkinson s disease. Neurobiol Dis 62:416–425. doi:10.1016/j.nbd.2013.10.008

Graves RA, Tontonoz P, Spiegelman BM (1992) Analysis of a tissue-specific enhancer: ARF6 regulates adipogenic gene expression. Mol Cell Biol 12(3):1202–1208

Haile CN, Kosten TA (2017) The peroxisome proliferator-activated receptor alpha agonist fenofibrate attenuates alcohol self-administration in rats. Neuropharmacology 116:364–370. doi:10.1016/j.neuropharm.2017.01.007

Hampson AJ, Hill WA, Zan-Phillips M, Makriyannis A, Leung E, Eglen RM, Bornheim LM (1995) Anandamide hydroxylation by brain lipoxygenase: metabolite structures and potencies at the cannabinoid receptor. Biochim Biophys Acta 1259(2):173–179

Hansen HS (2010) Palmitoylethanolamide and other anandamide congeners. Proposed role in the diseased brain. Exp Neurol 224(1):48–55. doi:10.1016/j.expneurol.2010.03.022

Hansen HS (2014) Role of anorectic N-acylethanolamines in intestinal physiology and satiety control with respect to dietary fat. Pharmacol Res 86:18–25. doi:10.1016/j.phrs.2014.03.006

Hansen HS, Diep TA (2009) N-acylethanolamines, anandamide and food intake. Biochem Pharmacol. doi:10.1016/j.bcp.2009.04.024

Hansen HS, Moesgaard B, Hansen HH, Petersen G (2000) N-Acylethanolamines and precursor phospholipids – relation to cell injury. Chem Phys Lipids 108(1–2):135–150. S0009308400001924 [pii]

Heneka MT, Sastre M, Dumitrescu-Ozimek L, Hanke A, Dewachter I, Kuiperi C, O'Banion K, Klockgether T, Van Leuven F, Landreth GE (2005) Acute treatment with the PPARgamma agonist pioglitazone and ibuprofen reduces glial inflammation and Abeta1-42 levels in APPV717I transgenic mice. Brain 128(Pt 6):1442–1453. doi:10.1093/brain/awh452

Hess R, Staubli W, Riess W (1965) Nature of the hepatomegalic effect produced by ethyl-chlorophenoxy-isobutyrate in the rat. Nature 208(5013):856–858

Huganir RL, Greengard P (1990) Regulation of neurotransmitter receptor desensitization by protein phosphorylation. Neuron 5(5):555–567

Issemann I, Green S (1990) Activation of a member of the steroid hormone receptor superfamily by peroxisome proliferators. Nature 347(6294):645–650. doi:10.1038/347645a0

Jiang Q, Heneka M, Landreth GE (2008) The role of peroxisome proliferator-activated receptor-gamma (PPARgamma) in Alzheimer's disease: therapeutic implications. CNS Drugs 22 (1):1–14

Jiang B, Wang YJ, Wang H, Song L, Huang C, Zhu Q, Wu F, Zhang W (2017) Antidepressant-like effects of fenofibrate in mice via the hippocampal brain-derived neurotrophic factor signalling pathway. Br J Pharmacol 174(2):177–194. doi:10.1111/bph.13668

Jin P, HL Y, Tian L, Zhang F, Quan ZS (2015) Antidepressant-like effects of oleoylethanolamide in a mouse model of chronic unpredictable mild stress. Pharmacol Biochem Behav 133:146–154. doi:10.1016/j.pbb.2015.04.001

Justinova Z, Mangieri RA, Bortolato M, Chefer SI, Mukhin AG, Clapper JR, King AR, Redhi GH, Yasar S, Piomelli D, Goldberg SR (2008) Fatty acid amide hydrolase inhibition heightens anandamide signaling without producing reinforcing effects in primates. Biol Psychiatry 64 (11):930–937. doi:10.1016/j.biopsych.2008.08.008

Justinova Z, Panlilio LV, Moreno-Sanz G, Redhi GH, Auber A, Secci ME, Mascia P, Bandiera T, Armirotti A, Bertorelli R, Chefer SI, Barnes C, Yasar S, Piomelli D, Goldberg SR (2015) Effects of fatty acid amide hydrolase (FAAH) inhibitors in non-human primate models of nicotine reward and relapse. Neuropsychopharmacology 40(9):2185–2197. doi:10.1038/npp.2015.62

Kapur A, Zhao P, Sharir H, Bai Y, Caron MG, Barak LS, Abood ME (2009) Atypical responsiveness of the orphan receptor GPR55 to cannabinoid ligands. J Biol Chem 284(43):29817–29827. doi:10.1074/jbc.M109.050187

Kemp DE, Schinagle M, Gao K, Conroy C, Ganocy SJ, Ismail-Beigi F, Calabrese JR (2014) PPAR-gamma agonism as a modulator of mood: proof-of-concept for pioglitazone in bipolar depression. CNS Drugs 28(6):571–581. doi:10.1007/s40263-014-0158-2

Kilaru A, Isaac G, Tamura P, Baxter D, Duncan SR, Venables BJ, Welti R, Koulen P, Chapman KD (2010) Lipid profiling reveals tissue-specific differences for ethanolamide lipids in mice lacking fatty acid amide hydrolase. Lipids 45(9):863–875. doi:10.1007/s11745-010-3457-5

Kozak KR, Crews BC, Morrow JD, Wang LH, Ma YH, Weinander R, Jakobsson PJ, Marnett LJ (2002) Metabolism of the endocannabinoids, 2-arachidonylglycerol and anandamide, into prostaglandin, thromboxane, and prostacyclin glycerol esters and ethanolamides. J Biol Chem 277(47):44877–44885. doi:10.1074/jbc.M206788200

Kucinski A, Syposs C, Wersinger S, Bencherif M, Stachowiak MK, Stachowiak EK (2012) Alpha7 neuronal nicotinic receptor agonist (TC-7020) reverses increased striatal dopamine release during acoustic PPI testing in a transgenic mouse model of schizophrenia. Schizophr Res 136 (1–3):82–87

Kuehl FA, Jacob TA, Ganley OH, Ormond RE, Meisinger MAP (1957) The identification of N-(2-hydroxyethyl)-palmitamide as a naturally occurring anti-inflammatory agent. J Am Chem Soc 79(20):5577–5578. doi:10.1021/ja01577a066

Lambert DM, Vandevoorde S, Diependaele G, Govaerts SJ, Robert AR (2001) Anticonvulsant activity of N-palmitoylethanolamide, a putative endocannabinoid, in mice. Epilepsia 42 (3):321–327

Landreth G, Jiang Q, Mandrekar S, Heneka M (2008) PPARgamma agonists as therapeutics for the treatment of Alzheimer's disease. Neurotherapeutics 5(3):481–489. doi:10.1016/j.nurt.2008.05.003. S1933-7213(08)00092-5 [pii]

Laudet V, Hanni C, Coll J, Catzeflis F, Stehelin D (1992) Evolution of the nuclear receptor gene superfamily. EMBO J 11(3):1003–1013

Lefebvre P, Chinetti G, Fruchart JC, Staels B (2006) Sorting out the roles of PPAR alpha in energy metabolism and vascular homeostasis. J Clin Invest 116(3):571–580. doi:10.1172/JCI27989

Leucht S, Corves C, Arbter D, Engel RR, Li C, Davis JM (2009) Second-generation versus first-generation antipsychotic drugs for schizophrenia: a meta-analysis. Lancet 373(9657):31–41. doi:10.1016/S0140-6736(08)61764-X

Leung D, Saghatelian A, Simon GM, Cravatt BF (2006) Inactivation of N-acyl phosphatidylethanolamine phospholipase D reveals multiple mechanisms for the biosynthesis of endocannabinoids. Biochemistry (Mosc) 45(15):4720–4726. doi:10.1021/bi0601631

Lo Verme J, Fu J, Astarita G, La Rana G, Russo R, Calignano A, Piomelli D (2005) The nuclear receptor peroxisome proliferator-activated receptor-alpha mediates the anti-inflammatory actions of palmitoylethanolamide. Mol Pharmacol 67(1):15–19. doi:10.1124/mol.104.006353

Loria F, Petrosino S, Mestre L, Spagnolo A, Correa F, Hernangomez M, Guaza C, Di Marzo V, Docagne F (2008) Study of the regulation of the endocannabinoid system in a virus model of

multiple sclerosis reveals a therapeutic effect of palmitoylethanolamide. Eur J Neurosci 28 (4):633–641. doi:10.1111/j.1460-9568.2008.06377.x
Losel R, Wehling M (2003) Nongenomic actions of steroid hormones. Nat Rev Mol Cell Biol 4 (1):46–56
Losel RM, Falkenstein E, Feuring M, Schultz A, Tillmann HC, Rossol-Haseroth K, Wehling M (2003) Nongenomic steroid action: controversies, questions, and answers. Physiol Rev 83 (3):965–1016
Luchicchi A, Lecca S, Carta S, Pillolla G, Muntoni AL, Yasar S, Goldberg SR, Pistis M (2010) Effects of fatty acid amide hydrolase inhibition on neuronal responses to nicotine, cocaine and morphine in the nucleus accumbens shell and ventral tegmental area: involvement of PPAR-alpha nuclear receptors. Addict Biol 15(3):277–288. doi:10.1111/j.1369-1600.2010.00222.x
Mandard S, Muller M, Kersten S (2004) Peroxisome proliferator-activated receptor alpha target genes. Cell Mol Life Sci 61(4):393–416. doi:10.1007/s00018-003-3216-3
Mascia P, Pistis M, Justinova Z, Panlilio LV, Luchicchi A, Lecca S, Scherma M, Fratta W, Fadda P, Barnes C, Redhi GH, Yasar S, Le Foll B, Tanda G, Piomelli D, Goldberg SR (2011) Blockade of nicotine reward and reinstatement by activation of alpha-type peroxisome proliferator-activated receptors. Biol Psychiatry 69(7):633–641. doi:10.1016/j.biopsych.2010.07.009
Mattace Raso G, Russo R, Calignano A, Meli R (2014) Palmitoylethanolamide in CNS health and disease. Pharmacol Res 86:32–41. doi:10.1016/j.phrs.2014.05.006
Mazzola C, Medalie J, Scherma M, Panlilio LV, Solinas M, Tanda G, Drago F, Cadet JL, Goldberg SR, Yasar S (2009) Fatty acid amide hydrolase (FAAH) inhibition enhances memory acquisition through activation of PPAR-alpha nuclear receptors. Learn Mem 16 (5):332-337. doi:16/5/332 [pii] 10.1101/lm.1145209
Melis M, Pistis M (2014) Targeting the interaction between fatty acid ethanolamides and nicotinic receptors: therapeutic perspectives. Pharmacol Res 86:42–49. doi:10.1016/j.phrs.2014.03.009
Melis M, Pillolla G, Luchicchi A, Muntoni AL, Yasar S, Goldberg SR, Pistis M (2008) Endogenous fatty acid ethanolamides suppress nicotine-induced activation of mesolimbic dopamine neurons through nuclear receptors. J Neurosci 28(51):13985–13994
Melis M, Carta S, Fattore L, Tolu S, Yasar S, Goldberg SR, Fratta W, Maskos U, Pistis M (2010) Peroxisome proliferator-activated receptors-alpha modulate dopamine cell activity through nicotinic receptors. Biol Psychiatry 68(3):256–264. doi:10.1016/j.biopsych.2010.04.016
Melis M, Carta G, Pistis M, Banni S (2013a) Physiological role of peroxisome proliferator-activated receptors type alpha on dopamine systems. CNS Neurol Disord Drug Targets 12 (1):70–77
Melis M, Scheggi S, Carta G, Madeddu C, Lecca S, Luchicchi A, Cadeddu F, Frau R, Fattore L, Fadda P, Ennas MG, Castelli MP, Fratta W, Schilstrom B, Banni S, De Montis MG, Pistis M (2013b) PPARalpha regulates cholinergic-driven activity of midbrain dopamine neurons via a novel mechanism involving alpha7 nicotinic acetylcholine receptors. J Neurosci 33 (14):6203–6211. doi:10.1523/JNEUROSCI.4647-12.2013
Mineur YS, Picciotto MR (2010) Nicotine receptors and depression: revisiting and revising the cholinergic hypothesis. Trends Pharmacol Sci 31(12):580–586. doi:10.1016/j.tips.2010.09.004
Moraes LA, Swales KE, Wray JA, Damazo A, Gibbins JM, Warner TD, Bishop-Bailey D (2007) Nongenomic signaling of the retinoid X receptor through binding and inhibiting Gq in human platelets. Blood 109(9):3741–3744
Moran BM, Abdel-Wahab YH, Flatt PR, McKillop AM (2014) Activation of GPR119 by fatty acid agonists augments insulin release from clonal beta-cells and isolated pancreatic islets and improves glucose tolerance in mice. Biol Chem 395(4):453–464. doi:10.1515/hsz-2013-0255
Moreno S, Farioli-Vecchioli S, Ceru MP (2004) Immunolocalization of peroxisome proliferator-activated receptors and retinoid X receptors in the adult rat CNS. Neuroscience 123 (1):131–145
Morishita J, Okamoto Y, Tsuboi K, Ueno M, Sakamoto H, Maekawa N, Ueda N (2005) Regional distribution and age-dependent expression of N-acylphosphatidylethanolamine-hydrolyzing

phospholipase D in rat brain. J Neurochem 94(3):753–762. doi:10.1111/j.1471-4159.2005.03234.x
Nadalin S, Giacometti J, Buretic-Tomljanovic A (2014) PPARalpha-L162V polymorphism is not associated with schizophrenia risk in a Croatian population. Prostaglandins Leukot Essent Fatty Acids 91(5):221–225. doi:10.1016/j.plefa.2014.07.003
Nestler EJ, Carlezon WA Jr (2006) The mesolimbic dopamine reward circuit in depression. Biol Psychiatry 59(12):1151–1159
Nyilas R, Dudok B, Urban GM, Mackie K, Watanabe M, Cravatt BF, Freund TF, Katona I (2008) Enzymatic machinery for endocannabinoid biosynthesis associated with calcium stores in glutamatergic axon terminals. J Neurosci 28(5):1058–1063. doi:10.1523/JNEUROSCI.5102-07.2008
Oka S, Nakajima K, Yamashita A, Kishimoto S, Sugiura T (2007) Identification of GPR55 as a lysophosphatidylinositol receptor. Biochem Biophys Res Commun 362(4):928–934. doi:10.1016/j.bbrc.2007.08.078
Okamoto Y, Morishita J, Tsuboi K, Tonai T, Ueda N (2004) Molecular characterization of a phospholipase D generating anandamide and its congeners. J Biol Chem 279(7):5298–5305
Owen MJ, Sawa A, Mortensen PB (2016) Schizophrenia. Lancet doi:10.1016/S0140-6736(15)01121-6
Panlilio LV, Justinova Z, Mascia P, Pistis M, Luchicchi A, Lecca S, Barnes C, Redhi GH, Adair J, Heishman SJ, Yasar S, Aliczki M, Haller J, Goldberg SR (2012) Novel use of a lipid-lowering fibrate medication to prevent nicotine reward and relapse: preclinical findings. Neuropsychopharmacology 37(8):1838–1847. doi:10.1038/npp.2012.31
Pazos MR, Nunez E, Benito C, Tolon RM, Romero J (2004) Role of the endocannabinoid system in Alzheimer's disease: new perspectives. Life Sci 75(16):1907–1915. doi:10.1016/j.lfs.2004.03.026. S0024-3205(04)00560-0 [pii]
Petrosino S, Iuvone T, Di Marzo V (2010) N-palmitoyl-ethanolamine: Biochemistry and new therapeutic opportunities. Biochimie 92(6):724–727. doi:10.1016/j.biochi.2010.01.006
Picciotto MR, Lewis AS, van Schalkwyk GI, Mineur YS (2015) Mood and anxiety regulation by nicotinic acetylcholine receptors: a potential pathway to modulate aggression and related behavioral states. Neuropharmacology 96(Pt B):235–243. doi:10.1016/j.neuropharm.2014.12.028
Pichat P, Bergis OE, Terranova JP, Urani A, Duarte C, Santucci V, Gueudet C, Voltz C, Steinberg R, Stemmelin J, Oury-Donat F, Avenet P, Griebel G, Scatton B (2007) SSR180711, a novel selective alpha7 nicotinic receptor partial agonist: (II) efficacy in experimental models predictive of activity against cognitive symptoms of schizophrenia. Neuropsychopharmacology 32(1):17–34. doi:10.1038/sj.npp.1301188
Pietr M, Kozela E, Levy R, Rimmerman N, Lin YH, Stella N, Vogel Z, Juknat A (2009) Differential changes in GPR55 during microglial cell activation. FEBS Lett 583(12):2071–2076. doi:10.1016/j.febslet.2009.05.028
Pinto M, Nissanka N, Peralta S, Brambilla R, Diaz F, Moraes CT (2016) Pioglitazone ameliorates the phenotype of a novel Parkinson's disease mouse model by reducing neuroinflammation. Mol Neurodegener 11:25. doi:10.1186/s13024-016-0090-7
Piomelli D (2013) A fatty gut feeling. Trends Endocrinol Metab 24(7):332–341. doi:10.1016/j.tem.2013.03.001
Pistis M, Melis M (2010) From surface to nuclear receptors: the endocannabinoid family extends its assets. Curr Med Chem 17(14):1450–1467
Porta N, Vallee L, Lecointe C, Bouchaert E, Staels B, Bordet R, Auvin S (2009) Fenofibrate, a peroxisome proliferator-activated receptor-alpha agonist, exerts anticonvulsive properties. Epilepsia 50(4):943–948. doi:10.1111/j.1528-1167.2008.01901.x
Pringsheim T, Jette N, Frolkis A, Steeves TD (2014) The prevalence of Parkinson's disease: a systematic review and meta-analysis. Mov Disord 29(13):1583–1590. doi:10.1002/mds.25945

Puligheddu M, Pillolla G, Melis M, Lecca S, Marrosu F, De Montis MG, Scheggi S, Carta G, Murru E, Aroni S, Muntoni AL, Pistis M (2013) PPAR-alpha agonists as novel antiepileptic drugs: preclinical findings. PloS One 8(5):e64541. doi:10.1371/journal.pone.0064541
Rahman IA, Tsuboi K, Uyama T, Ueda N (2014) New players in the fatty acyl ethanolamide metabolism. Pharmacol Res 86:1–10. doi:10.1016/j.phrs.2014.04.001
Randy LH, Guoying B (2007) Agonism of peroxisome proliferator receptor-gamma may have therapeutic potential for neuroinflammation and Parkinson's disease. Curr Neuropharmacol 5 (1):35–46
Rasgon NL, Kenna HA, Williams KE, Powers B, Wroolie T, Schatzberg AF (2010) Rosiglitazone add-on in treatment of depressed patients with insulin resistance: a pilot study. Sci World J 10:321–328. doi:10.1100/tsw.2010.32
Reed GM, Rebello TJ, Pike KM, Medina-Mora ME, Gureje O, Zhao M, Dai Y, Roberts MC, Maruta T, Matsumoto C, Krasnov VN, Kulygina M, Lovell AM, Stona AC, Sharan P, Robles R, Gaebel W, Zielasek J, Khoury B, de Jesus Mari J, Luis Ayuso-Mateos J, Evans SC, Kogan CS, Saxena S (2015) WHO's global clinical practice network for mental health. Lancet Psychiatry 2(5):379–380. doi:10.1016/S2215-0366(15)00183-2
Reguero L, Puente N, Elezgarai I, Ramos-Uriarte A, Gerrikagoitia I, Bueno-Lopez JL, Donate F, Grandes P (2014) Subcellular localization of NAPE-PLD and DAGL-alpha in the ventromedial nucleus of the hypothalamus by a preembedding immunogold method. Histochem Cell Biol 141(5):543–550. doi:10.1007/s00418-013-1174-x
Reitz C, Brayne C, Mayeux R (2011) Epidemiology of Alzheimer disease. Nat Rev Neurol 7 (3):137–152. doi:10.1038/nrneurol.2011.2
Risner ME, Saunders AM, Altman JF, Ormandy GC, Craft S, Foley IM, Zvartau-Hind ME, Hosford DA, Roses AD (2006) Efficacy of rosiglitazone in a genetically defined population with mild-to-moderate Alzheimer's disease. Pharmacogenomics J 6(4):246–254. doi:10.1038/sj.tpj.6500369
Rodriguez de Fonseca F, Navarro M, Gomez R, Escuredo L, Nava F, Fu J, Murillo-Rodriguez E, Giuffrida A, LoVerme J, Gaetani S, Kathuria S, Gall C, Piomelli D (2001) An anorexic lipid mediator regulated by feeding. Nature 414(6860):209–212
Rolland B, Marche K, Cottencin O, Bordet R (2012) The PPARalpha agonist fenofibrate reduces prepulse inhibition disruption in a neurodevelopmental model of Schizophrenia. Schizophr Res Treat 2012:839853. doi:10.1155/2012/839853
Rolland B, Deguil J, Jardri R, Cottencin O, Thomas P, Bordet R (2013) Therapeutic prospects of PPARs in psychiatric disorders: a comprehensive review. Curr Drug Targets 14(7):724–732
Ropero AB, Juan-Pico P, Rafacho A, Fuentes E, Bermudez-Silva FJ, Roche E, Quesada I, de Fonseca FR, Nadal A (2009) Rapid non-genomic regulation of Ca2+ signals and insulin secretion by PPAR alpha ligands in mouse pancreatic islets of Langerhans. J Endocrinol 200 (2):127–138
Rowley JW, Oler AJ, Tolley ND, Hunter BN, Low EN, Nix DA, Yost CC, Zimmerman GA, Weyrich AS (2011) Genome-wide RNA-seq analysis of human and mouse platelet transcriptomes. Blood 118(14):e101–e111. doi:10.1182/blood-2011-03-339705
Russo SJ, Nestler EJ (2013) The brain reward circuitry in mood disorders. Nat Rev Neurosci 14 (9):609–625. doi:10.1038/nrn3381
Ryberg E, Larsson N, Sjogren S, Hjorth S, Hermansson NO, Leonova J, Elebring T, Nilsson K, Drmota T, Greasley PJ (2007) The orphan receptor GPR55 is a novel cannabinoid receptor. Br J Pharmacol 152(7):1092–1101. doi:10.1038/sj.bjp.0707460
Sadaghiani MS, Javadi-Paydar M, Gharedaghi MH, Fard YY, Dehpour AR (2011) Antidepressant-like effect of pioglitazone in the forced swimming test in mice: the role of PPAR-gamma receptor and nitric oxide pathway. Behav Brain Res 224(2):336–343. doi:10.1016/j.bbr.2011.06.011
Saha L, Bhandari S, Bhatia A, Banerjee D, Chakrabarti A (2014) Anti-kindling effect of bezafibrate, a peroxisome proliferator-activated receptors alpha agonist, in pentylenetetrazole induced kindling seizure model. J Epilepsy Res 4(2):45–54

Salehi-Sadaghiani M, Javadi-Paydar M, Gharedaghi MH, Zandieh A, Heydarpour P, Yousefzadeh-Fard Y, Dehpour AR (2012) NMDA receptor involvement in antidepressant-like effect of pioglitazone in the forced swimming test in mice. Psychopharmacology (Berl) 223(3):345–355. doi:10.1007/s00213-012-2722-0

Saricicek A, Esterlis I, Maloney KH, Mineur YS, Ruf BM, Muralidharan A, Chen JI, Cosgrove KP, Kerestes R, Ghose S, Tamminga CA, Pittman B, Bois F, Tamagnan G, Seibyl J, Picciotto MR, Staley JK, Bhagwagar Z (2012) Persistent beta2*-nicotinic acetylcholinergic receptor dysfunction in major depressive disorder. Am J Psychiatry 169(8):851–859. doi:10.1176/appi.ajp.2012.11101546

Sawzdargo M, Nguyen T, Lee DK, Lynch KR, Cheng R, Heng HH, George SR, O'Dowd BF (1999) Identification and cloning of three novel human G protein-coupled receptor genes GPR52, PsiGPR53 and GPR55: GPR55 is extensively expressed in human brain. Brain Res Mol Brain Res 64(2):193–198

Scheggi S, Melis M, De Felice M, Aroni S, Muntoni AL, Pelliccia T, Gambarana C, De Montis MG, Pistis M (2016) PPARalpha modulation of mesolimbic dopamine transmission rescues depression-related behaviors. Neuropharmacology 110(Pt A):251–259. doi:10.1016/j.neuropharm.2016.07.024

Scherma M, Panlilio LV, Fadda P, Fattore L, Gamaleddin I, Le Foll B, Justinova Z, Mikics E, Haller J, Medalie J, Stroik J, Barnes C, Yasar S, Tanda G, Piomelli D, Fratta W, Goldberg SR (2008) Inhibition of anandamide hydrolysis by URB597 reverses abuse-related behavioral and neurochemical effects of nicotine in rats. J Pharmacol Exp Ther. doi:10.1124/jpet.108.142224

Scherma M, Muntoni AL, Melis M, Fattore L, Fadda P, Fratta W, Pistis M (2016) Interactions between the endocannabinoid and nicotinic cholinergic systems: preclinical evidence and therapeutic perspectives. Psychopharmacology (Berl). doi:10.1007/s00213-015-4196-3

Schintu N, Frau L, Ibba M, Caboni P, Garau A, Carboni E, Carta AR (2009) PPAR-gamma-mediated neuroprotection in a chronic mouse model of Parkinson's disease. Eur J Neurosci 29 (5):954–963

Schmid HH, Schmid PC, Natarajan V (1990) N-acylated glycerophospholipids and their derivatives. Prog Lipid Res 29(1):1–43

Schmidt A, Endo N, Rutledge SJ, Vogel R, Shinar D, Rodan GA (1992) Identification of a new member of the steroid hormone receptor superfamily that is activated by a peroxisome proliferator and fatty acids. Mol Endocrinol 6(10):1634–1641

Scuderi C, Steardo L (2013) Neuroglial roots of neurodegenerative diseases: therapeutic potential of palmitoylethanolamide in models of Alzheimer's disease. CNS Neurol Disord Drug Targets 12(1):62–69

Scuderi C, Esposito G, Blasio A, Valenza M, Arietti P, Steardo L, Jr., Carnuccio R, De Filippis D, Petrosino S, Iuvone T, Di Marzo V, Steardo L (2011) Palmitoylethanolamide counteracts reactive astrogliosis induced by beta-amyloid peptide. J Cell Mol Med 15 (12):2664-2674. doi:10.1111/j.1582-4934.2011.01267.x

Scuderi C, Valenza M, Stecca C, Esposito G, Carratu MR, Steardo L (2012) Palmitoylethanolamide exerts neuroprotective effects in mixed neuroglial cultures and organotypic hippocampal slices via peroxisome proliferator-activated receptor-alpha. J Neuroinflamm 9:49. doi:10.1186/1742-2094-9-21

Sepanjnia K, Modabbernia A, Ashrafi M, Modabbernia MJ, Akhondzadeh S (2012) Pioglitazone adjunctive therapy for moderate-to-severe major depressive disorder: randomized double-blind placebo-controlled trial. Neuropsychopharmacology 37(9):2093–2100. doi:10.1038/npp.2012.58

Sheerin AH, Zhang X, Saucier DM, Corcoran ME (2004) Selective antiepileptic effects of N-palmitoylethanolamide, a putative endocannabinoid. Epilepsia 45(10):1184–1188. doi:10.1111/j.0013-9580.2004.16604.x

Sher T, Yi HF, McBride OW, Gonzalez FJ (1993) cDNA cloning, chromosomal mapping, and functional characterization of the human peroxisome proliferator activated receptor. Biochemistry (Mosc) 32(21):5598–5604

Sheu MY, Fowler AJ, Kao J, Schmuth M, Schoonjans K, Auwerx J, Fluhr JW, Man MQ, Elias PM, Feingold KR (2002) Topical peroxisome proliferator activated receptor-alpha activators reduce inflammation in irritant and allergic contact dermatitis models. J Invest Dermatol 118 (1):94–101. doi:10.1046/j.0022-202x.2001.01626.x

Shohami E, Mechoulam R (2006) Multiple sclerosis may disrupt endocannabinoid brain protection mechanism. Proc Natl Acad Sci U S A 103(16):6087–6088

Simon GM, Cravatt BF (2010) Characterization of mice lacking candidate N-acyl ethanolamine biosynthetic enzymes provides evidence for multiple pathways that contribute to endocannabinoid production in vivo. Mol Biosyst 6(8):1411–1418. doi:10.1039/c000237b

Sipe JC, Chiang K, Gerber AL, Beutler E, Cravatt BF (2002) A missense mutation in human fatty acid amide hydrolase associated with problem drug use. Proc Natl Acad Sci U S A 99 (12):8394–8399. doi:10.1073/pnas.08223579999/12/8394. [pii]

Staton PC, Hatcher JP, Walker DJ, Morrison AD, Shapland EM, Hughes JP, Chong E, Mander PK, Green PJ, Billinton A, Fulleylove M, Lancaster HC, Smith JC, Bailey LT, Wise A, Brown AJ, Richardson JC, Chessell IP (2008) The putative cannabinoid receptor GPR55 plays a role in mechanical hyperalgesia associated with inflammatory and neuropathic pain. Pain 139 (1):225–236. doi:10.1016/j.pain.2008.04.006

Steinlein OK (2004) Genetic mechanisms that underlie epilepsy. Nat Rev Neurosci 5 (5):400-408. doi:10.1038/nrn1388 nrn1388 [pii]

Steinlein OK, Magnusson A, Stoodt J, Bertrand S, Weiland S, Berkovic SF, Nakken KO, Propping P, Bertrand D (1997) An insertion mutation of the CHRNA4 gene in a family with autosomal dominant nocturnal frontal lobe epilepsy. Hum Mol Genet 6(6):943–947

Straus DS, Glass CK (2007) Anti-inflammatory actions of PPAR ligands: new insights on cellular and molecular mechanisms. Trends Immunol 28(12):551–558. doi:10.1016/j.it.2007.09.003

Suarez J, Bermudez-Silva FJ, Mackie K, Ledent C, Zimmer A, Cravatt BF, de Fonseca FR (2008) Immunohistochemical description of the endogenous cannabinoid system in the rat cerebellum and functionally related nuclei. J Comp Neurol 509(4):400–421. doi:10.1002/cne.21774

Sutor B, Zolles G (2001) Neuronal nicotinic acetylcholine receptors and autosomal dominant nocturnal frontal lobe epilepsy: a critical review. Pflugers Arch 442(5):642–651

Terrazzino S, Berto F, Dalle Carbonare M, Fabris M, Guiotto A, Bernardini D, Leon A (2004) Stearoylethanolamide exerts anorexic effects in mice via down-regulation of liver stearoyl-coenzyme A desaturase-1 mRNA expression. FASEB J 18(13):1580–1582. doi:10.1096/fj.03-1080fje

Thomsen MS, Hansen HH, Timmerman DB, Mikkelsen JD (2012) Cognitive improvement by activation of alpha7 nicotinic acetylcholine receptors: from animal models to human pathophysiology. Curr Pharm Des 16(3):323–343

Thorp JM, Waring WS (1962) Modification of metabolism and distribution of lipids by ethyl chlorophenoxyisobutyrate. Nature 194:948–949

Thurman DJ, Beghi E, Begley CE, Berg AT, Buchhalter JR, Ding D, Hesdorffer DC, Hauser WA, Kazis L, Kobau R, Kroner B, Labiner D, Liow K, Logroscino G, Medina MT, Newton CR, Parko K, Paschal A, Preux PM, Sander JW, Selassie A, Theodore W, Tomson T, Wiebe S (2011) Standards for epidemiologic studies and surveillance of epilepsy. Epilepsia 52(Suppl 7):2–26. doi:10.1111/j.1528-1167.2011.03121.x

Tizabi Y, Rezvani AH, Russell LT, Tyler KY, Overstreet DH (2000) Depressive characteristics of FSL rats: involvement of central nicotinic receptors. Pharmacol Biochem Behav 66(1):73–77

Tsuboi K, Sun YX, Okamoto Y, Araki N, Tonai T, Ueda N (2005) Molecular characterization of N-acylethanolamine-hydrolyzing acid amidase, a novel member of the choloylglycine hydrolase family with structural and functional similarity to acid ceramidase. J Biol Chem 280 (12):11082–11092. doi:10.1074/jbc.M413473200

Tsuboi K, Takezaki N, Ueda N (2007a) The N-acylethanolamine-hydrolyzing acid amidase (NAAA). Chem Biodivers 4(8):1914–1925. doi:10.1002/cbdv.200790159

Tsuboi K, Zhao LY, Okamoto Y, Araki N, Ueno M, Sakamoto H, Ueda N (2007b) Predominant expression of lysosomal N-acylethanolamine-hydrolyzing acid amidase in macrophages revealed by immunochemical studies. Biochim Biophys Acta 1771(5):623–632. doi:10.1016/j.bbalip.2007.03.005

Tsuboi K, Okamoto Y, Ikematsu N, Inoue M, Shimizu Y, Uyama T, Wang J, Deutsch DG, Burns MP, Ulloa NM, Tokumura A, Ueda N (2011) Enzymatic formation of N-acylethanolamines from N-acylethanolamine plasmalogen through N-acylphosphatidylethanolamine-hydrolyzing phospholipase D-dependent and -independent pathways. Biochim Biophys Acta 1811 (10):565–577. doi:10.1016/j.bbalip.2011.07.009

Tyndale RF, Payne JI, Gerber AL, Sipe JC (2007) The fatty acid amide hydrolase C385A (P129T) missense variant in cannabis users: studies of drug use and dependence in Caucasians. Am J Med Genet B Neuropsychiatr Genet 144B(5):660–666. doi:10.1002/ajmg.b.30491

Ueda N, Yamamoto K, Yamamoto S, Tokunaga T, Shirakawa E, Shinkai H, Ogawa M, Sato T, Kudo I, Inoue K et al (1995) Lipoxygenase-catalyzed oxygenation of arachidonylethanolamide, a cannabinoid receptor agonist. Biochim Biophys Acta 1254 (2):127–134

Ueda N, Liu Q, Yamanaka K (2001) Marked activation of the N-acylphosphatidylethanolamine-hydrolyzing phosphodiesterase by divalent cations. Biochim Biophys Acta 1532(1-2):121–127

Ueda N, Tsuboi K, Uyama T (2010a) Enzymological studies on the biosynthesis of N-acylethanolamines. Biochim Biophys Acta (BBA) – Mol Cell Biol Lipids 1801 (12):1274–1285. doi:10.1016/j.bbalip.2010.08.010

Ueda N, Tsuboi K, Uyama T (2010b) N-acylethanolamine metabolism with special reference to N-acylethanolamine-hydrolyzing acid amidase (NAAA). Prog Lipid Res 49(4):299–315. doi:10.1016/j.plipres.2010.02.003

Umathe SN, Manna SS, Jain NS (2011) Involvement of endocannabinoids in antidepressant and anti-compulsive effect of fluoxetine in mice. Behav Brain Res 223(1):125–134. doi:10.1016/j.bbr.2011.04.031

Varvel NH, Jiang J, Dingledine R (2015) Candidate drug targets for prevention or modification of epilepsy. Annu Rev Pharmacol Toxicol 55:229–247. doi:10.1146/annurev-pharmtox-010814-124607

Volkow ND, Koob GF, McLellan AT (2016) Neurobiologic advances from the brain disease model of addiction. N Engl J Med 374(4):363–371. doi:10.1056/NEJMra1511480

Wangensteen T, Akselsen H, Holmen J, Undlien D, Retterstøl L (2010) A common haplotype in NAPEPLD is associated with severe obesity in a norwegian population-based cohort (the HUNT study). Obesity 19(3):612–617. doi:10.1038/oby.2010.219

Watson GS, Cholerton BA, Reger MA, Baker LD, Plymate SR, Asthana S, Fishel MA, Kulstad JJ, Green PS, Cook DG, Kahn SE, Keeling ML, Craft S (2005) Preserved cognition in patients with early Alzheimer disease and amnestic mild cognitive impairment during treatment with rosiglitazone: a preliminary study. Am J Geriatr Psychiatry 13(11):950–958. doi:10.1176/appi.ajgp.13.11.950

Whiteford HA, Ferrari AJ, Baxter AJ, Charlson FJ, Degenhardt L (2013) How did we arrive at burden of disease estimates for mental and illicit drug use disorders in the Global Burden of Disease Study 2010? Curr Opin Psychiatry 26(4):376–383. doi:10.1097/YCO.0b013e328361e60f

World Health Organization (2003) Investing in mental health

Xu J, Xiao G, Trujillo C, Chang V, Blanco L, Joseph SB, Bassilian S, Saad MF, Tontonoz P, Lee WN, Kurland IJ (2002) Peroxisome proliferator-activated receptor alpha (PPARalpha) influences substrate utilization for hepatic glucose production. J Biol Chem 277(52):50237–50244. doi:10.1074/jbc.M201208200

Yang R, Wang P, Chen Z, Hu W, Gong Y, Zhang W, Huang C (2017) WY-14643, a selective agonist of peroxisome proliferator-activated receptor-alpha, ameliorates lipopolysaccharide-induced depressive-like behaviors by preventing neuroinflammation and oxido-nitrosative stress in mice. Pharmacol Biochem Behav 153:97–104. doi:10.1016/j.pbb.2016.12.010

Yu M, Ives D, Ramesha CS (1997) Synthesis of prostaglandin E2 ethanolamide from anandamide by cyclooxygenase-2. J Biol Chem 272(34):21181–21186

Yu HL, Deng XQ, Li YJ, Li YC, Quan ZS, Sun XY (2011) N-palmitoylethanolamide, an endocannabinoid, exhibits antidepressant effects in the forced swim test and the tail suspension test in mice. Pharmacol Rep 63(3):834–839

Yu HL, Sun LP, Li MM, Quan ZS (2015) Involvement of norepinephrine and serotonin system in antidepressant-like effects of oleoylethanolamide in the mice models of behavior despair. Neurosci Lett 593:24–28. doi:10.1016/j.neulet.2015.03.019

Zanaletti R, Bettinetti L, Castaldo C, Cocconcelli G, Comery T, Dunlop J, Gaviraghi G, Ghiron C, Haydar SN, Jow F, Maccari L, Micco I, Nencini A, Scali C, Turlizzi E, Valacchi M (2012) Discovery of a novel alpha-7 nicotinic acetylcholine receptor agonist series and characterization of the potent, selective, and orally efficacious agonist 5-(4-acetyl[1,4]diazepan-1-yl) pentanoic acid [5-(4-methoxyphenyl)-1H-pyrazol-3-yl] amide (SEN15924, WAY-361789). J Med Chem 55(10):4806–4823

Mast Cells and Glia as Targets for the Anandamide Congener Palmitoylethanolamide: an Anti-inflammatory and Neuroprotective Lipid Signaling Molecule

Stephen D. Skaper

Abstract Glia and microglia in particular elaborate pro-inflammatory molecules which play key roles in central nervous system (CNS) disorders from neuropathic pain and epilepsy to neurodegenerative diseases. Microglia respond also to pro-inflammatory signals released from other nonneuronal cells, mainly those of immune origin such as mast cells. The latter can be found in most tissues, are CNS-resident, and traverse the blood–spinal cord and blood–brain barriers when barrier compromise results from CNS pathology. The existence of multiple lines of communication between mast cells and glia may provide new avenues for the development of therapies which target neuroinflammation by differentially modulating activation of those nonneuronal cell populations controlling neuronal sensitization—both peripherally and centrally. Mast cells and glia have "built in" homeostatic mechanisms/molecules that come into play as a result of tissue damage or stimulation of inflammatory responses. Such molecules include the N-acylethanolamine family. One such member, N-palmitoylethanolamine, is proposed to have a key role in the maintenance of cellular homeostasis in the face of external stressors provoking, for example, inflammation. N-Palmitoylethanolamine has been proven efficacious in mast cell-mediated experimental models of acute and neurogenic inflammation. This review will give an overview of current knowledge relating to the pathobiology of neuroinflammation, the role of microglia and mast cells, and the proposal that mast cell–microglia cross talk may exacerbate acute symptoms of chronic neurodegenerative disease and accelerate disease progression, as well as promote pain transmission pathways. We will conclude by considering the therapeutic potential of treating systemic inflammation or the blocking of signaling pathways from the periphery to the brain.

S.D. Skaper (✉)
Department of Pharmaceutical and Pharmacological Sciences, Largo "Egidio Meneghetti" 2, University of Padua, 35131 Padua, Italy
e-mail: stephen.skaper@unipd.it

M. Melis (ed.), *Endocannabinoids and Lipid Mediators in Brain Functions*,
DOI 10.1007/978-3-319-57371-7_12

1 Introduction

Inflammation is the body's attempt at self-protection to remove harmful stimuli (Nathan and Ding 2010) and is part of the immune response. Yet, there are settings when the inflammatory response itself damages host tissue and causes organ dysfunction, for example, when there is an overly robust acute or subacute inflammatory response to pathogens or debris from damaged host cells (Castellheim et al. 2009). In their recent review, Nathan and Ding (2010) clearly point out that the core problem with inflammation is not how often it starts but how often it fails to subside. In fact, few would dispute that non-resolving inflammation is one of the principal contributors to the medical burden in industrialized societies. Inflammation is particularly insidious where the peripheral and central nervous systems are involved ("neuroinflammation"), playing an important role in the pathogenesis of chronic pain and neuropathic pain (Myers et al. 2006), chronic neurodegenerative diseases (Freeman and Ting 2015; Iadecola and Anrather 2011; McGeer and McGeer 2013), neuropsychiatric illness (Castanon et al. 2015; Najjar et al. 2013; Theoharides et al. 2015a), and probably even autism spectrum disorder (Noriega and Savelkoul 2014; Theoharides et al. 2015b). It has even been suggested that neuroinflammation may raise the brain's sensitivity to stress (Rivat et al. 2010).

One of the most significant advances in the neurosciences over the past several decades has been the revelation that an extensive bidirectional communication exists between the immune system and the central nervous system (CNS). Pro-inflammatory cytokines occupy a key position in this highway, acting to regulate host responses to infection, inflammation, and reactions to stress or trauma. Principal protagonists in these actions are glia (astrocytes and microglia—the brain's resident macrophages) and mast cells. Both cell populations constitute important sources of inflammatory mediators and may have cardinal roles in conditions ranging from chronic pain to neurodegenerative diseases and neuropsychiatric disorders (Amor et al. 2014; Appel et al. 2011; Cunningham 2013; Harcha et al. 2015; Lull and Block 2010; Silver and Curley 2013; Skaper et al. 2014a, b; Thacker et al. 2007; Theoharides et al. 2015a, b). This review is intended to provide the reader with an overview of neuroinflammation's role in neurological diseases, followed by a consideration of approaches to counteract neuroinflammation based on endogenous defense mechanisms and lipid signaling molecules.

2 Glia and Mast Cells: Partners in Neuroinflammation

A number of cell types participate in neuroinflammation, including glia and immune system-derived cells, both tissue-resident and blood-borne. This review will focus on glia and mast cells, a choice supported by a large and ever-growing literature. Not only do glial cell activations contribute to neuropathology, but microglia and astrocytes also respond to pro-inflammatory signals released from

Table 1 Mast cells in a nutshell

Origin and classification
• Paul Ehrlich in 1878 first observed these cells on the basis of their unique staining characteristics and large cytoplasmic granules
• Very close to basophil granulocytes in blood; however, appear to be generated by different precursor cells in the bone marrow
• Thought to originate from bone marrow precursors expressing CD34; a distinct subset of mast cells can also be induced upon host responses to inflammation
• Hematopoietic lineage development of tissue mast cells is unique compared to other myeloid-derived cells because it is early lineage progenitors, undetectable by histochemistry, that leave the bone marrow to enter the circulation
• Immature lineage mast cells undergo transendothelial recruitment into peripheral tissues; appearance of secretory granules with a particular protease phenotype is regulated by the peripheral tissue
• Two types of mast cells are recognized, those from connective tissue and a distinct set of mucosal mast cells. Activities of the latter are dependent on T cells
• Present in most tissues close to blood vessels; are especially prominent near boundaries between the body's external environment and the internal milieu, e.g., the skin, mucosa of lungs and digestive tract, as well as in mouth, conjunctiva, and nose
• Found within the nervous system, including the meninges, brain parenchyma, and peripheral nerves
Physiology
• Play a key role in the inflammatory process
• Activation leads to rapid release of granules into the interstitium
• Degranulation can be caused by direct injury (e.g., physical or chemical), cross-linking of IgE receptors, or activated complement proteins
• Capable of elaborating a vast array of important cytokines and other inflammatory mediators
• Express multiple "pattern recognition receptors" thought to be involved in recognizing broad classes of pathogens
• Granules "preloaded" with bioactive chemicals, proteoglycans, serine proteases, and neuropeptides; these can be transferred to adjacent cells of the immune system and neurons via transgranulation and their pseudopodia
Role in disease
• Allergic reactions
• Anaphylactic shock
• Chronic and neuropathic pain
• Acute and chronic neurodegenerative disorders

other cells of immune origin. Mast cells represent a potentially important—and still relatively underappreciated—peripheral immune signaling link to the brain in an inflammatory setting (Table 1). These effector cells of the innate immune system, once activated, secrete numerous vasoactive, neurosensitizing, and pro-inflammatory mediators, which include biogenic amines, cytokines, enzymes, lipid metabolites, ATP, neuropeptides, growth factors, and nitric oxide (Table 2) (Silver and Curley 2013). By nature of their immune regulatory role, mast cells participate in IgE switching by B cells (Gauchat et al. 1993) and the release of chemoattractants that recruit eosinophils (Wardlaw et al. 1986) and monocytes.

Table 2 Mast cell mediators

• Biogenic amines
–Biogenic amines (histamine (2–5 pg/cell), serotonin)
• Cytokines
–Interleukins 1, 4, 6
–Leukemia inhibitory factor
–Tumor necrosis factor-α
–Interferon-γ
–Transforming growth factor-β
–Granulocyte–macrophage colony-stimulating factor
• Enzymes
–Acid hydrolases
–Chymase
–Phospholipases
–Rat mast cell protease I and II
–Tryptase
• Lipid metabolites
–Prostaglandin D2
–Leukotriene C4
–Chemokines (CCXL8, CCL2)
–Platelet-activating factor
• Other bioactive molecules
–Neuropeptides (e.g., vasoactive intestinal peptide, substance P)
–Proteoglycans, mainly heparin (active as an anticoagulant)
–Nerve growth factor
–ATP
–Nitric oxide

Given their oftentimes proximity at sites of inflammation and propensity for reciprocal interaction, the following discussion will integrate glia and mast cells when describing neuropathologies.

2.1 *Neuropathic Pain*

Neuropathic pain results from damage, degeneration, or dysfunction of the sensory nervous system and represents a substantial and growing unmet medical need. Central neuropathic pain accompanies spinal cord injury, multiple sclerosis, and some strokes, while painful peripheral neuropathies frequently occur with diabetes and other metabolic conditions. Neuropathic pain can result also as a direct result of cancer on peripheral nerves or as a side effect of chemotherapy. The triggering and maintenance of neuropathic pain states depends on Schwann cells, spinal microglia, and astrocytes, together with elements of the peripheral immune system (DeLeo and Yezierski 2001). Spinal microglia release interleukin-1β (IL-1β) which may induce phosphorylation of the glutamatergic N-methyl-D-aspartate receptor NR1 subunit to strengthen painful signal transmission (Wei et al. 2007). Under pathological conditions dorsal horn microglia become activated with the upregulation of purinergic receptors (Chessell et al. 2005; Tsuda et al. 2009) which participate in neuropathic pain (Chessell et al. 2005; Ballini et al. 2011).

Mast cells, upon degranulation, release molecules which activate or sensitize nociceptors, thereby contributing directly to neuropathic pain (Xanthos et al. 2011). Peripheral nerve mast cells constitute the first line of activation at the site of damage and promote the recruitment of neutrophils and macrophages (Zuo et al. 2003). Degranulation of mast cells activates also trigemino-cervical and lumbosacral pain pathways to elicit widespread tactile pain hypersensitivity (Levy et al. 2012), possibly mediated by a sensitizing effect of histamine on nociceptors. Further, the rapid release of nerve growth factor (NGF) by mast cells can sensitize nociceptors by engaging the latter's high-affinity TrkA receptors (and indirectly via other peripheral cell types) (Koda and Mizumura 2002). Mast cells themselves respond to NGF in a paracrine/autocrine manner (Leon et al. 1994). Collectively, these events facilitate T-cell recruitment to reinforce and maintain inflammatory reactions. The released mediators/factors can induce activity in axons and/or undergo retrograde transport to the cell body of spinal ganglion neurons and influence gene expression. In addition, mast cells may enhance recruitment of other key immune cell types which, in turn, release pro-nociceptive mediators, such as IL-6 (Leal-Berumen et al. 1994). Glucocorticoid therapy reduces pain and tumor necrosis factor-α (TNF-α)-positive mast cell numbers in rats with chronic constrictive nerve injury, strengthening a role for mast cells in chronic pain states (Hayashi et al. 2011). Mast cells are important mediators of chronic visceral pain, as well (Wood 2011).

2.2 *Acute Brain Injury*

Stroke and traumatic brain injury are characterized by an inflammatory response in which microglia activation and macrophage/neutrophil infiltration are well-recognized players (Wang et al. 2007). Left unchecked, this can ultimately lead to secondary injury. In certain instances, however, attenuating microglial activation can actually be beneficial (Hanisch and Kettenmann 2007). It is thus not surprising that a great deal of effort has been expended on strategies to inhibit the consequences of blood-borne neutrophil and phagocyte infiltration in ischemia. Consequently, brain-resident cell types able to mount an immediate host response in the cerebral parenchyma and meninges (e.g., mast cells) have received much less attention (Silver and Curley 2013). As with peripheral nerve damage—and contrary to long-held beliefs (Chew et al. 2006)—mast cell activation is clearly the "first responder" in this injury (Jin et al. 2009). While CNS microglia/macrophages and endothelial cells may produce TNF-α in response to stimuli, mast cells are "primed" to initiate acute inflammation with their stores of *preformed* TNF-α (Gordon and Galli 1991). Mast cell-stabilizing agents limit brain damage caused by perinatal hypoxia–ischemia and transient focal ischemia (Jin et al. 2007, 2009; Strbian et al. 2007). With their complement of vasoactive and matrix-degrading components and proteases able to activate matrix metalloproteinases, mast cells are well-positioned as an early response element in the regulation of acute blood–brain barrier changes after cerebral ischemia and hemorrhage (Lindsberg et al. 2010). Cerebral mast cells, through their regulation of acute microvascular gelatinase activation, can effect blood–brain barrier disruption following transient cerebral ischemia (Mattila et al. 2011). For a discussion of microglia and mast cell involvement in chronic neurodegenerative diseases, see recent reviews by Skaper et al. (2012, 2014a, b).

3 Microglia and Mast Cells: The Other Side of the Story

Acute CNS injuries are characterized by a prolonged inflammatory response involving microglial activation and infiltration of macrophages and neutrophils. Conceivably, microglia accumulation at the lesion site and penumbra may serve a neuroprotective role. Genetic ablation of microglia aggravates infarct size after middle cerebral artery occlusion (Lalancette-Hébert et al. 2007). Injection of microglia into the bloodstream of Mongolian gerbils is "redirected" to an ischemic hippocampal lesion (probably aided by a compromised blood–brain barrier) and improves neuron cell survival (Imai et al. 2007). Microglia may protect hippocampal neurons from excitotoxicity (Vinet et al. 2012). Resting microglia respond/repair subclinical abnormalities of the brain without a complete activation transformation (Hanisch and Kettenmann 2007) and likely also have a key role in so-called developmental "synaptic pruning" (Paolicelli et al. 2011) as well as adult neurogenesis (Ribeiro Xavier et al. 2015).

Apart from well-recognized roles of activated microglia in disease progression (Kim and de Vellis 2005), they also deliver trophic factors (Kotter et al. 2005), remove myelin debris (Filbin 2003), and recruit oligodendrocyte precursor cells to the lesion site (Fan et al. 2007), thereby supporting myelin regeneration. In the case of Alzheimer disease, Fan et al. (2007) reported that anti-inflammatory drugs which suppress microglial inflammatory response attenuate neurological symptoms in a mouse model; however, other authors have found that reducing or ablating resident microglia fails to alter amyloid plaque load in these transgenic Alzheimer disease mouse models (Grathwohl et al. 2009). Deleting the microglial chemokine receptor CCR2 (which mediates the accumulation of mononuclear phagocytes at sites of inflammation) accelerated early disease progression and impaired microglial accumulation in an Alzheimer disease mouse model (El Khoury et al. 2007). Although Toll-like receptors (TLRs) are generally associated with inflammation, Song et al. (2011) claimed that microglia activation via TLR4 signaling partially reduced amyloid β-peptide deposits and preserved cognitive functions otherwise compromised by amyloid β-peptide-mediated neurotoxicity.

In certain settings, mast cells may exert a beneficial action. For example, mast cells of human origin express and store angiogenin within their granules, which is released upon stimulation by FcεRI-mediated signals, Toll-like receptor ligands, and NGF (Kulka et al. 2009), with NGF being more efficacious compared with FcεRI cross-linking in the release of angiogenin. That human angiogenin is reported to be neuroprotective and to promote the survival and neuritogenesis of motor neurons (Subramanian et al. 2008) is of interest, as recent studies associate angiogenin gene mutations with amyotrophic lateral sclerosis (Gellera et al. 2008) and suggest a possible disease link. Mast cells are a source of serotonin in the hippocampus, which can contribute to behavioral and physiological functions in the hippocampus (Nautiyal et al. 2012). Additionally, in cutaneous tissue mast cells may contribute to wound healing (Weller et al. 2006).

4 Microglia and Mast Cells: A Binary Act

The contribution of mast cells and microglia to neuroinflammation has the potential to be aggravated by the ability of these two cell types to interact with each other. This aspect of nonneuronal cell involvement in neuroinflammation has been discussed in detail earlier (Skaper et al. 2012); here we will only describe several representative examples of such interactions. Consider, for example, TLRs. The latter comprise one of the principal classes of pathogen-associated molecular patterns, that is, molecules associated with groups of pathogens recognized by innate immune system cells (such as microglia and mast cells). Ligand engagement of mast cell surface TLR2/TLR4 triggers cytokine release which recruits immune cells to the sites of injury; microglia recruitment depends on signaling pathways involving TLR2/TLR4 (Pietrzak et al. 2011). Mast cell activation upregulates also chemokine expression (including CCL5/RANTES); these chemokines can induce a pro-inflammatory response in microglia. Further, microglia-derived IL-6 and CCL5 may, in turn, affect mast cell expression of TLR2/TLR4. ATP, while generally thought of as an energy source, may act as an autocrine/paracrine factor for mast cells: ATP released from one mast cell (e.g., FcεR1 cross-linking, stress) is capable of diffusing several hundred micrometers to provoke a rise in Ca^{2+} in neighboring cells (Osipchuk and Cahalan 1992). ATP, by binding to P2 receptors, stimulates the release of IL-33 from microglia pre-activated ("primed") with pathogen-associated molecular patterns (Chakraborty et al. 2010), which then induces mast cells to secrete IL-6, IL-13, and CCL2 which then modulate microglia activity. In effect, this creates a continual cycle. Mast cell tryptase cleavage/activation of proteinase-activated receptor 2 (Bushell 2007) on microglia upregulates $P2X_4$ receptor and brain-derived neurotrophic factor release (Trang et al. 2009), while TNF-α and IL-6 from microglia upregulate proteinase-activated receptor 2 expression by mast cells, leading to mast cell activation and TNF-α release (Zhang et al. 2010a, b). Elements

Table 3 Potential avenues of mast cell–glia communication

	Biological actions		
Effector	Microglia/astrocytes	Mast cells	References
ATP/P2 receptors	ATP stimulates IL-33 release from microglia pre-activated with pathogen-associated molecular patterns via TLRs	IL-33 binds to its receptor on mast cells and induces secretion of IL-6, IL-13, and monocyte chemoattractant protein 1 which, in turn, could modulate microglia activity	Burnstock et al. (2011), Osipchuk and Cahalan (1992)
Proteinase-activated receptor 2 (PAR2)	Mast cell tryptase cleaves/activates PAR2 on microglia, resulting in $P2X_4$ upregulation and release of brain-derived neurotrophic factor	IL-6 and TNF-α from microglia upregulate mast cell expression of PAR2, leading to mast cell activation and TNF-α release	Zhang et al. (2010a, b, 2012), Osipchuk and Cahalan (1992)
TLR2, TLR4	IL-6 and CCL5 released from microglia affect surface expression of TLR2/TLR4 on mast cells	Upregulation of cytokine/chemokine release; CCL5/RANTES induces pro-inflammatory profile in microglia; recruitment of immune cells to site of injury	Buchanan et al. (2010), Kim et al. (2007), Liu et al. (2012), Orinska et al. (2005), Pietrzak et al. (2011), Skuljec et al. (2011), Tanga et al. (2005)
CXCR4/ CXCL12	Promotes microglia migration and activation; CXCR4 and CXCL12 are both upregulated in hypoxia/ischemia	CXCR4 acts as mast cell chemotaxin	Juremalm et al. (2000), Knerlich-Lukoschus et al. (2011), Yang et al. (2010)
C5a receptor (C5aR)	C5aR upregulated by microglia activation; C5a peptide released in neuroinflammation; cross talk between C5a and TLR4	C5aR upregulated upon activation; a strong mast cell chemoattractant signal toward C5a peptide; cross talk between C5a and TLR4	Gasque et al. (1997), Griffin et al. (2007), Soruri et al. (2008)
CD40/ CD40L	Astrocyte-enhanced expression of CD40—cross talk with CD40L leads to production of inflammatory cytokines; astrocyte-derived cytokines/chemokines trigger mast cell degranulation	Enhanced surface expression of CD40L—cross talk with CD40 leads to production of inflammatory cytokines	Kim et al. (2011)

(continued)

Table 3 (continued)

	Biological actions		
Effector	Microglia/astrocytes	Mast cells	References
Translocator protein (TSPO)	Upregulated in retinal microglia in retinal inflammation/injury	The TSPO endogenous ligand diazepam-binding inhibitor (DBI) upregulated in microglia; DBI–TSPO signaling can promote microglia–microglia interactions which influence mast cell activity	Wang et al. (2014)

IL interleukin, *PAR* proteinase-activated receptor, *TLR* Toll-like receptor, *TNF-α* tumor necrosis factor-α

of microglia–mast cell cross talk appear to engage even the complement system. The chemoattractant C5a and its receptor are upregulated on reactive astrocytes and microglia in inflamed CNS tissue (Gasque et al. 1997), and C5a receptor on activated mast cells is a strong chemoattractant signal toward C5a peptide. Moreover, there is evidence to suggest mast cell–astrocyte communication at the level of CD40L/CD40 (Kim et al. 2011), as well as microglia–astrocyte interaction at the level of translocator protein, a marker of gliosis in neurodegeneration (Wang et al. 2014). Conceivably, these last findings propose an additional element whereby microglia and mast cells may work in concert to promote neuroinflammation (Table 3).

5 Microglia and Mast Cells as Therapeutic Targets for Neuroinflammation

Pharmacological attenuation of mast cell and microglial activation is viewed as an attractive therapeutic approach for neuropathic pain and other neuropathological conditions associated with inflammation (Gosselin et al. 2010; Mika 2008; Oliveira et al. 2011; Suk and Ock 2011; Xanthos et al. 2011). Attention has been directed also to therapeutic strategies aimed at inhibiting neurotoxic glial cell activation (Ralay Ranaivo et al. 2006).

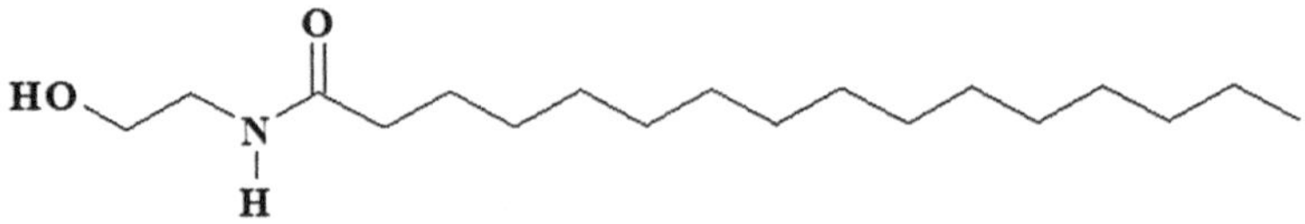

Fig. 1 Chemical structure of N-palmitoylethanolamine

5.1 Palmitoylethanolamide: A Wide-Acting Anti-inflammatory and Neuroprotective N-Acylethanolamine

In addition to more "traditional" medicinal chemistry approaches to develop molecules for controlling neuroinflammation, we should be aware of natural agents which are part and parcel of endogenous protective mechanisms activated following tissue damage or stimulation of inflammatory responses. Indeed, a number of molecules have so far been identified which take part in these protective mechanisms, whereby chronic inflammatory processes may trigger a program of resolution that encompasses the production of lipid mediators with the capacity to switch off inflammation (Buckley et al. 2013; Piomelli and Sasso 2014). One family of such molecules worth taking note of are the N-acylethanolamines, a class of naturally occurring lipid signaling molecules composed of a fatty acid and ethanolamine—the fatty acid ethanolamides (FAEs). Principal FAE family members include the endocannabinoid N-arachidonoylethanolamine (anandamide) and its congeners N-stearoylethanolamine, N-oleoylethanolamine, and N-palmitoylethanolamine (PEA) (chemical name: N-(2-hydroxyethyl)hexadecanamide) (Pacher et al. 2006). The PEA chemical structure is shown in Fig. 1. It is intriguing that, apart from its abundance in mammalian brain, PEA (and other NAEs) has been found even in some marine species (Bisogno et al. 1997b; Sepe et al. 1998) as well as in the CNS of the leech *Hirudo medicinalis* (Matias et al. 2001).

PEA and its congeners can be generated from N-acylated phosphatidylethanolamine (NAPE) by more than one enzymatic pathway (Rahman et al. 2014; Ueda et al. 2013), although this principally involves a membrane-associated NAPE-phospholipase D which produces the respective NAE and phosphatidic acid (Fig. 2) (Leung et al. 2006) and includes the conversion of N-palmitoyl-phosphatidylethanolamine into PEA. In the mammalian brain, NAEs are hydrolyzed by (1) an endoplasmic reticulum fatty acid amide hydrolase (FAAH) which breaks down NAEs into the corresponding fatty acid and ethanolamine (Cravatt et al. 1996) and (2) a lysosomal NAE-hydrolyzing acid amidase (NAAA) (Tsuboi et al. 2007) (Fig. 2). NAAA is mainly located in macrophages, where it hydrolyzes NAEs shorter than 18 carbon atoms like PEA. In contrast, FAAH hydrolyzes PEA, N-stearoylethanolamine, and N-oleoylethanolamine.

The idea that PEA plays a role in maintaining cellular homeostasis by acting as mediator of resolution of inflammatory processes draws support from a number of studies. For example, consider that PEA is produced/hydrolyzed by microglia and

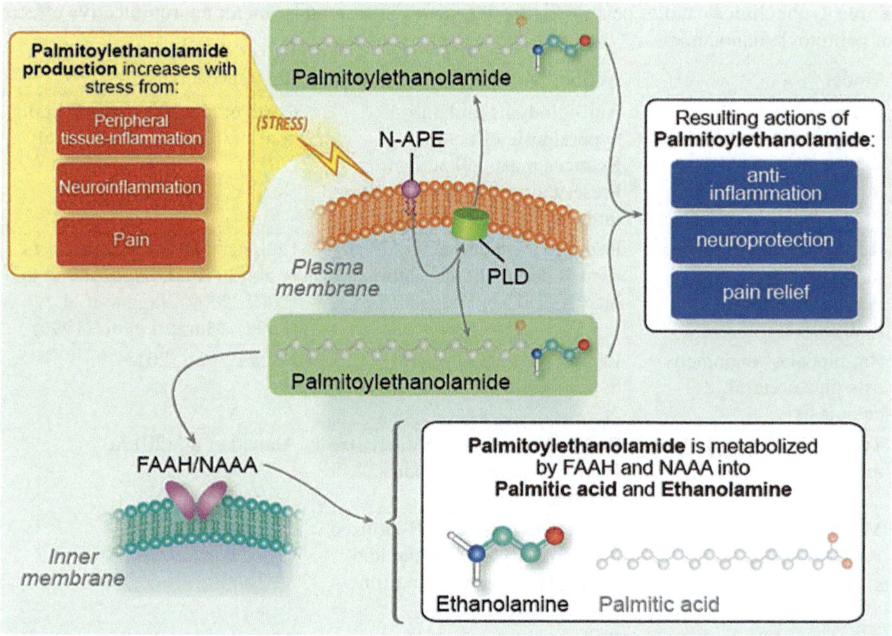

Fig. 2 Palmitoylethanolamide synthesis and metabolism. A plasma membrane-associated N-acylated phosphatidylethanolamine phospholipase D (PLD) converts N-palmitoyl-phosphatidylethanolamine (NAPE) into palmitoylethanolamide and phosphatidic acid. Palmitoylethanolamide is metabolized to palmitic acid and ethanolamine by both a relatively nonselective fatty acid amide hydrolase (FAAH) and the more selective N-acylethanolamine-hydrolyzing acid amidase (NAAA). Tissue levels of palmitoylethanolamide increase in stressful settings such as peripheral tissue inflammation, neuroinflammation, and pain [Reprinted from *Immunology, 141(3)*, S.D. Skaper, L. Facci, P. Giusti, Mast cells, glia and neuroinflammation: partners in crime?, 314–327 (Fig. 2). Copyright (2014), with permission from John Wiley and Sons]

mast cells (Bisogno et al. 1997a; Muccioli and Stella 2008), PEA down-modulates mast cell activation (Facci et al. 1995), and it controls microglial cell behaviors (Franklin et al. 2003; Luongo et al. 2013). Further, PEA levels are raised in brain areas involved in nociception and in spinal cord following induction of neuropathic pain (Petrosino et al. 2007), in conditions associated with pain development (Franklin et al. 2003; Jhaveri et al. 2008), in spinal cord of spastic mice with chronic relapsing experimental allergic encephalomyelitis (Baker et al. 2001), and in interstitium of trapezius muscle of women with chronic widespread pain and chronic neck–shoulder pain (Ghafouri et al. 2013).

The last two decades have seen an ever-expanding database dedicated to the anti-neuroinflammatory and neuroprotective actions of PEA at the preclinical level (Alhouayek and Muccioli 2014; Esposito et al. 2014; Fidaleo et al. 2014; Mattace Raso et al. 2014; Skaper et al. 1996). The principle findings reported in Table 4 clearly illustrate the diversity of models and neuropathological conditions where

Table 4 Preclinical studies demonstrating anti-neuroinflammatory and/or neuroprotective effects of palmitoylethanolamide

Model	Action	References
Chronic constriction injury in sciatic nerve	Anti-allodynic and anti-hyperalgesic effects Reduces mast cell activation Preservation of nerve structural integrity	Costa et al. (2008), De Filippis et al. (2011), Di Cesare et al. (2013)
Acute inflammation (formalin, dextran, carrageenan injection in rat hindpaw)	Reduces mast cell activation, tissue edema, inflammatory/ mechanical hyperalgesia	Calignano et al. (1998), Costa et al. (2002), D'Agostino et al. (2007, 2009), Jaggar et al. (1998), Mazzari et al. (1996)
Rat model of endometriosis plus ureteral calculosis	Reduces viscero-visceral hyperalgesia	Iuvone et al. (2015)
Traumatic brain injury in mice	Reduces edema and infarct size Improves neurobehavior functions	Ahmad et al. (2012a)
MPTP mouse model of Parkinson disease	Protects against MPTP-induced neurotoxicity, microglial and astrocyte activation, and functional deficits	Esposito et al. (2012)
Stroke (middle cerebral artery occlusion in rats)	Reduces edema and infarct size Improves neurobehavior functions	Ahmad et al. (2012b)
β-Amyloid peptide injection in rat brain	Counteracts reactive gliosis Reduces behavior impairments	Scuderi et al. (2011)
β-Amyloid-induced astrogliosis	Regulates production of pro-angiogenic mediators	Cipriano et al. (2015)
WAG/Rij rat model of absence epilepsy	Antiepileptic action	Citraro et al. (2013)

MPTP 1-methyl-4-phenyl-1,2,3,6-tetrahydropyridine

PEA was found to be pharmacologically efficacious. The case of chronic pelvic pain in women related to endometriosis is of particular interest, as this presents an example of what one might describe as "reverse" translational science: a newly published study by Iuvone et al. (2015) demonstrates the ability of ultramicronized PEA to significantly reduce viscero-visceral hyperalgesia in a rat model of endometriosis plus ureteral calculosis with mast cell involvement—in confirmation of clinical studies carried out several years before (Cobellis et al. 2011; Lo Monte et al. 2013).

The bulk of PEA studies in man have been conducted to address the molecule's effectiveness against chronic pain and neuropathic pain associated with a variety of etiopathogeneses. This review will not treat this aspect of PEA action, and the reader is referred instead to several recent reports (Hesselink and Hekker 2012; Paladini et al. 2016; Skaper et al. 2014a, b). In particular, the report by Paladini et al. (2016) presents a pooled data meta-analysis of 1484 patients with chronic pain, 1188 of which were treated with micronized or ultramicronized PEA for periods of 21–60 days with daily doses ranging from 300 to 1200 mg. None of the clinical trials reported treatment-related adverse events.

Inhibiting PEA degradation by targeting its principal catabolic enzyme, NAE-hydrolyzing acid amidase (NAAA) represents a potential alternative route in the treatment of neuroinflammation. A chemically diverse range of selective NAAA inhibitors have been published (Solorzano et al. 2009; Saturnino et al. 2010; Duranti et al. 2012; Li et al. 2012; Yamano et al. 2012; Yang et al. 2015; Sasso et al. 2013; Vitale et al. 2014) including systemically active compounds which are able to modulate responses induced by inflammatory stimuli in vivo and in vitro (Solorzano et al. 2009; Li et al. 2012; Sasso et al. 2013; Ribeiro et al. 2015) and also elevate PEA levels (Solorzano et al. 2009; Li et al. 2012; Yang et al. 2015). The most recent NAAA inhibitors report single-digit nanomolar intracellular activity ($IC_{50} = 7$ nM) on both the rat and human enzyme (Sasso et al. 2013).

Pharmacological block of FAAH and/or NAAA with nonphysiological agents as a strategy to increase endogenous N-acylethanolamine levels, in particular PEA, may present certain drawbacks from a metabolic point of view. One needs to keep in mind that NAAA substrates such as PEA are produced on demand, rather than being constitutive. As such, enzymes like NAAA are probably intended to *modulate* substrate availability. Indeed, several studies suggest that genetic deletion or pharmacological block of FAAH may actually result in paradoxical effects (Siegmund et al. 2013; Hoyer et al. 2014), including β-amyloid peptide exacerbation of inflammation in astrocytes lacking FAAH (Benito et al. 2012) and altered neural proliferation, apoptosis, and gliosis in a brain region-dependent manner, in a negative energy context (Rivera et al. 2015). This could be a consequence of the fact that, in order to exert a regulatory effect on nonneuronal cells (e.g., glia and mast cells) implicated in neuroinflammation, PEA pleiotropic effects (Skaper and Facci 2012) should thus be tightly controlled by a mechanism allowing for their inactivation. Conceivably, protracted block of FAAH might also alter endocannabinoid system tone by reducing levels of 2-arachidonoylglycerol while increasing those of anandamide. We have thus proposed that pharmacologically modulating—and not blocking—the specific amidases for N-acylamides and in particular NAAA could be a viable strategy which preserves the PEA role in maintaining cellular homeostasis through its rapid on-demand synthesis and equally rapid degradation (Skaper et al. 2015).

PEA is an agonist for peroxisome proliferator-activated receptor (PPAR)α, a member of a group of lipid-activated nuclear receptor proteins that function as transcription factors exerting numerous functions in development and metabolism (Alhouayek and Muccioli 2014; Fidaleo et al. 2014; Lo Verme et al. 2005). PPARα- and γ-isoforms especially appear to be associated with inflammatory events. PEA anti-inflammatory, antinociceptive/anti-neuropathic, and neuroprotective actions are either absent in PPARα-null mice or blocked by PPARα antagonists (Fidaleo et al. 2014; Skaper et al. 2014a, b). It has also been suggested that PEA exerts a so-called entourage effect, whereby the lipid mediator enhances the anti-inflammatory and antinociceptive activity of other endogenous compounds by potentiating their affinity for a receptor or by inhibiting their metabolic degradation (Smart et al. 2002). In this regard anandamide is a candidate molecule, given its anti-inflammatory and antinociceptive effects. Anandamide and its congeners like PEA may act also through the transient receptor potential vanilloid type 1 (TRPV1) receptor, a nonselective cation channel expressed in small diameter sensory neurons which is activated by noxious heat, low pH, and capsaicin. Anandamide is a TRPV1 receptor agonist, while PEA enhances anandamide stimulation of the human TRPV1 receptor (De Petrocellis et al. 2001) in a cannabinoid CB2 receptor antagonist-sensitive fashion. Because PEA has no appreciable affinity for either CB1 or CB2 receptors, this latter effect could result from PEA acting indirectly by potentiating anandamide actions (Calignano et al. 1998). Mast cells (Bíró et al. 1998) and microglia (Kim et al. 2006) are reported to express TRPV1 receptors.

6 Concluding Remarks

Inflammatory signaling molecules derived from the innate and adaptive immune systems, together with CNS glia, can profoundly affect a wide range of CNS functions. Microglia, acting as sensors for disturbed brain tissue homeostasis, accumulate locally in response to neuronal cell injury or entry of foreign material into the brain. We still know relatively little about resident brain cell types capable of mounting immediate host responses in the brain and meninges. Mast cells, rather than microglia, represent the "first responders" to injury rather than microglia (Jin et al. 2009). Degranulation/mediator release by the former is very rapid, while longer-lasting activation elaborates de novo-formed mediators. In spite of research efforts to date, gaps remain in our knowledge of glial and mast cell changes in human chronic pain, especially a clear demonstration that glial and mast cell activation occurs in hypersensitized patients. We lack also systematic studies which provide a correlation between the magnitude of glial and/or mast cell markers in cerebrospinal fluid or spinal tissue and the intensity of pain in patients.

In spite of the fact that we have at our disposition a well-stocked "medicine cabinet" (antidepressants, anticonvulsants, sodium channel blockers, glutamate receptor antagonists, opioids) to combat chronic and neuropathic pain, these agents treat the symptoms rather than the underlying pathophysiology. Unfortunately, at best they provide transient relief in only a fraction of neuropathic pain patients and can lead to serious CNS side effects. Incomplete understanding of mechanisms underlying the induction and maintenance of neuropathic pain has hindered more effective treatments, including elucidating the role of mast cells and how they might interact with microglia.

Without doubt, we still have much to learn about signaling mechanisms that regulate neuroinflammation. Targeting endogenous regulators of neuroinflammation may be one strategy for impinging on a diverse array of nervous system disorders. In this context it may be useful also to investigate the role of mast cells in inflammatory diseases as a network, which would call for a critical examination of tissue localization, function, and dynamic interaction with endogenous cells.

The production of PEA during inflammatory conditions supports its role as part of a complex homeostatic system controlling the basal threshold of both inflammation and pain. That selective inhibition of PEA degradation is anti-inflammatory further supports the view of PEA involvement in the control of pain and inflammation. One might also ask if we are not missing important therapeutic avenues by studying glia and mast cells as isolated players in neuroinflammation. PEA, its analogues, and its agents that inhibit its degradation are likely to further the development of novel therapeutic strategies for the treatment of neuropathological conditions.

Acknowledgments This study was supported in part by MIUR, PON "Ricerca e Competitività 2007–2013" project PON01_02512 and by Regione Veneto project protocol 103173COF/14/LR52001C2/000051.

Conflict of Interest The authors report no conflicts of interest.

References

Ahmad A, Crupi R, Impellizzeri D, Campolo M, Marino A, Esposito E, Cuzzocrea S (2012a) Administration of palmitoylethanolamide (PEA) protects the neurovascular unit and reduces secondary injury after traumatic brain injury in mice. Brain Behav Immun 26:1310–1321

Ahmad A, Genovese T, Impellizzeri D, Crupi R, Velardi E, Marino A, Esposito E, Cuzzocrea S (2012b) Reduction of ischemic brain injury by administration of palmitoylethanolamide after transient middle cerebral artery occlusion in rats. Brain Res 1477:45–58

Alhouayek M, Muccioli GG (2014) Harnessing the anti-inflammatory potential of palmitoylethanolamide. Drug Discov Today 19:1632–1639

Amor S, Peferoen LA, Vogel DY, Breur M, van der Valk P, Baker D, van Noort JM (2014) Innate and adaptive immune responses in neurodegeneration and repair. Immunology 141:287–291

Appel SH, Zhao W, Beers DR, Henkel JS (2011) The microglial-motoneuron dialogue in ALS. Acta Myol 30:4–8

Baker D, Pryce G, Croxford JL, Brown P, Pertwee RG, Makriyannis A, Khanolkar A, Layward L, Fezza F, Bisogno T, Di Marzo V (2001) Endocannabinoids control spasticity in a multiple sclerosis model. FASEB J 15:300–302

Ballini E, Virginio C, Medhurst SJ, Summerfield SG, Aldegheri L, Buson A, Carignani C, Chen YH, Giacometti A, Lago I, Powell AJ, Jarolimek W (2011) Characterization of three diaminopyrimidines as potent and selective antagonists of P2X3 and P2X2/3 receptors with in vivo efficacy in a pain model. Br J Pharmacol 163:1315–1325

Benito C, Tolón RM, Castillo AI, Ruiz-Valdepeñas L, Martínez-Orgado JA, Fernández-Sánchez FJ, Vázquez C, Cravatt BF, Romero J (2012) β-Amyloid exacerbates inflammation in astrocytes lacking fatty acid amide hydrolase through a mechanism involving PPAR-α, PPAR-γ and TRPV1, but not CB_1 or CB_2 receptors. Br J Pharmacol 166:1474–1489

Bíró T, Maurer M, Modarres S, Lewin E, Brodie C, Acs G, Acs P, Paus R, Blumberg PM (1998) Characterization of functional vanilloid receptors expressed by mast cells. Blood 91:1332–1340

Bisogno T, Maurelli S, Melck D, De Petrocellis L, Di Marzo V (1997a) Biosynthesis, uptake, and degradation of anandamide and palmitoylethanolamide in leukocytes. J Biol Chem 272:3315–3323

Bisogno T, Ventriglia M, Milone A, Mosca M, Cimino G, Di Marzo V (1997b) Occurrence and metabolism of anandamide and related acyl-ethanolamides in ovaries of the sea urchin Paracentrotus lividus. Biochim Biophys Acta 1345:338–348

Buchanan MM, Hutchinson M, Watkins LR, Yin H (2010) Toll-like receptor 4 in CNS pathologies. J Neurochem 114:13–27

Buckley CD, Gilroy DW, Serhan CN, Stockinger B, Tak PP (2013) The resolution of inflammation. Nat Rev Immunol 13:59–66

Burnstock G, Krügel U, Abbracchio MP, Illes P (2011) Purinergic signalling: from normal behaviour to pathological brain function. Prog Neurobiol 95:229–274

Bushell T (2007) The emergence of proteinase-activated receptor-2 as a novel target for the treatment of inflammation-related CNS disorders. J Physiol 581(Pt 1):7–16

Calignano A, La Rana G, Giuffrida A, Piomelli D (1998) Control of pain initiation by endogenous cannabinoids. Nature 394:277–281

Castanon N, Luheshi G, Layé S (2015) Role of neuroinflammation in the emotional and cognitive alterations displayed by animal models of obesity. Front Neurosci 9:229. doi:10.3389/fnins.2015.00229

Castellheim A, Brekke OL, Espevik T, Harboe M, Mollnes TE (2009) Innate immune responses to danger signals in systemic inflammatory response syndrome and sepsis. Scand J Immunol 69:479–491

Chakraborty S, Kaushik DK, Gupta M, Basu A (2010) Inflammasome signaling at the heart of central nervous system pathology. J Neurosci Res 88:1615–1631

Chessell IP, Hatcher JP, Bountra C, Michel AD, Hughes JP, Green P, Egerton J, Murfin M, Richardson J, Peck WL, Grahames CB, Casula MA, Yiangou Y, Birch R, Anand P, Buell GN (2005) Disruption of the P2X7 purinoceptor gene abolishes chronic inflammatory and neuropathic pain. Pain 114:386–396

Chew LJ, Takanohashi A, Bell M (2006) Microglia and inflammation: impact on developmental brain injuries. Ment Retard Dev Disabil Res Rev 12:105–112

Cipriano M, Esposito G, Negro L, Capoccia E, Sarnelli G, Scuderi C, Filippis DD, Steardo L, Iuvone T (2015) Palmitoylethanolamide regulates production of pro-angiogenic mediators in a model of β amyloid-induced astrogliosis in vitro. CNS Neurol Disord Drug Targets 14:828–837

Citraro R, Russo E, Scicchitano F, van Rijn CM, Cosco D, Avagliano C, Russo R, D'Agostino G, Petrosino S, Guida F, Gatta L, van Luijtelaar G, Maione S, Di Marzo V, Calignano A, De Sarro G (2013) Antiepileptic action of N-palmitoylethanolamine through CB1 and PPAR-α receptor activation in a genetic model of absence epilepsy. Neuropharmacology 69:115–126

Cobellis L, Castaldi MA, Giordano V, Trabucco E, De Franciscis P, Torella M, Colacurci N (2011) Effectiveness of the association micronized N-Palmitoylethanolamine (PEA)-transpolydatin in the treatment of chronic pelvic pain related to endometriosis after laparoscopic assessment: a pilot study. Eur J Obstet Gynecol Reprod Biol 158:82–86

Costa B, Conti S, Giagnoni G, Colleoni M (2002) Therapeutic effect of the endogenous fatty acid amide, palmitoylethanolamide, in rat acute inflammation: inhibition of nitric oxide and cyclo-oxygenase systems. Br J Pharmacol 137:413–120

Costa B, Comelli F, Bettoni I, Colleoni M, Giagnoni G (2008) The endogenous fatty acid amide, palmitoylethanolamide, has anti-allodynic and anti-hyperalgesic effects in a murine model of neuropathic pain: involvement of CB1, TRPV1 and PPARγ receptors and neurotrophic factors. Pain 139:541–550

Cravatt BF, Giang DK, Mayfield SP, Boger DL, Lerner RA, Gilula NB (1996) Molecular characterization of an enzyme that degrades neuromodulatory fatty-acid amides. Nature 384:83–87

Cunningham C (2013) Microglia and neurodegeneration: the role of systemic inflammation. Glia 61:71–90

D'Agostino G, La Rana G, Russo R, Sasso O, Iacono A, Esposito E, Raso GM, Cuzzocrea S et al (2007) Acute intracerebroventricular administration of palmitoylethanolamide, an endogenous peroxisome proliferator-activated receptor-alpha agonist, modulates carrageenan-induced paw edema in mice. J Pharmacol Exp Ther 322:1137–1143

D'Agostino G, La Rana G, Russo R, Sasso O, Iacono A, Esposito E, Mattace Raso G, Cuzzocrea S, Loverme J, Piomelli D, Meli R, Calignano A (2009) Central administration of palmitoylethanolamide reduces hyperalgesia in mice via inhibition of NF-κB nuclear signalling in dorsal root ganglia. Eur J Pharmacol 613:54–59

De Filippis D, Luongo L, Cipriano M, Palazzo E, Cinelli MP, de Novellis V, Maione S, Iuvone T (2011) Palmitoylethanolamide reduces granuloma-induced hyperalgesia by modulation of mast cell activation in rats. Mol Pain 10:3. doi:10.1186/1744-8069-7-3

De Petrocellis L, Davis JB, Di Marzo V (2001) Palmitoylethanolamide enhances anandamide stimulation of human vanilloid VR1 receptors. FEBS Lett 506:253–256

DeLeo JA, Yezierski RP (2001) The role of neuroinflammation and neuroimmune activation in persistent pain. Pain 90:1–6

Di Cesare ML, D'Agostino G, Pacini A, Russo R, Zanardelli M, Ghelardini C, Calignano A (2013) Palmitoylethanolamide is a disease-modifying agent in peripheral neuropathy: pain relief and neuroprotection share a PPAR-alpha-mediated mechanism. Mediators Inflamm 2013:328797. doi:10.1155/2013/328797

Duranti A, Tontini A, Antonietti F, Vacondio F, Fioni A, Silva C, Lodola A, Rivara S, Solorzano C, Piomelli D, Tarzia G, Mor M (2012) N-(2-oxo-3-oxetanyl)carbamic acid esters as N-acylethanolamine acid amidase inhibitors: synthesis and structure-activity and structure-property relationships. J Med Chem 55:4824–4836

El Khoury J, Toft M, Hickman SE, Means TK, Terada K, Geula C, Luster AD (2007) Ccr2 deficiency impairs microglial accumulation and accelerates progression of Alzheimer-like disease. Nat Med 13:432–438

Esposito E, Impellizzeri D, Mazzon E, Paterniti I, Cuzzocrea S (2012) Neuroprotective activities of palmitoylethanolamide in an animal model of Parkinson's disease. PLoS One 7(8):e41880. doi:10.1371/journal.pone.0041880

Esposito E, Cordaro M, Cuzzocrea S (2014) Roles of fatty acid ethanolamides (FAE) in traumatic and ischemic brain injury. Pharmacol Res 86:26–31

Facci L, Dal Toso R, Romanello S, Buriani A, Skaper SD, Leon A (1995) Mast cells express a peripheral cannabinoid receptor with differential sensitivity to anandamide and palmitoylethanolamide. Proc Natl Acad Sci USA 92:3376–3380

Fan R, Xu F, Previti ML, Davis J, Grande AM, Robinson JK, Van Nostrand WE (2007) Minocycline reduces microglial activation and improves behavioral deficits in a transgenic model of cerebral microvascular amyloid. J Neurosci 27:3057–3063

Fidaleo M, Fanelli F, Ceru MP, Moreno S (2014) Neuroprotective properties of peroxisome proliferator-activated receptor alpha (PPARα) and its lipid ligands. Curr Med Chem 21:2803–2821

Filbin MT (2003) Myelin-associated inhibitors of axonal regeneration in the adult mammalian CNS. Nat Rev Neurosci 4:703–713

Franklin A, Parmentier-Batteur S, Walter L, Greenberg DA, Stella N (2003) Palmitoylethanolamide increases after focal cerebral ischemia and potentiates microglial cell motility. J Neurosci 23:7767–7775

Freeman LC, Ting JP (2015) The pathogenic role of the inflammasome in neurodegenerative diseases. J Neurochem. doi:10.1111/jnc.13217

Gasque P, Singhrao SK, Neal JW, Götze O, Morgan BP (1997) Expression of the receptor for complement C5a (CD88) is up-regulated on reactive astrocytes, microglia, and endothelial cells in the inflamed human central nervous system. Am J Pathol 150:31–41

Gauchat JF, Henchoz S, Mazzei G, Aubry JP, Brunner T, Blasey H, Life P, Talabot D, Flores-Romo L, Thompson J, Kishi K, Butterfield J, Dahinden C, Bonnefoy J-Y (1993) Induction of human IgE synthesis in B cells by mast cells and basophils. Nature 365:340–343

Gellera C, Colombrita C, Ticozzi N, Castellotti B, Bragato C, Ratti A, Taroni F, Silani V (2008) Identification of new ANG gene mutations in a large cohort of Italian patients with amyotrophic lateral sclerosis. Neurogenetics 9:33–40

Ghafouri N, Ghafouri B, Larsson B, Stensson N, Fowler CJ, Gerdle B (2013) Palmitoylethanolamide and stearoylethanolamide levels in the interstitium of the trapezius muscle of women with chronic widespread pain and chronic neck-shoulder pain correlate with pain intensity and sensitivity. Pain 154:1649–1658

Gordon JR, Galli SJ (1991) Release of both preformed and newly synthesized tumor necrosis factor alpha (TNF-alpha)/cachectin by mouse mast cells stimulated via the Fc epsilon RI. A mechanism for the sustained action of mast cell-derived TNF-alpha during IgE-dependent biological responses. J Exp Med 174:103–107

Gosselin RD, Suter MR, Ji RR, Decosterd I (2010) Glial cells and chronic pain. Neuroscientist 16:519–531

Grathwohl SA, Kälin RE, Bolmont T, Prokop S, Winkelmann G, Kaeser SA, Odenthal J, Radde R, Eldh T, Gandy S, Aguzzi A, Staufenbiel M, Mathews PM, Wolburg H, Heppner FL, Jucker M (2009) Formation and maintenance of Alzheimer's disease β-amyloid plaques in the absence of microglia. Nat Neurosci 12:1361–1363

Griffin RS, Costigan M, Brenner GJ, Ma CH, Scholz J, Moss A, Allchorne AJ, Stahl GL et al (2007) Complement induction in spinal cord microglia results in anaphylatoxin C5a-mediated pain hypersensitivity. J Neurosci 27:8699–8708

Hanisch U-K, Kettenmann H (2007) Microglia: active sensor and versatile effector cells in the normal and pathologic brain. Nat Neurosci 10:1387–1394

Harcha PA, Vargas A, Yi C, Koulakoff AA, Giaume C, Sáez JC (2015) Hemichannels are required for amyloid β-peptide-induced degranulation and are activated in brain mast cells of APPswe/PS1dE9 mice. J Neurosci 35:9526–9538

Hayashi R, Xiao W, Kawamato M, Yuge O, Bennett GJ (2011) Systemic glucocorticoid therapy reduces pain and the number of endoneurial tumor necrosis factor-alpha (TNFα)-positive mast cells in rats with a painful peripheral neuropathy. J Pharmacol Sci 106:559–565

Hesselink JM, Hekker TA (2012) Therapeutic utility of palmitoylethanolamide in the treatment of neuropathic pain associated with various pathological conditions: a case series. J Pain Res 5:437–442

Hoyer FF, Khoury M, Slomka H, Kebschull M, Lerner R, Lutz B, Schott H, Lütjohann D, Wojtalla A, Becker A, Zimmer A, Nickenig G (2014) Inhibition of endocannabinoid-degrading enzyme fatty acid amide hydrolase increases atherosclerotic plaque vulnerability in mice. J Mol Cell Cardiol 66:126–1232

Iadecola C, Anrather J (2011) The immunology of stroke: from mechanisms to translation. Nat Med 17:796–808

Imai F, Suzuki H, Oda J, Ninomiya T, Ono K, Sano H, Sawada M (2007) Neuroprotective effect of exogenous microglia in global brain ischemia. J Cereb Blood Flow Metab 27:488–500

Iuvone T, Affaitati G, De Filippis D, Lopopolo M, Grassia G, Lapenna D, Negro L, Costantini R, Vaia M, Cipollone F, Ialenti A, Giamberardino MA (2015) Ultramicronized palmitoylethanolamide reduce viscero-visceral hyperalgesia in a rat model of endometriosis plus ureteral calculosis: role of mast cells. Pain. doi:10.1097/j.pain.0000000000000220

Jaggar SI, Hasnie FS, Sellaturay S, Rice AS (1998) The antihyperalgesic actions of the cannabinoid anandamide and the putative CB2 receptor agonist palmitoylethanolamide in visceral and somatic inflammatory pain. Pain 76:189–199

Jhaveri MD, Richardson D, Robinson I, Garle MJ, Patel A, Sun Y, Sagar DR, Bennett AJ, Alexander SP, Kendall DA, Barrett DA, Chapman V (2008) Inhibition of fatty acid amide hydrolase and cyclooxygenase-2 increases levels of endocannabinoid related molecules and

produces analgesia via peroxisome proliferator-activated receptor-alpha in a model of inflammatory pain. Neuropharmacology 55:85–93
Jin Y, Silverman AJ, Vannucci SJ (2007) Mast cell stabilization limits hypoxic-ischemic brain damage in the immature rat. Dev Neurosci 29:373–384
Jin Y, Silverman AJ, Vannucci SJ (2009) Mast cells are early responders after hypoxia-ischemia in immature rat brain. Stroke 40:3107–3112
Juremalm M, Hjertson M, Olsson N, Harvima I, Nilsson K, Nilsson G (2000) The chemokine receptor CXCR4 is expressed within the mast cell lineage and its ligand stromal cell-derived factor-1alpha acts as a mast cell chemotaxin. Eur J Immunol 30:3614–3622
Kim SU, de Vellis J (2005) Microglia in health and disease. J Neurosci Res 81:302–313
Kim SR, Kim SU, Oh U, Jin BK (2006) Transient receptor potential vanilloid subtype 1 mediates microglial cell death in vivo and in vitro via Ca2+-mediated mitochondrial damage and cytochrome c release. J Immunol 177:4322–4329
Kim D, Kim MA, Cho IH, Kim MS, Lee S, Jo EK, Choi SY, Park K, Kim JS, Akira S, Na HS, Oh SB, Lee SJ (2007) A critical role of toll-like receptor 2 in nerve injury-induced spinal cord glial cell activation and pain hypersensitivity. J Biol Chem 282:14975–14983
Kim DY, Hong GU, Ro JY (2011) Signal pathways in astrocytes activated by cross-talk between of astrocytes and mast cells through CD40-CD40L. J Neuroinflammation 8:25. doi:10.1186/1742-2094-8-25
Knerlich-Lukoschus F, von der Ropp-Brenner B, Lucius R, Mehdorn HM, Held-Feindt J (2011) Spatiotemporal CCR1, CCL3(MIP-1α), CXCR4, CXCL12(SDF-1α) expression patterns in a rat spinal cord injury model of posttraumatic neuropathic pain. J Neurosurg Spine 14:583–597
Koda H, Mizumura K (2002) Sensitization to mechanical stimulation by inflammatory mediators and by mild burn in canine visceral nociceptors in vitro. J Neurophysiol 87:2043–2051
Kotter MR, Zhao C, van Rooijen N, Franklin RJM (2005) Macrophage-depletion induced impairment of experimental CNS remyelination is associated with a reduced oligodendrocyte progenitor cell response and altered growth factor expression. Neurobiol Dis 18:166–175
Kulka M, Fukuishi N, Metcalfe DD (2009) Human mast cells synthesize and release angiogenin, a member of the ribonuclease A (RNase A) superfamily. J Leukoc Biol 86:1217–1226
Lalancette-Hébert M, Gowing G, Simard A, Weng YC, Kriz J (2007) Selective ablation of proliferating microglial cells exacerbates ischemic injury in the brain. J Neurosci 27:2596–2605
Leal-Berumen I, Conlon P, Marshall JS (1994) IL-6 production by rat peritoneal mast cells is not necessarily preceded by histamine release and can be induced by bacterial lipopolysaccharide. J Immunol 152:5468–5476
Leon A, Buriani A, Dal Toso R, Fabris M, Romanello S, Aloe L, Levi-Montalcini R (1994) Mast cells synthesize, store, and release nerve growth factor. Proc Natl Acad Sci USA 91:3739–3743
Leung D, Saghatelian A, Simon GM, Cravatt BF (2006) Inactivation of N-acyl phosphatidylethanolamine phospholipase D reveals multiple mechanisms for the biosynthesis of endocannabinoids. Biochemistry 45:4720–4726
Levy D, Kainz V, Burstein R, Strassman AM (2012) Mast cell degranulation distinctly activates trigemino-cervical and lumbosacral pain pathways and elicits widespread tactile pain hypersensitivity. Brain Behav Immun 26:311–317
Li Y, Yang L, Chen L, Zhu C, Huang R, Zheng X, Qiu Y, Fu J (2012) Design and synthesis of potent N-acylethanolamine-hydrolyzing acid amidase (NAAA) inhibitor as anti-inflammatory compounds. PLoS One 7(8):e43023. doi:10.1371/journal.pone.0043023
Lindsberg PJ, Strbian D, Karjalainen-Lindsberg ML (2010) Mast cells as early responders in the regulation of acute blood-brain barrier changes after cerebral ischemia and hemorrhage. J Cereb Blood Flow Metab 30:689–702
Liu S, Liu Y, Hao W, Wolf L, Kiliaan AJ, Penke B, Rübe CE, Walter J et al (2012) TLR2 is a primary receptor for Alzheimer's amyloid β-peptide to trigger neuroinflammatory activation. J Immunol 188:1098–1107

Lo Monte G, Soave I, Marci R (2013) Administration of micronized palmitoylethanolamide (PEA)-transpolydatin in the treatment of chronic pelvic pain in women affected by endometriosis: preliminary results. Minerva Ginecol 65:453–463. [Italian]

Lo Verme J, Fu J, Astarita G, La Rana G, Russo R, Calignano A, Piomelli D (2005) The nuclear receptor peroxisome proliferator-activated receptor-alpha mediates the anti-inflammatory actions of palmitoylethanolamide. Mol Pharmacol 67:15–19

Lull ME, Block ML (2010) Microglial activation and chronic neurodegeneration. Neurotherapeutics 7:354–365

Luongo L, Guida F, Boccella S, Bellini G, Gatta L, Rossi F, de Novellis V, Maione S (2013) Palmitoylethanolamide reduces formalin-induced neuropathic-like behaviour through spinal glial/microglial phenotypical changes in mice. CNS Neurol Disord Drug Targets 12:45–54

Matias I, Bisogno T, Melck D, Vandenbulcke F, Verger-Bocquet M, De Petrocellis L, Sergheraert C, Breton C, Di Marzo V, Salzet M (2001) Evidence for an endocannabinoid system in the central nervous system of the leech Hirudo medicinalis. Mol Brain Res 87:145–159

Mattace Raso G, Russo R, Calignano A, Meli R (2014) Palmitoylethanolamide in CNS health and disease. Pharmacol Res 86:32–41

Mattila OS, Strbian D, Saksi J, Pikkarainen TO, Rantanen V, Tatlisumak T, Lindsberg PJ (2011) Cerebral mast cells mediate blood-brain barrier disruption in acute experimental ischemic stroke through perivascular gelatinase activation. Stroke 42:3600–3605

Mazzari S, Canella R, Petrelli L, Marcolongo G, Leon L (1996) N-(2-hydroxyethyl) hexadecanamide is orally active in reducing edema formation and inflammatory hyperalgesia by downmodulating mast cell activation. Eur J Pharmacol 300:227–236

McGeer PL, McGeer EG (2013) The amyloid cascade-inflammatory hypothesis of Alzheimer disease: implications for therapy. Acta Neuropathol 126:479–497

Mika J (2008) Modulation of microglia can attenuate neuropathic pain symptoms and enhance morphine tolerance. Pharmacol Rep 60:297–307

Muccioli GG, Stella N (2008) Microglia produce and hydrolyze palmitoylethanolamide. Neuropharmacology 54:16–22

Myers RR, Campana WM, Shubayev VI (2006) The role of neuroinflammation in neuropathic pain: mechanisms and therapeutic targets. Drug Discov Today 11:8–20

Najjar S, Pearlman DM, Alper K, Najjar A, Devinsky O (2013) Neuroinflammation and psychiatric illness. J Neuroinflammation 10:43. doi:10.1186/1742-2094-10-43

Nathan C, Ding A (2010) Nonresolving inflammation. Cell 140:871–882

Nautiyal KM, Dailey CA, Jahn JL, Rodriquez E, Son NH, Sweedler JV, Silver R (2012) Serotonin of mast cell origin contributes to hippocampal function. Eur J Neurosci 36:2347–2359

Noriega DB, Savelkoul HF (2014) Immune dysregulation in autism spectrum disorder. Eur J Pediatr 173:33–43

Oliveira SM, Drewes CC, Silva CR, Trevisan G, Boschen SL, Moreira CG, de Almeida Cabrini D, Da Cunha C, Ferreira J (2011) Involvement of mast cells in a mouse model of postoperative pain. Eur J Pharmacol 672:88–95

Orinska Z, Bulanova E, Budagian V, Metz M, Maurer M, Bulfone-Paus S (2005) TLR3-induced activation of mast cells modulates CD8+ T-cell recruitment. Blood 106:978–987

Osipchuk Y, Cahalan M (1992) Cell-to-cell spread of calcium signals mediated by ATP receptors in mast cells. Nature 359:241–244

Pacher P, Bátkai S, Kunos G (2006) The endocannabinoid system as an emerging target of pharmacotherapy. Pharmacol Rev 58:389–462

Paladini A, Fusco M, Cenacchi T, Schievano C, Piroli A, Varrassi G (2016) Palmitoylethanolamide, a special food for medical purposes, in the treatment of chronic pain: a pooled data meta-analysis. Pain Physician 19(2):11–24

Paolicelli RC, Bolasco G, Pagani F, Maggi L, Scianni M, Panzanelli P, Giustetto M, Ferreira TA, Guiducci E, Dumas L, Ragozzino D, Gross CT (2011) Synaptic pruning by microglia is necessary for normal brain development. Science 333:1456–1458

Petrosino S, Palazzo E, de Novellis V, Bisogno T, Rossi F, Maione S, Di Marzo V (2007) Changes in spinal and supraspinal endocannabinoid levels in neuropathic rats. Neuropharmacology 52:415–422

Pietrzak A, Wierzbicki M, Wiktorska M, Brzezińska-Błaszczyk E (2011) Surface TLR2 and TLR4 expression on mature rat mast cells can be affected by some bacterial components and proinflammatory cytokines. Mediators Inflamm 2011:427473. doi:10.1155/2011/427473

Piomelli D, Sasso O (2014) Peripheral gating of pain signals by endogenous lipid mediators. Nat Neurosci 17:164–174

Rahman IA, Tsuboi K, Uyama T, Ueda N (2014) New players in the fatty acyl ethanolamide metabolism. Pharmacol Res 86:1–10

Ralay Ranaivo H, Craft JM, Hu W, Guo L, Wing LK, Van Eldik LJ, Watterson DM (2006) Glia as a therapeutic target: selective suppression of human amyloid-beta-induced upregulation of brain proinflammatory cytokine production attenuates neurodegeneration. J Neurosci 26:662–670

Ribeiro Xavier AL, Kress BT, Goldman SA, Lacerda de Menezes JR, Nedergaard M (2015) A distinct population of microglia supports adult neurogenesis in the subventricular zone. J Neurosci 35:11848–11861

Ribeiro A, Pontis S, Mengatto L, Armirotti A, Chiurchiù V, Capurro V, Fiasella A, Nuzzi A, Romeo E, Moreno-Sanz G, Maccarrone M, Reggiani A, Tarzia G, Mor M, Bertozzi F, Bandiera T, Piomelli D (2015) A potent systemically active N-acylethanolamine acid amidase inhibitor that suppresses inflammation and human macrophage activation. ACS Chem Biol 10:1838–1846

Rivat C, Becker C, Blugeot A, Zeau B, Mauborgne A, Pohl M, Benoliel JJ (2010) Chronic stress induces transient spinal neuroinflammation, triggering sensory hypersensitivity and long-lasting anxiety-induced hyperalgesia. Pain 150:358–368

Rivera P, Bindila L, Pastor A, Pérez-Martín M, Pavón FJ, Serrano A, de la Torre R, Lutz B, Rodríguez de Fonseca F, Suárez J (2015) Pharmacological blockade of the fatty acid amide hydrolase (FAAH) alters neural proliferation, apoptosis and gliosis in the rat hippocampus, hypothalamus and striatum in a negative energy context. Front Cell Neurosci 9:98. doi:10.3389/fncel.2015.00098

Sasso O, Moreno-Sanz G, Martucci C, Realini N, Dionisi M, Mengatto L, Duranti A, Tarozzo G, Tarzia G, Mor M, Bertorelli R, Reggiani A, Piomelli D (2013) Antinociceptive effects of the N-acylethanolamine acid amidase inhibitor ARN077 in rodent pain models. Pain 154:350–360

Saturnino C, Petrosino S, Ligresti A, Palladino C, De Martino G, Bisogno T, Di Marzo V (2010) Synthesis and biological evaluation of new potential inhibitors of N-acylethanolamine hydrolyzing acid amidase. Bioorg Med Chem Lett 20:1210–1213

Scuderi C, Esposito G, Blasio A, Valenza M, Arietti P, Steardo L Jr, Carnuccio R, De Filippis D, Petrosino S, Iuvone T, Di Marzo V, Steardo L (2011) Palmitoylethanolamide counteracts reactive astrogliosis induced by β-amyloid peptide. J Cell Mol Med 15:2664–2674

Sepe N, De Petrocellis L, Montanaro F, Cimino G, Di Marzo V (1998) Bioactive long chain N-acylethanolamines in five species of edible bivalve molluscs. Possible implications for mollusc physiology and sea food industry. Biochim Biophys Acta 1389:101–111

Siegmund SV, Wojtalla A, Schlosser M, Zimmer A, Singer MV (2013) Fatty acid amide hydrolase but not monoacyl glycerol lipase controls cell death induced by the endocannabinoid 2-arachidonoyl glycerol in hepatic cell populations. BiochemBiophys Res Commun 437:48–54

Silver R, Curley JP (2013) Mast cells on the mind: new insights and opportunities. Trends Neurosci 36:513–521

Skaper SD, Facci L (2012) Mast cell-glia axis in neuroinflammation and therapeutic potential of the anandamide congener palmitoylethanolamide. Philos Trans R Soc Lond B Biol Sci 367:3312–3325

Skaper SD, Buriani A, Dal Toso R, Petrelli L, Romanello S, Facci L, Leon A (1996) The ALIAmide palmitoylethanolamide and cannabinoids, but not anandamide, are protective in a

delayed postglutamate paradigm of excitotoxic death in cerebellar granule neurons. Proc Natl Acad Sci USA 93:3984–3989
Skaper SD, Giusti P, Facci L (2012) Microglia and mast cells: two tracks on the road to neuroinflammation. FASEB J 26:3103–3117
Skaper SD, Facci L, Fusco M, Della Valle MF, Zusso M, Costa B, Giusti P (2014a) Palmitoylethanolamide, a naturally occurring disease-modifying agent in neuropathic pain. Inflammopharmacology 22:79–94
Skaper SD, Facci L, Giusti P (2014b) Mast cells, glia and neuroinflammation: partners in crime? Immunology 141:314–327
Skaper SD, Facci L, Barbierato M, Zusso M, Bruschetta G, Impellizzeri D, Cuzzocrea S, Giusti P (2015) N-Palmitoylethanolamine and neuroinflammation: a novel therapeutic strategy of resolution. Mol Neurobiol 52:1034–1042
Skuljec J, Sun H, Pul R, Bénardais K, Ragancokova D, Moharregh-Khiabani D, Kotsiari A, Trebst C et al (2011) CCL5 induces a pro-inflammatory profile in microglia in vitro. Cell Immunol 270:164–171
Smart D, Jonsson KO, Vandevoorde S, Lambert DM, Fowler CJ (2002) 'Entourage' effects of N-acyl ethanolamines at human vanilloid receptors. Comparison of effects upon anandamide-induced vanilloid receptor activation and upon anandamide metabolism. Brit J Pharmacol 136:452–458
Solorzano C, Zhu C, Battista N, Astarita G, Lodola A, Rivara S, Mor M, Russo R, Maccarrone M, Antonietti F, Duranti A, Tontini A, Cuzzocrea S, Tarzia G, Piomelli D (2009) Selective N-acylethanolamine-hydrolyzing acid amidase inhibition reveals a key role for endogenous palmitoylethanolamide in inflammation. Proc Natl Acad Sci USA 106:20966–20971
Song M, Jin J, Lim JE, Kou J, Pattanayak A, Rehman JA, Kim HD, Tahara K, Lalonde R, Fukuchi K (2011) TLR4 mutation reduces microglial activation, increases Aβ deposits and exacerbates cognitive deficits in a mouse model of Alzheimer's disease. J Neuroinflammation 8:92. doi:10.1186/1742-2094-8-92
Soruri A, Grigat J, Kiafard Z, Zwirner J (2008) Mast cell activation is characterized by upregulation of a functional anaphylatoxin C5a receptor. BMC Immunol 9:29. doi:10.1186/1471-2172-9-29
Strbian D, Karjalainen-Lindsberg ML, Kovanen PT, Tatlisumak T, Lindsberg PJ (2007) Mast cell stabilization reduces hemorrhage formation and mortality after administration of thrombolytics in experimental ischemic stroke. Circulation 116:411–418
Subramanian V, Crabtree B, Acharya KR (2008) Human angiogenin is a neuroprotective factor and amyotrophic lateral sclerosis associated angiogenin variants affect neurite extension/pathfinding and survival of motor neurons. Hum Mol Genet 17:130–149
Suk K, Ock J (2011) Chemical genetics of neuroinflammation: natural and synthetic compounds as microglial inhibitors. Inflammopharmacology 20:151–158
Tanga FY, Nutile-McMenemy N, DeLeo JA (2005) The CNS role of Toll-like receptor 4 in innate neuroimmunity and painful neuropathy. Proc Natl Acad Sci USA 102:5856–58561
Thacker MA, Clark AK, Marchand F, McMahon SB (2007) Pathophysiology of peripheral neuropathic pain: immune cells and molecules. Anesth Analg 105:838–847
Theoharides TC, Stewart JM, Hatziagelaki E, Kolaitis G (2015a) Brain "fog," inflammation and obesity: key aspects of neuropsychiatric disorders improved by luteolin. Front Neurosci. 9:225. doi:10.3389/fnins.2015.00225
Theoharides TC, Stewart JM, Panagiotidou S, Melamed I (2015b) Mast cells, brain inflammation and autism. Eur J Pharmacol. doi:10.1016/j.ejphar.2015.03.086
Trang T, Beggs S, Wan X, Salter MW (2009) P2X4-receptor-mediated synthesis and release of brain-derived neurotrophic factor in microglia is dependent on calcium and p38-mitogen-activated protein kinase activation. J Neurosci 29:3518–3528
Tsuboi K, Takezaki N, Ueda N (2007) The N-acylethanolamine-hydrolyzing acid amidase (NAAA). Chem Biodivers 4:1914–1925

Tsuda M, Kuboyama K, Inoue T, Nagata K, Tozaki-Saitoh H, Inoue K (2009) Behavioral phenotypes of mice lacking purinergic P2X4 receptors in acute and chronic pain assays. Mol Pain 5:28. doi:10.1186/1744-8069-5-28

Ueda N, Tsuboi K, Uyama T (2013) Metabolism of endocannabinoids and related N-acylethanolamines: Canonical and alternative pathways. FEBS J 280:1874–1894

Vinet J, van Weering HR, Heinrich A, Kälin RE, Wegner A, Brouwer N, Heppner FL, van Rooijen N, Boddeke HW, Biber K (2012) Neuroprotective function for ramified microglia in hippocampal excitotoxicity. J Neuroinflammation 9:27. doi:10.1186/1742-2094-9-27

Vitale R, Ottonello G, Petracca R, Bertozzi SM, Ponzano S, Armirotti A, Berteotti A, Dionisi M, Cavalli A, Piomelli D, Bandiera T, Bertozzi F (2014) Synthesis, structure-activity, and structure-stability relationships of 2-substituted-N-(4-oxo-3-oxetanyl) N-acylethanolamine acid amidase (NAAA) inhibitors. ChemMedChem 9:323–336

Wang Q, Tang XN, Yenari MA (2007) The inflammatory response in stroke. J Neuroimmunol 184:53–68

Wang M, Wang X, Zhao L, Ma W, Rodriguez IR, Fariss RN, Wong WT (2014) Macroglia-microglia interactions via TSPO signaling regulates microglial activation in the mouse retina. J Neurosci 34:3793–3806

Wardlaw AJ, Moqbel R, Cromwell O, Kay AB (1986) Platelet-activating factor. A potent chemotactic and chemokinetic factor for human eosinophils. J Clin Invest 78:1701–1706

Wei F, Dubner R, Ren K (2007) Glial-cytokine-neuronal interactions underlying the mechanisms of persistent pain. J Neurosci 27:6006–6018

Weller K, Foitzik K, Paus R, Syska W, Maurer M (2006) Mast cells are required for normal healing of skin wounds in mice. FASEB J 20:2366–2368

Wood D (2011) Visceral pain: spinal afferents, enteric mast cells, enteric nervous system and stress. Curr Pharm Des 17:1573–1575

Xanthos DN, Gaderer S, Drdla R, Nuro E, Abramova A, Ellmeier W, Sandkühler J (2011) Central nervous system mast cells in peripheral inflammatory nociception. Mol Pain 7:42. doi:10.1186/1744-8069-7-42

Yamano Y, Tsuboi K, Hozaki Y, Takahashi K, Jin XH, Ueda N, Wada A (2012) Lipophilic amines as potent inhibitors of N-acylethanolamine-hydrolyzing acid amidase. Bioorg Med Chem 20:3658–3665

Yang H, Wei J, Zhang H, Song W, Wei W, Zhang L, Qian K, He S (2010) Upregulation of Toll-like receptor (TLR) expression and release of cytokines from mast cells by IL-12. Cell Physiol Biochem 26:337–346

Yang L, Li L, Chen L, Li Y, Chen H, Li Y, Ji G, Lin D, Liu Z, Qiu Y (2015) Potential analgesic effects of a novel N-acylethanolamine acid amidase inhibitor F96 through PPAR-α. Sci Rep 5:13565. doi:10.1038/srep13565

Zhang H, Lin L, Yang H, Zhang Z, Yang X, Zhang L, He S (2010a) Induction of IL-13 production and upregulation of gene expression of protease activated receptors in P815 cells by IL-6. Cytokine 50:138–145

Zhang H, Yang H, He S (2010b) TNF increases expression of IL-4 and PARs in mast cells. Cell Physiol Biochem 26:327–336

Zhang S, Zeng X, Yang H, Hu G, He S (2012) Mast cell tryptase induces microglia activation via protease-activated receptor 2 signaling. Cell Physiol Biochem 29:931–940

Zuo Y, Perkins NM, Tracey DJ, Geczy CL (2003) Inflammation and hyperalgesia induced by nerve injury in the rat: a key role of mast cells. Pain 105:467–479

Printed by Printforce, the Netherlands